Hilbert Schemes of Points and Infinite Dimensional Lie Algebras

Mathematical Surveys and Monographs

Volume 228

Hilbert Schemes of Points and Infinite Dimensional Lie Algebras

Zhenbo Qin

AMERICAN MATHEMATICAL SOCIETY
Providence, Rhode Island

The author was supported in part by a Collaboration Grant for Mathematicians (Award Number: 268702) from the Simons Foundation and by a Research Council Grant (Award Number: URC-17-084-n) from the University of Missouri.

2010 *Mathematics Subject Classification.* Primary 14C05, 17B65;
Secondary 14F43, 14J60, 14N35, 17B69.

For additional information and updates on this book, visit
www.ams.org/bookpages/surv-228

Library of Congress Cataloging-in-Publication Data

Names: Qin, Zhenbo, author.
Title: Hilbert schemes of points and infinite dimensional lie algebras / Zhenbo Qin.
Description: Providence, Rhode Island: American Mathematical Society, [2018] | Series: Mathematical surveys and monographs; volume 228 | Includes bibliographical references and index.
Identifiers: LCCN 2017036491 | ISBN 9781470441883 (alk. paper)
Subjects: LCSH: Hilbert schemes. | Schemes (Algebraic geometry) | Lie algebras. | AMS: Algebraic geometry – Cycles and subschemes – Parametrization (Chow and Hilbert schemes). msc | Nonassociative rings and algebras – Lie algebras and Lie superalgebras – Infinite-dimensional Lie (super)algebras. msc | Algebraic geometry – (Co)homology theory – Other algebro-geometric (co)homologies (e.g., intersection, equivariant, Lawson, Deligne (co)homologies). msc | Algebraic geometry – Surfaces and higher-dimensional varieties – Vector bundles on surfaces and higher-dimensional varieties, and their moduli. msc | Algebraic geometry – Projective and enumerative geometry – Gromov-Witten invariants, quantum cohomology, Gopakumar-Vafa invariants, Donaldson-Thomas invariants. msc | Nonassociative rings and algebras – Lie algebras and Lie superalgebras – Vertex operators; vertex operator algebras and related structures. msc
Classification: LCC QA564 .Q56 2018 | DDC 516.3/5–dc23
LC record available at https://lccn.loc.gov/2017036491

∞ The paper used in this book is acid-free and falls within the guidelines established to ensure permanence and durability.
Visit the AMS home page at `http://www.ams.org/`

10 9 8 7 6 5 4 3 2 1 23 22 21 20 19 18

Contents

Preface

Since the pioneering work of Grothendieck [**Grot**], Hilbert schemes, which parametrize subschemes in algebraic varieties, have been studied extensively as evidenced in the survey [**Iar2**]. Viewed also as the simplest moduli spaces of sheaves, they are fundamental objects in algebraic geometry, play essential roles in various important enumerative problems, and provide testing grounds for many interesting conjectures in algebraic geometry and its interplay with theoretical physics. Apart from the Hilbert schemes of curves on 3-folds over which the Donaldson-Thomas theory is defined and investigated [**MNOP1**, **MNOP2**], much attention has been concentrated on Hilbert schemes of points, i.e., those Hilbert schemes parametrizing 0-dimensional closed subschemes. When the dimension of the variety is at least three, the corresponding Hilbert schemes of points are in general singular. When the variety is a smooth curve, the corresponding Hilbert schemes of points coincide with the symmetric products of the curve. When X is a smooth algebraic surface, the Hilbert scheme $X^{[n]}$ parametrizing length-n 0-dimensional closed subschemes of X is irreducible and smooth [**Bri**, **Fog1**, **Iar1**]. Moreover, the Hilbert-Chow morphism ρ_n from $X^{[n]}$ to the symmetric product $X^{(n)}$, which sends an element in $X^{[n]}$ to its support (counted with multiplicities) in $X^{(n)}$, is a crepant desingularization of $X^{(n)}$. Many fundamental aspects of $X^{[n]}$ such as its cohomology groups, Chow groups, motive, cobordism class, and relations with algebraic combinatorics, the McKay correspondence and integrable systems have been analyzed [**CoG**, **dCM**, **EGL**, **ES1**, **Got1**, **Hai1**, **Hai2**, **Mar1**].

In the seminal papers [**Groj**, **Nak3**] which were motivated by [**Nak1**, **Nak2**, **Nak4**] regarding the construction of representations of affine Lie algebras on the homology groups of the moduli spaces of instantons on ALE spaces (equivalently, on the homology groups of quiver varieties), Grojnowski and Nakajima geometrically constructed Heisenberg algebra actions on the cohomology of the Hilbert schemes $X^{[n]}$, where X denotes a smooth algebraic surface. Their geometric constructions started a whole new chapter investigating interplays between the Hilbert schemes of points and infinite dimensional Lie algebras. Subsequently, using the Hilbert schemes $X^{[n]}$, Lehn [**Leh1**] geometrically constructed the Virasoro algebras and the boundary operator, W.-P. Li, W. Wang and the author [**LQW1**, **LQW4**] constructed the Chern character operators and the $\mathcal{W}$ algebras, and Carlsson and Okounkov [**Car1**, **Car2**, **CO**] constructed the Ext vertex operators. As noted in [**FW**, Introduction] and [**Mat**, Section 5], Lehn's boundary operator is a version of the bosonized Calogero-Sutherland operator. The Chern character operators are vertex operators of higher conformal weights, and the $\mathcal{W}$ algebras are higher-spin generalizations of Lehn's Virasoro algebras. The Ext vertex operators of Carlsson and Okounkov are motivated by the study of Nekrasov partition functions (which arise from supersymmetric quantum gauge theory) in the setting of Hilbert schemes of points.

All of these operators and algebras are powerful tools in understanding the finer geometric properties of the Hilbert schemes $X^{[n]}$. Indeed, the Heisenberg algebras serve as a fundamental language in describing the homology and cohomology classes of $X^{[n]}$, which is crucial in studying the Gromov-Witten theory of $X^{[n]}$ [**Cheo2**, **ELQ**, **HLQ**, **LL**, **LQ1**, **LQ5**, **OP3**, **OP4**]. The Virasoro operators and the Chern character operators determine the cup products and ring structures of the cohomology of $X^{[n]}$ [**CoG**, **LQW1**, **LQW2**, **LQW3**, **LQW5**, **LS1**, **LS2**]. The Ext vertex operators provide a very useful approach in understanding the intersection theory of $X^{[n]}$ when the tangent bundle of $X^{[n]}$ is involved [**Car1**, **Car2**, **QY**].

Two excellent books on Hilbert schemes of points exist. The first one is Göttsche's book [**Got2**] which includes fundamental facts about $X^{[n]}$, the Betti numbers of $X^{[n]}$ when X is a surface, and the computation of the homology and Chow rings of Hilbert schemes. The second is Nakajima's book [**Nak5**] which contains his Heisenberg algebra constructions, symmetric products (in the Hilbert scheme $X^{[n]}$) of an embedded curve in a surface X, and interactions with singularities, symplectic geometry and the ring of symmetric functions. We also refer to [**EG**, **Leh2**] for surveys on Hilbert schemes of points, the Heisenberg algebras and the Virasoro algebras, and to [**Nak7**] for a survey on the equivariant cohomology of the Hilbert schemes, the Heisenberg algebras, the Virasoro algebras and some interesting applications to algebraic combinatorics.

The purpose of this book is to present a detailed survey of the developments, in the context of interactions between Hilbert schemes of points and infinite dimensional Lie algebras, that appeared after the book [**Nak5**]. This book contains 5 parts consisting of 16 chapters. Part 1 deals with the basics of the Hilbert schemes of points and some geometry of the Hilbert schemes of points on the projective plane. Part 2 is devoted to the constructions of various infinite dimensional Lie algebra actions on the cohomology of the Hilbert schemes of points on surfaces, and to the connections with multiple q-zeta values. It includes Nakajima's affine Lie algebra actions on the homology of quiver varieties as a motivation and background material, the Heisenberg algebras of Grojnowski and Nakajima, the boundary operator and the Virasoro algebras of Lehn, the Ext vertex operators of Carlsson and Okounkov, and the Chern character operators and the $\mathcal{W}$ algebras of W.-P. Li, W. Wang and the author. Part 3 studies the cohomology ring structure of these Hilbert schemes, such as the ring generators and the ideals, the approach of Lehn and Sorger when the canonical divisor of the surface is numerically trivial, the approach of Costello and Grojnowski in terms of the Calogero-Sutherland operators and the Dunkl-Cherednik operators, the integral basis for the cohomology group, and the orbifold cohomology ring of the symmetric product of a surface. Part 4 is about the equivariant cohomology of the Hilbert schemes via Jack polynomials, the Hilbert/Gromov-Witten correspondence between the equivariant cohomology of the Hilbert schemes of points on the affine plane and the Gromov-Witten theory of curves, and applications to the Hurwitz numbers of the projective line. Part 5 includes the cosection localization technique of Kiem and J. Li, the Gromov-Witten theory of the Hilbert schemes of points, the equivariant quantum corrected boundary operator and the quantum differential equations of Okounkov and Pandharipande, the quantum corrected boundary operator of J. Li and W.-P. Li, and the proof of Ruan's Cohomological Crepant Resolution Conjecture.

This book is suitable for graduate students and researchers in algebraic geometry, representation theory, algebraic combinatorics, topology, number theory and theoretical physics. Furthermore, a semester of an advanced course on Hilbert schemes of points and infinite dimensional Lie algebras may be organized from selected chapters in this book, e.g., Chapter 1, Chapter 3, Chapter 4, Chapter 7, Chapter 8, Chapter 11 and Chapter 12.

The author would like to thank all of his collaborators on works related to the Hilbert schemes of points, without whom this book would be impossible: Dan Edidin, Jianxun Hu, Wei-Ping Li, Yuping Tu, Weiqiang Wang, Fei Yu and Qi Zhang. In addition, thanks are due to Lie Fu, Wei-Ping Li, Hiraku Nakajima, Boiar Qin, Wenzer Qin, Weiqiang Wang and the two anonymous referees for carefully reading the manuscript and providing useful comments which have greatly improved the exposition of the book.

The author also thanks Sergei Gelfand, Christine M. Thivierge and the Editorial Committee of the American Mathematical Society for their editorial guidance.

December 14, 2017

Zhenbo Qin

Part 1

Hilbert schemes of points on surfaces

CHAPTER 1

Basic results on Hilbert schemes of points

This chapter collects basic definitions and results regarding partitions, the ring of symmetric functions, symmetric products, Hilbert schemes of points, and incidence Hilbert schemes. Details may be found in Chapter I of [**Mac2**] and Chapters 1 and 2 of [**Got2**]. The focuses of this chapter are various symmetric functions and the Hilbert schemes of points on a smooth projective complex surface. Throughout the book, all the homology and cohomology are with $\mathbb{C}$-coefficients, unless otherwise specified.

1.1. Partitions

Let $n \geq 1$. A *partition* λ of n, denoted by $\lambda \vdash n$, is an infinite sequence

$$\lambda = (\lambda_1 \geq \lambda_2 \geq \lambda_3 \geq \ldots)$$

of nonnegative integers λ_i, or equivalently a finite sequence $\lambda = (\lambda_1 \geq \ldots \geq \lambda_\ell)$ of positive integers λ_i, such that $n = \sum_i \lambda_i$. In this case, ℓ is the *length* of λ and denoted by $\ell(\lambda)$, and n is the *size* of λ and denoted by $|\lambda|$. A partition may be rewritten as

$$\lambda = (1^{m_1} 2^{m_2} 3^{m_3} \cdots).$$

Put

$$\mathfrak{z}_\lambda = \prod_{i \geq 1} (i^{m_i} m_i!). \tag{1.1}$$

Definition 1.1. Let $\lambda = (\lambda_1 \geq \cdots \geq \lambda_k > 0)$ and $\tilde{\lambda} = (\tilde{\lambda}_1 \geq \cdots \geq \tilde{\lambda}_\ell > 0)$ be partitions. We define $\lambda \prec \tilde{\lambda}$ if one of the three following conditions holds:

(i) $|\lambda| < |\tilde{\lambda}|$.
(ii) $|\lambda| = |\tilde{\lambda}|$ and $k < \ell$,
(iii) $|\lambda| = |\tilde{\lambda}|$, $k = \ell$, and $\lambda_r < \tilde{\lambda}_r$ for some r but $\lambda_i = \tilde{\lambda}_i$ for all $i = r+1, \ldots, k$.

This is a total ordering on the set of all partitions. Another useful ordering is the dominance (partial) ordering $\leq$.

Definition 1.2. Let $\lambda = (\lambda_1 \geq \lambda_2 \geq \cdots)$ and $\tilde{\lambda} = (\tilde{\lambda}_1 \geq \tilde{\lambda}_2 \geq \cdots)$ be partitions. We define $\lambda \leq \tilde{\lambda}$ if

$$\lambda_1 + \ldots + \lambda_i \leq \tilde{\lambda}_1 + \ldots + \tilde{\lambda}_i$$

for all $i \geq 1$.

A partition is also referred to as a 2-dimensional partition.

Definition 1.3. (i) Let $d \geq 2$ and $n \geq 0$. *A d-dimensional partition* λ of n, denoted by $\lambda \vdash n$, is an array

$$\lambda = (\lambda_{i_1, \ldots, i_{d-1}})_{i_1, \ldots, i_{d-1}}$$

of nonnegative integers $\lambda_{i_1,\ldots,i_{d-1}}$ indexed by the tuples $(i_1, \ldots, i_{d-1}) \in (\mathbb{Z}_{\geq 0})^{d-1}$ such that

$$\sum_{i_1,\ldots,i_{d-1}} \lambda_{i_1,\ldots,i_{d-1}} = n,$$

and $\lambda_{i_1,\ldots,i_{d-1}} \geq \lambda_{j_1,\ldots,j_{d-1}}$ if $i_1 \leq j_1, \ldots, i_{d-1} \leq j_{d-1}$.

(ii) We define $P_d(n)$ to be the number of d-dimensional partitions of n.

One sees immediately that the generating function for $P_2(n)$ is given by

$$\sum_{n=0}^{+\infty} P_2(n)q^n = \prod_{n=1}^{+\infty} \frac{1}{1-q^n} \tag{1.2}$$

where q is a formal variable. A well-known result of McMahon [**And**] states that

$$\sum_{n=0}^{+\infty} P_3(n)q^n = \prod_{n=1}^{+\infty} \frac{1}{(1-q^n)^n}. \tag{1.3}$$

There is no analogous formula for $P_d(n)$ when $d > 3$.

Geometrically, a (2-dimensional) partition can be represented by a Young diagram. Here we adopt the convention of drawing diagrams upside down (so equivalently, a partition can be represented by stacking unit squares in the first quadrant).

DEFINITION 1.4. Let D_λ denote the Young diagram associated to a partition λ. For a cell $\square$ in D_λ, let $\ell(\square)$, $\ell'(\square)$, $a(\square)$ and $a'(\square)$ be respectively the numbers of cells in D_λ to the south, north, east and west of $\square$ (excluding $\square$ itself). The numbers $\ell(\square)$ and $\ell'(\square)$ are called the *leg length* and the *leg colength* respectively, and $a(\square)$ and $a'(\square)$ are called the *arm length* and the *arm colength* respectively. The *hook length* $h(\square)$ and the *content* $c(\square)$ are defined to be $a(\square) + \ell(\square) + 1$ and $a'(\square) - \ell'(\square)$ respectively.

EXAMPLE 1.5. Let $\lambda = (5, 5, 5, 2, 1) = (1^1 2^1 5^3)$. Then λ is a partition of 18. Its Young diagram D_λ

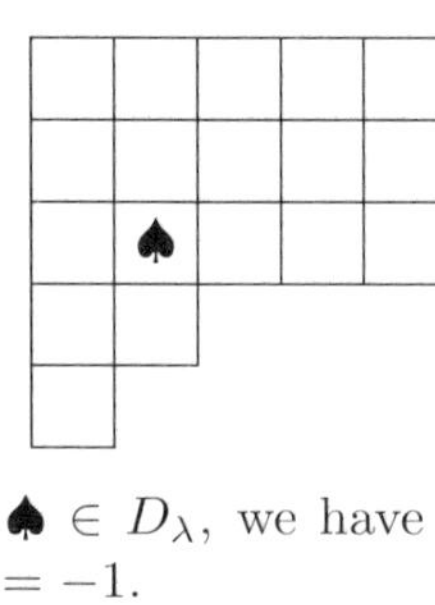

contains 18 cells. For the cell $\spadesuit \in D_\lambda$, we have $\ell(\spadesuit) = 1$, $\ell'(\spadesuit) = 2$, $a(\spadesuit) = 3$, $a'(\spadesuit) = 1$, $h(\spadesuit) = 5$, and $c(\spadesuit) = -1$.

Similarly, a 3-dimensional partition can be represented by stacking unit cubes at a corner.

DEFINITION 1.6. A *generalized partition* of an integer n is of the form

$$\lambda = (\cdots (-2)^{m_{-2}} (-1)^{m_{-1}} 1^{m_1} 2^{m_2} \cdots)$$

such that part $i \in \mathbb{Z}$ has multiplicity m_i, $m_i \neq 0$ for only finitely many i's, and $n = \sum_i im_i$. Define

$$\begin{aligned} \ell(\lambda) &= \sum_i m_i, \\ |\lambda| &= \sum_i im_i, \\ s(\lambda) &= \sum_i i^2 m_i, \\ \lambda^! &= \prod_i m_i!. \end{aligned}$$

Let $-\lambda$ be the generalized partition whose multiplicity of $i \in \mathbb{Z}$ is m_{-i}. The set of all generalized partitions is denoted by $\widetilde{\mathcal{P}}$.

1.2. The ring of symmetric functions

Let $x_1, \ldots, x_n$ be formal variables. The n-th symmetric group S_n acts on the polynomial ring $\mathbb{Z}[x_1, \ldots, x_n]$ by permuting the variables $x_1, \ldots, x_n$. Let

$$\Lambda_n = \mathbb{Z}[x_1, \ldots, x_n]^{S_n}$$

be the subring of symmetric polynomials in $x_1, \ldots, x_n$. For $r \geq 0$, let Λ_n^r be the additive group of homogeneous symmetric polynomials of degree r. Then,

$$\Lambda_n = \bigoplus_{r=0}^{+\infty} \Lambda_n^r.$$

By setting $x_{n+1} = 0$, there is a natural surjective homomorphism $\Lambda_{n+1} \to \Lambda_n$ which maps Λ_{n+1}^r surjectively to Λ_n^r. Let

$$\Lambda^r = \varprojlim_n \Lambda_n^r.$$

Then the *ring of symmetric functions* in countably many independent variables $x_1, x_2, \ldots$ is defined to be

$$\Lambda = \bigoplus_{r=0}^{+\infty} \Lambda^r.$$

Note that Λ is not the inverse limit of the rings Λ_n relative to the homomorphisms $\Lambda_{n+1} \to \Lambda_n$. For instance, the infinite product $\prod_{i=1}^{+\infty}(1 + x_i)$ is not contained in Λ since it is not a finite sum of homogeneous symmetric functions.

For each integer $r \geq 0$, the r-th *elementary symmetric function* e_r is the sum of all products of r distinct variables x_i, so that $e_0 = 1$ and

$$e_r = \sum_{i_1 < \ldots < i_r} x_{i_1} \cdots x_{i_r}$$

for $r \geq 1$. For a partition $\lambda = (\lambda_1 \geq \lambda_2 \geq \ldots)$, define the *elementary symmetric function* e_λ by

$$e_\lambda = e_{\lambda_1} e_{\lambda_2} \cdots .$$

For a sequence $\alpha = (\alpha_1, \alpha_2, \ldots)$ of nonnegative integers, define

$$x^\alpha = x_1^{\alpha_1} x_2^{\alpha_2} \cdots .$$

For a partition $\lambda = (\lambda_1 \geq \lambda_2 \geq \ldots)$, define the *monomial symmetric function* by

$$m_\lambda = \sum_\alpha x^\alpha$$

where α runs over all *distinct* permutations of the sequence $(\lambda_1, \lambda_2, \ldots)$.

For each integer $r \geq 1$, the r-th *power sum* is defined to be

$$p_r = \sum_{i=1}^{+\infty} x_i^r.$$

For a partition $\lambda = (\lambda_1 \geq \lambda_2 \geq \ldots)$, define the *power symmetric function* p_λ by

$$p_\lambda = p_{\lambda_1} p_{\lambda_2} \cdots .$$

Since the monomial symmetric function $m_{(1^n)}$ is the same as the n-th elementary symmetric function, we conclude from formula (2.14′) in Chapter I of [**Mac2**] that

$$m_{(1^n)} = \sum_{|\mu|=n} \frac{(-1)^{n-\ell(\mu)} p_\mu}{\mathfrak{z}_\mu}. \tag{1.4}$$

Moreover, it is a classical result that

$$p_i m_\lambda = \sum_\mu a_{\lambda,\mu} m_\mu \tag{1.5}$$

where the summation is over certain partitions μ of $|\lambda| + i$, which are obtained as follows:

(i) add i to a part in λ, say λ_k (possibly 0), and then
(ii) arrange the parts in descending order.

The coefficient $a_{\lambda,\mu}$ is equal to the number of elements in $\{\ell \mid \mu_\ell = \lambda_k + i\}$.

To define the Schur functions, we will work in $\mathbb{Z}[x_1, \ldots, x_n]$ for the moment. For a sequence $\alpha = (\alpha_1, \ldots, \alpha_n)$ of nonnegative integers, let

$$a_\alpha = \det(x_i^{\alpha_j})_{1 \leq i,j \leq n}.$$

When $\delta = (n-1, n-2, \ldots, 1, 0)$, a_δ is the Vandermonde determinant and equal to $\prod_{i<j}(x_i - x_j)$. Note that $a_\alpha = 0$ if two of the integers α_i are equal. So up to sign, we may assume that $\alpha_1 > \ldots > \alpha_n \geq 0$. For $1 \leq i \leq n$, define λ_i by $\alpha_i = \lambda_i + n - i$. Put $\lambda = (\lambda_1, \ldots, \lambda_n)$. Then λ is a partition of length at most n, and $\alpha = \lambda + \delta$. The determinant $a_\alpha = a_{\lambda+\delta}$ is divisible by a_δ in the polynomial ring $\mathbb{Z}[x_1, \ldots, x_n]$, and

$$s_\lambda(x_1, \ldots, x_n) = a_{\lambda+\delta}/a_\delta$$

is a symmetric homogeneous polynomial of degree $|\lambda|$. Moreover,

$$s_\lambda(x_1, \ldots, x_n, 0) = s_\lambda(x_1, \ldots, x_n).$$

Therefore, there exists a unique element

$$s_\lambda \in \Lambda$$

that reduces to $s_\lambda(x_1, \ldots, x_n)$ when $x_{n+1}, x_{n+2}, \ldots$ are all set equal to zero, for every $n \geq \ell(\lambda)$. This symmetric function s_λ is the *Schur function* corresponding to λ.

Next, we review the Jack symmetric functions. Fix a positive real number α. Let $\mathbb{Q}(\alpha)$ be the subfield of $\mathbb{R}$ generated by α over $\mathbb{Q}$. Let $\langle \cdot, \cdot \rangle_\alpha$ be the inner product on $\Lambda \otimes_{\mathbb{Z}} \mathbb{Q}(\alpha)$ defined by declaring

$$\langle p_\mu, p_\nu \rangle_\alpha = \alpha^{\ell(\mu)}\, \mathfrak{z}_\mu\, \delta_{\mu,\nu}.$$

For a fixed positive real number α, the *Jack symmetric functions* $P_\lambda^{(\alpha)}$ form a basis of $\Lambda \otimes_{\mathbb{Z}} \mathbb{Q}(\alpha)$ indexed by partitions λ and are characterized by the following properties (1.6) and (1.7):

$$P_\lambda^{(\alpha)} = m_\lambda + \sum_{\mu<\lambda} u_{\lambda,\mu}^{(\alpha)} m_\mu \tag{1.6}$$

where $u_{\lambda,\mu}^{(\alpha)}$ are real numbers and $<$ denotes the dominance ordering, and

$$\langle P_\lambda^{(\alpha)}, P_\mu^{(\alpha)} \rangle_\alpha = 0 \tag{1.7}$$

if $\lambda \neq \mu$.

It is well-known that each of the sets $\{e_\lambda\}_\lambda$, $\{m_\lambda\}_\lambda$ and $\{s_\lambda\}_\lambda$ is a $\mathbb{Z}$-linear basis of Λ. In addition, the set

$$\{e_i = m_{(1^i)}\}_{i\geq 1}$$

generates the integral ring Λ. The set $\{p_\lambda\}_\lambda$ is a $\mathbb{Q}$-linear basis of $\Lambda \otimes_{\mathbb{Z}} \mathbb{Q}$, but not a $\mathbb{Z}$-linear basis of Λ. For a fixed positive real number α, $\{P_\lambda^{(\alpha)}\}_\lambda$ is a linear basis of $\Lambda \otimes_{\mathbb{Z}} \mathbb{Q}(\alpha)$.

Finally, there is an involution ω on Λ satisfying

$$\omega(p_\lambda) = (-1)^{|\lambda|-\ell(\lambda)} p_\lambda. \tag{1.8}$$

The *forgotten symmetric function* f_λ associated to a partition λ is defined by

$$f_\lambda = \omega(m_\lambda). \tag{1.9}$$

The set $\{f_\lambda\}_\lambda$ is another $\mathbb{Z}$-linear basis of Λ.

1.3. Symmetric products

Let X be a compact polyhedron, and X^n be the n-th Cartesian product of X. The n-th symmetric group S_n acts on X^n by permuting the factors. The n-th *symmetric product* $X^{(n)}$ is the quotient space X^n/S_n. An element in X^n is denoted by a tuple $(x_1, \ldots, x_n)$. An element in $X^{(n)}$ is denoted by $x_1 + \ldots + x_n$ where the points $x_1, \ldots, x_n$ are not necessarily distinct. Equivalently, an element in $X^{(n)}$ can also be denoted by $n_1 x_1 + \ldots + n_k x_k$ where the points $x_1, \ldots, x_k \in X$ are distinct, $n_1, \ldots, n_k$ are positive, and $n_1 + \ldots + n_k = n$.

Let $H_i(X)$ be the i-th homology group of X with complex coefficients, and $b_i(X) = \dim H_i(X)$ be the i-th Betti number of X. The Poincaré polynomial of X is defined to be

$$P_t(X) = \sum_{i=0}^{+\infty} b_i(X)\, t^i.$$

Macdonald [**Mac1**] proved that the generating function for the Poincaré polynomial of $X^{(n)}$ is given by

$$\sum_{n=0}^{+\infty} P_t(X^{(n)})\, q^n = \frac{(1+tq)^{b_1(X)}(1+t^3q)^{b_3(X)} \cdots}{(1-q)^{b_0(X)}(1-t^2q)^{b_2(X)}(1-t^4q)^{b_4(X)} \cdots}. \tag{1.10}$$

Now let X be a quasi-projective complex variety. To see whether the symmetric product $X^{(n)}$ exists as an algebraic variety, we introduce the following.

DEFINITION 1.7. Let Y be a quasi-projective complex variety with an action of a group G. A variety Z together with a surjective morphism $\pi : Y \to Z$ is defined to be a *quotient* of Y by G if

(i) the fibers of π are the orbits of G,
(ii) $\tilde{\pi} : Y \to W$ factors through π whenever $\tilde{\pi}$ is a G-equivariant morphism and maps every G-orbit in Y to a single point in W.

If a quotient exists, then it is unique up to isomorphisms and denoted by Y/G. In general, it is not easy to see whether a quotient exists, and the question is somewhat related to geometric invariant theory. On the other hand, when the group G is finite, the quotient Y/G always exists.

THEOREM 1.8. *Let Y be a quasi-projective complex variety with an action of a finite group G. Then the quotient Y/G exists as a quasi-projective variety.*

PROOF. First of all, we assume that Y is affine. Let

$$Y/G = \mathrm{Spec}\big(A[Y]^G\big)$$

where $A[Y]$ denotes the affine coordinate ring of Y. To prove that $A[Y]^G$ is finitely generated over $\mathbb{C}$, note that we can write $A[Y]$ in the form $\mathbb{C}[x_1, \ldots, x_m]/I(Y)$ where G acts on $x_1, \ldots, x_m$ as permutations. Moreover, we have a surjection

$$\mathbb{C}[x_1, \ldots, x_m]^G \to \big(\mathbb{C}[x_1, \ldots, x_m]/I(Y)\big)^G.$$

Indeed, an element $\overline{\alpha} \in \big(\mathbb{C}[x_1, \ldots, x_m]/I(Y)\big)^G$ corresponds to an element $\alpha \in \mathbb{C}[x_1, \ldots, x_m]$ congruent to every element $g^*(\alpha)$ mod $I(Y)$. Then the average

$$\frac{1}{|G|} \sum_{g \in G} g^*(\alpha) \in \mathbb{C}[x_1, \ldots, x_m]$$

is G-invariant and will be mapped to $\overline{\alpha}$. Thus it suffices to prove that $\mathbb{C}[x_1, \ldots, x_m]^G$ is finitely generated over $\mathbb{C}$. We have

$$\mathbb{C}[x_1, \ldots, x_m]^{S_m} \subset \mathbb{C}[x_1, \ldots, x_m]^G \subset \mathbb{C}[x_1, \ldots, x_m].$$

By the fundamental theorem on symmetric polynomials,

$$\mathbb{C}[x_1, \ldots, x_m]^{S_m} = \mathbb{C}[\tilde{e}_1, \ldots, \tilde{e}_m]$$

where $\tilde{e}_i = e_i(x_1, \ldots, x_m, 0, \ldots)$ with $1 \le i \le m$ are the elementary symmetric polynomials with the m variables $x_1, \ldots, x_m$. Since $\mathbb{C}[x_1, \ldots, x_m]$ is a finitely generated module over $\mathbb{C}[\tilde{e}_1, \ldots, \tilde{e}_m]$, so is $\mathbb{C}[x_1, \ldots, x_m]^G$. It follows that $\mathbb{C}[x_1, \ldots, x_m]^G$ is a finitely generated algebra over $\mathbb{C}$.

Next we prove that the points of Y/G correspond to the G-orbits on Y. To see that $\pi(p) \ne \pi(q)$ for two points $p, q \in Y$ not in the same G-orbit, observe that we can find a function $f \in A[Y]$ such that $f(p) = 0$ but $f(g(q)) \ne 0$ for every $g \in G$. The product $\prod_{g \in G} g^*(f)$ gives rise to a function in $A[Y]^G$ which vanishes at $\pi(p)$ but not at $\pi(q)$. To see that π is surjective, let $\overline{\mathfrak{m}} = (h_1, \ldots, h_k)$ be a maximal ideal in $A[Y]^G$. We claim that $\overline{\mathfrak{m}} \cdot A[Y] \ne A[Y]$. If $\overline{\mathfrak{m}} \cdot A[Y] = A[Y]$, then

$$1 = a_1 h_1 + \ldots + a_k h_k$$

for some $a_1, \ldots, a_k \in A[Y]$. Applying g^* for every $g \in G$ and summing up yield

$$|G| = \left(\sum_{g \in G} g^*(a_1)\right) h_1 + \ldots + \left(\sum_{g \in G} g^*(a_k)\right) h_k.$$

Thus $1 \in \overline{\mathbf{m}}$ which is absurd. Now let $\mathbf{m}$ be a maximal ideal of $A[Y]$ containing $\overline{\mathbf{m}} \cdot A[Y]$. Then $\overline{\mathbf{m}} = \mathbf{m} \cap A[Y]^G$.

To prove the universal property of π, we may assume that W is affine. Then $\tilde{\pi} : Y \to W$ corresponds to a homomorphism $\tilde{\alpha} : A[W] \to A[Y]$. It suffices to prove that $\tilde{\alpha}(A[W]) \subset A[Y]^G$. Let $f \in A[W]$. For every $g \in G$ and $p \in Y$, we have

$$\big(g^*(\tilde{\alpha}(f))\big)(p) = (\tilde{\alpha}(f))(g(p)) = f\big(\tilde{\pi}(g(p))\big) = f\big(\tilde{\pi}(p)\big) = (\tilde{\alpha}(f))(p)$$

by the assumptions on $\tilde{\pi}$. So $g^*(\tilde{\alpha}(f)) = \tilde{\alpha}(f)$ for every $g \in G$, and $\tilde{\alpha}(f) \in A[Y]^G$.

Finally, if Y is not affine, then there exists an open cover of Y by G-invariant affine open subsets Y_i, and Y/G is the gluing of the quasi-projective varieties Y_i/G. We refer to [**Harr**, p. 127] for the proof that Y/G is quasi-projective. □

By Theorem 1.8, if X is a quasi-projective complex variety, then so is the symmetric product $X^{(n)}$. Note that the symmetric product $X^{(n)}$ may not be smooth even if X itself is smooth.

EXAMPLE 1.9. (i) Note that $\mathbb{C}[x_1, \ldots, x_n]^{S_n} = \mathbb{C}[\tilde{e}_1, \ldots, \tilde{e}_n]$ where

$$\tilde{e}_i = e_i(x_1, \ldots, x_n, 0, \ldots)$$

with $1 \le i \le n$ are the elementary symmetric polynomials with the n variables $x_1, \ldots, x_n$. It follows that $(\mathbb{A}^1)^{(n)} = \mathbb{A}^n$. Similarly, $(\mathbb{P}^1)^{(n)} = \mathbb{P}^n$.

(ii) If X is a smooth curve, then locally in the étale topology X is isomorphic to $\mathbb{A}^1$. In view of (i), the symmetric product $X^{(n)}$ is smooth.

EXAMPLE 1.10. The symmetric product $(\mathbb{A}^2)^{(2)}$ is not smooth. In fact,

$$(\mathbb{A}^2)^{(2)} \cong \mathbb{A}^2 \times \operatorname{Spec} \frac{\mathbb{C}[u, v, w]}{\langle uw - v^2 \rangle}. \tag{1.11}$$

To prove (1.11), let x_1, y_1 and x_2, y_2 be the coordinates on the two factors of $\mathbb{A}^2 \times \mathbb{A}^2$. Let $s = x_1 - x_2$, $t = y_1 - y_2$, $s' = x_1 + x_2$ and $t' = y_1 + y_2$. Denote the transposition in S_2 by τ. Then, $\tau(s) = -s$, $\tau(t) = -t$, $\tau(s') = s'$, and $\tau(t') = t'$. So

$$\mathbb{C}[x_1, y_1, x_2, y_2]^{S_2} = \mathbb{C}[s, t, s', t']^{S_2} = \big(\mathbb{C}[s, t]^{S_2}\big)[s', t'].$$

Note that $\tau(s^i t^j) = (-1)^{i+j} s^i t^j$. Thus, $\mathbb{C}[s, t]^{S_2}$ is the subalgebra generated by s^2, st, t^2. Putting $u = s^2$, $v = st$ and $w = t^2$, we see that

$$\mathbb{C}[s, t]^{S_2} \cong \frac{\mathbb{C}[u, v, w]}{\langle uw - v^2 \rangle}.$$

Hence

$$(\mathbb{A}^2)^{(2)} = \operatorname{Spec}\big(\mathbb{C}[x_1, y_1, x_2, y_2]^{S_2}\big) \cong \mathbb{A}^2 \times \operatorname{Spec} \frac{\mathbb{C}[u, v, w]}{\langle uw - v^2 \rangle}.$$

1.4. Hilbert schemes of points

Let X be a projective complex scheme and $\mathcal{O}_X(1)$ be an ample line bundle on X. Consider the contravariant functor

$$\mathcal{H}ilb_X : [\text{Schemes}] \to [\text{Sets}]$$

from the category of complex schemes to the category of sets defined by

$$\mathcal{H}ilb_X(U) = \{Z \subset U \times X | Z \text{ is a closed subscheme and flat over } U\}.$$

Namely, $\mathcal{H}ilb_X$ is the functor associating a complex scheme U to the set of flat families of closed subschemes in X parametrized by U. Let $\pi : Z \to U$ be the restriction of the natural projection $U \times X \to U$. For $u \in U$, we define the Hilbert polynomial

$$P_u(m) = \chi\big(\mathcal{O}_{Z_u} \otimes \mathcal{O}_X(m)\big)$$

where $Z_u = \pi^*(u)$. For each polynomial P, let $\mathcal{H}ilb_X^P$ be the subfunctor of $\mathcal{H}ilb_X$ associating a complex scheme U to the set of flat families of closed subschemes in X parametrized by U such that P is the Hilbert polynomial of the closed subschemes. Grothendieck [**Grot**] proved the following fundamental result.

THEOREM 1.11. *The functor $\mathcal{H}ilb_X^P$ is representable by a projective scheme* Hilb_X^P.

In particular, there exists a closed subscheme $\mathcal{Z} \subset \text{Hilb}_X^P \times X$ such that $\mathcal{Z} \to \text{Hilb}_X^P$ is a universal family in the following sense: if Z is a flat family of closed subschemes in X parametrized by U such that P is the Hilbert polynomial of the closed subschemes, then there exists a unique morphism $f : U \to \text{Hilb}_X^P$ such that

$$Z = (f \times \text{Id}_X)^* \mathcal{Z}.$$

Furthermore, if Y is an open subscheme of X, then there is a corresponding open subscheme Hilb_Y^P of Hilb_X^P parametrizing closed subschemes in Y. In particular, Hilb_Y^P is defined for a quasi-projective scheme Y.

If $[Z] \in \text{Hilb}_X^P$ is a point corresponding to a closed subscheme Z of X, then the Zariski tangent space of Hilb_X^P at $[Z]$ is canonically isomorphic to

$$\text{Hom}(I_Z, \mathcal{O}_Z). \tag{1.12}$$

Since every homomorphism $I_Z \to \mathcal{O}_Z$ of $\mathcal{O}_X$-modules factors uniquely through I_Z/I_Z^2, the Zariski tangent space of Hilb_X^P at $[Z]$ is also canonically isomorphic to

$$\text{Hom}(I_Z/I_Z^2, \mathcal{O}_Z). \tag{1.13}$$

DEFINITION 1.12. Let n be a positive integer. We define $X^{[n]}$ to be Hilb_X^P where P is the constant polynomial given by $P(m) = n$ for all $m \in \mathbb{Z}$, and call it the *Hilbert scheme of n points* in X. An element $\xi \in X^{[n]}$ represents a 0-dimensional closed subscheme of X with length $\ell(\xi) = h^0(X, \mathcal{O}_\xi) = n$. For convenience, we define $X^{[0]}$ to be a reduced point.

EXAMPLE 1.13. (i) Note that a length-1 0-dimensional closed subscheme of a smooth scheme X is simply a point of X. So $X^{[1]} = X$. The universal closed subscheme in $X^{[1]} \times X = X \times X$ is the diagonal in $X \times X$.

(ii) Let ξ be the closed subscheme of a smooth scheme X consisting of distinct points $x_1, \ldots, x_n \in X$, denoted by $\xi = x_1 + \ldots + x_n$. Then, $\xi \in X^{[n]}$. These elements

ξ form a smooth open (but not necessarily dense) subset of $X^{[n]}$ of dimension $\dim(X) \cdot n$.

(iii) A length-n 0-dimensional closed subscheme of $\mathbb{A}^d$ corresponds to an ideal $I \subset \mathbb{C}[x_1, \ldots, x_d]$ of *colength-n*, i.e.,

$$\dim \mathbb{C}[x_1, \ldots, x_d]/I = n.$$

Therefore, as sets,

$$(\mathbb{A}^d)^{[n]} = \{I \subset \mathbb{C}[x_1, \ldots, x_d] | \, I \text{ is a colength-}n \text{ ideal of } \mathbb{C}[x_1, \ldots, x_d]\}.$$

THEOREM 1.14. *Let X be a smooth irreducible variety with* $\dim(X) \leq 2$. *Then $X^{[n]}$ is a smooth irreducible variety of dimension* $\dim(X) \cdot n$.

PROOF. Since X is connected, $X^{[n]}$ is connected as well. In view of Example 1.13 (ii), it remains to prove that for every element $\xi \in X^{[n]}$, the dimension of the Zariski tangent space $T_{\xi, X^{[n]}}$ of $X^{[n]}$ at ξ is at most $\dim(X) \cdot n$. By (1.12), we have

$$T_{\xi, X^{[n]}} = \operatorname{Hom}(I_\xi, \mathcal{O}_\xi).$$

When $\dim(X) = 1$, we obtain $I_\xi = \mathcal{O}_X(-\xi)$ and $\operatorname{Hom}(I_\xi, \mathcal{O}_\xi) \cong H^0(X, \mathcal{O}_\xi)$. So $\dim T_{\xi, X^{[n]}} = \ell(\xi) = n$ and we are done.

Let $\dim(X) = 2$. Applying the functor $\operatorname{Hom}(\cdot, \mathcal{O}_\xi)$ to the exact sequence

$$0 \to I_\xi \to \mathcal{O}_X \to \mathcal{O}_\xi \to 0$$

yields an exact sequence

$$0 \to \operatorname{Hom}(\mathcal{O}_\xi, \mathcal{O}_\xi) \to \operatorname{Hom}(\mathcal{O}_X, \mathcal{O}_\xi) \to \operatorname{Hom}(I_\xi, \mathcal{O}_\xi) \to \operatorname{Ext}^1(\mathcal{O}_\xi, \mathcal{O}_\xi).$$

Since

$$\dim \operatorname{Hom}(\mathcal{O}_\xi, \mathcal{O}_\xi) = \ell(\xi) = n = \dim \operatorname{Hom}(\mathcal{O}_X, \mathcal{O}_\xi),$$

we have an injection

$$0 \to \operatorname{Hom}(I_\xi, \mathcal{O}_\xi) \to \operatorname{Ext}^1(\mathcal{O}_\xi, \mathcal{O}_\xi).$$

So it remains to prove that

$$\dim \operatorname{Ext}^1(\mathcal{O}_\xi, \mathcal{O}_\xi) = 2n. \tag{1.14}$$

We have

$$\dim \operatorname{Ext}^2(\mathcal{O}_\xi, \mathcal{O}_\xi) = \dim \operatorname{Hom}(\mathcal{O}_\xi, \mathcal{O}_\xi \otimes K_X) = n$$

by the Serre duality and $\dim \operatorname{Hom}(\mathcal{O}_\xi, \mathcal{O}_\xi) = n$. By the Riemann-Roch Theorem,

$$\sum_{i=0}^{2} (-1)^i \cdot \dim \operatorname{Ext}^i(\mathcal{O}_\xi, \mathcal{O}_\xi) = \chi(\mathcal{O}_\xi, \mathcal{O}_\xi) = 0.$$

It follows that $\dim \operatorname{Ext}^1(\mathcal{O}_\xi, \mathcal{O}_\xi) = 2n$. This proves (1.14). □

REMARK 1.15. When $\dim X = 2$, Theorem 1.14 is due to Fogarty [**Fog1**, Theorem 2.4]. In a different direction, one can also show that if X is a smooth variety, then $X^{[n]}$ is smooth and irreducible for all $n \leq 3$ [**FG2**, Remark 7.2.5].

EXAMPLE 1.16. Let X be a smooth quasi-projective variety. An element $\xi \in X^{[2]}$ is either a set $\{x_1, x_2\}$ of distinct points in X or a length-2 closed subscheme supported at one point $x \in X$. In the latter case, let m be the ideal of x. Then, $m^2 \subset I_\xi \subset m$. So I_ξ is given by a codimension-1 subspace of m/m^2, i.e., by a point in $\mathbb{P}(T_{x,X}) = \mathbb{P}^{d-1}$ where d denotes the dimension of X. Hence an element in $X^{[2]}$

is either a set of two distinct points or a point together with a tangent direction at that point. In fact, the Hilbert scheme $X^{[2]}$ can be described globally as follows. Let $\widetilde{X^2}$ be the blow-up of X^2 along the diagonal. The action of S_2 on X^2 extends to $\widetilde{X^2}$. Then $X^{[2]} = \widetilde{X^2}/S_2$.

Let X be a smooth quasi-projective variety. Sending an element $\xi \in X^{[n]}$ to the element

$$\sum_{x \in X} \ell(\mathcal{O}_{\xi,x})\, x \in X^{(n)}$$

defines the *Hilbert-Chow morphism*

$$\rho_n : X^{[n]} \to X^{(n)}. \tag{1.15}$$

A detailed construction of ρ_n can be found in Section 5.4 of [**MF**].

DEFINITION 1.17. (i) For a closed subset S of X, define $M_n(S) \subset X^{[n]}$ to be the closed subset

$$M_n(S) = \bigcup_{x \in S} (\rho_n)^{-1}(nx)$$

with the reduced induced subscheme structure.

(ii) The *punctual Hilbert scheme* at a point $x \in X$ is defined to be $M_n(x)$.

Analytically, the punctual Hilbert scheme $M_n(x)$ is independent of the point $x \in X$ and the smooth variety X as long as the dimension of X is fixed. In fact, let $d = \dim X$,

$$R = \mathbb{C}[[x_1, \ldots, x_d]]$$

be the local ring of formal power series in the variables $x_1, \ldots, x_d$, and $\mathbf{m} = (x_1, \ldots, x_d)$ be the maximal ideal of R. Then,

$$M_n(x) \cong \mathrm{Hilb}^n(R)$$

which denotes the Hilbert scheme parametrizing all colength-n ideals in $R/\mathbf{m}^n$. The punctual Hilbert scheme $\mathrm{Hilb}^n(R)$ has been studied extensively (e.g. [**Bri, Iar1, Iar2**]). The following is due to Briancon [**Bri**].

THEOREM 1.18. *The punctual Hilbert scheme* $\mathrm{Hilb}^n(\mathbb{C}[[x,y]])$ *is irreducible of dimension* $(n-1)$. *Moreover, generic elements in* $\mathrm{Hilb}^n(\mathbb{C}[[x,y]])$ *correspond to curvi-linear ideals of the form*

$$\langle y - a_1 x - \ldots - a_{n-1} x^{n-1}, x^n \rangle$$

where $a_1, \ldots, a_{n-1} \in \mathbb{C}$.

REMARK 1.19. Let X be a smooth surface, and let $n_1 x_1 + \ldots + n_\ell x_\ell \in X^{(n)}$ where $x_1, \ldots, x_\ell$ are distinct and $n_1, \ldots, n_\ell > 0$. Then

$$\begin{aligned} (\rho_n)^{-1}(n_1 x_1 + \ldots + n_\ell x_\ell) &\cong \prod_{i=1}^{\ell} (\rho_{n_i})^{-1}(n_i x_i) \\ &\cong \prod_{i=1}^{\ell} \mathrm{Hilb}^{n_i}(\mathbb{C}[[x,y]]). \end{aligned}$$

By Theorem 1.18, $(\rho_n)^{-1}(n_1 x_1 + \ldots + n_\ell x_\ell)$ is irreducible of dimension $(n - \ell)$.

REMARK 1.20. Fix a point $x \in X$. For $m \geq 0$ and $n > 0$, define

$$M_{m,m+n}(x) = \{(\xi, \xi')| \, \xi \subset \xi'\} \subset M_m(x) \times M_{m+n}(x). \tag{1.16}$$

Ellingsrud and Strømme [**ES3**] proved that $M_{m,m+1}(x)$ is irreducible with

$$\dim M_{m,m+1}(x) = m. \tag{1.17}$$

PROPOSITION 1.21. *Let X be a smooth quasi-projective variety.*

(i) *If $\dim X = 1$, then the Hilbert-Chow morphism ρ_n is an isomorphism.*
(ii) *If $\dim X = 2$, then ρ_n is a resolution of singularities.*

PROOF. (i) Since $\dim X = 1$, every ideal sheaf is invertible. Thus for every $x \in X$ and every positive integer k, there exists a unique length-k 0-dimensional closed subscheme ξ supported at x. This unique subscheme ξ is defined by the ideal sheaf $\mathcal{O}_X(-kx)$, and $\rho_k(\xi) = kx$. It follows that the Hilbert-Chow morphism $\rho_n : X^{[n]} \to X^{(n)}$ is bijective. By Theorem 1.14 and Example 1.9 (ii), both $X^{[n]}$ and $X^{(n)}$ are smooth of dimension n. Hence ρ_n is an isomorphism in view of Zariski's Main Theorem.

(ii) Let $X_0^{(n)}$ be the open subset of $X^{(n)}$ consisting of sums of n distinct points. Then $X_0^{(n)}$ is smooth, and ρ_n is an isomorphism over $X_0^{(n)}$. By Theorem 1.14, $X^{[n]}$ is smooth, irreducible and of dimension $2n$. Since $X^{(n)}$ is irreducible and of dimension $2n$, it remains to prove that the singular locus $\mathrm{Sing}\big(X^{(n)}\big)$ of $X^{(n)}$ is $X^{(n)} \backslash X_0^{(n)}$.

Let $\pi : X^n \to X^{(n)}$ be the quotient map. The big diagonal of X^n is

$$\Delta = \bigcup_{1 \leq i < j \leq n} \Delta_{i,j}$$

where

$$\Delta_{i,j} = \{(x_1, \ldots, x_n) \in X^n | x_i = x_j\}.$$

Let Δ_0 be the open subset of Δ where precisely two of the x_i's coincide. Let $p \in \Delta_0$. Without loss of generality, we may assume that $p \in \Delta_{1,2}$. By Example 1.10, a formal neighborhood of the point $\pi(p)$ in $X^{(n)}$ (i.e., the complete local ring of $X^{(n)}$ at $\pi(p)$) is isomorphic to

$$\mathrm{Spec}\, \frac{\mathbb{C}[[u, v, w, s', t', x_3, y_3, \ldots, x_n, y_n]]}{\langle uw - v^2 \rangle}.$$

Thus all the points in $\pi(\Delta_0)$ are singular points of $X^{(n)}$. So $\pi(\Delta_0)$ is contained in the singular locus $\mathrm{Sing}\big(X^{(n)}\big)$. Since $\mathrm{Sing}\big(X^{(n)}\big)$ is closed, the closure of $\pi(\Delta_0)$ is contained in $\mathrm{Sing}\big(X^{(n)}\big)$. Since the closure of $\pi(\Delta_0)$ is $X^{(n)} \backslash X_0^{(n)}$ and

$$\mathrm{Sing}\big(X^{(n)}\big) \subset X^{(n)} \backslash X_0^{(n)},$$

we conclude that $\mathrm{Sing}\big(X^{(n)}\big)$ is equal to $X^{(n)} \backslash X_0^{(n)}$. □

The generating function for the Euler characteristic $\chi\big(X^{[n]}\big)$ of the Hilbert scheme $X^{[n]}$ was calculated in [**Che1**] (the case $d = 1$ also follows from Proposition 1.21 (i) and (1.10) by setting $t = -1$, and the case $d = 2$ is due to Göttsche and follows from Theorem 1.23 below by setting $t = -1$).

PROPOSITION 1.22. *Let X be a smooth quasi-projective variety of dimension d, and let $M_n(O)$ be the punctual Hilbert scheme at the origin $O \in \mathbb{A}^d$. Then,*

$$\sum_{n=0}^{+\infty} \chi(X^{[n]})q^n = \left(\sum_{n=0}^{+\infty} \chi(M_n(O))q^n\right)^{\chi(X)} = \left(\sum_{n=0}^{+\infty} P_d(n)q^n\right)^{\chi(X)} \tag{1.18}$$

where $P_d(n)$ denotes the number of d-dimensional partitions of n.

PROOF. The first equality follows from the standard stratifications of $X^{[n]}$ and $X^{(n)}$ together with a careful analysis of the fibers of the Hilbert-Chow morphism ρ_n. We refer to [**Che1**] for the details. For the second equality in (1.18), it suffices to prove

$$\chi\big(M_n(O)\big) = P_d(n). \tag{1.19}$$

Note that the d-dimensional torus $\mathbb{T} = (\mathbb{C}^*)^d$ acts on $\mathbb{A}^d$ via

$$(t_1, \ldots, t_d)(x_1, \ldots, x_d) = (t_1x_1, \ldots, t_dx_d) \tag{1.20}$$

where $x_1, \ldots, x_d$ denote the affine coordinates of $\mathbb{A}^d$, and $(t_1, \ldots, t_d) \in \mathbb{T}$. This action lifts to a $\mathbb{T}$-action on $(\mathbb{A}^d)^{[n]}$, and $M_n(O)$ is invariant. By the Theorem 4.1 in [**LY**],

$$\chi\big(M_n(O)\big) = \chi\big(M_n(O)^{\mathbb{T}}\big) \tag{1.21}$$

where $M_n(O)^{\mathbb{T}}$ denotes the set of fixed points under the $\mathbb{T}$-action. Using Example 1.13 (iii), the elements in $M_n(O)^{\mathbb{T}}$ are in one-to-one correspondence with $\mathbb{T}$-invariant colength-n ideals I of $\mathbb{C}[x_1, \ldots, x_d]$. These ideals must be generated by monomials, and are in one-to-one correspondence with d-dimensional partitions of n. Indeed, given a d-dimensional partition $(\lambda_{i_1,\ldots,i_{d-1}})_{i_1,\ldots,i_{d-1}\ge 0}$ of n, the ideal of $\mathbb{C}[x_1, \ldots, x_d]$ generated by the monomials

$$x_1^{i_1} \cdots x_{d-1}^{i_{d-1}} x_d^{n_{i_1,\ldots,i_{d-1}}}$$

has colength-n. Conversely, given a colength-n ideal I of $\mathbb{C}[x_1, \ldots, x_d]$ generated by monomials, we obtain a d-dimensional partition $(\lambda_{i_1,\ldots,i_{d-1}})_{i_1,\ldots,i_{d-1}\ge 0}$ of n by putting

$$\lambda_{i_1,\ldots,i_{d-1}} = \min\Big\{i_d \big|\, x_1^{i_1} \cdots x_{d-1}^{i_{d-1}} x_d^{i_d} \in I\Big\}.$$

Therefore, $\chi\big(M_n(O)\big) = P_d(n)$. This completes the proof of (1.19). □

In particular, when $\dim X = 2$, we see from (1.18) and (1.2) that

$$\sum_{n=0}^{+\infty} \chi(X^{[n]})q^n = \prod_{n=1}^{+\infty} \frac{1}{(1-q^n)^{\chi(X)}}. \tag{1.22}$$

When $\dim X = 3$, we see from (1.18) and (1.3) that

$$\sum_{n=0}^{+\infty} \chi(X^{[n]})q^n = M(q)^{\chi(X)} \tag{1.23}$$

where

$$M(q) = \prod_{n=1}^{+\infty} \frac{1}{(1-q^n)^n}.$$

For the finer invariants of $X^{[n]}$ such as its Betti numbers, we have the following elegant formula [**Got1**, **GS**] when X is a surface.

THEOREM 1.23. *Let X be a smooth quasi-projective surface. Then,*

$$\sum_{n=0}^{+\infty} P_t(X^{[n]})q^n = \prod_{n=1}^{+\infty}\prod_{i=0}^{4}\left(\frac{1}{1-(-t)^{2n-2+i}q^n}\right)^{(-1)^i b_i(X)}.$$

It will be clear from Chapter 3 that Theorem 1.23 is important in the interplay between the Hilbert schemes of points and infinite dimensional Lie algebras.

Let $\mathcal{Z}_n \subset X^{[n]} \times X$ be the universal codimension-2 closed subscheme:

$$\mathcal{Z}_n = \big\{(\xi, x) \subset X^{[n]} \times X \,|\, x \in \mathrm{Supp}\,(\xi)\big\}. \tag{1.24}$$

Let $p : \mathcal{Z}_n \to X^{[n]}$ and $q : \mathcal{Z}_n \to X$ be the natural morphisms. For a sheaf F on X, let

$$F^{[n]} = p_* q^* F. \tag{1.25}$$

Since p is a degree-n flat morphism, if F is locally free of rank r, then $F^{[n]}$ is locally free of rank rn; in this case, $F^{[n]}$ is called a *tautological bundle* over $X^{[n]}$. We define the subset B_n of the Hilbert scheme $X^{[n]}$ by putting

$$B_n = \Big\{\xi \in X^{[n]} |\, |\mathrm{Supp}(\xi)| < n\Big\}. \tag{1.26}$$

Let X be a projective surface. Then B_n is codimension-1 in $X^{[n]}$ and called the *boundary* of the Hilbert scheme $X^{[n]}$. For a real-surface C in X, define

$$D_C = \Big\{\xi \in X^{[n]} |\ \mathrm{Supp}(\xi) \cap C \neq \emptyset\Big\}. \tag{1.27}$$

Note that D_C is a codimension-1 subvariety of $X^{[n]}$ when C is a smooth algebraic curve in X. We extend the notion D_C to all divisors C in X by linearity.

THEOREM 1.24. *Let $n > 1$ and X be a smooth projective surface. Then,*

(i) $\det\big((\mathcal{O}_X(C))^{[n]}\big) = D_C - B_n/2$ *for every divisor C in X. Moreover, if C is a smooth curve in X and $C^{[n]}$ is regarded as a closed subscheme of $X^{[n]}$, then*

$$c_n\big((\mathcal{O}_X(C))^{[n]}\big) = \big[C^{[n]}\big], \tag{1.28}$$

$$N_{C^{[n]}\subset X^{[n]}} = (\mathcal{O}_X(C))^{[n]}|_{C^{[n]}} \tag{1.29}$$

where $N_{C^{[n]}\subset X^{[n]}}$ denotes the normal bundle of $C^{[n]}$ in $X^{[n]}$.

(ii) $K_{X^{[n]}} = \rho_n^* K_{X^{(n)}} = D_{K_X}$.

(iii) $\mathrm{Pic}(X^{[n]}) \cong (\mathrm{Pic}(X^n))^{S_n} \oplus (B_n/2)\,\mathbb{Z}$.

Theorem 1.24 (i), (ii) and (iii) were proved in [**AIK**, **BSG**], [**Bea1**] and [**BSG**, **Fog2**] respectively. When $H^1(X, \mathcal{O}_X) = 0$, it follows from Theorem 1.24 (iii) that

$$\mathrm{Pic}(X^{[n]}) \cong \mathrm{Pic}(X) \oplus (B_n/2)\,\mathbb{Z}. \tag{1.30}$$

Under this isomorphism, the divisor $D_C \in \mathrm{Pic}(X^{[n]})$ corresponds to $C \in \mathrm{Pic}(X)$.

REMARK 1.25. Let $n > 1$ and X be a smooth projective surface. By Proposition 1.21 (ii) and Theorem 1.24 (ii), the Hilbert-Chow morphism $\rho_n : X^{[n]} \to X^{(n)}$ is a crepant resolution.

DEFINITION 1.26. Let L be a line bundle on a smooth projective surface X. Then,

(i) L is *nef* if $L \cdot C \ge 0$ for every curve C in X.
(ii) L is *nef and big* if it is nef and $L^2 > 0$.
(iii) L is *k-very ample* if the restriction

$$H^0(X, L) \to H^0(X, \mathcal{O}_\xi \otimes L)$$

is surjective for every $\xi \in X^{[k+1]}$.

The concept of k-very ampleness relates the Hilbert scheme $X^{[k+1]}$ to a Grassmannian as follows. The surjective map in Definition 1.26 (iii) represents an element in $\text{Grass}(H^0(X, L), k+1)$. So we get a morphism

$$\varphi_k(L): \ X^{[k+1]} \to \text{Grass}(H^0(X, L), k+1) \tag{1.31}$$

when L is k-very ample. Let $h = h^0(X, L)$ and

$$\mathfrak{p}: \ \text{Grass}(\mathbb{C}^h, k+1) \to \mathbb{P}((\wedge^{h-(k+1)}\mathbb{C}^h)^*)$$

be the Plücker embedding. Then by the Appendix of [**BSG**], we have

$$(\mathfrak{p} \circ \varphi_k(L))^*\mathcal{H} = \mathcal{O}_{X^{[k+1]}}\big(D_{c_1(L)} - B_{k+1}/2\big) \tag{1.32}$$

where $\mathcal{H}$ is the hyperplane line bundle over $\mathbb{P}((\wedge^{h-(k+1)}\mathbb{C}^h)^*)$.

THEOREM 1.27. *Let L be a line bundle on a smooth projective surface X.*

(i) *If L is nef and big, and if $L^2 \ge 4k+5$, then either $K_X + L$ is k-very ample, or there exists an effective divisor D such that $L - 2D$ is $\mathbb{Q}$-effective, D contains some 0-cycle of degree $\le (k+1)$ where the k-very ampleness fails, and*

$$L \cdot D - k - 1 \le D^2 < L \cdot D/2 < (k+1).$$

(ii) *If L is $(k+1)$-very ample, then the morphism $\varphi_k(L)$ in (1.31) is an embedding.*

Definition 1.26 (iii) and Theorem 1.27 (i) are from [**BSG**], and Theorem 1.27 (ii) is from [**CG**].

REMARK 1.28. Let $k \ge 1$ be an integer. By Theorem 1.27 (i), there exists a very ample divisor C on X such that the line bundle $L = \mathcal{O}_X(C)$ is $(k+1)$-very ample. Then by Theorem 1.27 (ii), the morphism $\varphi_k(L)$ is an embedding. Therefore, by (1.32), the divisor $(D_C - B_{k+1}/2)$ on $X^{[k+1]}$ is very ample.

Let X be a smooth projective surface. A basic intersection formula ([**ES3**, Theorem 1.1] and [**Nak5**, Theorem 9.20]) for $X^{[n]}$ states that

$$[M_n(A_1)] \cdot [M_n(A_2)] = (-1)^{n-1} n(A_1 \cdot A_2) \tag{1.33}$$

where A_1 and A_2 are subvarieties of X with $\dim(A_1) + \dim(A_2) = 2$.

For $n \ge 2$, let $\beta_n \in H_2(X^{[n]}; \mathbb{Z})$ be the homology class in $H_2(X^{[n]})$ represented by a curve of the form:

$$M_2(x) + \eta \tag{1.34}$$

where $x \in X$ and $\eta \in X^{[n-2]}$ are fixed such that $x \notin \mathrm{Supp}(\eta)$. Then,

$$\langle \beta_n, B_n \rangle = [M_2(x)] \cdot [M_2(X)] = -2 \tag{1.35}$$

where we have used (1.33) in the last step.

REMARK 1.29. Using (1.35), we now prove the formula

$$K_{X^{[n]}} = \rho_n^* K_{X^{(n)}}$$

in Theorem 1.24 (ii). Since $\rho_n : X^{[n]} \to X^{(n)}$ is the desingularization of $X^{(n)}$ and the exceptional divisor B_n is irreducible,

$$K_{X^{[n]}} = \rho_n^* K_{X^{(n)}} + aB_n$$

for some rational number a. So it suffices to show $a = 0$. To see this, we have

$$a = -K_{X^{[n]}} \cdot \beta_n/2 = -K_{X^{[2]}} \cdot \beta_2/2$$

by (1.35). Recall from Example 1.16 that $X^{[2]} = \widetilde{X^2}/S_2$ where $\widetilde{X^2}$ is the blow-up of X^2 along the diagonal Δ. Let E be the exceptional divisor in $\widetilde{X^2}$, and let f be a fiber of the $\mathbb{P}^1$-bundle $E \to \Delta$. Let $\pi_1 : \widetilde{X^2} \to X^2$ be the blowing-up morphism, and let $\pi_2 : \widetilde{X^2} \to X^{[2]}$ be the quotient map. Then, $K_{\widetilde{X^2}} = \pi_1^* K_{X^2} + E$ and $K_{\widetilde{X^2}} = \pi_2^* K_{X^{[2]}} + E$. Thus, $\pi_1^* K_{X^2} = \pi_2^* K_{X^{[2]}}$. Finally, since $\pi_{2*} f = \beta_2$, we obtain

$$a = -K_{X^{[2]}} \cdot \pi_{2*} f/2 = -\pi_2^* K_{X^{[2]}} \cdot f/2 = -\pi_1^* K_{X^2} \cdot f/2 = 0$$

where the projection formula has been used twice.

1.5. Incidence Hilbert schemes

In this section, let X be a smooth projective surface. The incidence Hilbert scheme $X^{[n,n+1]}$ is defined by

$$X^{[n,n+1]} = \{(\xi, \eta) \,|\, \xi \subset \eta\} \subset X^{[n]} \times X^{[n+1]}. \tag{1.36}$$

It is known [**Che2**, **Tik**] that $X^{[n,n+1]}$ is irreducible, smooth and of dimension $2(n+1)$. Define

$$M_{n,n+1}(X) = \{(\xi, \eta) \in X^{[n,n+1]} |\ \mathrm{Supp}(\eta) = \{x\} \text{ for some } x \in X\}. \tag{1.37}$$

NOTATION 1.30. *We fix the following notations for various morphisms:*

$$\begin{aligned}
f_{n+1} &: X^{[n,n+1]} \to X^{[n]} \text{ with } f_{n+1}(\xi,\eta) = \xi. \\
g_{n+1} &: X^{[n,n+1]} \to X^{[n+1]} \text{ with } g_{n+1}(\xi,\eta) = \eta. \\
\psi_{n+1} &: X^{[n,n+1]} \to \mathcal{Z}_{n+1} \text{ with } \psi_{n+1}(\xi,\eta) = \big(\eta, \mathrm{Supp}(I_\xi/I_\eta)\big). \\
\rho &: X^{[n,n+1]} \to X \text{ with } \rho(\xi,\eta) = \mathrm{Supp}(I_\xi/I_\eta).
\end{aligned}$$

The following theorem was proved in [**ES3**].

THEOREM 1.31. (i) *The morphism $\psi_{n+1} : X^{[n,n+1]} \to \mathcal{Z}_{n+1}$ is canonically isomorphic to the projectification $\mathbb{P}(\omega_{\mathcal{Z}_{n+1}}) \to \mathcal{Z}_{n+1}$ where $\omega_{\mathcal{Z}_{n+1}}$ is the dualizing sheaf of $\mathcal{Z}_{n+1}$.*

(ii) *The morphism $(f_{n+1}, \rho)\colon X^{[n,n+1]} \to X^{[n]} \times X$ is canonically isomorphic to the blowing-up of $X^{[n]} \times X$ along $\mathcal{Z}_n$. The exceptional locus E_{n+1} is equal to*

$$\big\{(\xi, \eta) \in X^{[n,n+1]} \,|\, \mathrm{Supp}(\xi) = \mathrm{Supp}(\eta)\big\}.$$

(iii) *As cycles in the Chow group,*

$$\frac{1}{n+1}\, g_{n+1}^*[M_{n+1}(X)] = [M_{n,n+1}(X)].$$

Put $f_X = f_{n+1} \times \mathrm{Id}_X$, $g_X = g_{n+1} \times \mathrm{Id}_X$ and $\rho_X = \rho \times \mathrm{Id}_X$. Let Δ_X be the diagonal in $X \times X$. Then there exists an exact sequence over $X^{[n,n+1]} \times X$:

$$0 \to \rho_X^* \mathcal{O}_{\Delta_X} \otimes p_1^* \mathcal{O}_{X^{[n,n+1]}}(-E_{n+1}) \to g_X^* \mathcal{O}_{\mathcal{Z}_{n+1}} \to f_X^* \mathcal{O}_{\mathcal{Z}_n} \to 0 \tag{1.38}$$

where $p_1 : X^{[n,n+1]} \times X \to X^{[n,n+1]}$ is the first projection. Let F be a locally free sheaf on X. Applying the functor $p_{1*}(\cdot \otimes p_1^* \rho^* F)$ to (1.38) yields an exact sequence:

$$0 \to \rho^* F \otimes \mathcal{O}_{X^{[n,n+1]}}(-E_{n+1}) \to g_{n+1}^* F^{[n+1]} \to f_{n+1}^* F^{[n]} \to 0. \tag{1.39}$$

In particular, setting $F = \mathcal{O}_X$ and using Theorem 1.24 (i), we obtain

$$g_{n+1}^* B_{n+1} - f_{n+1}^* B_n = 2E_{n+1}. \tag{1.40}$$

The generating function for the Poincaré polynomials of $X^{[n,n+1]}$ has a closed formula due to Cheah [**Che3**, Section 0.3]:

$$\begin{aligned} &\sum_{n=0}^{+\infty} P_t(X^{[n,n+1]}) q^n \\ = &\frac{P_t(X)}{1-t^2 q} \cdot \prod_{n=1}^{+\infty} \prod_{i=0}^{4} \left(\frac{1}{1-(-t)^{2n-2+i} q^n} \right)^{(-1)^i b_i(X)} \\ = &\frac{P_t(X)}{1-t^2 q} \cdot \sum_{n=0}^{+\infty} P_t(X^{[n]}) q^n. \end{aligned} \tag{1.41}$$

Formula (1.41) will be important in Chapter 6 which deals with the relations between the incidence Hilbert schemes of points and infinite dimensional Lie algebras.

We refer to [**Can**, **Che2**, **Che3**] for generalizations from incidence Hilbert schemes to nested Hilbert schemes of points on surfaces, and to [**GSY1**, **GSY2**] for generalizations to nested Hilbert schemes of points *and curves* on surfaces which are related to localized Donaldson-Thomas theory.

CHAPTER 2

The nef cone and flip structure of $(\mathbb{P}^2)^{[n]}$

The Hilbert schemes of points on the complex projective plane $\mathbb{P}^2$ provide excellent and simple examples illustrating some geometric aspects of the Hilbert schemes of points on general surfaces. In this chapter, we will study the nef cone and birational geometry of the Hilbert schemes $(\mathbb{P}^2)^{[n]}$. We will begin with the description of certain curves which are contracted by the Hilbert-Chow morphism (1.15) for a simply connected surface X. Then for the Hilbert scheme $(\mathbb{P}^2)^{[n]}$, we will describe its nef cone and the cone spanned by all the effective curves, and study the curves having minimal degree with respect to a certain very ample divisor. We will end with the construction of a flip structure on $(\mathbb{P}^2)^{[n]}$ for $n \geq 3$. The main results in this chapter appeared in the announcement [**LQZ**], and have been partially generalized in [**QT**]. We refer to [**ABCH, LZ**] for further birational geometry of $(\mathbb{P}^2)^{[n]}$ in connection with the Bridgeland stability.

2.1. Curves homologous to β_n

In this section, for simplicity, X stands for a simply connected smooth projective surface. We will determine all the curves in the Hilbert scheme $X^{[n]}$ homologous to β_n defined in (1.34), and study their moduli space. The understanding of these curves plays an important role in studying the extremal Gromov-Witten invariants of the Hilbert scheme $X^{[n]}$ which will be dealt with in Chapters 14 and 15.

First of all, we determine bases of $H_2(X^{[n]})$ and $H_{4n-2}(X^{[n]})$. For a real-surface C in X, define β_C to be the homology class in $H_2(X^{[n]})$ represented by a curve of the form:

$$\{x + \eta | \, x \in C\} \tag{2.1}$$

where $\eta \in X^{[n-1]}$ is fixed such that $C \cap \mathrm{Supp}(\eta) = \emptyset$. Recall the notations B_n and D_C from (1.26) and (1.27) respectively.

LEMMA 2.1. *Let $n \geq 2$ and X be a simply connected smooth projective surface. Let $C_1, \ldots, C_s$ be real-surfaces in X such that their homology classes form a basis of $H_2(X)$. Then,*

(i) *a basis of $H_2(X^{[n]})$ consists of the homology classes*

$$\beta_n, \beta_{C_1}, \ldots, \beta_{C_s};$$

(ii) *a basis of $H_{4n-2}(X^{[n]})$ consists of the homology classes*

$$B_n, D_{C_1}, \ldots, D_{C_s}.$$

PROOF. Let $\widetilde{C}_1, \ldots, \widetilde{C}_s$ be real-surfaces in X such that

$$\widetilde{C}_1 \sim C_1, \ldots, \widetilde{C}_s \sim \widetilde{C}_s,$$

where the symbol "$A_1 \sim A_2$" means that A_1 and A_2 are homologous. We further assume that $C_1, \ldots, C_s, \widetilde{C}_1, \ldots, \widetilde{C}_s$ are in general position. It follows that

$$\langle \beta_{C_i}, D_{\widetilde{C}_j} \rangle = \langle C_i, \widetilde{C}_j \rangle \tag{2.2}$$

for all $i, j = 1, \ldots, s$. By (1.35), $\langle \beta_n, B_n \rangle = -2$. Also, it is clear that for all $j = 1, \ldots, s$, we have

$$\langle \beta_n, D_{\widetilde{C}_j} \rangle = 0. \tag{2.3}$$

Thus, the matrix consisting of the intersection pairings between $\beta_n, \beta_{C_1}, \ldots, \beta_{C_s}$ and $B_n, D_{C_1}, \ldots, D_{C_s}$ is nondegenerate. By Theorem 1.23, the second Betti number of $X^{[n]}$ is equal to $1 + b_2(X) = 1 + s$. It follows that $\beta_n, \beta_{C_1}, \ldots, \beta_{C_s}$ form a basis of $H_2(X^{[n]})$ and $B_n, D_{C_1}, \ldots, D_{C_s}$ form a basis of $H_{4n-2}(X^{[n]})$. □

Next, we will study the curves γ in the punctual Hilbert scheme $M_n(x) \subset X^{[n]}$ such that $\gamma \sim \beta_n$. We will establish some notations, and prove three technical lemmas. Throughout this section, we identify $M_n(x)$ with the punctual Hilbert scheme $M_n(O)$ of $\mathbb{C}^2$ at the origin O. Let

$$R = \mathcal{O}_{\mathbb{C}^2, O}$$

be the local ring of $\mathbb{C}^2$ at the origin, and $\mathfrak{m} = \langle u, v \rangle$ be the maximal ideal of R where u, v form a coordinate system of $\mathbb{C}^2$ at the origin. Let $\eta \in M_n(O)$. It is known from [**Iar1**] that $I_{\eta, O} \supset \mathfrak{m}^n$ and there is an embedding

$$\tau\colon M_n(O) \to \mathrm{Grass}(R/\mathfrak{m}^n, n) \tag{2.4}$$

where $R/\mathfrak{m}^n$ is considered as a $\mathbb{C}$-vector space of dimension $\binom{n+1}{2}$, and τ maps $\eta \in M_n(O)$ to the quotient

$$R/\mathfrak{m}^n \to R/I_{\eta, O} = \mathcal{O}_{\eta, O} \to 0.$$

For simplicity, put

$$\begin{aligned} \mathbb{G} &= \mathrm{Grass}(R/\mathfrak{m}^n, n), \\ \ell_1 &= \binom{n+1}{2}. \end{aligned}$$

Then over $\mathbb{G}$, there are a universal quotient bundle $\mathcal{Q}$, a universal subbundle $\mathcal{S}$ and an exact sequence:

$$0 \to \mathcal{S} \to (R/\mathfrak{m}^n) \otimes \mathcal{O}_{\mathbb{G}} \to \mathcal{Q} \to 0. \tag{2.5}$$

It is known that $\det(\mathcal{Q})$ induces the Plücker embedding

$$\mathfrak{p} : \mathbb{G} \to \mathbb{P}((\wedge^{\ell_1 - n} \mathbb{C}^{\ell_1})^*). \tag{2.6}$$

So if $\mathcal{H}$ is the hyperplane line bundle over $\mathbb{P}((\wedge^{\ell_1 - n} \mathbb{C}^{\ell_1})^*)$, then $\det(\mathcal{Q}) = \mathfrak{p}^* \mathcal{H}$.

LEMMA 2.2. *Let $n \geq 2$. Fix a point $x \in X$ and a curve $\gamma \subset M_n(x)$. Regard $\gamma \subset M_n(O)$ via the identification $M_n(x) \cong M_n(O)$. Then $\gamma \sim \beta_n$ if and only if $\gamma \cdot B_n = -2$ if and only if $\mathfrak{p} \circ \tau(\gamma)$ is a line.*

PROOF. Consider the rank-n tautological bundle $\mathcal{O}_X^{[n]}$ over $X^{[n]}$. Following the similar argument as in the proof of the Claim 2 in [**BSG**, Appendix], we can show that $\mathcal{O}_X^{[n]}|_{M_n(x)}$ is a quotient of the trivial bundle $(\mathcal{O}_{M_n(x)})^{\oplus \ell_1}$. The fiber of $\mathcal{O}_X^{[n]}|_{M_n(x)}$ over $\eta \in M_n(x) \subset X^{[n]}$ is $H^0(X, \mathcal{O}_\eta) = \mathcal{O}_{\eta, x}$. Since the fiber of $\tau^* \mathcal{Q}$ over $\eta \in M_n(O)$ is $\mathcal{O}_{\eta, O}$, we see from the universality of the quotient bundle that $\mathcal{O}_X^{[n]}|_{M_n(x)} \cong \tau^*\mathcal{Q}$. By Theorem 1.24 (i), $c_1(\mathcal{O}_X^{[n]}) = -B_n/2$. Combining with $\det(\mathcal{Q}) = \mathfrak{p}^*\mathcal{H}$ yields

$$\begin{aligned} \gamma \cdot B_n &= -2\gamma \cdot c_1(\mathcal{O}_X^{[n]}) \\ &= -2\gamma \cdot c_1(\mathcal{O}_X^{[n]}|_{M_n(x)}) \\ &= -2\gamma \cdot c_1(\tau^*\mathcal{Q}) \\ &= -2\gamma \cdot \tau^*\mathfrak{p}^* c_1(\mathcal{H}) \\ &= -2(\mathfrak{p}\circ\tau)(\gamma)\cdot c_1(\mathcal{H}). \end{aligned} \tag{2.7}$$

Next, since $\gamma \subset M_n(x)$, we have $\gamma \cdot D_C = 0 = \beta_n \cdot D_C$ for every real-surface C in X. By Lemma 2.1, $\gamma \sim \beta_n$ if and only if $\gamma \cdot B_n = -2 = \beta_n \cdot B_n$. By (2.7), $\gamma \cdot B_n = -2$ if and only if $\mathfrak{p}\circ\tau(\gamma)$ is a line. □

LEMMA 2.3. *Let $k = n(n-1)/2$, and $\mathbb{L} \subset \mathbb{G} = \mathrm{Grass}(R/\mathfrak{m}^n, n)$ be a curve. Then, $\mathbb{L}$ is mapped to a line by $\mathfrak{p}$ if and only if $\mathbb{L}$ is of the form*

$$\mathbb{L} = \left\{\mathrm{Span}(\overline{f}_1, \ldots, \overline{f}_{k-1}, \lambda\overline{f}_k + \mu\overline{f}_{k+1}) \mid \lambda, \mu \in \mathbb{C}, |\lambda| + |\mu| \neq 0\right\} \tag{2.8}$$

where $f_1, \ldots, f_{k+1} \in R$ are polynomials such that $\overline{f}_1, \ldots, \overline{f}_{k+1}$ are linearly independent in the vector space $R/\mathfrak{m}^n \cong \mathbb{C}^{n+k}$.

PROOF. From the definition of $\mathfrak{p}$, it is clear that if $\mathbb{L}$ is of the form (2.8), then $\mathfrak{p}(\mathbb{L})$ is a line. Conversely, assume that $\mathfrak{p}(\mathbb{L})$ is a line. We see from $\det(\mathcal{Q}) = \mathfrak{p}^*\mathcal{H}$ that $\det(\mathcal{Q}|_{\mathbb{L}}) \cong \mathcal{O}_{\mathbb{L}}(1)$. Restricting (2.5) to $\mathbb{L}$, we obtain an exact sequence:

$$0 \to \mathcal{S}|_{\mathbb{L}} \to (R/\mathfrak{m}^n) \otimes \mathcal{O}_{\mathbb{L}} \to \mathcal{Q}|_{\mathbb{L}} \to 0. \tag{2.9}$$

So $\det(\mathcal{S}|_{\mathbb{L}}) \cong \mathcal{O}_{\mathbb{L}}(-1)$. From the injection $\mathcal{S}|_{\mathbb{L}} \to (R/\mathfrak{m}^n) \otimes \mathcal{O}_{\mathbb{L}}$, we obtain

$$\mathcal{S}|_{\mathbb{L}} \cong \bigoplus_{i=1}^{k} \mathcal{O}_{\mathbb{L}}(a_i)$$

with $\sum_{i=1}^k a_i = -1$ and $a_i \le 0$. Thus

$$\mathcal{S}|_{\mathbb{L}} \cong \mathcal{O}_{\mathbb{L}}^{\oplus(k-1)} \oplus \mathcal{O}_{\mathbb{L}}(-1).$$

Now the injection $\mathcal{O}_{\mathbb{L}}^{\oplus(k-1)} \to (R/\mathfrak{m}^n) \otimes \mathcal{O}_{\mathbb{L}}$ gives rise to linearly independent elements $\overline{f}_1, \ldots, \overline{f}_{k-1}$ in $R/\mathfrak{m}^n$. The injection $\mathcal{O}_{\mathbb{L}}(-1) \to (R/\mathfrak{m}^n) \otimes \mathcal{O}_{\mathbb{L}}$ gives rise to two linearly independent elements $\overline{f}_k$ and $\overline{f}_{k+1}$ in $R/\mathfrak{m}^n$ such that

$$\overline{f}_1, \ldots, \overline{f}_{k-1}, \lambda\overline{f}_k + \mu\overline{f}_{k+1}$$

are linearly independent when $|\lambda| + |\mu| \neq 0$. So $\mathbb{L}$ is of the form (2.8). □

LEMMA 2.4. *Let $x \in X$ and $\eta \in M_{n+1}(x)$. Let $\xi_1, \xi_2 \in M_n(x)$ and $\xi_1 \neq \xi_2$. If $\xi_1, \xi_2 \subset \eta$, then $I_{\eta,x} \supset \mathfrak{m}_x^n$ where $\mathfrak{m}_x$ is the maximal ideal of $\mathcal{O}_{X,x}$.*

PROOF. We have $I_{\eta,x} \subset I_{\xi_1,x} \cap I_{\xi_2,x} \subset I_{\xi_2,x}$ and $I_{\xi_1,x} \cap I_{\xi_2,x} \neq I_{\xi_2,x}$. Since the colength of $I_{\eta,x}$ is $n+1$ and the colength of $I_{\xi_2,x}$ is n, we obtain $I_{\eta,x} = I_{\xi_1,x} \cap I_{\xi_2,x}$. Since $\mathfrak{m}_x^n \subset I_{\xi_1,x}$ and $\mathfrak{m}_x^n \subset I_{\xi_2,x}$, we get $\mathfrak{m}_x^n \subset I_{\xi_1,x} \cap I_{\xi_2,x} = I_{\eta,x}$. □

Now we are able to characterize all the curves γ in the punctual Hilbert scheme $M_n(x)$ which is homologous to β_n. By Theorem 1.31 (i), the fiber $(\psi_{n+1})^{-1}(\eta, x)$ over $(\eta, x) \in \mathcal{Z}_{n+1} \subset X^{[n+1]} \times X$ is isomorphic to $\mathbb{P}(\omega_{\eta_x} \otimes \mathbb{C}_x)$.

PROPOSITION 2.5. *Let $n \geq 2$, and X be simply connected. Let $x \in X$, and γ be a curve in $M_n(x)$. Then, $\gamma \sim \beta_n$ if and only if $\gamma = f_{n+1}(C)$ where C is a line in the fiber $(\psi_{n+1})^{-1}(\eta, x) \cong \mathbb{P}(\omega_\eta \otimes \mathbb{C}_x)$ for some $\eta \in M_{n+1}(x)$.*

PROOF. (i) Fix the identification $M_n(x) \cong M_n(O)$. Let C be a line in

$$(\psi_{n+1})^{-1}(\eta, x) \cong \mathbb{P}(\omega_\eta \otimes \mathbb{C}_x)$$

for some $\eta \in M_{n+1}(x)$. By Lemma 2.4, we have $I_{\eta,x} \supset \mathfrak{m}^n$. Choose polynomials $f_1, \ldots, f_{k-1} \in I_\eta$ such that $\overline{f}_1, \ldots, \overline{f}_{k-1}$ form a basis of the $(k-1)$-dimensional subspace $I_{\eta,x}/\mathfrak{m}^n \subset R/\mathfrak{m}^n$. Now, $I_{\eta,x} = \mathbb{C}f_1 + \ldots + \mathbb{C}f_{k-1} + \mathfrak{m}^n$. So

$$I_\eta = I_{\eta,x} \cap \mathbb{C}[u,v] = \mathbb{C}f_1 + \ldots + \mathbb{C}f_{k-1} + \langle u, v\rangle^n.$$

Next, since $\mathbb{P}(\omega_\eta \otimes \mathbb{C}_x) \cong \mathbb{P}(\mathrm{Hom}(\mathbb{C}_x, \mathcal{O}_\eta)^*)$, the line C in

$$(\psi_{n+1})^{-1}(\eta, x) \cong \mathbb{P}(\mathrm{Hom}(\mathbb{C}_x, \mathcal{O}_\eta)^*)$$

is generated by two different points $[\epsilon_1], [\epsilon_2] \in \mathbb{P}(\mathrm{Hom}(\mathbb{C}_x, \mathcal{O}_\eta)^*)$ (so that the two morphisms $\epsilon_1, \epsilon_2 \in \mathrm{Hom}(\mathbb{C}_x, \mathcal{O}_\eta) = \mathrm{Hom}_R(R/\mathfrak{m}, \mathcal{O}_\eta)$ are linearly independent). Choose $f_k, f_{k+1} \in R$ such that $[f_k] = \epsilon_1(1)$ and $[f_{k+1}] = \epsilon_2(1)$ in $R/I_{\eta,x}$. Note that $\overline{f}_1, \ldots, \overline{f}_{k+1} \in R/\mathfrak{m}^n$ are linearly independent.

Let $\xi \in \gamma = f_{n+1}(C)$, and assume that the point $(\xi, \eta) \in C$ corresponds to $[\lambda\epsilon_1 + \mu\epsilon_2] \in \mathbb{P}(\mathrm{Hom}(\mathbb{C}_x, \mathcal{O}_\eta)^*)$. Then we have the exact sequence:

$$0 \longrightarrow \mathbb{C}_x \stackrel{\lambda\epsilon_1+\mu\epsilon_2}{\longrightarrow} \mathcal{O}_\eta \longrightarrow \mathcal{O}_\xi \longrightarrow 0.$$

So

$$\begin{aligned} I_\xi &= I_\eta + \mathbb{C}(\lambda f_k + \mu f_{k+1}) \\ &= \mathbb{C}f_1 + \ldots + \mathbb{C}f_{k-1} + \mathbb{C}(\lambda f_k + \mu f_{k+1}) + \langle u, v\rangle^n. \end{aligned}$$

Thus $\tau(\gamma) = \mathbb{L}$ which is of the form (2.8). By Lemma 2.3 and Lemma 2.2, $\gamma \sim \beta_n$.

(ii) Conversely, assume $\gamma \sim \beta_n$. By Lemma 2.2 and Lemma 2.3, $\tau(\gamma) = \mathbb{L}$ which is of the form (2.8). Let $u\colon R/\mathfrak{m}^n \to R/\mathfrak{m}^n$ be the map defined by multiplication with u. Let

$$\varphi_u\colon R/\mathfrak{m}^n \otimes \mathcal{O}_\mathbb{L} \to R/\mathfrak{m}^n \otimes \mathcal{O}_\mathbb{L}$$

be the induced morphism, and consider the restriction of φ_u to the subbundle $\mathcal{S}|_\mathbb{L} \cong \mathcal{O}_\mathbb{L}(-1) \oplus \mathcal{O}_\mathbb{L}^{\oplus(k-1)}$ in (2.9). Since $\mathbb{L} = \tau(\gamma)$,

$$\mathrm{Span}(\overline{f}_1, \ldots, \overline{f}_{k-1}, \lambda\overline{f}_k + \mu\overline{f}_{k+1})$$

is an ideal of $R/\mathfrak{m}^n$ whenever $|\lambda| + |\mu| \neq 0$. So we see from the proof of Lemma 2.3 that φ_u maps $\mathcal{O}_\mathbb{L}(-1) \oplus \mathcal{O}_\mathbb{L}^{\oplus(k-1)}$ to itself. Since $\mathrm{Hom}(\mathcal{O}_\mathbb{L}, \mathcal{O}_\mathbb{L}(-1)) = 0$, φ_u maps $\mathcal{O}_\mathbb{L}^{\oplus(k-1)}$ to $\mathcal{O}_\mathbb{L}^{\oplus(k-1)}$. The same statement holds when we replace u by v. Hence, $\mathrm{Span}(\overline{f}_1, \ldots, \overline{f}_{k-1})$ is an ideal of $R/\mathfrak{m}^n$, and $\mathbb{C}f_1 + \ldots + \mathbb{C}f_{k-1} + \mathfrak{m}^n$ is an ideal of R with colength $n+1$. Let $\eta \in M_{n+1}(x)$ be defined by the ideal

$$I_\eta = \mathbb{C}f_1 + \ldots + \mathbb{C}f_{k-1} + \langle u, v\rangle^n.$$

Since $\mathrm{Hom}(\mathcal{O}_\mathbb{L}(-1), \mathcal{O}_\mathbb{L}(-1)) = \mathbb{C} \cdot \mathrm{Id}_{\mathcal{O}_\mathbb{L}(-1)}$, we have $\varphi_u|_{\mathcal{O}_\mathbb{L}(-1)} = c \cdot \mathrm{Id}_{\mathcal{O}_\mathbb{L}(-1)}$ for some $c \in \mathbb{C}$. Then $(\varphi_u|_{\mathcal{O}_\mathbb{L}(-1)})^n = c^n \cdot \mathrm{Id}_{\mathcal{O}_\mathbb{L}(-1)}$. However, $(\varphi_u)^n = 0$ since $u^n \in \mathfrak{m}^n$. So $c = 0$, and φ_u maps $\mathcal{O}_\mathbb{L}(-1) \oplus \mathcal{O}_\mathbb{L}^{\oplus(k-1)}$ to $\mathcal{O}_\mathbb{L}^{\oplus(k-1)}$. In particular, $u\overline{f}_k, u\overline{f}_{k+1} \in \mathrm{Span}(\overline{f}_1, \ldots, \overline{f}_{k-1})$. Hence $uf_k, uf_{k+1} \in I_\eta$. Similarly, $vf_k, vf_{k+1} \in I_\eta$.

Define two R-morphisms $R \to \mathcal{O}_\eta$ by sending $1 \in R$ to $[f_k], [f_{k+1}] \in \mathcal{O}_\eta$ respectively. They induce two R-morphisms $\epsilon_1, \epsilon_2 \colon \mathbb{C}_x = R/\mathfrak{m} \to \mathcal{O}_\eta$ since uf_k, uf_{k+1}, vf_k, $vf_{k+1} \in I_\eta$. Since $\overline{f}_1, \ldots, \overline{f}_{k-1}, \overline{f}_k, \overline{f}_{k+1} \in R/\mathfrak{m}^n$ are linearly independent, so are the two morphisms $\epsilon_1, \epsilon_2 \in \mathrm{Hom}(\mathbb{C}_x, \mathcal{O}_\eta)$. Let C be the line in

$$(\psi_{n+1})^{-1}(\eta, x) \cong \mathbb{P}(\mathrm{Hom}(\mathbb{C}_x, \mathcal{O}_\eta)^*)$$

generated by the points $[\epsilon_1], [\epsilon_2] \in \mathbb{P}(\mathrm{Hom}(\mathbb{C}_x, \mathcal{O}_\eta)^*)$. As in the last paragraph in part (i) above, the points in $f_{n+1}(C)$ correspond to the ideals

$$\mathbb{C}f_1 + \ldots + \mathbb{C}f_{k-1} + \mathbb{C}(\lambda f_k + \mu f_{k+1}) + \langle u, v\rangle^n, \qquad \lambda, \mu \in \mathbb{C}$$

with $|\lambda| + |\mu| \neq 0$. These same ideals define the points in γ since $\tau(\gamma) = \mathbb{L}$ is given by (2.8). So $\gamma = f_{n+1}(C)$. $\square$

Next, we study an arbitrary curve $\gamma \subset X^{[n]}$ homologous to β_n.

LEMMA 2.6.

(i) *A curve $\gamma \subset X^{[n]}$ homologous to β_n is a smooth rational curve;*

(ii) *A curve $\gamma \subset X^{[n]}$ is homologous to β_n if and only if $\gamma = \gamma_k + \xi$ where $2 \le k \le n$, γ_k is a curve in the punctual Hilbert scheme $M_k(x)$ for some $x \in X$, $\gamma_k \sim \beta_k$ in $X^{[k]}$, and $\xi \in X^{[n-k]}$ is a fixed point with $x \notin \mathrm{Supp}(\xi)$.*

PROOF. (i) By Remark 1.28, $(D_C - B_n/2)$ is very ample for some divisor C on X. Since $\gamma \cdot (D_C - B_n/2) = \beta_n \cdot (D_C - B_n/2) = 1$, γ is a smooth rational curve.

(ii) First of all, assume $\gamma \sim \beta_n$. Since γ is irreducible and $\gamma \cdot D_C = \beta_n \cdot D_C = 0$, $\rho_n(\gamma)$ is a point in $X^{(n)}$. Let $\rho_n(\gamma) = n_1x_1 + \ldots + n_mx_m$ where $n_i \ge 1$, $\sum_{i=1}^m n_i = n$, and $x_1, \ldots, x_m$ are distinct points. Then γ is contained in the product $\prod_{i=1}^m M_{n_i}(x_i) \subset X^{[n]}$. Since $M_{n_i}(x_i) \cong M_{n_i}(O)$, $H_1(M_{n_i}(x_i); \mathbb{Z}) = 0$ by the Theorem 1.1 (i) of [**ES1**]. By the Künneth formula, $\gamma \sim \sum_{i=1}^m (p_i(\gamma) + \xi_i)$ in $\prod_{i=1}^m M_{n_i}(x_i)$, where p_i is the i-th projection of $\prod_{i=1}^m M_{n_i}(x_i)$, and $\xi_i \in X^{[n-n_i]}$ with $\mathrm{Supp}(\xi_i) \cap \mathrm{Supp}\,\big(p_i(\gamma)\big) = \emptyset$. So

$$1 = (D_C - B_n/2) \cdot \gamma = \sum_{i=1}^m (D_C - B_n/2) \cdot (p_i(\gamma) + \xi_i).$$

Since $(D_C - B_n/2)$ is very ample, there exists exactly one i such that $p_i(\gamma)$ is a curve in $X^{[n_i]}$ and $p_j(\gamma)$ is a point in $X^{[n_j]}$ for all $j \neq i$. We may assume $i = 1$. Let $k = n_1$ and $x = x_1$. Then $\gamma = \gamma_k + \xi$ where $\gamma_k \subset M_k(x)$ and $\xi \in X^{[n-k]}$ is fixed with $x \notin \mathrm{Supp}(\xi)$. It remains to prove $\gamma_k \sim \beta_k$ in $X^{[k]}$.

Claim. *Let $\tilde{x} \in X$ be a fixed point, and $\alpha \subset X^{[n-1]}$ be a curve such that $\tilde{x} \notin \mathrm{Supp}(\xi)$ for every $\xi \in \alpha$. Then, $(\alpha + \tilde{x}) \cdot B_n = \alpha \cdot B_{n-1}$.*

PROOF. Define

$$\widetilde{\alpha} = \{\, (\xi, \xi + \tilde{x}) \,|\, \xi \in \alpha\} \subset X^{[n-1,n]}.$$

Since $\tilde{x} \notin \mathrm{Supp}(\alpha)$, the restriction $g_n|_{\widetilde{\alpha}} : \widetilde{\alpha} \to \alpha + \tilde{x}$ is an isomorphism. So $(g_n)_*\widetilde{\alpha} = \alpha + \tilde{x}$. Similarly, $(f_n)_*\widetilde{\alpha} = \alpha$. Let

$$E_n = \{(\xi, \eta) \in X^{[n-1,n]} \mid \mathrm{Supp}(\xi) = \mathrm{Supp}(\eta), \xi \subset \eta\}.$$

Since $\tilde{x} \notin \mathrm{Supp}(\alpha)$, we obtain $\widetilde{\alpha} \cap E_n = \emptyset$. By (1.40),

$$(g_n)^*B_n - (f_n)^*B_{n-1} = 2E_n.$$

So $\widetilde{\alpha}\cdot((g_n)^*B_n-(f_n)^*B_{n-1})=0$. Therefore, we obtain

$$(\alpha+\tilde{x})\cdot B_n=(g_n)_*\widetilde{\alpha}\cdot B_n=\widetilde{\alpha}\cdot(g_n)^*B_n=\widetilde{\alpha}\cdot(f_n)^*B_{n-1}.$$

By the projection formula,

$$(\alpha+\tilde{x})\cdot B_n=(f_n)_*\widetilde{\alpha}\cdot B_{n-1}=\alpha\cdot B_{n-1}.$$

This proves the claim. □

By deforming ξ in $X^{[n-k]}$, we can assume that ξ is $\tilde{x}_1+\ldots+\tilde{x}_{n-k}$ where $\tilde{x}_1,\ldots,\tilde{x}_{n-k}$ are distinct points different from x. From the Claim, we get

$$\begin{aligned}(\gamma_k+\tilde{x}_1+\ldots+\tilde{x}_{n-k})\cdot B_n &= (\gamma_k+\tilde{x}_1+\cdots+\tilde{x}_{n-k-1})\cdot B_{n-1}\\ &= \ldots\\ &= \gamma_k\cdot B_k.\end{aligned}\tag{2.10}$$

So $\gamma_k\cdot B_k=\gamma\cdot B_n=\beta_n\cdot B_n=-2$. Since $\gamma_k\subset M_k(x)$, $\gamma_k\sim\beta_k$ by Lemma 2.2.

Conversely, suppose $\gamma=\gamma_k+\xi$ where $2\le k\le n$, $\gamma_k\subset M_k(x)$ for some $x\in X$, $\gamma_k\sim\beta_k$ in $X^{[k]}$, and $\xi\in X^{[n-k]}$ is fixed with $x\notin\operatorname{Supp}(\xi)$. We may assume that ξ consists of $(n-k)$ distinct points. By (2.10),

$$\gamma\cdot B_n=\gamma_k\cdot B_k=\beta_k\cdot B_k=-2.$$

Then the same argument as in the proof of Lemma 2.2 yields $\gamma\sim\beta_n$. □

THEOREM 2.7. *Let $n\ge 2$, and X be a simply connected smooth projective surface. Then, a curve γ in $X^{[n]}$ is homologous to β_n if and only if $\gamma=f_{n+1}(C)$ where C is a line in the projective space $(\psi_{n+1})^{-1}(\eta,x)$ for some $(\eta,x)\in\mathcal{Z}_{n+1}$. Moreover, in this case, (η,x) and C are uniquely determined by γ.*

PROOF. (i) First, assume $\gamma\sim\beta_n$. By Lemma 2.6 (ii), we know that $\gamma=\gamma_k+\xi$ where $\gamma_k\subset M_k(x)$ for some $x\in X$, $\xi\in X^{[n-k]}$ is fixed with $x\notin\operatorname{Supp}(\xi)$, and $\gamma_k\sim\beta_k$ in $X^{[k]}$. By Proposition 2.5, $\gamma_k=f_{k+1}(\widetilde{C})$ where $\widetilde{C}$ is a line in $(\psi_{k+1})^{-1}(\eta_{k+1},x)$ for some $\eta_{k+1}\in M_{k+1}(x)$. Since $\operatorname{Supp}(\eta_{k+1})=\{x\}\not\subset\operatorname{Supp}(\xi)$, we can define $\eta=\eta_{k+1}+\xi\in X^{[n+1]}$. Then, $(\eta,x)\in\mathcal{Z}_{n+1}$. By the definition of ψ_{n+1}, we have

$$\begin{aligned}(\psi_{n+1})^{-1}(\eta,x) &= \{(\zeta,\eta)\in X^{[n,n+1]}\mid\zeta\subset\eta\text{ and }\operatorname{Supp}(I_\zeta/I_\eta)=x\}\\ &= \{(\zeta_k+\xi,\eta_{k+1}+\xi)\in X^{[n,n+1]}\mid\zeta_k\subset\eta_{k+1}\}\\ &\cong \{(\zeta_k,\eta_{k+1})\in X^{[k,k+1]}\mid\zeta_k\subset\eta_{k+1}\}\\ &= (\psi_{k+1})^{-1}(\eta_{k+1},x).\end{aligned}$$

Put

$$\begin{aligned}C &= \{(\zeta_k+\xi,\eta_{k+1}+\xi)|(\zeta_k,\eta_{k+1})\in\widetilde{C}\}\\ &= \{(\zeta_k+\xi,\eta)|(\zeta_k,\eta_{k+1})\in\widetilde{C}\},\end{aligned}$$

i.e., let C be the line in $(\psi_{n+1})^{-1}(\eta,x)$ corresponding to the line $\widetilde{C}$ in

$$(\psi_{k+1})^{-1}(\eta_{k+1},x)$$

via the above isomorphism $(\psi_{n+1})^{-1}(\eta,x)\cong(\psi_{k+1})^{-1}(\eta_{k+1},x)$. Then, we see that

$$\gamma=\gamma_k+\xi=f_{k+1}(\widetilde{C})+\xi=f_{n+1}(C).$$

Conversely, suppose $\gamma = f_{n+1}(C)$ where C is a line in $(\psi_{n+1})^{-1}(\eta, x)$ for some $(\eta, x) \in \mathcal{Z}_{n+1}$. Assume $\ell(\eta_x) = k+1$. Let $\eta_{k+1} = \eta_x \in X^{[k+1]}$. Reversing the argument in the preceding paragraph and using Lemma 2.6 (ii), we get $\gamma \sim \beta_n$.

(ii) To show the uniqueness of (η, x) and C, let $\xi_1, \xi_2 \in \gamma$ with $\xi_1 \neq \xi_2$. Then, $\xi_1, \xi_2 \subset \eta$. Since $\ell(\eta) = 1 + \ell(\xi_1) = 1 + \ell(\xi_2)$, we get $I_\eta = I_{\xi_1} \cap I_{\xi_2}$. So η and hence the point x are uniquely determined by γ. Since

$$f_{n+1}|_{(\psi_{n+1})^{-1}(\eta,x)} : \quad (\psi_{n+1})^{-1}(\eta, x) \to f_{n+1}((\psi_{n+1})^{-1}(\eta, x))$$

is an isomorphism, C is also uniquely determined by γ. □

Our next goal is to give a global description of the moduli space $\mathfrak{M}(\beta_n)$ of all the curves in $X^{[n]}$ homologous to β_n. Note that $\mathfrak{M}(\beta_n)$ is the union of certain irreducible components in the Hilbert scheme of curves in $X^{[n]}$. We will use the concept of the relative Grassmannian scheme from [**Sim**].

THEOREM 2.8. *Let $n \geq 2$, X be a simply connected smooth projective surface, and $\omega_{\mathcal{Z}_{n+1}}$ be the dualizing sheaf of $\mathcal{Z}_{n+1} \subset X^{[n+1]} \times X$. Then, there exists a bijective morphism from the relative Grassmannian* $\mathrm{Grass}(\omega_{\mathcal{Z}_{n+1}}, 2)$ *over $\mathcal{Z}_{n+1}$ to the moduli space $\mathfrak{M}(\beta_n)$ of all the curves in $X^{[n]}$ homologous to β_n.*

PROOF. Put $\widetilde{\mathbb{G}} = \mathrm{Grass}(\omega_{\mathcal{Z}_{n+1}}, 2)$ for simplicity. Define

$$\Gamma = \Big\{([C], (\xi, \eta)) \in \widetilde{\mathbb{G}} \times X^{[n,n+1]} | \ C \text{ is a line in } (\psi_{n+1})^{-1}(\eta, x)$$
$$\text{and } (\xi, \eta) \in C \text{ with } \mathrm{Supp}(I_\xi/I_\eta) = \{x\}\Big\}.$$

By Theorem 1.31 (i), $\mathbb{P}(\omega_{\mathcal{Z}_{n+1}}) \cong X^{[n,n+1]}$ over $\mathcal{Z}_{n+1}$. So Γ is the universal subscheme in $\widetilde{\mathbb{G}} \times_{\mathcal{Z}_{n+1}} \mathbb{P}(\omega_{\mathcal{Z}_{n+1}})$ which is flat over $\widetilde{\mathbb{G}}$. Consider

$$\alpha : \widetilde{\mathbb{G}} \times_{\mathcal{Z}_{n+1}} X^{[n,n+1]} \subset \widetilde{\mathbb{G}} \times X^{[n,n+1]} \xrightarrow{\mathrm{Id}_{\widetilde{\mathbb{G}}} \times f_{n+1}} \widetilde{\mathbb{G}} \times X^{[n]}.$$

Note that the fiber of $\widetilde{\mathbb{G}}$ over $(\eta, x) \in \mathcal{Z}_{n+1}$ is $\widetilde{\mathbb{G}}|_{(\eta,x)} = \mathrm{Grass}(\omega_{\eta_x} \otimes \mathbb{C}_x, 2)$. So

$$\begin{aligned}
(2.11) \qquad \Gamma|_{(\eta,x)} &\subset \widetilde{\mathbb{G}}|_{(\eta,x)} \times \mathbb{P}(\omega_{\mathcal{Z}_{n+1}})|_{(\eta,x)} \\
&= \mathrm{Grass}(\omega_{\eta_x} \otimes \mathbb{C}_x, 2) \times \mathbb{P}(\omega_{\eta_x} \otimes \mathbb{C}_x) \\
&= \mathrm{Grass}(\omega_{\eta_x} \otimes \mathbb{C}_x, 2) \times (\psi_{n+1})^{-1}(\eta, x).
\end{aligned}$$

Since f_{n+1} maps $(\psi_{n+1})^{-1}(\eta, x)$ isomorphically to $f_{n+1}((\psi_{n+1})^{-1}(\eta, x))$, we see that $\alpha|_\Gamma : \Gamma \to \alpha(\Gamma)$ is bijective. Moreover, assuming that $q \in \widetilde{\mathbb{G}}$ is mapped to (η, x) via the natural projection $\widetilde{\mathbb{G}} \to \mathcal{Z}_{n+1}$ and that q corresponds to the line

$$\{q\} \times C = \Gamma|_q \subset \{q\} \times \mathbb{P}(\omega_{\eta_x} \otimes \mathbb{C}_x) = \{q\} \times (\psi_{n+1})^{-1}(\eta, x)$$

via the universal subscheme $\Gamma \subset \widetilde{\mathbb{G}} \times_{\mathcal{Z}_{n+1}} \mathbb{P}(\omega_{\mathcal{Z}_{n+1}})$, then we see from (2.11) that the same point $q \in \widetilde{\mathbb{G}}$ also corresponds to the curve

$$\alpha(\Gamma)|_q = \alpha(\Gamma|_q) = \alpha(\{q\} \times C) = \{q\} \times f_{n+1}(C) \subset \{q\} \times f_{n+1}((\psi_{n+1})^{-1}(\eta, x))$$

via the subscheme $\alpha(\Gamma) \subset \widetilde{\mathbb{G}} \times X^{[n]}$. By Theorem 2.7, we conclude that via the subscheme $\alpha(\Gamma) \subset \widetilde{\mathbb{G}} \times X^{[n]}$ which is flat over $\widetilde{\mathbb{G}}$, the relative Grassmannian $\widetilde{\mathbb{G}}$ parameterizes curves in $X^{[n]}$ homologous to β_n. By the universal property of the Hilbert scheme of curves, we obtain a morphism $\Lambda : \widetilde{\mathbb{G}} \to \mathfrak{M}(\beta_n)$. Moreover, we see from Theorem 2.7 that the morphism Λ is bijective. □

Next, we study an open subset in $\mathrm{Grass}(\omega_{\mathcal{Z}_{n+1}}, 2)$. Let $x \in X$ and

$$\xi = x_1 + \ldots + x_{n-2} \in X^{[n-2]}$$

be fixed such that $x, x_1, \ldots, x_{n-2}$ are distinct points. Then the curve $M_2(x) + \xi$ is homologous to β_n. The set of all such curves has dimension

$$\#(\text{moduli of } x) + \#(\text{moduli of } \xi) = 2 + 2(n-2) = 2n - 2.$$

We now give a more detailed geometric description of this set as follows. Note that $\mathfrak{m}_x^2$ is the only colength-3 ideal in $\mathcal{O}_{X,x}$ whose minimal number of generators is 3. Let $\theta(x)$ be the unique length-3 0-dimensional subscheme of X corresponding to $\mathfrak{m}_x^2$. Let

$$\Delta_{n-1} = \{(x, x_1, \ldots, x_{n-2}) \in X^{n-1} | x, x_1, \ldots, x_{n-2} \text{ are not distinct}\}.$$

Define a map $\Psi : X^{n-1} - \Delta_{n-1} \to X^{[n+1]}$ by putting

$$\Psi(x, x_1, \ldots, x_{n-2}) = \theta(x) + \sum_{i=1}^{n-2} x_i.$$

Let $\theta(x) + \xi \in \mathrm{Im}(\Psi)$ where ξ denotes $x_1 + \ldots + x_{n-2}$. Then, $(\pi_{n+1})^{-1}(\theta(x) + \xi)$ consists of $(n-1)$ distinct points in $\mathcal{Z}_{n+1}$:

$$(\theta(x) + \xi, x), \quad (\theta(x) + \xi, x_i), 1 \le i \le n-2.$$

Note that

$$\mathrm{Grass}(\omega_{\mathcal{Z}_{n+1}}, 2)|_{(\theta(x)+\xi,x)} = \mathrm{Grass}(\omega_{\theta(x)} \otimes \mathbb{C}_x, 2) \cong \mathrm{Grass}(\mathbb{C}^2, 2) = pt.$$

In fact, $\mathrm{Grass}(\omega_{\mathcal{Z}_{n+1}}, 2)|_{(\theta(x)+\xi,x)}$ is the point in $\mathrm{Grass}(\omega_{\mathcal{Z}_{n+1}}, 2)$ corresponding to the curve $M_2(x) + \xi$ in $X^{[n]}$. Also,

$$\mathrm{Grass}(\omega_{\mathcal{Z}_{n+1}}, 2)|_{(\theta(x)+\xi,x_i)} \cong \mathrm{Grass}(\mathbb{C}, 2) = \emptyset.$$

So $\mathrm{Grass}(\omega_{\mathcal{Z}_{n+1}}, 2)|_{(\pi_{n+1})^{-1}(\mathrm{Im}(\Psi))}$ is exactly the subset of $\mathrm{Grass}(\omega_{\mathcal{Z}_{n+1}}, 2)$ described at the beginning of this paragraph. This proves (i) of the following proposition.

PROPOSITION 2.9. (i) *The open subset* $\mathrm{Grass}(\omega_{\mathcal{Z}_{n+1}}, 2)|_{(\pi_{n+1})^{-1}(\mathrm{Im}(\Psi))}$ *of* $\mathrm{Grass}(\omega_{\mathcal{Z}_{n+1}}, 2)$ *has dimension* $(2n-2)$;

(ii) *Let* $\gamma = M_2(x) + x_1 + \ldots + x_{n-2}$ *where* $x, x_1, \ldots, x_{n-2}$ *are distinct points in* X, *and let* $N_{\gamma \subset X^{[n]}}$ *be the normal bundle of* γ *in* $X^{[n]}$. *Then,*

$$\begin{aligned} N_{\gamma \subset X^{[n]}} &\cong \mathcal{O}_\gamma^{\oplus(2n-2)} \oplus \mathcal{O}_\gamma(-2), \\ T_{X^{[n]}}|_\gamma &\cong \mathcal{O}_\gamma^{\oplus(2n-2)} \oplus \mathcal{O}_\gamma(-2) \oplus \mathcal{O}_\gamma(2). \end{aligned}$$

PROOF. To prove (ii), note that in an analytic open neighborhood of γ, $X^{[n]}$ splits into $X^{[2]} \times X^{[n-2]}$. So

$$N_{\gamma \subset X^{[n]}} \cong N \oplus \mathcal{O}_\gamma^{\oplus(2n-4)}$$

where N is isomorphic to the normal bundle $\widetilde{N}$ of $M_2(x)$ in $X^{[2]}$ via $\gamma \cong M_2(x)$. Since $\widetilde{N} \cong \mathcal{O}_{M_2(x)}^{\oplus 2} \oplus \mathcal{O}_{M_2(x)}(-2)$, we obtain

$$N_{\gamma \subset X^{[n]}} \cong \mathcal{O}_\gamma^{\oplus(2n-2)} \oplus \mathcal{O}_\gamma(-2).$$

Next, we have an exact sequence

$$0 \to T_\gamma \to T_{X^{[n]}}|_\gamma \to N_{\gamma \subset X^{[n]}} \to 0.$$

Since $T_\gamma = \mathcal{O}_\gamma(2)$, we obtain the exact sequence

$$0 \to \mathcal{O}_\gamma(2) \to T_{X^{[n]}}|_\gamma \to \mathcal{O}_\gamma^{\oplus(2n-2)} \oplus \mathcal{O}_\gamma(-2) \to 0$$

which must split. Therefore, we conclude that

$$T_{X^{[n]}}|_\gamma \cong \mathcal{O}_\gamma^{\oplus(2n-2)} \oplus \mathcal{O}_\gamma(-2) \oplus \mathcal{O}_\gamma(2).$$

This completes the proof of the proposition. □

By Proposition 2.9 (i) and Theorem 2.8, the dimension of the moduli space $\mathfrak{M}(\beta_n)$ is at least $(2n-2)$, one more than the expected dimension

$$c_1(T_{X^{[n]}}) \cdot \beta_n + \dim X^{[n]} - \dim \mathrm{Aut}(\mathbb{P}^1) = (2n-3).$$

Therefore the moduli space of curves in $X^{[n]}$ homologous to β_n is obstructed. The obstruction may be understood via Proposition 2.9 (ii). In Lemma 15.4 below, we will use Proposition 2.9 (ii) to determine the obstruction sheaf of the 1-point genus-0 extremal Gromov-Witten theory of the Hilbert scheme $X^{[n]}$.

Finally, we construct some special curves in the punctual Hilbert schemes.

LEMMA 2.10. *Fix $n \geq 2$. Let $M_n(O)$ be the punctual Hilbert scheme at the origin $O \in \mathbb{C}^2$, and u, v be the coordinates of $\mathbb{C}^2$. Then,*

$$H_2(M_n(O); \mathbb{Z}) \cong \mathbb{Z}.$$

Moreover, a generator of $H_2(M_n(O); \mathbb{Z})$ is given by

$$\sigma_n = \{(\lambda u + \mu v^{n-1}, u^2, uv, v^n) \mid \lambda, \mu \in \mathbb{C} \text{ with } |\lambda| + |\mu| \neq 0\}. \tag{2.12}$$

PROOF. The first statement follows from [**ES1**, Theorem 1.1]. To prove the second statement, take a $\mathbb{C}^*$-action on $\mathbb{C}^2$ given by

$$t \cdot (u, v) = (t^{-\alpha} u, t^{-\beta} v)$$

with $\beta \gg \alpha$. For $\xi \in M_n(O)$, we use the ideal $I_\xi \subset \mathbb{C}[u, v]$ to represent ξ. Then the $\mathbb{C}^*$-invariant ideal in $\mathbb{C}[u, v]$ corresponding to a generator σ_n of $H_2(M_n(O); \mathbb{Z})$ is (v^{n-1}, uv, u^2). Therefore σ_n is the closure of the cell

$$\begin{aligned} & \Big\{I \in \mathbb{C}[u, v] \mid \ell(\mathbb{C}[u, v]/I) = n, \quad \lim_{t \to 0} (t \cdot I) = (v^{n-1}, uv, u^2)\Big\} \\ = \; & \{(v^{n-1} + au, uv, u^2) \mid a \in \mathbb{C}\} \\ \cong \; & \mathbb{C}. \end{aligned}$$

Finally, notice that if $a \neq 0$, then $(v^{n-1} + au, uv, u^2) = (v^{n-1} + au, v^n)$. So letting $a \to \infty$, we see that the ideal (u, v^n) is also contained in σ_n. Thus,

$$\sigma_n = \{(v^{n-1} + au, uv, u^2) \mid a \in \mathbb{C}\} \cup \{(u, v^n)\}$$

which is the same as $\{(\lambda u + \mu v^{n-1}, u^2, uv, v^n) \mid \lambda, \mu \in \mathbb{C} \text{ with } |\lambda| + |\mu| \neq 0\}$. □

PROPOSITION 2.11. *Identify $M_n(x)$ with $M_n(O)$, and regard σ_n as a curve in $M_n(x) \subset X^{[n]}$. Then as a curve in $X^{[n]}$, σ_n is homologous to β_n.*

PROOF. By Lemma 2.2, it suffices to show that the image $(\mathfrak{p} \circ \tau)(\sigma_n)$ is a line. Fix a basis for the $\mathbb{C}$-vector space $R/\mathfrak{m}^n$:

$$\overline{1}, \overline{u}, \overline{u}^2, \overline{u}\,\overline{v}, \overline{u}^3, \overline{u}^2\,\overline{v}, \overline{u}\,\overline{v}^2, \ldots, \overline{u}^{n-1}, \overline{u}^{n-2}\overline{v}, \ldots, \overline{u}\,\overline{v}^{n-2}, \overline{v}, \ldots, \overline{v}^{n-1}.$$

Note the special ordering of this basis. Recall from (2.12) that for any $\eta \in \sigma_n \subset M_n(O)$,

$$I_{\eta,0} = (\lambda u + \mu v^{n-1}, u^2, uv, v^n)$$

for some $\lambda, \mu \in \mathbb{C}$ with $|\lambda| + |\mu| \neq 0$. So a basis for the subspace $I_{\eta,0}/\mathfrak{m}^n \subset R/\mathfrak{m}^n$ can be chosen as

$$\lambda \overline{u} + \mu \overline{v}^{n-1}, \overline{u}^2, \overline{u}\,\overline{v}, \overline{u}^3, \overline{u}^2\overline{v}, \overline{u}\,\overline{v}^2, \ldots, \overline{u}^{n-1}, \overline{u}^{n-2}\overline{v}, \ldots, \overline{u}\,\overline{v}^{n-2},$$

and the matrix representation of $I_{\eta,0}/\mathfrak{m}^n$ is given by the $\binom{n}{2} \times \binom{n+1}{2}$-matrix:

$$\begin{bmatrix} 0 & \lambda & 0 & \ldots & 0 & 0 & \ldots & 0 & \mu \\ 0 & 0 & 1 & \ldots & 0 & 0 & \ldots & 0 & 0 \\ \vdots & \vdots & \vdots & \vdots & \vdots & \vdots & \ldots & \vdots & \vdots \\ 0 & 0 & 0 & \ldots & 1 & 0 & \ldots & 0 & 0 \end{bmatrix}.$$

Thus, $(\mathfrak{p} \circ \tau)(\eta) = [0, \ldots, 0, \lambda, 0, \ldots, 0, \mu, 0, \ldots, 0]$ where the positions of λ and μ are independent of $\eta \in \sigma_n$. So the image $(\mathfrak{p} \circ \tau)(\sigma_n)$ is a line. $\square$

Note that the flat limits of the elements $(\lambda u + v, v^n)$, $\lambda \in \mathbb{C}^*$ in $M_n(O)$ as $\lambda \to 0$ and $\lambda \to \infty$ are equal to (v, u^n) and (u, v^n) respectively. So in the punctual Hilbert scheme $M_n(O)$, we have the projective curve:

$$\tilde{\sigma}_n = \{(\lambda u + v, v^n) \mid \lambda \in \mathbb{C}^*\} \cup \{(v, u^n),\ (u, v^n)\}. \tag{2.13}$$

LEMMA 2.12. *As a curve in* $X^{[n]}$, $\tilde{\sigma}_n$ *is homologous to* $\binom{n}{2}\beta_n$.

PROOF. It suffices to show that $\tilde{\sigma}_n$ is homologous to $\binom{n}{2}\sigma_n$ in $H_2(M_n(O); \mathbb{Z})$. By (2.13), if $\eta \in \tilde{\sigma}_n - \{(v, u^n),\ (u, v^n)\}$, then a basis for $I_{\eta,0}/\mathfrak{m}^n \subset R/\mathfrak{m}^n$ consists of

$$\lambda \overline{u} + \overline{v}, \lambda \overline{u}^2 + \overline{u}\,\overline{v}, \lambda \overline{u}\,\overline{v} + \overline{v}^2, \ldots,$$
$$\lambda \overline{u}^{n-1} + \overline{u}^{n-2}\overline{v}, \lambda \overline{u}^{n-2}\overline{v} + \overline{u}^{n-3}\overline{v}^2, \ldots, \lambda \overline{u}\,\overline{v}^{n-2} + \overline{v}^{n-1}.$$

As in the proof of Lemma 2.11, we see that the degree of

$$(\mathfrak{p} \circ \tau)\Big(\tilde{\sigma}_n - \{(v, u^n),\ (u, v^n)\}\Big)$$

is $\binom{n}{2}$. So $(\mathfrak{p} \circ \tau)(\tilde{\sigma}_n)$ has degree $\binom{n}{2}$. By Lemma 2.10, there exists an integer d such that $\tilde{\sigma}_n \sim d\sigma_n$ in $H_2(M_n(O); \mathbb{Z})$. Since $(\mathfrak{p} \circ \tau)(\sigma_n)$ is a line, $d = \binom{n}{2}$. $\square$

2.2. The nef cone of $(\mathbb{P}^2)^{[n]}$

In this section, we will determine the nef cone of the Hilbert scheme $(\mathbb{P}^2)^{[n]}$ and the cone $\mathrm{NE}\big((\mathbb{P}^2)^{[n]}\big)$ spanned by all the effective curves in $(\mathbb{P}^2)^{[n]}$. We remark that as a general fact, these two cones are dual to each other.

First of all, recall the concept of k-very ampleness from Definition 1.26 (iii). This concept will be applied to produce nef divisors and very ample divisors on the Hilbert scheme $(\mathbb{P}^2)^{[n]}$.

LEMMA 2.13. *Let* $X = \mathbb{P}^2$. *Then the line bundle* $\mathcal{O}_X(n)$ *is* n*-very ample.*

PROOF. Let ℓ be a line on $\mathbb{P}^2$, $C = (n+3)\ell$, and $k = n$. Then, $K_X + C = n\ell$ and $C \cdot C = (n+3)^2 > 4n + 5$. Suppose $\mathcal{O}_X(n) = \mathcal{O}_X(K_X + C)$ is not n-very ample. By Theorem 1.27 (i), there exists an effective divisor D on X such that

$$C \cdot D - n - 1 \le D \cdot D < C \cdot D/2 < n + 1.$$

Let $D \sim d\ell$. Then, $d \ge 0$ and

$$(n+3)d - n - 1 \le d^2 < (n+3)d/2 < n + 1.$$

Since $d^2 < (n+3)d/2$, $d > 0$. Now, $(n+3)d/2 < (n+1)$ implies $d < 2$. So $d = 1$ which contradicts $(n+3)d - n - 1 \le d^2$. Hence, $\mathcal{O}_X(n)$ is n-very ample. □

Note that for $k \ge 1$, k-very ampleness implies $(k-1)$-very ampleness. Therefore, by Lemma 2.13, we get the following two morphisms:

$$\varphi_{n-1}\big(\mathcal{O}_X(n-1)\big)\colon\ X^{[n]} \to \operatorname{Grass}\big(H^0\big(X, \mathcal{O}_X(n-1)\big), n\big),$$
$$\varphi_{n-1}\big(\mathcal{O}_X(n)\big)\colon\ X^{[n]} \to \operatorname{Grass}\big(H^0\big(X, \mathcal{O}_X(n)\big), n\big).$$

Put

$$\varphi_1 = \varphi_{n-1}\big(\mathcal{O}_X(n-1)\big)\colon\ X^{[n]} \to \operatorname{Grass}\big(H^0\big(X, \mathcal{O}_X(n-1)\big), n\big), \tag{2.14}$$
$$\varphi_2 = \varphi_{n-1}\big(\mathcal{O}_X(n)\big)\colon\ X^{[n]} \to \operatorname{Grass}\big(H^0\big(X, \mathcal{O}_X(n)\big), n\big). \tag{2.15}$$

For $i = 1$ or 2, let

$$\ell_i = h^0\big(X, \mathcal{O}_X(n-2+i)\big) = \binom{n+i}{2},$$

$N_i = \dim(\wedge^{\ell_i - n}\mathbb{C}^{\ell_i}) - 1$, and

$$\mathfrak{p}_i\colon\ \operatorname{Grass}(\mathbb{C}^{\ell_i}, n) \to \mathbb{P}^{N_i}$$

be the Plücker embedding. Let

$$\Phi_i = \mathfrak{p}_i \circ \varphi_i\colon\ X^{[n]} \to \mathbb{P}^{N_i},$$

and $\mathcal{H}_i$ be the hyperplane line bundle over $\mathbb{P}^{N_i}$. By (1.32),

$$(\Phi_1)^*\mathcal{H}_1 = \mathcal{O}_{X^{[n]}}\big((n-1)D_\ell - B_n/2\big),$$
$$(\Phi_2)^*\mathcal{H}_2 = \mathcal{O}_{X^{[n]}}(nD_\ell - B_n/2). \tag{2.16}$$

LEMMA 2.14. *Let $X = \mathbb{P}^2$. Then the morphism φ_2 in (2.15) is an embedding of the Hilbert scheme $X^{[n]}$, and the divisor $(nD_\ell - B_n/2)$ is very ample.*

PROOF. By Lemma 2.13, $\mathcal{O}_X(n)$ is n-very ample. So by Theorem 1.27 (ii), φ_2 is an embedding. By (2.16), $(nD_\ell - B_n/2)$ is very ample. □

Next, we study $((n-1)D_\ell - B_n/2)$. We begin with the construction of a curve which lies on the boundary of the cone $\operatorname{NE}\big((\mathbb{P}^2)^{[n]}\big)$. Let

$$\xi = x_1 + \ldots + x_{n-1} \in X^{[n-1]}$$

where $x_1, \ldots, x_{n-1}$ are distinct points in the line $\ell \subset X$. Let $(\ell + \xi)$ be the closure of $\big(\ell - \operatorname{Supp}(\xi)\big) + \xi$ in $X^{[n]}$. Alternatively, consider:

$$\begin{array}{ccccccc}
\tilde{\ell}_\xi & \subset & \widetilde{X}_\xi & \subset & X^{[n-1,n]} & \overset{g_n}{\to} & X^{[n]} \\
\downarrow & & \downarrow & & \downarrow \phi_n & & \\
\xi \times \ell & \subset & \xi \times X & \subset & X^{[n-1]} \times X & &
\end{array} \tag{2.17}$$

where $\tilde{\ell}_\xi$ and $\widetilde{X}_\xi$ are the strict transforms of $\xi \times \ell$ and $\xi \times X$ in $X^{[n-1,n]}$ respectively. Since ϕ_n is the blowup of $X^{[n-1]} \times X$ along $\mathcal{Z}_{n-1}$, $\widetilde{X}_\xi$ is isomorphic to the blowup of $\xi \times X \cong X$ at $x_1, \ldots, x_{n-1}$. Let E_i be the exceptional divisor in $\widetilde{X}_\xi$ over x_i. Then,

$$(\phi_n|_{\widetilde{X}_\xi})^*(\xi \times \ell) = \tilde{\ell}_\xi + \sum_{i=1}^{n-1} E_i$$

in the Chow group $A_1(\widetilde{X}_\xi)$. Notice that $g_n(\tilde{\ell}_\xi) = (\ell + \xi)$ and

$$g_n(E_i) = M_2(x_i) + x_1 \ldots + x_{i-1} + x_{i+1} + \ldots + x_{n-1}.$$

In fact, since $g_n|_{\widetilde{X}_\xi} : \widetilde{X}_\xi \to g_n(\widetilde{X}_\xi)$ is an isomorphism, we have

$$(2.18) \qquad (g_n|_{\widetilde{X}_\xi})_*(\tilde{\ell}_\xi) = (\ell + \xi) \quad \text{and} \quad (g_n|_{\widetilde{X}_\xi})_*(E_i) = \beta_n$$

in $A_1(X^{[n]})$.

LEMMA 2.15. *With the above notations, then*

$$(\ell + \xi) = \beta_\ell - (n-1)\beta_n$$

in $A_1(X^{[n]})$.

PROOF. Choose another line ℓ' such that $\mathrm{Supp}(\xi) \cap \ell' = \emptyset$. Then in $A_1(X^{[n]})$,

$$(g_n|_{\widetilde{X}_\xi})_*(\phi_n|_{\widetilde{X}_\xi})^*(\xi \times \ell) = (g_n|_{\widetilde{X}_\xi})_*(\phi_n|_{\widetilde{X}_\xi})^*(\xi \times \ell') = (\ell' + \xi) = \beta_{\ell'} = \beta_\ell,$$
$$(g_n|_{\widetilde{X}_\xi})_*(\phi_n|_{\widetilde{X}_\xi})^*(\xi \times \ell) = (g_n|_{\widetilde{X}_\xi})_*(\tilde{\ell}_\xi + \textstyle\sum_{i=1}^{n-1} E_i) = (\ell + \xi) + (n-1)\beta_n$$

where we have used (2.18). So $(\ell + \xi) = \beta_\ell - (n-1)\beta_n$ in $A_1(X^{[n]})$. □

PROPOSITION 2.16. *Let* $X = \mathbb{P}^2$, *and* φ_1 *be the morphism defined in (2.14).*

(i) *If* $n \geq 2$, *then the divisor* $((n-1)D_\ell - B_n/2)$ *on* $X^{[n]}$ *is nef but not ample.*
(ii) *If* $n = 2$, *then*

$$\varphi_1 : X^{[2]} \to \mathrm{Grass}\left(H^0(X, \mathcal{O}_X(1)), 2\right) = (\mathbb{P}^2)^*$$

is isomorphic to the relative Hilbert scheme $\mathrm{Hilb}^2(\mathbb{P}((T_{(\mathbb{P}^2)^*})^*)/(\mathbb{P}^2)^*)$ *over the dual space* $(\mathbb{P}^2)^*$.

PROOF. (i) We have the following intersection numbers:

$$(2.19) \qquad \beta_\ell \cdot D_\ell = 1, \quad \beta_\ell \cdot (-B_n/2) = 0, \quad \beta_n \cdot D_\ell = 0, \quad \beta_n \cdot (-B_n/2) = 1.$$

It follows that

$$((n-1)D_\ell - B_n/2) \cdot (\beta_\ell - (n-1)\beta_n) = 0.$$

By Lemma 2.15, $(\beta_\ell - (n-1)\beta_n)$ is rationally equivalent to an effective curve. So $((n-1)D_\ell - B_n/2)$ is not ample. By (2.16), the divisor $((n-1)D_\ell - B_n/2)$ is nef.

(ii) When $n = 2$, we have

$$\varphi_1 = \varphi_1(\mathcal{O}_X(1)) : \quad X^{[2]} \to \mathrm{Grass}\left(H^0(X, \mathcal{O}_X(1)), 2\right) = (\mathbb{P}^2)^*.$$

An element $\xi \in X^{[2]}$ determines a unique line ℓ_ξ in X and $\varphi_1(\xi) = \ell_\xi \in (\mathbb{P}^2)^*$. So φ_1 is a surjection whose fiber over a line $\ell \in (\mathbb{P}^2)^*$ is $\ell^{[2]} \cong \ell^{(2)} \cong \mathbb{P}^2$. Note that

$$\{(x, \ell) \in \mathbb{P}^2 \times (\mathbb{P}^2)^* | \ x \in \ell\} \cong \mathbb{P}((T_{(\mathbb{P}^2)^*})^*).$$

So $\varphi_1 : X^{[2]} \to (\mathbb{P}^2)^*$ is isomorphic to the relative Hilbert scheme

$$\mathrm{Hilb}^2(\mathbb{P}((T_{(\mathbb{P}^2)^*})^*)/(\mathbb{P}^2)^*)$$

over $(\mathbb{P}^2)^*$. □

THEOREM 2.17. *Let* $n \geq 2$ *and* ℓ *be a line in* $X = \mathbb{P}^2$. *Let* $\mathrm{NE}(X^{[n]})$ *be the cone spanned by all the effective curves on* $X^{[n]}$.

(i) *The nef cone of* $X^{[n]}$ *is the cone spanned by* D_ℓ *and* $((n-1)D_\ell - B_n/2)$;
(ii) *The cone* $\mathrm{NE}(X^{[n]})$ *is spanned by* β_n *and* $(\beta_\ell - (n-1)\beta_n)$;

(iii) *Let β be the homology class of an effective curve in $X^{[n]}$. Then,*

$$\beta \cdot (nD_\ell - B_n/2) = 1$$

if and only if either $\beta = \beta_n$ or $\beta = (\beta_\ell - (n-1)\beta_n)$.

PROOF. (i) Note that D_ℓ is nef and big, but not ample. By Proposition 2.16, $((n-1)D_\ell - B_n/2)$ is also nef but not ample. By Theorem 1.24 (iii), $(-B_n/2)$ and D_ℓ form a basis for $A_{2n-1}(X^{[n]}) \cong H_{4n-2}(X^{[n]};\mathbb{Z})$. So the nef cone of the Hilbert scheme $X^{[n]}$ is the cone spanned by D_ℓ and $((n-1)D_\ell - B_n/2)$.

(ii) Since $(-B_n/2)$ and D_ℓ form a basis for $H_{4n-2}(X^{[n]};\mathbb{Z})$, we see from (2.19) and the Poincaré duality that $\{\beta_\ell, \beta_n\}$ is a basis for $H_2(X^{[n]};\mathbb{Z})$. By Lemma 2.15, both β_n and $(\beta_\ell - (n-1)\beta_n)$ are numerical equivalence classes of effective curves. By (2.19) again,

$$\begin{aligned} \beta_n \cdot D_\ell &= 0, \\ (\beta_\ell - (n-1)\beta_n) \cdot D_\ell &= 1, \\ \beta_n \cdot ((n-1)D_\ell - B_n/2) &= 1, \\ (\beta_\ell - (n-1)\beta_n) \cdot ((n-1)D_\ell - B_n/2) &= 0. \end{aligned}$$

It follows from (i) that the cone $\mathrm{NE}(X^{[n]})$ is spanned by β_n and $(\beta_\ell - (n-1)\beta_n)$.

(iii) Let $\beta = a\beta_n + b(\beta_\ell - (n-1)\beta_n)$. Intersecting β with the nef divisors D_ℓ and $((n-1)D_\ell - B_n/2)$, we see that a and b are nonnegative integers. Since $(nD_\ell - B_n/2)$ is very ample and

$$\beta_n \cdot (nD_\ell - B_n/2) = (\beta_\ell - (n-1)\beta_n) \cdot (nD_\ell - B_n/2) = 1,$$

we conclude that $\beta \cdot (nD_\ell - B_n/2) = 1$ if and only if $\beta = \beta_n$ or $\beta = (\beta_\ell - (n-1)\beta_n)$. ☐

Theorem 2.17 can be generalized to the following, and we refer to [**QT**] for further details.

THEOREM 2.18. *Let $n \geq 2$, and let X be a smooth projective surface X satisfying $h^1(X, \mathcal{O}_X) = 0$. Assume that the nef cone of X is the span of the divisors $F_1, \ldots, F_t$, and the cone $\mathrm{NE}(X)$ is the span of the curves $C_1, \ldots, C_t$ with $F_i \cdot C_j = \delta_{i,j}$ for all i and j. Assume further that*

$$\mathcal{O}_X\big((n-1)\sum_{i=1}^{t} F_i\big)$$

is an $(n-1)$-very ample line bundle. Then,

(i) *the nef cone of the Hilbert scheme $X^{[n]}$ is spanned by*

$$D_{F_1}, \quad \ldots, \quad D_{F_t}, \quad (n-1)\sum_{i=1}^{t} D_{F_i} - B_n/2; \tag{2.20}$$

(ii) *the cone $\mathrm{NE}(X^{[n]})$ is spanned by the classes*

$$\beta_{C_1} - (n-1)\beta_n, \quad \ldots, \quad \beta_{C_t} - (n-1)\beta_n, \quad \beta_n. \tag{2.21}$$

2.3. Curves homologous to $\beta_\ell - (n-1)\beta_n$

In this section, we will continue to take $X = \mathbb{P}^2$. We will characterize all the curves in $X^{[n]}$ homologous to the class $\beta_\ell - (n-1)\beta_n$. In addition, we will compute the normal bundles of these curves in $X^{[n]}$, and describe their moduli space.

First of all, we describe some curves in $X^{[n]}$ homologous to $\beta_\ell - (n-1)\beta_n$.

LEMMA 2.19. *Let C be a line in $X = \mathbb{P}^2$, and consider $C^{(n)} = C^{[n]} \subset X^{[n]}$.*

(i) *Every line in $C^{(n)} \cong \mathbb{P}^n$ is homologous to $\beta_\ell - (n-1)\beta_n$;*

(ii) $\mathcal{O}_{X^{[n]}}(D_\ell)|_{C^{(n)}} = \mathcal{O}_{C^{(n)}}(1)$.

PROOF. (i) Let $x_1, \ldots, x_{n-1} \in C$ be distinct, and put $\gamma = C + (x_1 + \ldots + x_{n-1})$. Then γ is a line in the projective space $C^{(n)} \cong \mathbb{P}^n$. By Lemma 2.15, $\gamma \sim \beta_\ell - (n-1)\beta_n$. So every line in $C^{(n)} \cong \mathbb{P}^n$ is homologous to $\beta_\ell - (n-1)\beta_n$.

(ii) Follows from $\gamma \cdot D_\ell|_{C^{(n)}} = \gamma \cdot D_\ell = (\beta_\ell - (n-1)\beta_n) \cdot D_\ell = 1$. □

Next we will prove the converse of Lemma 2.19. Let γ be a curve in $X^{[n]}$ homologous to $\beta_\ell - (n-1)\beta_n$. Then, we have

$$\gamma \cdot (nD_\ell - B_n/2) = (\beta_\ell - (n-1)\beta_n) \cdot (nD_\ell - B_n/2) = 1.$$

By Lemma 2.14, $(nD_\ell - B_n/2)$ is very ample. So γ is a smooth rational curve. Let p_1 be the projection from X^n to the first factor. Let C_γ be the union of all the curves in $p_1\big(\upsilon_n^{-1}(\rho_n(\gamma))\big)$ where $\upsilon_n : X^n \to X^{(n)}$ is the quotient map. The properties of C_γ will be studied in the next two lemmas.

LEMMA 2.20. *With the notations as above, C_γ is a line in $X = \mathbb{P}^2$.*

PROOF. Suppose $\deg(C_\gamma) \geq 2$. Take a point $\xi \in \gamma$ and a smooth point $x \in C_\gamma$ such that $x \notin \mathrm{Supp}(\xi)$. Since $x \in C_\gamma \subset p_1\big(\upsilon_n^{-1}(\rho_n(\gamma))\big)$, there exists $\xi_x \in \gamma$ such that $\rho_n(\xi_x) = n_x x + \eta_x$ where $n_x \geq 1$, $\eta_x \in X^{(n-n_x)}$, and $x \notin \mathrm{Supp}(\eta_x)$. Choose a line ℓ in X missing $\mathrm{Supp}(\eta_x) \cup \mathrm{Supp}(\xi)$, passing through x, and intersecting C_γ transversally. Then the intersection $\ell \cap C_\gamma$ is a finite set. Since $\deg(C_\gamma) \geq 2$, $\ell \cap C_\gamma$ contains one more point $y \neq x$. Hence there exists $\xi_y \in \gamma$ with $y \in \mathrm{Supp}(\xi_y)$. Thus $\xi_x, \xi_y \in \gamma \cap D_\ell$. Since $y \neq x$, $y \in \ell$ and ℓ misses $\mathrm{Supp}(\eta_x)$, we have

$$y \notin \{x\} \cup \mathrm{Supp}(\eta_x) = \mathrm{Supp}(\xi_x).$$

So $\xi_x \neq \xi_y$. Since $\mathrm{Supp}(\xi) \cap \ell = \emptyset$, we have $\xi \notin D_\ell$. Since $\xi \in \gamma$, γ is not contained in D_ℓ. So $\gamma \cap D_\ell$ is a finite set of points. Since $\xi_x, \xi_y \in \gamma \cap D_\ell$ and $\xi_x \neq \xi_y$, we obtain $\gamma \cdot D_\ell \geq 2$, a contradiction to $\gamma \cdot D_\ell = 1$. □

LEMMA 2.21. *With the notations as above, let p_γ be a linear homogeneous polynomial defining the line $C_\gamma \subset X = \mathbb{P}^2$. Then, any degree-$(n-1)$ homogeneous polynomial divisible by p_γ must vanish at ξ for every $\xi \in \gamma$.*

PROOF. Let V_{p_γ} be the subspace of $H^0\big(X, \mathcal{O}_X(n-1)\big)$ consisting of all the degree-$(n-1)$ homogeneous polynomials divisible by p_γ. Recall from (2.14) that φ_1 sends $\xi \in X^{[n]}$ to the $n(n-1)/2$-dimensional subspace $H^0(X, \mathcal{O}_X(n-1) \otimes I_\xi)$ of $H^0\big(X, \mathcal{O}_X(n-1)\big)$. Since

$$\gamma \cdot \big((n-1)D_\ell - B_n/2\big) = (\beta_\ell - (n-1)\beta_n) \cdot \big((n-1)D_\ell - B_n/2\big) = 0,$$

γ is contracted by φ_1. So the subspaces $H^0(X, \mathcal{O}_X(n-1) \otimes I_\xi) \subset H^0\big(X, \mathcal{O}_X(n-1)\big)$ are independent of $\xi \in \gamma$. Let $V_\gamma = H^0(X, \mathcal{O}_X(n-1) \otimes I_\xi)$ where $\xi \in \gamma$. It suffices to prove that $V_\gamma = V_{p_\gamma}$ as subspaces of $H^0\big(X, \mathcal{O}_X(n-1)\big)$.

Let $s \in V_\gamma$. Then s vanishes at every $\xi \in \gamma$ by definition. Since $C_\gamma \subset p_1\big(\upsilon_n^{-1}\big(\rho_n(\gamma)\big)\big)$, s vanishes at every point in C_γ. So the polynomial s is divisible by p_γ. Thus, $s \in V_{p_\gamma}$, and $V_\gamma \subset V_{p_\gamma}$. Since

$$\dim V_{p_\gamma} = h^0(X, \mathcal{O}_X(n-2)) = n(n-1)/2 = h^0(X, \mathcal{O}_X(n-1) \otimes I_\xi) = \dim V_\gamma$$

where $\xi \in \gamma$, we have $V_\gamma = V_{p_\gamma}$. $\square$

THEOREM 2.22. *Let $n \geq 2$, and ℓ be a line in X. Then, a curve $\gamma \subset X^{[n]}$ is homologous to $\beta_\ell - (n-1)\beta_n$ if and only if there is a line C in X such that γ is a line in $C^{[n]} \subset X^{[n]}$. Moreover, C is uniquely determined by the curve γ.*

PROOF. By Lemma 2.19 (i), it remains to prove the "only if" part.

Let $C = C_\gamma$ where C_γ is from Lemma 2.20. First, we claim $\gamma \subset C^{[n]}$. Indeed, let $\xi \in \gamma$. With the same notations as those in the proof of Lemma 2.21, we choose $s = p_\gamma \cdot q \in V_{p_\gamma}$ where $q \in H^0\big(X, \mathcal{O}_X(n-2)\big)$ and q doesn't vanish at any point in $\mathrm{Supp}(\xi)$. By Lemma 2.21, s vanishes at ξ. So p_γ must vanish at ξ. Therefore, $\xi \subset C_\gamma = C$. Hence, $\xi \in C^{[n]} = C^{(n)}$ for every $\xi \in \gamma$. So $\gamma \subset C^{[n]} \subset X^{[n]}$. Next,

$$\mathcal{O}_{X^{[n]}}(D_\ell)|_{C^{[n]}} = \mathcal{O}_{C^{[n]}}(1)$$

by Lemma 2.19 (ii). So viewing γ as a curve in $C^{[n]}$, we obtain

$$\gamma \cdot c_1(\mathcal{O}_{C^{[n]}}(1)) = \gamma \cdot D_\ell = 1.$$

Thus γ is a line in $C^{[n]}$.

Finally, the uniqueness of the line C follows from the observation that if $\xi \in X^{[n]}$ and $n \geq 2$, then ξ is contained in at most one line in $X = \mathbb{P}^2$. $\square$

In the following, we will study the normal bundle of a curve γ in $X^{[n]}$ homologous to $\beta_\ell - (n-1)\beta_n$. By Theorem 2.22, there exists a line C in $X = \mathbb{P}^2$ such that γ is a line in $C^{(n)} = C^{[n]} \subset X^{[n]}$. In particular,

$$N_{\gamma \subset C^{(n)}} \cong \mathcal{O}_\gamma(1)^{\oplus(n-1)}.$$

So we have the following exact sequence of normal bundles:

$$0 \to \mathcal{O}_\gamma(1)^{\oplus(n-1)} \to N_{\gamma \subset X^{[n]}} \to N_{C^{(n)} \subset X^{[n]}}|_\gamma \to 0. \tag{2.22}$$

PROPOSITION 2.23. *Let $n \geq 2$, and C and ℓ be lines in $X = \mathbb{P}^2$. Let $\gamma \subset X^{[n]}$ be a curve homologous to $\beta_\ell - (n-1)\beta_n$. Then,*

(i) $N_{C^{(n)} \subset X^{[n]}} \cong \mathcal{O}_X(1)^{[n]}|_{C^{(n)}} \cong \mathcal{O}_{C^{(n)}}^{\oplus 2} \oplus \mathcal{O}_{C^{(n)}}(-1)^{\oplus(n-2)}$;

(ii) $N_{\gamma \subset X^{[n]}} = \mathcal{O}_\gamma(1)^{\oplus(n-1)} \oplus \mathcal{O}_\gamma^{\oplus 2} \oplus \mathcal{O}_\gamma(-1)^{\oplus(n-2)}$;

(iii) *the moduli space $\mathfrak{M}(\beta_\ell - (n-1)\beta_n)$ of all the curves in $X^{[n]}$ homologous to $(\beta_\ell - (n-1)\beta_n)$ is unobstructed, i.e., is smooth with the expected dimension.*

PROOF. (i) The isomorphism

$$N_{C^{(n)} \subset X^{[n]}} \cong \mathcal{O}_X(1)^{[n]}|_{C^{(n)}}$$

follows from (1.29). To prove the second isomorphism, let $\widetilde{\mathcal{Z}}_n$ be the universal subscheme in $C^{(n)} \times C \cong \mathbb{P}^n \times \mathbb{P}^1$. Then, we have

$$\mathcal{O}_X(1)^{[n]}|_{C^{(n)}} \cong \tilde{\pi}_{n*}\big(\tilde{q}_n^* \mathcal{O}_{\mathbb{P}^1}(1)|_{\widetilde{\mathcal{Z}}_n}\big)$$

where $\tilde{\pi}_{n*} : \widetilde{\mathcal{Z}}_n \to \mathbb{P}^n$ and $\tilde{q}_n : \mathbb{P}^n \times \mathbb{P}^1 \to \mathbb{P}^1$ are the projections. It is known that $\widetilde{\mathcal{Z}}_n \subset \mathbb{P}^n \times \mathbb{P}^1$ is defined by the equation

$$a_0 U^n + a_1 U^{n-1} V + \ldots + a_n V^n = 0$$

where $a_0, a_1, \ldots, a_n$ and U, V are the homogeneous coordinates on $\mathbb{P}^n$ and $\mathbb{P}^1$ respectively. So the line bundle $\mathcal{O}_{\mathbb{P}^n \times \mathbb{P}^1}(\widetilde{\mathcal{Z}}_n)$ is of type $(1, n)$ in $\operatorname{Pic}(\mathbb{P}^n \times \mathbb{P}^1) \cong \mathbb{Z} \oplus \mathbb{Z}$. Let $\tilde{p}_n : \mathbb{P}^n \times \mathbb{P}^1 \to \mathbb{P}^n$ be the projection. Applying $\tilde{p}_{n*}$ to the exact sequence

$$0 \to \tilde{q}_n^* \mathcal{O}_{\mathbb{P}^1}(1) \otimes \mathcal{O}_{\mathbb{P}^n \times \mathbb{P}^1}(-\widetilde{\mathcal{Z}}_n) \to \tilde{q}_n^* \mathcal{O}_{\mathbb{P}^1}(1) \to \tilde{q}_n^* \mathcal{O}_{\mathbb{P}^1}(1)|_{\widetilde{\mathcal{Z}}_n} \to 0,$$

we obtain

$$0 \to \mathcal{O}_{C^{(n)}}^{\oplus 2} \to \tilde{\pi}_{n*}\big(\tilde{q}_n^* \mathcal{O}_{\mathbb{P}^1}(1)|_{\widetilde{\mathcal{Z}}_n}\big) \to \mathcal{O}_{C^{(n)}}(-1)^{\oplus(n-2)} \to 0.$$

Since this exact sequence splits, we conclude the second isomorphism.

(ii) Follows immediately from (2.22) and the isomorphisms in (i).

(iii) By (ii), $H^1(\gamma, N_{\gamma \subset X^{[n]}}) = 0$ for any curve γ homologous to $(\beta_\ell - (n-1)\beta_n)$. By deformation theory, $\mathfrak{M}(\beta_\ell - (n-1)\beta_n)$ is unobstructed. □

In the next theorem, we further identify $\mathfrak{M}(\beta_\ell - (n-1)\beta_n)$ with a certain Grassmannian bundle.

THEOREM 2.24. *Let $n \geq 2$, $X = \mathbb{P}^2$, and ℓ be a line in X. Then the moduli space $\mathfrak{M}(\beta_\ell - (n-1)\beta_n)$ of curves in $X^{[n]}$ homologous to $(\beta_\ell - (n-1)\beta_n)$ is isomorphic to the Grassmannian bundle* $\operatorname{Grass}(\operatorname{Sym}^n((T_{(\mathbb{P}^2)^*})^*), 2)$ *over* $(\mathbb{P}^2)^*$.

PROOF. First of all, we give a global description of the union W_n of all the subsets $C^{[n]} \subset X^{[n]}$ with C being lines in $X = \mathbb{P}^2$. Consider

$$F = \{(x, [C]) \in \mathbb{P}^2 \times (\mathbb{P}^2)^* \mid x \in C\} \subset \mathbb{P}^2 \times (\mathbb{P}^2)^*.$$

We know that $F \cong \mathbb{P}((T_{(\mathbb{P}^2)^*})^*)$. Then we have

$$\begin{aligned} \operatorname{Hilb}^n(\mathbb{P}((T_{(\mathbb{P}^2)^*})^*)/(\mathbb{P}^2)^*) &\cong \operatorname{Hilb}^n(F/(\mathbb{P}^2)^*) \\ &\subset \operatorname{Hilb}^n(\mathbb{P}^2 \times (\mathbb{P}^2)^*/(\mathbb{P}^2)^*) \\ &\cong X^{[n]} \times (\mathbb{P}^2)^* \end{aligned}$$

of relative Hilbert schemes over $(\mathbb{P}^2)^*$. Let

$$\pi\colon \operatorname{Hilb}^n(\mathbb{P}((T_{(\mathbb{P}^2)^*})^*)/(\mathbb{P}^2)^*) \to (\mathbb{P}^2)^*$$

be the projection. Consider the map

$$\alpha\colon \operatorname{Hilb}^n(\mathbb{P}((T_{(\mathbb{P}^2)^*})^*)/(\mathbb{P}^2)^*) \hookrightarrow X^{[n]} \times (\mathbb{P}^2)^* \to X^{[n]}.$$

Note that via the isomorphism

$$\operatorname{Hilb}^n(\mathbb{P}((T_{(\mathbb{P}^2)^*})^*)/(\mathbb{P}^2)^*) \cong \operatorname{Hilb}^n(F/(\mathbb{P}^2)^*),$$

the fiber of the map π over a point $[C] \in (\mathbb{P}^2)^*$ is isomorphic to $C^{[n]}$. So $\operatorname{Im}(\alpha) = W_n$. Moreover, since $n \geq 2$, an element $\xi \in X^{[n]}$ is contained in at most one line C in $X = \mathbb{P}^2$. So if $\alpha(\xi') = \alpha(\eta')$ for $\xi', \eta' \in \operatorname{Hilb}^n(\mathbb{P}((T_{(\mathbb{P}^2)^*})^*)/(\mathbb{P}^2)^*)$, then there exists a unique line C such that $\alpha(\xi') = \alpha(\eta') \subset C$, i.e., ξ' and η' are contained in the fiber $\pi^{-1}([C])$ over the point $[C] \in (\mathbb{P}^2)^*$. Since α maps this fiber isomorphically to $C^{[n]} \subset X^{[n]}$, we must have $\xi' = \eta'$. Also

$$\operatorname{Hilb}^n(\mathbb{P}((T_{(\mathbb{P}^2)^*})^*)/(\mathbb{P}^2)^*) \cong \mathbb{P}(\operatorname{Sym}^n((T_{(\mathbb{P}^2)^*})^*))$$

by the Theorem 4.1.11 in [**Got2**]. Thus, α gives the isomorphism

$$\mathbb{P}(\mathrm{Sym}^n((T_{(\mathbb{P}^2)^*})^*)) \cong W_n. \tag{2.23}$$

So the set of all the lines in $C^{[n]} \subset X^{[n]}$ with C being lines in $X = \mathbb{P}^2$ is parameterized by the Grassmannian bundle $\mathrm{Grass}(\mathrm{Sym}^n((T_{(\mathbb{P}^2)^*})^*), 2)$ over $(\mathbb{P}^2)^*$. By Theorem 2.22, from the universal properties of the Hilbert schemes and the Grassmannians, we obtain a bijective morphism

$$\Lambda : \mathrm{Grass}(\mathrm{Sym}^n((T_{(\mathbb{P}^2)^*})^*), 2) \to \mathfrak{M}(\beta_\ell - (n-1)\beta_n).$$

Since both $\mathrm{Grass}(\mathrm{Sym}^n((T_{(\mathbb{P}^2)^*})^*), 2)$ and $\mathfrak{M}(\beta_\ell - (n-1)\beta_n)$ are smooth, the bijectivity of Λ implies that Λ is an isomorphism. □

By Lemma 2.14 and (2.16), the morphism

$$\Phi_2 = \mathfrak{p}_2 \circ \varphi_2 \colon \; X^{[n]} \to \mathbb{P}^{N_2}$$

is an embedding associated to the complete linear system $|nD_\ell - B_n/2|$. By Theorem 2.17 (iii), $\gamma \subset X^{[n]}$ is mapped to a line by Φ_2 if and only if either $\gamma \sim \beta_n$ or $\gamma \sim (\beta_\ell - (n-1)\beta_n)$. Therefore, regarding $X^{[n]}$ as a closed subvariety of $\mathbb{P}^{N_2}$, then the Hilbert scheme of lines in $X^{[n]}$ is the disjoint union

$$\mathfrak{M}(\beta_n) \coprod \mathfrak{M}(\beta_\ell - (n-1)\beta_n).$$

2.4. A flip structure on $(\mathbb{P}^2)^{[n]}$ when $n \geq 3$

Let $X = \mathbb{P}^2$. This section is devoted to the birational geometry of the Hilbert scheme $(\mathbb{P}^2)^{[n]}$. We will study the structure of the morphism φ_1 defined in (2.14) when $n \geq 3$, and construct a flip for φ_1.

We recall from Proposition 2.16 (ii) that the structure of φ_1 has been determined for $n = 2$. For $n \geq 3$, we begin with a lemma.

LEMMA 2.25. *Let $n \geq 3$, $X = \mathbb{P}^2$, and W_n be the contraction locus of the morphism $\varphi_1 : X^{[n]} \to \varphi_1(X^{[n]})$ defined by (2.14). Then,*

(i) *a fiber of φ_1 is either a point or the subset $C^{(n)} \subset X^{[n]}$ for some line $C \subset X$;*
(ii) *the restriction $\varphi_1|_{W_n} : W_n \to \varphi_1(W_n)$ is isomorphic to the projective bundle $\mathbb{P}(\mathrm{Sym}^n((T_{(\mathbb{P}^2)^*})^*))$ over $(\mathbb{P}^2)^*$;*
(iii) *$N \cong (\varphi_1|_{W_n})^* M \otimes \mathcal{O}(-1)$ where $N = N_{W_n \subset X^{[n]}}$ is the normal bundle of W_n in $X^{[n]}$, M is some vector bundle over $\varphi_1(W_n) \cong (\mathbb{P}^2)^*$, and $\mathcal{O}(1)$ stands for the tautological line bundle over the projective bundle $\varphi_1|_{W_n} : W_n \to \varphi_1(W_n)$.*

PROOF. (i) Fix $\xi \in X^{[n]}$. By definition,

$$\varphi_1(\xi) \in \varphi_1(X^{[n]}) \subset \mathrm{Grass}\left(H^0(X, \mathcal{O}_X(n-1)), n\right)$$

corresponds to the $n(n-1)/2$-dimensional subspace $H^0\big(X, I_\xi \otimes \mathcal{O}_X(n-1)\big)$ in $H^0\big(X, \mathcal{O}_X(n-1)\big)$. So $\eta \in (\varphi_1)^{-1}(\varphi_1(\xi))$ if and only if

$$H^0\big(X, I_\eta \otimes \mathcal{O}_X(n-1)\big) = H^0\big(X, I_\xi \otimes \mathcal{O}_X(n-1)\big).$$

Let $\mathfrak{d} \subset |(n-1)\ell|$ be the linear system corresponding to the subspace $H^0\big(X, I_\xi \otimes \mathcal{O}_X(n-1)\big)$ in $H^0\big(X, \mathcal{O}_X(n-1)\big)$. Then, either $\mathfrak{d}$ cuts out a 0-dimensional subscheme ζ with $\zeta \supset \xi$, or $\mathfrak{d}$ has a fixed component C.

Assume that $\mathfrak{d}$ cuts out a 0-dimensional subscheme ζ with $\zeta \supset \xi$. Then,

$$H^0\big(X, I_\zeta \otimes \mathcal{O}_X(n-1)\big) = H^0\big(X, I_\xi \otimes \mathcal{O}_X(n-1)\big),$$

and $(\varphi_1)^{-1}(\varphi_1(\xi)) \subset \{\eta \in X^{[n]} | \ \eta \subset \zeta\}$. Conversely, if $\eta \in X^{[n]}$ and $\eta \subset \zeta$, then

$$\begin{aligned} H^0\big(X, I_\eta \otimes \mathcal{O}_X(n-1)\big) &\supset H^0\big(X, I_\zeta \otimes \mathcal{O}_X(n-1)\big) \\ &= H^0\big(X, I_\xi \otimes \mathcal{O}_X(n-1)\big). \end{aligned}$$

So

$$H^0\big(X, I_\eta \otimes \mathcal{O}_X(n-1)\big) = H^0\big(X, I_\xi \otimes \mathcal{O}_X(n-1)\big)$$

since they have the same dimension. Thus $\eta \in (\varphi_1)^{-1}(\varphi_1(\xi))$, and

$$\{\eta \in X^{[n]} | \ \eta \subset \zeta\} \subset (\varphi_1)^{-1}(\varphi_1(\xi)).$$

Hence $(\varphi_1)^{-1}(\varphi_1(\xi)) = \{\eta \in X^{[n]} | \ \eta \subset \zeta\}$. Now we claim

$$(\varphi_1)^{-1}(\varphi_1(\xi)) = \{\xi\}.$$

Indeed, if $(\varphi_1)^{-1}(\varphi_1(\xi))$ contains more than one point, then there is an irreducible projective curve Γ in $(\varphi_1)^{-1}(\varphi_1(\xi)) = \{\eta \in X^{[n]} | \ \eta \subset \zeta\}$ since the closed subset $\{\eta \in X^{[n]} | \ \eta \subset \zeta\}$ of $X^{[n]}$ is known to be connected. Moreover, the projective curve Γ is contracted to points by both the Hilbert-Chow morphism ρ_n and the morphism φ_1. So $\Gamma \cdot D_\ell = 0$, and $\Gamma \cdot B_n = 0$ in view of (2.16). By Theorem 2.17 (i), the intersection of Γ with every divisor on $X^{[n]}$ is zero. This is impossible. So $(\varphi_1)^{-1}(\varphi_1(\xi)) = \{\xi\}$.

Next, assume that $\mathfrak{d}$ has a fixed component C. Let $C \sim d\ell$ where $d \geq 1$. Then,

$$n(n-1)/2 - 1 = \dim(\mathfrak{d}) \leq \dim |(n-1-d)\ell| = (n+1-d)(n-d)/2 - 1.$$

It follows that $d = 1$, C is a line in $X = \mathbb{P}^2$, and $\xi \subset C$ (i.e. $\xi \in C^{(n)} \subset X^{[n]}$). So $(\varphi_1)^{-1}(\varphi_1(\xi)) \subset C^{(n)}$. On the other hand, letting $p(C)$ be a linear homogeneous polynomial defining the line C, we see that for every $\eta \in C^{(n)} \subset X^{[n]}$, the subspace $H^0\big(X, I_\eta \otimes \mathcal{O}_X(n-1)\big)$ in $H^0\big(X, \mathcal{O}_X(n-1)\big)$ consists of all the degree-$(n-1)$ homogeneous polynomials divisible by $p(C)$. In particular,

$$H^0\big(X, I_\eta \otimes \mathcal{O}_X(n-1)\big) = H^0\big(X, I_\xi \otimes \mathcal{O}_X(n-1)\big).$$

So $C^{(n)} \subset (\varphi_1)^{-1}(\varphi_1(\xi))$. Hence $(\varphi_1)^{-1}(\varphi_1(\xi)) = C^{(n)}$.

(ii) By (i), the contraction locus W_n is the union of all the subsets $C^{(n)} \subset X^{[n]}$ with C being lines in $X = \mathbb{P}^2$. So (ii) follows from (2.23) and (i).

(iii) Fix a line C in $X = \mathbb{P}^2$. By (ii), $C^{(n)}$ is a fiber in the fiber bundle $\varphi_1|_{W_n}$: $W_n \to \varphi_1(W_n)$. So we have $N_{C^{(n)} \subset W_n} = \mathcal{O}_{C^{(n)}}^{\oplus 2}$. Thus by Proposition 2.23 (i), we obtain an exact sequence of normal bundles:

$$0 \to \mathcal{O}_{C^{(n)}}^{\oplus 2} \to \mathcal{O}_{C^{(n)}}^{\oplus 2} \oplus \mathcal{O}_{C^{(n)}}(-1)^{\oplus(n-2)} \to (N_{W_n \subset X^{[n]}})|_{C^{(n)}} = N|_{C^{(n)}} \to 0.$$

It follows that $N|_{C^{(n)}} \cong \mathcal{O}_{C^{(n)}}(-1)^{\oplus(n-2)}$ for every line C in $X = \mathbb{P}^2$. Since these $C^{(n)}$, C being lines in $X = \mathbb{P}^2$, are precisely the fibers of the projective bundle $\varphi_1|_{W_n} : W_n \to \varphi_1(W_n)$, we conclude that $N \cong (\varphi_1|_{W_n})^* M \otimes \mathcal{O}(-1)$ for some rank-$(n-2)$ vector bundle M over $\varphi_1(W_n) \cong (\mathbb{P}^2)^*$. □

Let $n \geq 3$. By Lemma 2.25 (ii), we have

$$\dim(W_n) = (n+2) \leq \dim(X^{[n]}) - 1.$$

So $\varphi_1 : X^{[n]} \to \varphi_1(X^{[n]})$ is a birational morphism. Since $K_{X^{[n]}} = -3D_\ell$ where ℓ is a line in $X = \mathbb{P}^2$, $(\beta_\ell - (n-1)\beta_n) \cdot K_{X^{[n]}} = -3 < 0$. This says that

$$\mathbb{R}^+ \cdot (\beta_\ell - (n-1)\beta_n)$$

is an extremal ray since $(-K_{X^{[n]}})$ is nef and the Picard number of $X^{[n]}$ is 2. Thus φ_1 is the contraction of the ray $\mathbb{R}^+ \cdot (\beta_\ell - (n-1)\beta_n)$ since

$$((n-1)D_\ell - B_n/2) \cdot (\beta_\ell - (n-1)\beta_n) = 0.$$

Moreover, when $n \geq 4$, $\dim(W_n) \leq \dim(X^{[n]}) - 2$, so φ_1 is a small contraction.

THEOREM 2.26. *Let $n \geq 4$ and $X = \mathbb{P}^2$. Then, the flip for the small contraction $\varphi_1 : X^{[n]} \to \varphi_1(X^{[n]})$ defined via (2.14) exists.*

PROOF. We will explicitely construct a commutative diagram

$$\begin{array}{ccccc} X^{[n]} & & \overset{(\varphi_+)^{-1}\circ\varphi_1}{\dashrightarrow} & & (X^{[n]})_+ \\ & \searrow^{\varphi_1} & & \swarrow^{\varphi_+} & \\ & & \varphi_1(X^{[n]}) & & \end{array}$$

satisfying the Flip Conjecture 5-1-10 in [**KMM**], by proving that $(X^{[n]})_+$ is smooth, φ_+ is a birational morphism, and $K_{(X^{[n]})_+}$ is φ_+-ample. The proof is a little long, but may be divided into two steps. In step one, we construct two additional birational morphisms $\widetilde{\varphi}_-$ and $\widetilde{\varphi}_+$. In step two, we construct the birational morphism φ_+.

Let

$$\widetilde{\varphi}_- : \widetilde{X^{[n]}} \to X^{[n]}$$

be the blowing-up of $X^{[n]}$ along the contraction locus W_n of φ_1, and $\widetilde{W_n} \subset \widetilde{X^{[n]}}$ be the exceptional divisor. Let N be the normal bundle of W_n in $X^{[n]}$. Then over $\widetilde{W_n} = \mathbb{P}(N^*)$, we have the tautological surjection

$$(\widetilde{\varphi}_-|_{\widetilde{W_n}})^*(N^*) \to \mathcal{O}_{\widetilde{W_n}}(1) \to 0 \tag{2.24}$$

where $\mathcal{O}_{\widetilde{W_n}}(1)$ is the tautological line bundle over $\widetilde{W_n} = \mathbb{P}(N^*)$. In view of Lemma 2.25 (iii),

$$N \cong (\varphi_1|_{W_n})^* M \otimes \mathcal{O}_{W_n}(-1)$$

where M is some rank-$(n-2)$ vector bundle over $\varphi_1(W_n) \cong (\mathbb{P}^2)^*$, and $\mathcal{O}_{W_n}(1)$ stands for the tautological line bundle over the projective bundle $\varphi_1|_{W_n} : W_n \to \varphi_1(W_n)$. So from (2.24), we obtain a surjection:

$$(\varphi_1|_{W_n} \circ \widetilde{\varphi}_-|_{\widetilde{W_n}})^*(M^*) \to (\widetilde{\varphi}_-|_{\widetilde{W_n}})^*\mathcal{O}_{W_n}(-1) \otimes \mathcal{O}_{\widetilde{W_n}}(1) \to 0.$$

By the Proposition 7.12 in Chapter II of [**Hart**], the map

$$(\varphi_1|_{W_n} \circ \widetilde{\varphi}_-|_{\widetilde{W_n}}) : \quad \widetilde{W_n} \to \varphi_1(W_n)$$

factors through the natural projection $\mathbb{P}(M^*) \to \varphi_1(W_n)$. In fact, the induced morphism $\widetilde{\varphi} : \widetilde{W_n} \to \mathbb{P}(M^*)$ is a $\mathbb{P}^n$-bundle. So the fibers of

$$(\varphi_1|_{W_n} \circ \widetilde{\varphi}_-|_{\widetilde{W_n}}) : \quad \widetilde{W_n} \to \varphi_1(W_n)$$

are naturally $\mathbb{P}^n \times \mathbb{P}^{n-3}$. Moreover, for the projective spaces $\mathbb{P}^n$ in the fibers $\mathbb{P}^n \times \mathbb{P}^{n-3}$, we have $\mathcal{O}_{\widetilde{X^{[n]}}}(\widetilde{W_n})|_{\mathbb{P}^n} \cong \mathcal{O}_{\mathbb{P}^n}(-1)$, using the facts that $\mathcal{O}_{\widetilde{X^{[n]}}}(\widetilde{W_n})|_{\mathbb{P}^n \times \mathbb{P}^{n-3}} \cong \mathcal{O}(a,-1)$ for some integer a and that

$$\begin{aligned} \mathcal{O}_{\mathbb{P}^n}^{\oplus(n-2)} \otimes \mathcal{O}_{\mathbb{P}^n}(1) &\cong N^*|_{\mathbb{P}^n} \\ &\cong (\widetilde{\varphi}_-|_{\mathbb{P}^n \times \mathbb{P}^{n-3}})_* \left[\mathcal{O}_{\widetilde{X^{[n]}}}(-\widetilde{W_n})|_{\mathbb{P}^n \times \mathbb{P}^{n-3}}\right] \\ &\cong \mathcal{O}_{\mathbb{P}^n}^{\oplus(n-2)} \otimes \mathcal{O}_{\mathbb{P}^n}(-a). \end{aligned}$$

In other words, the restriction $\mathcal{O}_{\widetilde{X^{[n]}}}(-\widetilde{W_n})|_{\widetilde{W_n}}$ is a tautological line bundle of $\widetilde{\varphi} : \widetilde{W_n} \to \mathbb{P}(M^*)$. By the results in [**Naka, FN**], there exists a morphism

$$\widetilde{\varphi}_+ : \widetilde{X^{[n]}} \to (X^{[n]})_+$$

such that $(X^{[n]})_+$ is smooth and contains $\mathbb{P}(M^*)$. Moreover, $\widetilde{\varphi}_+|_{\widetilde{W_n}} = \widetilde{\varphi}$ and $\widetilde{\varphi}_+$ is the blowing-up of $(X^{[n]})_+$ along $\mathbb{P}(M^*)$.

Now we claim that

$$(\widetilde{\varphi}_-)^*((n-1)D_\ell - B_n/2) = (\widetilde{\varphi}_+)^* H_+$$

for some divisor H_+ on $(X^{[n]})_+$. Indeed, recall from (2.16) that $((n-1)D_\ell - B_n/2) = (\varphi_1)^* H$ for some very ample divisor H on $\varphi_1(X^{[n]})$. Consider the divisor

$$(\widetilde{\varphi}_-)^*((n-1)D_\ell - B_n/2) = (\varphi_1 \circ \widetilde{\varphi}_-)^* H \tag{2.25}$$

on $\widetilde{X^{[n]}}$. Since

$$\mathbb{Z} \cdot [\widetilde{W_n}] \oplus (\widetilde{\varphi}_-)^* \mathrm{Cl}(X^{[n]}) = \mathrm{Cl}(\widetilde{X^{[n]}}) = \mathbb{Z} \cdot [\widetilde{W_n}] \oplus (\widetilde{\varphi}_+)^* \mathrm{Cl}((X^{[n]})_+)$$

where $\mathrm{Cl}(\cdot)$ stands for the divisor class group, we conclude that

$$(\widetilde{\varphi}_-)^*((n-1)D_\ell - B_n/2) = \lambda \widetilde{W_n} + (\widetilde{\varphi}_+)^* H_+ \tag{2.26}$$

for some integer λ and some divisor H_+ on $(X^{[n]})_+$. Let $\widetilde{C} \subset \widetilde{X^{[n]}}$ be a line contained in a fiber $\mathbb{P}^n$ of the projective bundle $\widetilde{\varphi} : \widetilde{W_n} \to \mathbb{P}(M^*)$. Then, $\widetilde{\varphi}_+(\widetilde{C}) = \widetilde{\varphi}(\widetilde{C})$ is a point in $\mathbb{P}(M^*) \subset (X^{[n]})_+$, and $(\varphi_1 \circ \widetilde{\varphi}_-)(\widetilde{C})$ is a point in $\varphi_1(W_n) \subset \varphi_1(X^{[n]})$. So applying $\cdot \widetilde{C}$ to both sides of (2.26) and using (2.25), we obtain

$$\begin{aligned} 0 &= (\varphi_1 \circ \widetilde{\varphi}_-)^* H \cdot \widetilde{C} \\ &= (\widetilde{\varphi}_-)^*((n-1)D_\ell - B_n/2) \cdot \widetilde{C} \\ &= (\lambda \widetilde{W_n} + (\widetilde{\varphi}_+)^* H_+) \cdot \widetilde{C} \\ &= \lambda \cdot (\widetilde{W_n} \cdot \widetilde{C}) \\ &= \lambda \cdot (c_1(\mathcal{O}_{\mathbb{P}^n}(-1)) \cdot \widetilde{C}) \\ &= -\lambda \end{aligned}$$

where we have used the fact that $\mathcal{O}_{\widetilde{X^{[n]}}}(\widetilde{W_n})|_{\mathbb{P}^n} = \mathcal{O}_{\mathbb{P}^n}(-1)$ when $\mathbb{P}^n$ is a fiber of the projective bundle $\widetilde{\varphi} : \widetilde{W_n} \to \mathbb{P}(M^*)$. So $\lambda = 0$ and

$$(\widetilde{\varphi}_-)^*((n-1)D_\ell - B_n/2) = (\widetilde{\varphi}_+)^* H_+. \tag{2.27}$$

Since $\widetilde{\varphi}_+$ is the blowing-up morphism of the smooth variety $(X^{[n]})_+$, we see that $|(\widetilde{\varphi}_+)^*H_+| = (\widetilde{\varphi}_+)^*|H_+|$. So combining with (2.25) and (2.27), we obtain

$$\begin{aligned} |(\varphi_1 \circ \widetilde{\varphi}_-)^*H| &= |(\widetilde{\varphi}_-)^*((n-1)D_\ell - B_n/2)| \\ &= |(\widetilde{\varphi}_+)^*H_+| \\ &= (\widetilde{\varphi}_+)^*|H_+|. \end{aligned} \tag{2.28}$$

Since φ_1 is the contraction of the extremal ray $R = \mathbb{R}^+ \cdot (\beta_\ell - (n-1)\beta_n)$, it is known from [**KMM**, Remark 5-1-1] that $\varphi_1(X^{[n]})$ has at most rational singularities. So

$$|(\varphi_1 \circ \widetilde{\varphi}_-)^*H| = (\varphi_1 \circ \widetilde{\varphi}_-)^*|H|.$$

In view of (2.28),

$$(\varphi_1 \circ \widetilde{\varphi}_-)^*|H| = |(\varphi_1 \circ \widetilde{\varphi}_-)^*H| = (\widetilde{\varphi}_+)^*|H_+|.$$

Since H is a very ample divisor on $\varphi_1(X^{[n]})$, $(\varphi_1 \circ \widetilde{\varphi}_-)^*|H|$ is base-point-free. It follows from $(\widetilde{\varphi}_+)^*|H_+| = (\varphi_1 \circ \widetilde{\varphi}_-)^*|H|$ that $|H_+|$ is also base-point-free. So $|H_+|$ induces a morphism

$$\varphi_+ : (X^{[n]})_+ \to \varphi_+((X^{[n]})_+).$$

Moreover, since $(\widetilde{\varphi}_+)^*|H_+| = (\varphi_1 \circ \widetilde{\varphi}_-)^*|H|$, we conclude that $\varphi_1 \circ \widetilde{\varphi}_- = \varphi_+ \circ \widetilde{\varphi}_+$ up to an isomorphism between $\varphi_1(X^{[n]})$ and $\varphi_+((X^{[n]})_+)$. So putting $\varphi_- = \varphi_1$, we obtain a commutative diagram:

$$\begin{array}{ccccc} & & \widetilde{X^{[n]}} & & \\ & \swarrow^{\widetilde{\varphi}_-} & & \searrow^{\widetilde{\varphi}_+} & \\ X^{[n]} & & & & (X^{[n]})_+. \\ & \searrow^{\varphi_-} & & \swarrow^{\varphi_+} & \\ & & \varphi_1(X^{[n]}) & & \end{array} \tag{2.29}$$

Note that $\varphi_+ : (X^{[n]})_+ - \mathbb{P}(M^*) \to \varphi_1(X^{[n]}) - \varphi_1(W_n)$ is an isomorphism.

It is standard to check that $K_{(X^{[n]})_+}$ is φ_+-ample. So the commutative triangle

$$\begin{array}{ccccc} X^{[n]} & & \overset{\widetilde{\varphi}_+ \circ (\widetilde{\varphi}_-)^{-1}}{\dashrightarrow} & & (X^{[n]})_+ \\ & \searrow^{\varphi_-} & & \swarrow^{\varphi_+} & \\ & & \varphi_1(X^{[n]}) & & \end{array} \tag{2.30}$$

is the flip for $\varphi_- = \varphi_1$. □

REMARK 2.27. The proof of Theorem 2.26 shows that for $n = 3$, the morphism

$$\varphi_1 : X^{[n]} \to \varphi_1(X^{[n]})$$

is the blowing-up along $\varphi_1(W_n) \cong (\mathbb{P}^2)^*$, and W_n is the exceptional locus.

Part 2

Hilbert schemes and infinite dimensional Lie algebras

CHAPTER 3

Hilbert schemes and infinite dimensional Lie algebras

In this chapter, we will survey various infinite dimensional Lie algebra actions on the cohomology of the Hilbert schemes of points on a smooth projective complex surface. As a motivation and background material, we will begin with Nakajima's affine Lie algebra action on the homology of quiver varieties [**Nak1**, **Nak2**, **Nak4**]. We will review the Heisenberg algebras of Nakajima and Grojnowski [**Groj**, **Nak3**] which provide an important language for describing the homology and cohomology classes of the Hilbert schemes, determine the geometric interpretations of Heisenberg monomial classes, and study the homology classes of curves in the Hilbert schemes. Then we will present the Virasoro algebras and boundary operator of Lehn [**Leh1**] as well as the Ext vertex operators of Carlsson and Okounkov [**Car1**, **Car2**, **CO**]. As evidenced by [**LQW1**, **LQW2**, **LQW3**, **LQW5**, **LQW7**, **LS1**, **LS2**], Lehn's boundary operator and the Heisenberg algebras of Nakajima and Grojnowski play essential roles in determining the cohomology ring structures of the Hilbert schemes. The Ext vertex operators of Carlsson and Okounkov are very powerful tools in understanding the intersection theory of the Hilbert schemes coupled with the Chern classes of their tangent bundles, and in investigating the connection between the multiple q-zeta values and the Hilbert schemes of points [**Ok2**, **QY**]. These operators also have applications in studying the AGT correspondence [**AGT**, **Neg**]. In addition, we will prove a structure theorem (Theorem 3.31) for the higher order derivatives of the Heisenberg operators.

3.1. Affine Lie algebra action of Nakajima

In this section, we will briefly review Nakajima's construction [**Nak1**, **Nak2**, **Nak4**] of affine Lie algebra action on the homology of quiver varieties. When a complex surface X_0 is the minimal resolution of a Kleinian (or simple surface) singularity, the Hilbert schemes of points on it can be described as quiver varieties of affine type. Therefore, Nakajima's work provides a strong motivation for the construction [**Groj**, **Nak3**] of Heisenberg algebra action on the cohomology of the Hilbert schemes of points on complex surfaces. To avoid getting into technicalities, we will follow the presentations in [**Nag**] and concentrate on the case of the affine Kac-Moody algebra $\widetilde{\mathfrak{sl}}_\ell$.

Let (I, H) be a *quiver* , i.e, I is a set of vertices and H is a set of oriented edges. Assume that there is a subset $\Omega \subset H$ such that $\Omega \cup \overline{\Omega} = H$ and $\Omega \cap \overline{\Omega} = \emptyset$, where $\overline{\Omega}$ means reversing orientations of the edges in Ω. Assume further that Ω has no cycle. Fix a dimension vector $\mathbf{v} \in (\mathbb{Z}_+)^I$ and a framing vector $\mathbf{w} \in (\mathbb{Z}_{\geq 0})^I$. Let

$$\zeta = (\zeta_{\mathbb{C}}, \zeta_{\mathbb{R}}) \in \mathbb{C}^I \oplus \mathbb{R}^I.$$

Let V, W be I-graded vector spaces such that $\dim V_i = v_i$ and $\dim W_i = w_i$. Put

$$\mathrm{M}(\mathbf{v},\mathbf{w}) = \left(\bigoplus_{h\in H} \mathrm{Hom}(V_{\mathrm{out}(h)}, V_{\mathrm{in}(h)})\right) \oplus \left(\bigoplus_{i\in I} \mathrm{Hom}(W_i, V_i) \oplus \mathrm{Hom}(V_i, W_i)\right)$$

where $h \in H$ is drawn from $\mathrm{out}(h)$ to $\mathrm{in}(h)$. An element of $\mathrm{M}(\mathbf{v},\mathbf{w})$ is called an ADHM (Atiyah, Drinfel'd, Hitchin and Manin) datum. Let

$$\begin{aligned} \mathrm{GL}_{\mathbf{v}} &= \prod_{i\in I} \mathrm{GL}(V_i), \\ \mathfrak{gl}_{\mathbf{v}} &= \bigoplus_{i\in I} \mathfrak{gl}(V_i). \end{aligned}$$

Then $\mathrm{GL}_{\mathbf{v}}$ acts on $\mathrm{M}(\mathbf{v},\mathbf{w})$ by

$$(g_i)\cdot(B_h, a_i, b_i) = \left(g_{\mathrm{in}(h)} B_h g_{\mathrm{out}(h)}^{-1}, g_i a_i, b_i g_i^{-1}\right)$$

for $g_i \in \mathrm{GL}(V_i)$, $B_h \in \mathrm{Hom}(V_{\mathrm{out}(h)}, V_{\mathrm{in}(h)})$, $a_i \in \mathrm{Hom}(W_i, V_i)$, and $b_i \in \mathrm{Hom}(V_i, W_i)$. The moment map $\mu_{\mathbb{C}}$ is defined by

$$\mu_{\mathbb{C}}(B, a, b) = \left(\sum_{\mathrm{in}(h)=i} \epsilon(h) B_h B_{\overline{h}} + a_i b_i\right) \in \bigoplus_{i\in I} \mathfrak{gl}(V_i) = \mathfrak{gl}_{\mathbf{v}}$$

where $\epsilon(h) = 1$ and $\epsilon(\overline{h}) = -1$ for $h \in \Omega$. Note that $\mathbb{C}^I$ is canonically identified with the center of $\mathfrak{gl}_{\mathbf{v}}$. The element $\zeta_{\mathbb{C}} = (\zeta_{\mathbb{C},i}) \in \mathbb{C}^I$ is identified with $\oplus_{i\in I}\zeta_{\mathbb{C},i}\mathrm{Id}_{V_i} \in \mathfrak{gl}_{\mathbf{v}}$.

The *quiver variety* $\mathfrak{M}_\zeta(\mathbf{v},\mathbf{w})$ associated to $\mathbf{v},\mathbf{w}$ and ζ is defined to be

$$\left(\mu_{\mathbb{C}}^{-1}(\zeta_{\mathbb{C}})\right)^{\mathrm{ss}} // \mathrm{GL}_{\mathbf{v}}$$

where the index "ss" means the subset of $\zeta_{\mathbb{R}}$-semistable objects in $\mu_{\mathbb{C}}^{-1}(\zeta_{\mathbb{C}})$.

Let $A = (a_{i,j})_{i,j\in I}$ be the adjacency matrix of the quiver (I, H), that is to say $a_{i,j}$ is the number of oriented edges in H which are drawn from i to j. Let

$$C = 2 \cdot \mathrm{Id} - A$$

be the associated Cartan matrix . Then we consider the set of positive roots

$$R_+ = \{\theta \in (\mathbb{Z}_{\geq 0})^I | \theta^t C\theta \leq 2\} - \{0\}.$$

For $\mathbf{v} \in (\mathbb{Z}_+)^I$, define

$$R_+(\mathbf{v}) = \{\theta = (\theta_i) \in R_+ | \theta_i \leq v_i \ \forall i \in I\}.$$

An element $\zeta \in \mathbb{C}^I \oplus \mathbb{R}^I$ is called generic with respect to $\mathbf{v}$ if for any $\theta \in R_+(\mathbf{v})$,

$$\zeta \notin D_\theta \otimes \mathbb{R}^3 \subset \mathbb{R}^I \otimes \mathbb{R}^3 \cong \mathbb{C}^I \oplus \mathbb{R}^I$$

where

$$D_\theta = \left\{\eta = (\eta_i) \in \mathbb{R}^I | \sum_{i\in I} \eta_i\theta_i = 0\right\}.$$

Note that

$$\zeta_0 = (0, \zeta_{\mathbb{R}})$$

is generic if $\zeta_{\mathbb{R}} \in C_0$ where

$$C_0 = \{\eta = (\eta_i) \in \mathbb{R}^I | \eta_i < 0 \ \forall i \in I\}.$$

It is known that $\mathfrak{M}_\zeta(\mathbf{v},\mathbf{w})$ is smooth if ζ is generic with respect to $\mathbf{v}$. Since

$$\mathbb{R}^I \otimes \mathbb{R}^3 - \bigcup_{\theta \in R_+(\mathbf{v})} D_\theta \otimes \mathbb{R}^3$$

is connected, $\mathfrak{M}_\zeta(\mathbf{v},\mathbf{w})$ and $\mathfrak{M}_{\zeta'}(\mathbf{v},\mathbf{w})$ are diffeomorphic for generic ζ and ζ'. Moreover, for generic ζ, $\mathfrak{M}_\zeta(\mathbf{v},\mathbf{w})$ is a fine moduli space of representations of the quiver [**King**, Section 3].

In the rest of this section, let (I,H) be the quiver of type $\hat{A}_{\ell-1}$ with cyclic orientation. Fix $\mathbf{w}_0 = (1,0,\ldots,0) \in \mathbb{Z}^I$, and $\zeta_0 = (0,\zeta_\mathbb{R})$ with $\zeta_\mathbb{R} \in C_0$. Let $\mathbf{e}_i$ denote the i-th coordinate vector of $\mathbb{Z}^I$. For fixed i and $\mathbf{v}$, define the *Hecke correspondence* to be the following subvariety of $\mathfrak{M}_{\zeta_0}(\mathbf{v},\mathbf{w}_0) \times \mathfrak{M}_{\zeta_0}(\mathbf{v}+\mathbf{e}_i,\mathbf{w}_0)$:

$$\begin{aligned}(3.1)\qquad \mathcal{B}_i(\mathbf{v}) = \big\{(J_1,J_2) | J_1 \in \mathfrak{M}_{\zeta_0}(\mathbf{v},\mathbf{w}_0) \text{ is}& \\ \text{a subrepresentation of } J_2 \in \mathfrak{M}_{\zeta_0}(\mathbf{v}+\mathbf{e}_i,\mathbf{w}_0)\big\}&.\end{aligned}$$

By [**Nak4**, Theorem 5.7], $\mathcal{B}_i(\mathbf{v})$ is Lagrangian in $\mathfrak{M}_{\zeta_0}(\mathbf{v},\mathbf{w}_0) \times \mathfrak{M}_{\zeta_0}(\mathbf{v}+\mathbf{e}_i,\mathbf{w}_0)$.

Next, let $\mathfrak{sl}_\ell$ be the Lie algebra consisting of all complex $\ell \times \ell$-matrices of trace zero. It has Chevalley generators

$$e_i = E_{i,i+1}, \quad f_i = E_{i+1,i}, \quad h_i = E_{i,i} - E_{i+1,i+1}$$

with $1 \le i \le \ell - 1$, where $E_{i,j}$ denotes the matrix which has a 1 on the i-th row and j-th column and 0 elsewhere. The affine Kac-Moody algebra $\widetilde{\mathfrak{sl}}_\ell$ is given by

$$\widetilde{\mathfrak{sl}}_\ell = \mathfrak{sl}_\ell[t,t^{-1}] \oplus \mathbb{C}c \oplus \mathbb{C}d$$

[**Hern**, Subsection 4.2]. In [**Nak4**, Section 9], a geometric representation of $\widetilde{\mathfrak{sl}}_\ell$ on the homology of the quiver varieties $\mathfrak{M}_{\zeta_0}(\mathbf{v},\mathbf{w}_0)$ indexed by $\mathbf{v}$ is constructed. The images of h_i and d are certain multiples of the fundamental classes of the diagonals in the products $\mathfrak{M}_{\zeta_0}(\mathbf{v},\mathbf{w}_0) \times \mathfrak{M}_{\zeta_0}(\mathbf{v},\mathbf{w}_0)$. The image of e_i is the operator $\mathbf{E}_i$ defined by

$$\mathbf{E}_i(\alpha) = (-1)^{v_{i-1}+v_i} p_{1*}\big(p_2^*\alpha \cdot \mathcal{B}_i(\mathbf{v})\big)$$

where p_j is the j-th projection on $\mathfrak{M}_{\zeta_0}(\mathbf{v},\mathbf{w}_0) \times \mathfrak{M}_{\zeta_0}(\mathbf{v}+\mathbf{e}_i,\mathbf{w}_0)$, while the image of f_i is the operator $\mathbf{F}_i$ defined by

$$\mathbf{F}_i(\alpha) = (-1)^{v_i+v_{i+1}} p_{2*}\big(p_1^*\alpha \cdot \mathcal{B}_i(\mathbf{v})\big).$$

The quiver variety $\mathfrak{M}_{\zeta_0}(\mathbf{v},\mathbf{w}_0)$ is related to the Hilbert schemes of points as follows. Regard the cyclic group $\mathbb{Z}/\ell\mathbb{Z}$ as the subgroup of $\mathrm{SL}_2(\mathbb{C})$ consisting of the diagonal matrices $\mathrm{diag}(\epsilon^i,\epsilon^{-i}), 0 \le i < \ell$ where ϵ is a primitive ℓ-th root of unity. Then, $\mathbb{Z}/\ell\mathbb{Z}$ acts on the affine plane $\mathbb{C}^2$. Let X_0 denote the minimal resolution of $\mathbb{C}^2/(\mathbb{Z}/\ell\mathbb{Z})$. Then there exists an element

$$\zeta_\infty = (0,\zeta'_\mathbb{R}) \in \mathbb{C}^I \oplus \mathbb{R}^I$$

generic with respect to $\mathbf{v}$ such that $\mathfrak{M}_{\zeta_\infty}(\mathbf{v},\mathbf{w}_0)$ is naturally identified with the Hilbert scheme $(X_0)^{[n]}$ where $n = \dim\big(\mathfrak{M}_{\zeta_\infty}(\mathbf{v},\mathbf{w}_0)\big)/2$. Since both ζ_0 and ζ_∞ are generic with respect to $\mathbf{v}$, $\mathfrak{M}_{\zeta_\infty}(\mathbf{v},\mathbf{w}_0)$ is diffeomorphic to $\mathfrak{M}_{\zeta_0}(\mathbf{v},\mathbf{w}_0)$. Therefore, the above geometric action of the affine Lie algebra $\widetilde{\mathfrak{sl}}_\ell$ on the homology of the quiver varieties $\mathfrak{M}_{\zeta_0}(\mathbf{v},\mathbf{w}_0)$ with varying $\mathbf{v}$ induces naturally an action of $\widetilde{\mathfrak{sl}}_\ell$ on the homology of the Hilbert schemes $(X_0)^{[n]}, n \ge 0$.

The preceding discussion indicates strongly that for an arbitrary complex surface X, there should exist, via certain Hecke correspondence, geometric action of infinite dimensional Lie algebras on the (co)homology groups of the Hilbert schemes

$X^{[n]}, n \geq 0$ (here we have by-passed the identification of quiver varieties with moduli spaces of torsion-free sheaves on ALE spaces). Indeed, Nakajima and Grojnowski [**Groj**, **Nak3**] constructed action of Heisenberg algebras on the (co)homology of these Hilbert schemes. This leads us to the next section.

REMARK 3.1. It is unclear whether the Hecke correspondence $\mathcal{B}_i(\mathbf{v})$ defined in (3.1) for the quiver varieties could be transferred directly to the Hilbert schemes $(X_0)^{[n]}$. On the other hand, by the theorem in [**Nag**, Subsection 6.2], the above induced action of the affine Lie algebra $\widetilde{\mathfrak{sl}}_\ell$ on the homology of the Hilbert schemes $(X_0)^{[n]}$ is related to the action of the Heisenberg algebra (to be constructed in the next section) via the Frenkel-Kac construction [**Kac**, Section 5.6].

3.2. Heisenberg algebras of Nakajima and Grojnowski

In this section, we will define the Fock space as the sum of the cohomology groups of all the Hilbert schemes $X^{[n]}$ of points on a smooth projective complex surface X. Then we will construct the Heisenberg algebras of Nakajima and Grojnowski, and discuss the Heisenberg monomial classes.

By our convention, all homology and cohomology groups are with complex coefficients, unless otherwise specified.

DEFINITION 3.2. Let X be a smooth projective complex surface.

(i) Let $\mathbb{H}_X^{n,i}$ denote $H^i(X^{[n]})$, and

$$\mathbb{H}_X^n = H^*(X^{[n]}) = \bigoplus_{i=0}^{4n} \mathbb{H}_X^{n,i}.$$

Let

$$\mathbb{H}_X = \bigoplus_{n=0}^{+\infty} \mathbb{H}_X^n = \bigoplus_{n,i \geq 0} \mathbb{H}_X^{n,i}$$

denote the bi-graded vector space. The element 1 in $H^0(X^{[0]}) = \mathbb{C}$ is called the *vacuum vector* and denoted by $|0\rangle$;

(ii) Define the *degree* of $\alpha \in H^*(X^{[n]})$ to be $|\alpha| = s$ if $\alpha \in H^s(X^{[n]})$;

(iii) An operator $\mathfrak{f} \in \operatorname{End}(\mathbb{H}_X)$ is *homogeneous of bi-degree* (ℓ, m) if

$$\mathfrak{f}(\mathbb{H}_X^{n,i}) \subset \mathbb{H}_X^{n+\ell, i+m};$$

(iv) For linear operators $\mathfrak{f}$ and $\mathfrak{g} \in \operatorname{End}(\mathbb{H}_X)$ of bi-degrees (ℓ, m) and (ℓ_1, m_1) respectively, define the *Lie superalgebra bracket* $[\mathfrak{f}, \mathfrak{g}]$ by putting

$$[\mathfrak{f}, \mathfrak{g}] = \mathfrak{f}\mathfrak{g} - (-1)^{mm_1} \mathfrak{g}\mathfrak{f}.$$

(v) The *non-degenerate super-symmetric bilinear form* $\langle \cdot, \cdot \rangle$ on $\mathbb{H}_X$ is induced from the standard one on $\mathbb{H}_X^n = H^*(X^{[n]})$ defined by

$$\langle a, b \rangle = \int_{X^{[n]}} a \cdot b$$

for $a, b \in H^*(X^{[n]})$. For an operator $\mathfrak{f} \in \operatorname{End}(\mathbb{H}_X)$ of bi-degree (ℓ, m), we define its *adjoint operator* $\mathfrak{f}^\dagger \in \operatorname{End}(\mathbb{H}_X)$ by putting

$$\langle \mathfrak{f}(a), b \rangle = (-1)^{m \cdot |a|} \cdot \langle a, \mathfrak{f}^\dagger(b) \rangle.$$

Let $\mathfrak{f} \in \mathrm{End}(\mathbb{H}_X)$ be of bi-degree (ℓ, m). Then the bi-degree of $\mathfrak{f}^\dagger$ is $(-\ell, m-4\ell)$. Also, for $\mathfrak{g}, \mathfrak{h} \in \mathrm{End}(\mathbb{H}_X)$ of bi-degrees $(\ell_1, m_1), (\ell_2, m_2)$ respectively, we have

$$(\mathfrak{f}\mathfrak{g})^\dagger = (-1)^{mm_1} \cdot \mathfrak{g}^\dagger \mathfrak{f}^\dagger, \tag{3.2}$$

$$[\mathfrak{f}, \mathfrak{g}]^\dagger = (-1)^{mm_1} \cdot [\mathfrak{g}^\dagger, \mathfrak{f}^\dagger] = -[\mathfrak{f}^\dagger, \mathfrak{g}^\dagger], \tag{3.3}$$

$$[\mathfrak{f}, [\mathfrak{g}, \mathfrak{h}]] = [[\mathfrak{f}, \mathfrak{g}], \mathfrak{h}] + (-1)^{mm_1}[\mathfrak{g}, [\mathfrak{f}, \mathfrak{h}]], \tag{3.4}$$

$$[\mathfrak{f}, \mathfrak{g}\mathfrak{h}] = [\mathfrak{f}, \mathfrak{g}]\mathfrak{h} + (-1)^{mm_1}\mathfrak{g}[\mathfrak{f}, \mathfrak{h}]. \tag{3.5}$$

We recall Nakajima's definition of the Heisenberg operators [**Groj**, **Nak3**, **Nak5**]. For $m \geq 0$ and $n > 0$, let $Q^{[m,m]} = \emptyset$ and define $Q^{[m+n,m]}$ to be

$$\Big\{(\xi, x, \eta) \in X^{[m+n]} \times X \times X^{[m]} \mid \xi \supset \eta \text{ and } \mathrm{Supp}(I_\eta/I_\xi) = \{x\}\Big\}. \tag{3.6}$$

DEFINITION 3.3. Fix a cohomology class $\alpha \in H^*(X)$. Let $n > 0$. The *linear operator* $\mathfrak{a}_{-n}(\alpha) \in \mathrm{End}(\mathbb{H}_X)$ is defined by

$$\mathfrak{a}_{-n}(\alpha)(a) = \tilde{p}_{1*}\left([Q^{[m+n,m]}] \cdot \tilde{\rho}^*\alpha \cdot \tilde{p}_2^* a\right)$$

for $a \in H^*(X^{[m]})$, where $\tilde{p}_1, \tilde{\rho}, \tilde{p}_2$ are the projections of $X^{[m+n]} \times X \times X^{[m]}$ to $X^{[m+n]}, X, X^{[m]}$ respectively. Define $\mathfrak{a}_n(\alpha) \in \mathrm{End}(\mathbb{H}_X)$ to be $(-1)^n$ times the operator obtained from the definition of $\mathfrak{a}_{-n}(\alpha)$ by switching the roles of $\tilde{p}_1$ and $\tilde{p}_2$. We often refer to $\mathfrak{a}_{-n}(\alpha)$ (respectively, $\mathfrak{a}_n(\alpha)$) as the *creation* (respectively, *annihilation*) operator. We also set $\mathfrak{a}_0(\alpha) = 0$.

LEMMA 3.4. *Let $m \geq 0$ and $n > 0$. Then,* $\dim Q^{[m+n,m]} = 2m + n + 1$.

PROOF. Every element $(\xi, x, \eta) \in Q^{[m+n,m]}$ is of the form

$$\eta = \eta_x + \eta', \qquad \xi = \xi_x + \eta'$$

where $\mathrm{Supp}(\eta_x) = \mathrm{Supp}(\xi_x) = \{x\}$, $x \notin \mathrm{Supp}(\eta')$ and $\eta_x \subset \xi_x$. Fix the length $\ell = \ell(\eta_x)$. If $\ell = 0$, then these elements $(\xi, x, \eta) \in Q^{[m+n,m]}$ form an open subset of $Q^{[m+n,m]}$ with dimension equal to

$$\begin{aligned} \#\text{moduli}\{\eta'\} + \#\text{moduli}\{x\} + \#\text{moduli}\{\xi_x\} &= 2m + 2 + (n-1) \\ &= 2m + n + 1 \end{aligned}$$

where we have used $\dim M_n(x) = n - 1$ from Theorem 1.18. If $\ell > 0$, then the number of moduli of these elements $(\xi, x, \eta) \in Q^{[m+n,m]}$ is at most

$$\begin{aligned} & \#\text{moduli}\{\eta'\} + \#\text{moduli}\{x\} + \#\text{moduli}\{\eta_x\} + \#\text{moduli}\{\xi_x\} \\ \leq\ & 2(m-\ell) + 2 + (\ell - 1) + (n + \ell - 1) \\ =\ & 2m + n. \end{aligned}$$

It follows that the dimension of $Q^{[m+n,m]}$ is equal to $2m + n + 1$. □

COROLLARY 3.5. *Let $n \neq 0$. Then the bi-degree of $\mathfrak{a}_{-n}(\alpha)$ is $(n, 2n - 2 + |\alpha|)$.*

PROOF. Follows immediately from Lemma 3.4 and Definition 3.3. Note from the proof of Lemma 3.4 that $Q^{[m+n,m]}$ has a unique irreducible component of dimension $2m+n+1$. Although $Q^{[m+n,m]}$ may (or may not) contain other irreducible components of lower dimension, it does not matter since only the fundamental class $[Q^{[m+n,m]}]$ is used in Definition 3.3. □

REMARK 3.6. Let $n > 0$. The geometric meaning of the operator $\mathfrak{a}_{-n}(\alpha)$ is roughly the following. Let $a \in H^*\big(X^{[m]}\big)$. Represent α and a by the cycles $\Gamma_\alpha \subset X$ and $\Gamma_a \subset X^{[m]}$ respectively such that a generic element $x \in \Gamma_\alpha$ and a generic element $\eta \in \Gamma_a$ satisfy $x \notin \mathrm{Supp}(\eta)$. Then $\mathfrak{a}_{-n}(\alpha)$ is represent by the cycle $\Gamma \subset X^{[m+n]}$ whose generic elements are of the form $\eta + \xi_x$ where $\eta \in \Gamma_a$, $x \in \Gamma_\alpha$ but $x \notin \mathrm{Supp}(\eta)$, and $\xi_x \in M_n(x)$.

LEMMA 3.7. $\mathfrak{a}_{-n}(\alpha)^\dagger = (-1)^n \cdot \mathfrak{a}_n(\alpha)$.

PROOF. The argument is completely formal. We may assume that $n > 0$. Let $a \in H^*\big(X^{[m]}\big)$ and $b \in H^*\big(X^{[m+n]}\big)$. We must prove that

$$\big\langle \mathfrak{a}_{-n}(\alpha)(a), b\big\rangle = (-1)^{(2n-2+|\alpha|)\,|a|} \cdot \big\langle a, (-1)^n \cdot \mathfrak{a}_n(\alpha)(b)\big\rangle \tag{3.7}$$

noting that the bi-degree of $\mathfrak{a}_{-n}(\alpha)$ is $(n, 2n-2+|\alpha|)$.

Let notations be the same as in Definition 3.3. By Definition 3.3,

$$\begin{aligned} \big\langle \mathfrak{a}_{-n}(\alpha)(a), b\big\rangle &= \int_{X^{[m+n]}} \mathfrak{a}_{-n}(\alpha)(a) \cdot b \\ &= \int_{X^{[m+n]}} \tilde p_{1*}\Big([Q^{[m+n,m]}] \cdot \tilde\rho^*\alpha \cdot \tilde p_2^* a\Big) \cdot b \\ &= \int_{X^{[m+n]}} \tilde p_{1*}\Big([Q^{[m+n,m]}] \cdot \tilde\rho^*\alpha \cdot \tilde p_2^* a \cdot \tilde p_1^* b\Big) \end{aligned} \tag{3.8}$$

where we have used the projection formula in the last step. Similarly, for the right-hand-side of (3.7), we see from the definition of $\mathfrak{a}_n(\alpha)$ that

$$\begin{aligned} &(-1)^{(2n-2+|\alpha|)\,|a|} \cdot \big\langle a, (-1)^n \cdot \mathfrak{a}_n(\alpha)(b)\big\rangle \\ =\ & (-1)^{|\alpha|\,|a|} \cdot \int_{X^{[m]}} \tilde p_{2*}\Big(\tilde p_2^* a \cdot [Q^{[m+n,m]}] \cdot \tilde\rho^*\alpha \cdot \tilde p_1^* b\Big) \\ =\ & \int_{X^{[m]}} \tilde p_{2*}\Big([Q^{[m+n,m]}] \cdot \tilde\rho^*\alpha \cdot \tilde p_2^* a \cdot \tilde p_1^* b\Big). \end{aligned}$$

Combining this with (3.8), we conclude that (3.7) holds. □

The following theorem is proved by Nakajima [**Nak3**, Section 3] and Grojnowski [**Groj**, Section 3].

THEOREM 3.8. *The operators $\mathfrak{a}_n(\alpha) \in \mathrm{End}(\mathbb{H}_X)$ with $\alpha \in H^*(X)$ and $n \in \mathbb{Z}$ satisfy the following Heisenberg algebra commutation relation:*

$$[\mathfrak{a}_m(\alpha), \mathfrak{a}_n(\beta)] = -m \cdot \delta_{m,-n} \cdot (\alpha, \beta) \cdot \mathrm{Id}_{\mathbb{H}_X} \tag{3.9}$$

where we have used $\delta_{m,-n}$ to denote 1 if $m = -n$ and 0 otherwise. Moreover, the space $\mathbb{H}_X$ is an irreducible module over the Heisenberg algebra generated by the operators $\mathfrak{a}_n(\alpha)$ with a highest weight vector $|0\rangle = 1 \in H^0(X^{[0]})$.

It follows that $\mathbb{H}_X$ is linearly spanned by all the *Heisenberg monomial classes*:

$$\mathfrak{a}_{-n_1}(\alpha_1) \cdots \mathfrak{a}_{-n_k}(\alpha_k)|0\rangle \tag{3.10}$$

where $k \ge 0, n_1, \ldots, n_k > 0$, and the cohomology classes $\alpha_1, \ldots, \alpha_k$ run over a fixed linear basis of $H^*(X)$. In view of Corollary 3.5, we have

$$\mathfrak{a}_{-n_1}(\alpha_1) \cdots \mathfrak{a}_{-n_k}(\alpha_k)|0\rangle \in H^m(X^{[n]}) \tag{3.11}$$

where $n = n_1 + \ldots + n_k$ and $m = \sum_{i=1}^k (2n_i - 2 + |\alpha_i|)$.

Next, we will introduce the notion $\mathfrak{a}_{m_1} \cdots \mathfrak{a}_{m_k}(\tau_{k*}\alpha)$ and study its properties. This notion will be useful in writing down the Chern character operators and the higher derivatives of the Heisenberg operators.

DEFINITION 3.9. For $k \geq 1$, define $\tau_{k*} : H^*(X) \to H^*(X^k)$ to be the map induced by the diagonal embedding $\tau_k : X \to X^k$, and $\mathfrak{a}_{m_1} \cdots \mathfrak{a}_{m_k}(\tau_{k*}\alpha)$ to be

$$\sum_j \mathfrak{a}_{m_1}(\alpha_{j,1}) \cdots \mathfrak{a}_{m_k}(\alpha_{j,k})$$

when

$$\tau_{k*}\alpha = \sum_j \alpha_{j,1} \otimes \cdots \otimes \alpha_{j,k} \tag{3.12}$$

via the Künneth decomposition of $H^*(X^k)$.

LEMMA 3.10. *The notion $\mathfrak{a}_{m_1} \cdots \mathfrak{a}_{m_k}(\tau_{k*}\alpha)$ is well-defined in the sense that it is independent of the decomposition* (3.12).

PROOF. Assume that via the Künneth decomposition of $H^*(X^k)$, we have

$$\tau_{k*}\alpha = \sum_j \alpha_{j,1} \otimes \cdots \otimes \alpha_{j,k} = \sum_s \beta_{s,1} \otimes \cdots \otimes \beta_{s,k}. \tag{3.13}$$

Let $\{\gamma_\ell\}_\ell$ be a linear basis of $H^*(X)$. For j, s and $1 \leq t \leq k$, put

$$\alpha_{j,t} = \sum_\ell c_{j,t,\ell}\gamma_\ell, \quad \beta_{s,t} = \sum_\ell d_{s,t,\ell}\gamma_\ell$$

where $c_{j,t,\ell}, d_{s,t,\ell} \in \mathbb{C}$. In view of (3.13), we get

$$\begin{aligned} &\sum_{\ell_1,\ldots,\ell_k} \left(\sum_j c_{j,1,\ell_1} \cdots c_{j,k,\ell_k} \right) \gamma_{\ell_1} \otimes \cdots \otimes \gamma_{\ell_k} \\ = &\sum_{\ell_1,\ldots,\ell_k} \left(\sum_s d_{s,1,\ell_1} \cdots d_{s,k,\ell_k} \right) \gamma_{\ell_1} \otimes \cdots \otimes \gamma_{\ell_k}. \end{aligned}$$

Therefore,

$$\sum_j c_{j,1,\ell_1} \cdots c_{j,k,\ell_k} = \sum_s d_{s,1,\ell_1} \cdots d_{s,k,\ell_k}$$

for all $\ell_1, \ldots, \ell_k$. It follows that

$$\begin{aligned} &\sum_j \mathfrak{a}_{m_1}(\alpha_{j,1}) \cdots \mathfrak{a}_{m_k}(\alpha_{j,k}) \\ = &\sum_{\ell_1,\ldots,\ell_k} \left(\sum_j c_{j,1,\ell_1} \cdots c_{j,k,\ell_k} \right) \mathfrak{a}_{m_1}(\gamma_{\ell_1}) \cdots \mathfrak{a}_{m_k}(\gamma_{\ell_k}) \\ = &\sum_{\ell_1,\ldots,\ell_k} \left(\sum_s d_{s,1,\ell_1} \cdots d_{s,k,\ell_k} \right) \mathfrak{a}_{m_1}(\gamma_{\ell_1}) \cdots \mathfrak{a}_{m_k}(\gamma_{\ell_k}) \\ = &\sum_s \mathfrak{a}_{m_1}(\beta_{s,1}) \cdots \mathfrak{a}_{m_k}(\beta_{s,k}). \end{aligned}$$

Hence, $\mathfrak{a}_{m_1} \cdots \mathfrak{a}_{m_k}(\tau_{k*}\alpha)$ is independent of the decomposition (3.13). □

The following topological lemma is elementary, but very efficient in organizing the Heisenberg monomial classes.

LEMMA 3.11. *Let $k, u \geq 1$ and $\alpha, \beta \in H^*(X)$. Assume that*

$$\tau_{k*}(\alpha) = \sum_i \alpha_{i,1} \otimes \ldots \otimes \alpha_{i,k}$$

under the Künneth decomposition of $H^*(X^k)$. Then for $0 \leq j \leq k$, we have

$$\begin{aligned}
\tau_{k*}(\alpha\beta) &= \sum_i (-1)^{|\beta| \cdot \sum_{\ell=j+1}^k |\alpha_{i,\ell}|} \cdot \left(\otimes_{s=1}^{j-1} \alpha_{i,s}\right) \otimes (\alpha_{i,j}\beta) \otimes \left(\otimes_{t=j+1}^k \alpha_{i,t}\right) \\
\tau_{(k-1)*}(\alpha\beta) &= \sum_i (-1)^{|\beta| \sum_{\ell=j+1}^k |\alpha_{i,\ell}|} \int_X \alpha_{i,j}\beta \cdot \otimes_{1 \leq s \leq k, s \neq j} \alpha_{i,s} \\
\tau_{(k+u-1)*}(\alpha) &= \sum_i \left(\otimes_{s=1}^{j-1} \alpha_{i,s}\right) \otimes (\tau_{u*}\alpha_{i,j}) \otimes \left(\otimes_{t=j+1}^k \alpha_{i,t}\right).
\end{aligned}$$

PROOF. The basic idea is to use the projection formula. Let p_j be the projection of X^k to the j-th factor. Then, we have

$$\begin{aligned}
&\sum_i (-1)^{|\beta| \cdot \sum_{\ell=j+1}^k |\alpha_{i,\ell}|} \cdot \left(\otimes_{s=1}^{j-1} \alpha_{i,s}\right) \otimes (\alpha_{i,j}\beta) \otimes \left(\otimes_{t=j+1}^k \alpha_{i,t}\right) \\
= &\left(\sum_i \alpha_{i,1} \otimes \ldots \otimes \alpha_{i,k}\right) \cdot p_j^*(\beta) \\
= &\tau_{k*}(\alpha) \cdot p_j^*(\beta) \\
= &\tau_{k*}\left(\alpha \cdot (p_j \circ \tau_k)^*(\beta)\right) \\
= &\tau_{k*}(\alpha\beta).
\end{aligned}$$

This proves the first formula.

For the second formula, let $\tilde{p}_j : X^k \to X^{k-1}$ be the projection by ignoring the j-th factor. Then, we get

$$\begin{aligned}
&\sum_i (-1)^{|\beta| \sum_{\ell=j+1}^k |\alpha_{i,\ell}|} \int_X \alpha_{i,j}\beta \cdot \otimes_{1 \leq s \leq k, s \neq j} \alpha_{i,s} \\
= &\tilde{p}_{j*}\left(\sum_i (-1)^{|\beta| \cdot \sum_{\ell=j+1}^k |\alpha_{i,\ell}|} \cdot \left(\otimes_{s=1}^{j-1} \alpha_{i,s}\right) \otimes (\alpha_{i,j}\beta) \otimes \left(\otimes_{t=j+1}^k \alpha_{i,t}\right)\right) \\
= &\tilde{p}_{j*}\big(\tau_{k*}(\alpha\beta)\big) \\
= &\tau_{(k-1)*}(\alpha\beta)
\end{aligned}$$

where we have used the first formula in the second step. This proves the second formula.

For the third formula, let

$$\hat{p}_j = (\mathrm{Id}_X)^{j-1} \times \tau_u \times (\mathrm{Id}_X)^{k-j} : \quad X^k \to X^{k+u-1}.$$

Then, we obtain

$$\begin{aligned}
&\sum_i \left(\otimes_{s=1}^{j-1}\alpha_{i,s}\right) \otimes (\tau_{u*}\alpha_{i,j}) \otimes \left(\otimes_{t=j+1}^{k}\alpha_{i,t}\right) \\
= \;& \hat{p}_{j*}\left(\sum_i \alpha_{i,1} \otimes \ldots \otimes \alpha_{i,k}\right) \\
= \;& \hat{p}_{j*}\left(\tau_{k*}(\alpha)\right) \\
= \;& \tau_{(k+u-1)*}(\alpha).
\end{aligned}$$

This proves the third formula. □

In the lemma below and throughout the book, the products of Heisenberg operators are understood to be in the increasing order of the parametrizing indices from the left to the right, e.g.,

$$\prod_{1\le s<j} \mathfrak{a}_{n_s} = \mathfrak{a}_{n_1}\mathfrak{a}_{n_2}\cdots\mathfrak{a}_{n_{j-1}}$$

and

$$\prod_{1\le u\le k, u\ne t} \mathfrak{a}_{n_u} = \mathfrak{a}_{n_1}\cdots\mathfrak{a}_{n_{t-1}}\mathfrak{a}_{n_{t+1}}\cdots\mathfrak{a}_{n_k}.$$

LEMMA 3.12. *Let $k, s \ge 1$, $n_1, \ldots, n_k, m_1, \ldots, m_s \in \mathbb{Z}$ and $\alpha, \beta \in H^*(X)$.*

(i) *The commutator $[\mathfrak{a}_{n_1}\cdots\mathfrak{a}_{n_k}(\tau_{k*}\alpha), \mathfrak{a}_{m_1}\cdots\mathfrak{a}_{m_s}(\tau_{s*}\beta)]$ is equal to*

$$-\sum_{t=1}^{k}\sum_{j=1}^{s} n_t\delta_{n_t,-m_j}\cdot\left(\prod_{\ell=1}^{j-1}\mathfrak{a}_{m_\ell}\prod_{1\le u\le k, u\ne t}\mathfrak{a}_{n_u}\prod_{\ell=j+1}^{s}\mathfrak{a}_{m_\ell}\right)\left(\tau_{(k+s-2)*}(\alpha\beta)\right);$$

(ii) *Let j satisfy $1 \le j < k$. Then, $\mathfrak{a}_{n_1}\cdots\mathfrak{a}_{n_k}(\tau_{k*}\alpha)$ is equal to*

$$\begin{aligned}
&\left(\prod_{1\le s<j}\mathfrak{a}_{n_s}\cdot\mathfrak{a}_{n_{j+1}}\mathfrak{a}_{n_j}\cdot\prod_{j+1<s\le k}\mathfrak{a}_{n_s}\right)(\tau_{k*}\alpha) \\
- \;& n_j\delta_{n_j,-n_{j+1}}\prod_{\substack{1\le s\le k\\ s\ne j,j+1}}\mathfrak{a}_{n_s}\left(\tau_{(k-2)*}(e_X\alpha)\right)
\end{aligned}$$

where e_X is the Euler class of X.

PROOF. To simplify the signs, we will assume that all the cohomology classes have even degrees.

(i) Let

$$\begin{aligned}
\tau_{k*}\alpha &= \sum_a \alpha_{a,1}\otimes\cdots\otimes\alpha_{a,k}, \\
\tau_{s*}\beta &= \sum_b \beta_{b,1}\otimes\cdots\otimes\beta_{b,s}.
\end{aligned}$$

By (3.5), we have

$$
\begin{aligned}
(3.14)\qquad & [\mathfrak{a}_{n_1}\cdots\mathfrak{a}_{n_k}(\tau_{k*}\alpha), \mathfrak{a}_{m_1}\cdots\mathfrak{a}_{m_s}(\tau_{s*}\beta)] \\
&= \sum_b [\mathfrak{a}_{n_1}\cdots\mathfrak{a}_{n_k}(\tau_{k*}\alpha), \mathfrak{a}_{m_1}(\beta_{b,1})\cdots\mathfrak{a}_{m_s}(\beta_{b,s})] \\
&= \sum_b\sum_{j=1}^{s}\prod_{\ell=1}^{j-1}\mathfrak{a}_{m_\ell}(\beta_{b,\ell})\cdot[\mathfrak{a}_{n_1}\cdots\mathfrak{a}_{n_k}(\tau_{k*}\alpha),\mathfrak{a}_{m_j}(\beta_{b,j})]\cdot\prod_{\ell=j+1}^{s}\mathfrak{a}_{m_\ell}(\beta_{b,\ell}).
\end{aligned}
$$

By (3.5) again and Theorem 3.8, we get

$$
\begin{aligned}
& [\mathfrak{a}_{n_1}\cdots\mathfrak{a}_{n_k}(\tau_{k*}\alpha), \mathfrak{a}_{m_j}(\beta_{b,j})] \\
= \quad & \sum_a [\mathfrak{a}_{n_1}(\alpha_{a,1})\cdots\mathfrak{a}_{n_k}(\alpha_{a,k}), \mathfrak{a}_{m_j}(\beta_{b,j})] \\
= \quad & \sum_a\sum_{t=1}^{k}\prod_{u=1}^{t-1}\mathfrak{a}_{n_u}(\alpha_{a,u})\cdot[\mathfrak{a}_{n_t}(\alpha_{a,t}),\mathfrak{a}_{m_j}(\beta_{b,j})]\cdot\prod_{u=t+1}^{k}\mathfrak{a}_{n_u}(\alpha_{a,u}) \\
= \quad & -\sum_{t=1}^{k} n_t\delta_{n_t,-m_j}\cdot\sum_a\int_X\alpha_{a,t}\beta_{b,j}\cdot\prod_{1\le u\le k, u\ne t}\mathfrak{a}_{n_u}(\alpha_{a,u}).
\end{aligned}
$$

By the second formula in Lemma 3.11, we obtain

$$
\begin{aligned}
& [\mathfrak{a}_{n_1}\cdots\mathfrak{a}_{n_k}(\tau_{k*}\alpha), \mathfrak{a}_{m_j}(\beta_{b,j})] \\
= \quad & -\sum_{t=1}^{k} n_t\delta_{n_t,-m_j}\cdot\left(\prod_{1\le u\le k, u\ne t}\mathfrak{a}_{n_u}\right)\tau_{(k-1)*}(\alpha\beta_{b,j}).
\end{aligned}
$$

Combining with (3.14), we see that $[\mathfrak{a}_{n_1}\cdots\mathfrak{a}_{n_k}(\tau_{k*}\alpha), \mathfrak{a}_{m_1}\cdots\mathfrak{a}_{m_s}(\tau_{s*}\beta)]$ is equal to

$$
-\sum_{t=1}^{k}\sum_{j=1}^{s} n_t\delta_{n_t,-m_j}\cdot\sum_b\prod_{\ell=1}^{j-1}\mathfrak{a}_{m_\ell}(\beta_{b,\ell})
$$

$$
\cdot\left(\prod_{1\le u\le k, u\ne t}\mathfrak{a}_{n_u}\right)\tau_{(k-1)*}(\alpha\beta_{b,j})\cdot\prod_{\ell=j+1}^{s}\mathfrak{a}_{m_\ell}(\beta_{b,\ell}).
$$

By the first and third formulas in Lemma 3.11, we conclude that

$$
\begin{aligned}
& [\mathfrak{a}_{n_1}\cdots\mathfrak{a}_{n_k}(\tau_{k*}\alpha), \mathfrak{a}_{m_1}\cdots\mathfrak{a}_{m_s}(\tau_{s*}\beta)] \\
= \quad & -\sum_{t=1}^{k}\sum_{j=1}^{s} n_t\delta_{n_t,-m_j}\cdot\left(\prod_{\ell=1}^{j-1}\mathfrak{a}_{m_\ell}\prod_{1\le u\le k, u\ne t}\mathfrak{a}_{n_u}\prod_{\ell=j+1}^{s}\mathfrak{a}_{m_\ell}\right)(\tau_{(k+s-2)*}(\alpha\beta)).
\end{aligned}
$$

(ii) Let

$$
\begin{aligned}
\tau_{(k-1)*}\alpha &= \sum_a \alpha_{a,1}\otimes\cdots\otimes\alpha_{a,k-1}, \\
\tau_{2*}\alpha_{a,j} &= \sum_b \beta_{a,j,b,1}\otimes\beta_{a,j,b,2}
\end{aligned}
$$

where we may assume that b is independent of a and j. By the third formula in Lemma 3.11,

$$\tau_{k*}\alpha = \sum_{a,b} \big(\otimes_{s=1}^{j-1} \alpha_{a,s} \big) \otimes \beta_{a,j,b,1} \otimes \beta_{a,j,b,2} \otimes \big(\otimes_{s=j+1}^{k-1} \alpha_{a,s} \big).$$

Combining with Theorem 3.8, we obtain

$$\begin{aligned}
&\mathfrak{a}_{n_1} \cdots \mathfrak{a}_{n_k}(\tau_{k*}\alpha) \\
= &\sum_{a,b} \prod_{1 \le s < j} \mathfrak{a}_{n_s}(\alpha_{a,s}) \cdot \mathfrak{a}_{n_j}(\beta_{a,j,b,1}) \mathfrak{a}_{n_{j+1}}(\beta_{a,j,b,2}) \cdot \prod_{j < s \le k-1} \mathfrak{a}_{n_{s+1}}(\alpha_{a,s}) \\
= &\sum_{a,b} \prod_{1 \le s < j} \mathfrak{a}_{n_s}(\alpha_{a,s}) \cdot \mathfrak{a}_{n_{j+1}}(\beta_{a,j,b,2}) \mathfrak{a}_{n_j}(\beta_{a,j,b,1}) \cdot \prod_{j < s \le k-1} \mathfrak{a}_{n_{s+1}}(\alpha_{a,s}) \\
&- n_j \delta_{n_j, -n_{j+1}} \sum_{a,b} \int_X \beta_{a,j,b,1} \beta_{a,j,b,2} \cdot \prod_{1 \le s < j} \mathfrak{a}_{n_s}(\alpha_{a,s}) \cdot \prod_{j < s \le k-1} \mathfrak{a}_{n_{s+1}}(\alpha_{a,s}) \\
= &\left(\prod_{1 \le s < j} \mathfrak{a}_{n_s} \cdot \mathfrak{a}_{n_{j+1}} \mathfrak{a}_{n_j} \cdot \prod_{j+1 < s \le k} \mathfrak{a}_{n_s} \right) (\tau_{k*}\alpha) \\
&- n_j \delta_{n_j, -n_{j+1}} \sum_{a,b} \int_X \beta_{a,j,b,1} \beta_{a,j,b,2} \cdot \prod_{1 \le s < j} \mathfrak{a}_{n_s}(\alpha_{a,s}) \cdot \prod_{j < s \le k-1} \mathfrak{a}_{n_{s+1}}(\alpha_{a,s}).
\end{aligned}$$

Note that

$$\sum_b \beta_{a,j,b,1} \beta_{a,j,b,2} = e_X \alpha_{a,j}.$$

Therefore,

$$\begin{aligned}
&\mathfrak{a}_{n_1} \cdots \mathfrak{a}_{n_k}(\tau_{k*}\alpha) \\
= &\left(\prod_{1 \le s < j} \mathfrak{a}_{n_s} \cdot \mathfrak{a}_{n_{j+1}} \mathfrak{a}_{n_j} \cdot \prod_{j+1 < s \le k} \mathfrak{a}_{n_s} \right) (\tau_{k*}\alpha) \\
&- n_j \delta_{n_j, -n_{j+1}} \sum_{a} \int_X e_X \alpha_{a,j} \cdot \prod_{1 \le s < j} \mathfrak{a}_{n_s}(\alpha_{a,s}) \cdot \prod_{j < s \le k-1} \mathfrak{a}_{n_{s+1}}(\alpha_{a,s}) \\
= &\left(\prod_{1 \le s < j} \mathfrak{a}_{n_s} \cdot \mathfrak{a}_{n_{j+1}} \mathfrak{a}_{n_j} \cdot \prod_{j+1 < s \le k} \mathfrak{a}_{n_s} \right) (\tau_{k*}\alpha) \\
&- n_j \delta_{n_j, -n_{j+1}} \prod_{\substack{1 \le s \le k \\ s \ne j, j+1}} \mathfrak{a}_{n_s}(\tau_{(k-2)*}(e_X \alpha))
\end{aligned}$$

where at the last step, we have used the second formula in Lemma 3.11. □

In the following, we will define some convenient operation regarding Heisenberg monomial classes. It enables us to decompose cohomology classes. In Section 15.3, we will see that this operation plays an interesting role in describing the structure of the 3-point genus-0 extremal Gromov-Witten invariants of the Hilbert scheme $X^{[n]}$.

DEFINITION 3.13. Let $A = \mathfrak{a}_{-n_1}(\alpha_1) \cdots \mathfrak{a}_{-n_\ell}(\alpha_\ell)|0\rangle$ where $n_1, \ldots, n_\ell > 0$.

(i) If $B = \mathfrak{a}_{-m_1}(\beta_1) \cdots \mathfrak{a}_{-m_s}(\beta_s)|0\rangle$ with $m_1, \ldots, m_s > 0$, then we define

$$A \circ B = \mathfrak{a}_{-n_1}(\alpha_1) \cdots \mathfrak{a}_{-n_\ell}(\alpha_\ell) \mathfrak{a}_{-m_1}(\beta_1) \cdots \mathfrak{a}_{-m_s}(\beta_s)|0\rangle. \tag{3.15}$$

(ii) We use the symbol $B \subset A$ if $B = \mathfrak{a}_{-n_{i_1}}(\alpha_{i_1}) \cdots \mathfrak{a}_{-n_{i_s}}(\alpha_{i_s})|0\rangle$ with $1 \le i_1 < \ldots < i_s \le \ell$. In this case, we use A/B or AB^{-1} or $\dfrac{A}{B}$ to denote the cohomology class obtained from A by deleting the factors

$$\mathfrak{a}_{-n_{i_1}}(\alpha_{i_1}), \ldots, \mathfrak{a}_{-n_{i_s}}(\alpha_{i_s}).$$

Note that when $B \subset A$, then we have the identity $(A/B) \circ B = A$.

LEMMA 3.14. *Assume that the odd Betti numbers of X are equal to zero. Let $A = \mathfrak{a}_{-n_1}(\alpha_1) \cdots \mathfrak{a}_{-n_\ell}(\alpha_\ell)|0\rangle$ where $n_1, \ldots, n_\ell > 0$. Let p be a monomial of Heisenberg creation operators. Then,*

$$p^\dagger A = \sum_{B \subset A} \langle p|0\rangle, B\rangle \cdot \frac{A}{B}. \tag{3.16}$$

PROOF. Let

$$p = \mathfrak{a}_{-p_1}(\beta_{1,1}) \cdots \mathfrak{a}_{-p_1}(\beta_{1,k_1}) \cdots \mathfrak{a}_{-p_s}(\beta_{s,1}) \cdots \mathfrak{a}_{-p_s}(\beta_{s,k_s})$$

where $0 < p_1 < \ldots < p_s$. We use induction on s. Let $s = 1$. If $p_1 \ne n_i$ for every $i = 1, \ldots, \ell$, then both sides of (3.16) are equal to 0. Next, let $p_1 = n_i$ for some $i \in \{1, \ldots, \ell\}$. We may assume that $i = 1$ with

$$p_1 = n_1 = \ldots = n_t < n_{t+1} \le n_{t+2} \cdots \le n_\ell.$$

By Lemma 3.7 and Theorem 3.8,

$$\begin{aligned} & p^\dagger A \\ = \;& (-1)^{k_1 p_1} \mathfrak{a}_{p_1}(\beta_{1,1}) \cdots \mathfrak{a}_{p_1}(\beta_{1,k_1}) \mathfrak{a}_{-n_1}(\alpha_1) \cdots \mathfrak{a}_{-n_\ell}(\alpha_\ell)|0\rangle \\ = \;& (-1)^{k_1 p_1} \sum_{\substack{1 \le j_1 < \ldots, < j_{k_1} \le \ell \\ \sigma \in \mathrm{Per}\{j_1, \ldots, j_{k_1}\}}} \frac{(-n_1)^{k_1} \cdot \langle \beta_{1,1}, \alpha_{\sigma(j_1)}\rangle \cdots \langle \beta_{1,k_1}, \alpha_{\sigma(j_{k_1})}\rangle \cdot A}{\mathfrak{a}_{-n_{j_1}}(\alpha_{j_1}) \cdots \mathfrak{a}_{-n_{j_{k_1}}}(\alpha_{j_{k_1}})|0\rangle} \\ = \;& \sum_{B \subset A} \langle p|0\rangle, B\rangle \cdot \frac{A}{B} \end{aligned}$$

where $\mathrm{Per}\{j_1, \ldots, j_{k_1}\}$ denotes the permutation group of $\{j_1, \ldots, j_{k_1}\}$. Therefore, (3.16) is true when $s = 1$. Next, let $s > 1$ and assume that (3.16) holds for

$$\tilde{p} = \mathfrak{a}_{-p_1}(\beta_{1,1}) \cdots \mathfrak{a}_{-p_1}(\beta_{1,k_1}) \cdots \mathfrak{a}_{-p_{s-1}}(\beta_{s-1,1}) \cdots \mathfrak{a}_{-p_{s-1}}(\beta_{s-1,k_{s-1}}).$$

Note from (3.2) that $p^\dagger = \big(\mathfrak{a}_{-p_s}(\beta_{s,1}) \cdots \mathfrak{a}_{-p_s}(\beta_{s,k_s})\big)^\dagger \tilde{p}^\dagger$. By induction,

$$\begin{aligned} & p^\dagger A \\ = \;& \big(\mathfrak{a}_{-p_s}(\beta_{s,1}) \cdots \mathfrak{a}_{-p_s}(\beta_{s,k_s})\big)^\dagger \tilde{p}^\dagger A \\ = \;& \big(\mathfrak{a}_{-p_s}(\beta_{s,1}) \cdots \mathfrak{a}_{-p_s}(\beta_{s,k_s})\big)^\dagger \sum_{B_1 \subset A} \langle \tilde{p}|0\rangle, B_1\rangle \cdot \frac{A}{B_1} \\ = \;& \sum_{B_1 \subset A} \sum_{B_2 \subset A/B_1} \big\langle \mathfrak{a}_{-p_s}(\beta_{s,1}) \cdots \mathfrak{a}_{-p_s}(\beta_{s,k_s})|0\rangle, B_2\big\rangle \cdot \langle \tilde{p}|0\rangle, B_1\rangle \cdot \frac{A/B_1}{B_2} \end{aligned}$$

If $\big\langle \mathfrak{a}_{-p_s}(\beta_{s,1}) \cdots \mathfrak{a}_{-p_s}(\beta_{s,k_s})|0\rangle, B_2\big\rangle \cdot \langle \tilde{p}|0\rangle, B_1\rangle \ne 0$, then B_1 and B_2 do not share any common Heisenberg factors, and

$$\big\langle \mathfrak{a}_{-p_s}(\beta_{s,1}) \cdots \mathfrak{a}_{-p_s}(\beta_{s,k_s})|0\rangle, B_2\big\rangle \cdot \langle \tilde{p}|0\rangle, B_1\rangle = \langle p|0\rangle, B_1 \circ B_2\rangle.$$

Hence, putting $B = B_1 \circ B_2$, we obtain

$$p^\dagger A = \sum_{B \subset A} \langle p|0\rangle, B\rangle \cdot \frac{A}{B}.$$

This completes the proof of (3.16). $\square$

3.3. Geometric interpretations of Heisenberg monomial classes

By (3.10), the Heisenberg monomial classes linearly span the cohomology group of the Hilbert scheme $X^{[n]}$. In this section, we will describe the geometric representations of the Heisenberg monomial classes

$$\mathfrak{a}_{-n_1}(\alpha_1)\cdots\mathfrak{a}_{-n_k}(\alpha_k)|0\rangle \in H^*(X^{[n]})$$

where $n = n_1 + \ldots + n_k$.

We begin with the construction of a certain closed subset in the Hilbert scheme $X^{[n]}$ and the determination of its main component. Fix closed real cycles $X_1, \ldots, X_k$ of X in general position in the sense that any subset of the X_i's meet transversally in the expected dimension. Define

$$W(n_1, X_1; \ldots; n_k, X_k) \subset X^{[n]}$$

to be the closed subset consisting of all $\xi \in X^{[n]}$ which admit filtrations

$$\xi = \xi_k \supset \ldots \supset \xi_1 \supset \xi_0 = \emptyset$$

with $\ell(\xi_i) = \ell(\xi_{i-1}) + n_i$ and

$$\text{Supp}(I_{\xi_{i-1}}/I_{\xi_i}) = x_i \in X_i \tag{3.17}$$

for $1 \le i \le k$. Let

$$W(n_1, X_1; \ldots; n_k, X_k)^0 \subset W(n_1, X_1; \ldots; n_k, X_k)$$

be the open subset consisting of all $\xi \in W(n_1, X_1; \ldots; n_k, X_k)$ such that the points $x_1, \ldots, x_k$ in (3.17) are distinct.

LEMMA 3.15. *Denote $W(n_1, X_1; \ldots; n_k, X_k)^0$ and $W(n_1, X_1; \ldots; n_k, X_k)$ by W^0 and W respectively. Let $W^c = W - W^0$. Then,*

$$\dim_{\mathbb{R}}(W^c) < \dim_{\mathbb{R}}(W^0) = \sum_{i=1}^{k} \big(2n_i - 2 + \dim_{\mathbb{R}}(X_i)\big).$$

PROOF. Note that the points $\xi \in W^0$ are of the form

$$\xi = \eta_1 + \ldots + \eta_k$$

with $\ell(\eta_i) = n_i$ and $\text{Supp}(\eta_i) = x_i \in X_i$ for every i. So

$$\dim_{\mathbb{R}}(W^0) = \sum_{i=1}^{k} \big(2n_i - 2 + \dim_{\mathbb{R}}(X_i)\big).$$

Next, without loss of generality, we may assume that a point $\xi \in W^c$ satisfies

$$x_1 = \ldots = x_{k_1},$$
$$\ldots,$$
$$x_{k_1+\ldots+k_{r-1}+1} = \ldots = x_{k_1+\ldots+k_r}.$$

Here $r \geq 1$, $k_1 \geq 2$, $k_2, \ldots, k_r \geq 1$, $k_1 + \ldots + k_r = k$, and the points $y_1, \ldots, y_r$ are distinct where

$$y_1 = x_1 = \ldots = x_{k_1},$$
$$\ldots,$$
$$y_r = x_{k_1+\ldots+k_{r-1}+1} = \ldots = x_{k_1+\ldots+k_r}.$$

Since the cycles $X_1, \ldots, X_k$ are in general position, these ξ's either do not exist at all, or form a subset of dimension at most

$$\begin{aligned} &\left[2\left(\sum_{i=1}^{k_1} n_i - 1\right) + \sum_{i=1}^{k_1} \dim_{\mathbb{R}}(X_i) - 4(k_1-1)\right] + \ldots \\ &+ \left[2\left(\sum_{i=1}^{k_r} n_{k_1+\ldots+k_{r-1}+i} - 1\right) + \sum_{i=1}^{k_r} \dim_{\mathbb{R}}(X_{k_1+\ldots+k_{r-1}+i}) - 4(k_r-1)\right] \\ = &\sum_{i=1}^{k}(2n_i - 2 + \dim_{\mathbb{R}}(X_i)) - \sum_{i=1}^{r} 2(k_i-1) \\ < &\sum_{i=1}^{k}(2n_i - 2 + \dim_{\mathbb{R}}(X_i)). \end{aligned}$$

It follows that

$$\dim_{\mathbb{R}}(W^c) < \sum_{i=1}^{k}\big(2n_i - 2 + \dim_{\mathbb{R}}(X_i)\big) = \dim_{\mathbb{R}}(W^0).$$

This completes the proof of the lemma. □

We conclude from Lemma 3.15 that $W(n_1, X_1; \ldots; n_k, X_k)$ is a union of irreducible closed subsets among which the closure of $W(n_1, X_1; \ldots; n_k, X_k)^0$ is the only irreducible component having the largest dimension.

Proposition 3.16. *Let $\ell, k \geq 0$, $s_i \geq 0$ $(1 \leq i \leq \ell)$, $n_i > 0$ $(1 \leq i \leq k)$. Let*

$$\alpha_1, \ldots, \alpha_k \in \bigoplus_{i=1}^{4} H^i(X)$$

be homogeneous classes represented by the real cycles $X_1, \ldots, X_k \subset X$ respectively such that $X_1, \ldots, X_k$ are in general position. Then, the Heisenberg monomial class

$$\left(\prod_{i=1}^{\ell} \frac{\mathfrak{a}_{-i}(1_X)^{s_i}}{s_i!}\right)\left(\prod_{i=1}^{k} \mathfrak{a}_{-n_i}(\alpha_i)\right)|0\rangle \tag{3.18}$$

is represented by the closure of

$$W(\underbrace{1, X; \ldots; 1, X}_{s_1 \text{ times}}; \ldots; \underbrace{\ell, X; \ldots; \ell, X}_{s_\ell \text{ times}}; n_1, X_1; \ldots; n_k, X_k)^0. \tag{3.19}$$

Proof. We use induction on $\sum_{i=1}^{\ell} is_i + \sum_{i=1}^{k} n_i$. When

$$\sum_{i=1}^{\ell} is_i + \sum_{i=1}^{k} n_i = 0,$$

the conclusion is trivially true. In the following, assuming that the conclusion is true whenever $\sum_{i=1}^{\ell} is_i + \sum_{i=1}^{k} n_i < n$, we will prove that it still holds when

$$\sum_{i=1}^{\ell} is_i + \sum_{i=1}^{k} n_i = n.$$

There are two cases: $s_i > 0$ for some i, or $n_i > 0$ for some i. Since the proof of these two cases are similar, we only prove the first case.

So let $s_i > 0$ for some i. Without loss of generality, we may assume that $i = \ell$, i.e., $s_\ell > 0$. Denote (3.18) and (3.19) by A_{s_ℓ} and $W_{s_\ell}^0$ respectively. By the induction hypothesis, $A_{s_\ell - 1}$ is represented by the closure $\overline{W_{s_\ell-1}^0}$. By Definition 3.3, $\mathfrak{a}_{-\ell}(1_X)(A_{s_\ell-1})$ is represented by

$$p_{1*}\left([Q_X^{[n,n-\ell]}] \cap p_2^*(\overline{W_{s_\ell-1}^0})\right)$$

where p_1, p_2 are the projections of $X^{[n]} \times X^{[n-\ell]}$, and

$$Q_X^{[n,n-\ell]} = \left\{(\xi,\eta) \in X^{[n]} \times X^{[n-\ell]} \mid \xi \supset \eta \text{ and } \mathrm{Supp}(I_\eta/I_\xi) \text{ is a point}\right\}.$$

Since $\mathfrak{a}_{-\ell}(1_X)(A_{s_\ell-1}) = s_\ell A_{s_\ell}$, we see that A_{s_ℓ} is represented by

$$\frac{1}{s_\ell} \cdot p_{1*}\left([Q_X^{[n,n-\ell]}] \cap p_2^*(\overline{W_{s_\ell-1}^0})\right). \tag{3.20}$$

Note that the image $p_1\left(Q_X^{[n,n-\ell]} \cap p_2^{-1}(\overline{W_{s_\ell-1}^0})\right)$ is contained in the closed subset

$$W(\underbrace{1,X;\ldots;1,X}_{s_1 \text{ times}};\ldots;\underbrace{\ell,X;\ldots;\ell,X}_{s_\ell-1 \text{ times}};n_1,X_1;\ldots;n_k,X_k;\ell,X)$$

whose real dimension is precisely the cohomology degree of A_{s_ℓ}, and

$$W_{s_\ell}^0 = W(\underbrace{1,X;\ldots;1,X}_{s_1 \text{ times}};\ldots;\underbrace{\ell,X;\ldots;\ell,X}_{s_\ell-1 \text{ times}};n_1,X_1;\ldots;n_k,X_k;\ell,X)^0.$$

By Lemma 3.15, $p_{1*}\left([Q_X^{[n,n-\ell]}] \cap p_2^*(\overline{W_{s_\ell-1}^0})\right)$ is a multiple of $\overline{W_{s_\ell}^0}$. By (3.20), it remains to show that the multiplicity is equal to s_ℓ. Indeed, take a general element $\xi \in W_{s_\ell}^0$. Then

$$\xi = \sum_{i=1}^{\ell}\sum_{j=1}^{s_i} \xi_{i,j} + \sum_{i=1}^{k} \xi_i$$

where $\ell(\xi_{i,j}) = i$, $\ell(\xi_i) = n_i$, $\mathrm{Supp}(\xi_{i,j}) = x_{i,j}$, $\mathrm{Supp}(\xi_i) = x_i$, all the points $x_{i,j}, x_i$ are distinct, $x_i \in X_k$ if and only if $i = k$, and $x_{i,j} \notin X_k$ for all i,j,k. Now,

$$(p_1)^{-1}(\xi) \cap \left(Q_X^{[n,n-\ell]} \cap p_2^{-1}(\overline{W_{s_\ell-1}^0})\right)$$

consists of exactly s_ℓ different pairs $(\xi, \xi - \xi_{\ell,j})$, $1 \le j \le s_\ell$. $\square$

Let $1_{X^{[n]}}$ be the fundamental class of $X^{[n]}$. Recall the notations B_n and D_C from (1.26) and (1.27) respectively.

COROLLARY 3.17. *Let $n \geq 2$ and C be a real-surface on X. Then,*

$$1_{X^{[n]}} = \frac{1}{n!} \cdot \mathfrak{a}_{-1}(1_X)^n|0\rangle, \tag{3.21}$$

$$B_n = \frac{1}{(n-2)!} \cdot \mathfrak{a}_{-2}(1_X)\mathfrak{a}_{-1}(1_X)^{n-2}|0\rangle, \tag{3.22}$$

$$D_C = \frac{1}{(n-1)!} \cdot \mathfrak{a}_{-1}(C)\mathfrak{a}_{-1}(1_X)^{n-1}|0\rangle \tag{3.23}$$

where by abusing notations, we have also used C to denote its Poincaré dual in $H^2(X)$.

PROOF. These follows from Proposition 3.16, (1.26) and (1.27). □

DEFINITION 3.18. For a nonnegative integer n, we define the operator:

$$\mathbf{1}_{-n} = \frac{1}{n!} \cdot \mathfrak{a}_{-1}(1_X)^n$$

so that $1_{X^{[n]}} = \mathbf{1}_{-n}|0\rangle$. For simplicity, we set $\mathbf{1}_{-n} = 0$ when $n < 0$.

3.4. The homology classes of curves in Hilbert schemes

In this section, with the help of Heisenberg monomial classes, we will present two lemmas dealing with the homology classes of certain curves in the Hilbert schemes $X^{[n]}$. The results are useful in studying the Gromov-Witten theory of $X^{[n]}$.

First of all, we describe some elements in $H_2(X^{[n]})$. Let $\{\alpha_1, \ldots, \alpha_s\}$ be a basis of $H^2(X,)$ and $\{\alpha_{s+1}, \ldots, \alpha_{s+t}\}$ be a basis of $H^3(X)$, where every class α_i is homogeneous. By (3.10), a basis of the cohomology group $H^{4n-2}(X^{[n]})$ consists of the classes:

$$\mathfrak{a}_{-2}(x)\mathfrak{a}_{-1}(x)^{n-2}|0\rangle,$$
$$\mathfrak{a}_{-1}(\alpha_i)\mathfrak{a}_{-1}(x)^{n-1}|0\rangle,$$
$$\mathfrak{a}_{-1}(\alpha_j)\mathfrak{a}_{-1}(\alpha_k)\mathfrak{a}_{-1}(x)^{n-2}|0\rangle$$

where $|\alpha_i| = 2$, $|\alpha_j| = |\alpha_k| = 3$ with $j < k$, and by abusing notations, we have used x to denote the cohomology class Poincaré dual to a point $x \in X$. The homology class β_n defined by (1.34) is Poincaré dual to $\mathfrak{a}_{-2}(x)\mathfrak{a}_{-1}(x)^{n-2}|0\rangle$. Let

$$\widetilde{H}_2(X^{[n]}) \subset H_2(X^{[n]}) \tag{3.24}$$

be the linear span of the Poincaré duals of the basis classes

$$\mathfrak{a}_{-1}(\alpha_j)\mathfrak{a}_{-1}(\alpha_k)\mathfrak{a}_{-1}(x)^{n-2}|0\rangle$$

in the above list. Moreover, for a homology class $\gamma \in H_2(X)$, let $\beta_\gamma \in H_2(X^{[n]})$ be the Poincaré dual of $\mathfrak{a}_{-1}(\alpha)\mathfrak{a}_{-1}(x)^{n-1}|0\rangle$, where α is the Poincaré dual of γ. Note that when C is a real-surface in X, the homology class β_C is represented by (2.1).

Let Γ be an irreducible curve in the Hilbert scheme $X^{[n]}$. In view of the preceding paragraph,

$$[\Gamma] = \beta_\gamma + d\beta_n \pmod{\widetilde{H}_2(X^{[n]})} \tag{3.25}$$

for some homology class $\gamma \in H_2(X)$ and some integer d. To determine γ, recall the universal codimension-2 closed subscheme $\mathcal{Z}_n \subset X^{[n]} \times X$ from (1.24). Define

$$\mathcal{Z}_\Gamma = \Gamma \times_{X^{[n]}} \mathcal{Z}_n \quad \subset \mathcal{Z}_n \subset X^{[n]} \times X. \tag{3.26}$$

Let $\tilde{\pi}_1, \tilde{\pi}_2$ be the restrictions to $\mathcal{Z}_n$ of the two natural projections on $X^{[n]} \times X$. The following lemma and its proof are from [**Obe**, Lemma 1] (similar statement also appeared in [**HLQ**, Lemma 5.1]).

LEMMA 3.19. *Let $n \geq 2$. Let Γ be an irreducible curve in the Hilbert scheme $X^{[n]}$. Then,*

$$[\Gamma] = \beta_{\tilde{\pi}_{2*}[\mathcal{Z}_\Gamma]} + d\beta_n \pmod{\widetilde{H}_2(X^{[n]})} \tag{3.27}$$

for some integer d.

PROOF. Since $\tilde{\pi}_1$ is flat, $\tilde{\pi}_{2*}[\mathcal{Z}_\Gamma] = \tilde{\pi}_{2*}\tilde{\pi}_1^*[\Gamma]$. By (3.25), it remains to prove

$$\begin{aligned} \tilde{\pi}_{2*}\tilde{\pi}_1^*\beta_n &= 0, \\ \tilde{\pi}_{2*}\tilde{\pi}_1^*\beta_\gamma &= \gamma, \\ \tilde{\pi}_{2*}\tilde{\pi}_1^* \,\mathrm{PD}\left(\mathfrak{a}_{-1}(\alpha_j)\mathfrak{a}_{-1}(\alpha_k)\mathfrak{a}_{-1}(x)^{n-2}|0\rangle\right) &= 0 \end{aligned}$$

for $\gamma \in H_2(X)$ and $|\alpha_j| = |\alpha_k| = 3$. Representing β_n by (1.34), we see that $\tilde{\pi}_{2*}\tilde{\pi}_1^*\beta_n = 0$.

Next, to prove $\tilde{\pi}_{2*}\tilde{\pi}_1^*\beta_\gamma = \gamma$, represent the homology class γ by a real smooth surface S in X. Let $y_1, \ldots, y_{n-1} \in X$ be distinct with $\{y_1, \ldots, y_{n-1}\} \cap S = \emptyset$. Let

$$U = \{(x_1, \ldots, x_n) \in X^n | x_1, \ldots x_n \text{ are mutually distinct}\},$$

and let $\sigma : U \to X^{[n]}$ be the morphism sending $(x_1, \ldots, x_n)$ to $x_1 + \ldots + x_n$. Define $\widetilde{U} = U \times_{X^{[n]}} \mathcal{Z}_n$. Let $\tilde{\pi}_1' : \widetilde{U} \to U$ be the natural map. Then we have a diagram of morphisms:

$$\begin{array}{ccccc} \widetilde{U} & \rightarrow & \mathcal{Z}_n & \stackrel{\tilde{\pi}_2}{\rightarrow} & X \\ \big\downarrow \tilde{\pi}_1' & & \big\downarrow \tilde{\pi}_1 & & \\ U & \stackrel{\sigma}{\rightarrow} & X^{[n]}. & & \end{array}$$

Let $f : \widetilde{U} \to \mathcal{Z}_n \stackrel{\tilde{\pi}_2}{\rightarrow} X$ be the composition. Then, $\beta_\gamma = \sigma_*[y_1 \times \cdots \times y_{n-1} \times S]$. So

$$\begin{aligned} \tilde{\pi}_{2*}\tilde{\pi}_1^*\beta_\gamma &= \tilde{\pi}_{2*}\tilde{\pi}_1^*\sigma_*[y_1 \times \cdots \times y_{n-1} \times S] \\ &= f_*(\tilde{\pi}_1')^*[y_1 \times \cdots \times y_{n-1} \times S] \\ &= [S] \\ &= \gamma. \end{aligned}$$

Finally, to prove $\tilde{\pi}_{2*}\tilde{\pi}_1^* \,\mathrm{PD}\left(\mathfrak{a}_{-1}(\alpha_j)\mathfrak{a}_{-1}(\alpha_k)\mathfrak{a}_{-1}(x)^{n-2}|0\rangle\right) = 0$, represent α_j, α_k by real smooth curves $C_j, C_k \subset X$ respectively such that $C_j \cap C_k = \emptyset$. Let $y_1, \ldots, y_{n-2} \in X$ be distinct with $\{y_1, \ldots, y_{n-2}\} \cap (C_j \cup C_k) = \emptyset$. Then,

$$\begin{aligned} & \tilde{\pi}_{2*}\tilde{\pi}_1^* \,\mathrm{PD}\left(\mathfrak{a}_{-1}(\alpha_j)\mathfrak{a}_{-1}(\alpha_k)\mathfrak{a}_{-1}(x)^{n-2}|0\rangle\right) \\ = \; & \tilde{\pi}_{2*}\tilde{\pi}_1^*\sigma_*[y_1 \times \cdots \times y_{n-2} \times C_j \times C_k] \\ = \; & f_*(\tilde{\pi}_1')^*[y_1 \times \cdots \times y_{n-2} \times C_j \times C_k] \\ = \; & 0. \end{aligned}$$

This completes the proof of the lemma. $\square$

Next, we study the homology classes of curves in $C^{(n)} \subset X^{[n]}$ where C denotes a smooth curve in X. Let g_C be the genus of C. Assume that $g_C \geq 1$. We recall some standard facts about $C^{(n)}$ from [**ACGH**, **BT**]. For a fixed point $p \in C$, let Ξ denote the divisor $p + C^{(n-1)} \subset C^{(n)}$. Let

$$\mathrm{AJ} : C^{(n)} \to \mathrm{Jac}_n(C)$$

be the Abel-Jacobi map sending an element $\xi \in C^{(n)}$ to the corresponding degree-n divisor class in $\mathrm{Jac}_n(C)$. For an element $\delta \in \mathrm{Jac}_n(C)$, the fiber $\mathrm{AJ}^{-1}(\delta)$ is the complete linear system $|\delta|$. Let

$$\mathcal{Z}_n(C) \subset C^{(n)} \times C$$

be the universal divisor, and let $\tilde{\pi}_1, \tilde{\pi}_2$ be the two projections on $C^{(n)} \times C$. By the Lemma 2.5 on p.340 of [**ACGH**] and the Proposition 2.1 (iv) of [**BT**], we have

$$c_1\big(\tilde{\pi}_{1*}\mathcal{O}_{\mathcal{Z}_n(C)}\big) = (1 - g_C - n)\Xi + \Theta \tag{3.28}$$

where Θ is the pull-back via AJ of a Theta divisor on $\mathrm{Jac}_n(C)$.

LEMMA 3.20. *Let $n \ge 2$ and X be a smooth projective surface. Let $C \subset X$ be a smooth curve with genus $g_C \ge 1$, and $\Gamma \subset C^{(n)}$ be a curve. Then, $[\Gamma]$ is equal to*

$$(\Xi \cdot \Gamma)\beta_C + \big(-(n + g_C - 1)(\Xi \cdot \Gamma) + (\Theta \cdot \Gamma)\big)\beta_n \pmod{\widetilde{H}_2(X^{[n]})}. \tag{3.29}$$

In addition, for every line Γ_0 in a positive-dimensional fiber $\mathrm{AJ}^{-1}(\delta)$, we have

$$[\Gamma_0] = \beta_C - (n + g_C - 1)\beta_n \pmod{\widetilde{H}_2(X^{[n]})}. \tag{3.30}$$

PROOF. (i) Recall the subspace $\widetilde{H}_2(X^{[n]})$ from (3.24). By Lemma 3.19,

$$[\Gamma] = \beta_\gamma + d\beta_n \pmod{\widetilde{H}_2(X^{[n]})} \tag{3.31}$$

for some effective cycle γ and some integer d.

To determine d and γ, let $\mathcal{Z}_n$ be the universal codimension-2 subscheme of $X^{[n]} \times X$, and let π_1, π_2 be the two projections on $X^{[n]} \times X$. Then,

$$\big(\pi_{1*}\mathcal{O}_{\mathcal{Z}_n}\big)|_{C^{(n)}} = \tilde{\pi}_{1*}\mathcal{O}_{\mathcal{Z}_n(C)}.$$

By Theorem 1.24 (i), $c_1\big(\pi_{1*}\mathcal{O}_{\mathcal{Z}_n}\big) = -B_n/2$. Combining with (3.31) and (3.28), we obtain

$$\begin{aligned} d &= (-B_n/2) \cdot \Gamma \\ &= c_1\big(\pi_{1*}\mathcal{O}_{\mathcal{Z}_n}\big)|_{C^{(n)}} \cdot \Gamma \\ &= c_1\big(\tilde{\pi}_{1*}\mathcal{O}_{\mathcal{Z}_n(C)}\big) \cdot \Gamma \\ &= (1 - g_C - n)(\Xi \cdot \Gamma) + (\Theta \cdot \Gamma). \end{aligned} \tag{3.32}$$

Next, let $\alpha \in H^2(X)$. Then, $D_\alpha|_{C^{(n)}} = (C \cdot \alpha)\Xi$. Thus, we get

$$D_\alpha \cdot \Gamma = (C \cdot \alpha)(\Xi \cdot \Gamma).$$

By (3.31),

$$\alpha \cdot \gamma = D_\alpha \cdot \Gamma = (C \cdot \alpha)(\Xi \cdot \Gamma).$$

It follows that $\gamma = (\Xi \cdot \Gamma)\, C$.

(ii) Note that $\Theta \cdot \Gamma_0 = 0$. Also, it is known that $\Xi|_{\mathrm{AJ}^{-1}(\delta)} = \mathcal{O}_{\mathrm{AJ}^{-1}(\delta)}(1)$. So $\Xi \cdot \Gamma_0 = 1$. Now (3.30) follows from (3.29). ☐

REMARK 3.21. In view of the proof of Lemma 2.19 (i), we conclude that when the genus g_C of the smooth curve C is equal to 0, Lemma 3.20 still holds if we delete the term $(\Theta \cdot \Gamma)$ in (3.29).

3.5. Virasoro algebras of Lehn

There is a standard construction which associates a Virasoro algebra to a Heisenberg algebra. In this section, using the Heisenberg algebras constructed in Section 3.2, we will introduce the Virasoro algebras and boundary operator of Lehn [**Leh1**], and discuss their properties. Up to a factor of $(-1/2)$, the boundary operator is defined as the cup product by the boundary divisors of the Hilbert schemes $X^{[n]}$. These operators play an essential role in understanding the cohomology ring structure of the Hilbert scheme $X^{[n]}$. Lehn's work [**Leh1**] is the first step in this direction.

DEFINITION 3.22. The *normally ordered product* $:\mathfrak{a}_{m_1}\mathfrak{a}_{m_2}:$ is defined by

$$:\mathfrak{a}_{m_1}\mathfrak{a}_{m_2}: \quad = \quad \begin{cases} \mathfrak{a}_{m_1}\mathfrak{a}_{m_2}, & m_1 \leq m_2 \\ \mathfrak{a}_{m_2}\mathfrak{a}_{m_1}, & m_1 \geq m_2. \end{cases}$$

For $n \in \mathbb{Z}$, define the *Virasoro operator* $\mathfrak{L}_n : H^*(X) \to \mathrm{End}(\mathbb{H}_X)$ by putting

$$\mathfrak{L}_n = -\frac{1}{2} \cdot \sum_{m \in \mathbb{Z}} :\mathfrak{a}_m \mathfrak{a}_{n-m}: \tau_{2*}.$$

By (3.9), $[\mathfrak{a}_m(\alpha), \mathfrak{a}_{n-m}(\beta)] = 0$ if $n \neq 0$. It follows that

$$\mathfrak{L}_n = \begin{cases} -\frac{1}{2} \cdot \sum_{m \in \mathbb{Z}} \mathfrak{a}_m \mathfrak{a}_{n-m} \tau_{2*}, & \text{if } n \neq 0, \\ -\sum_{m > 0} \mathfrak{a}_{-m} \mathfrak{a}_m \tau_{2*}, & \text{if } n = 0. \end{cases} \tag{3.33}$$

Moreover, a geometric interpretation of the Virasoro operator $\mathfrak{L}_n$ with $n \geq 0$ is given by [**Leh1**, Theorem 3.5].

DEFINITION 3.23. Define the *boundary operator* $\mathfrak{d} \in \mathrm{End}(\mathbb{H}_X)$ by

$$\mathfrak{d} = \bigoplus_n (-B_n/2) \cup$$

where the class $(-B_n/2)$ acts on $\mathbb{H}_X^n = H^*(X^{[n]})$ by the cup product. For a linear operator $\mathfrak{f} \in \mathrm{End}(\mathbb{H}_X)$, define its *derivative* $\mathfrak{f}'$ by

$$\mathfrak{f}' \; = \; \mathrm{ad}(\mathfrak{f}) \; = \; [\mathfrak{d}, \mathfrak{f}].$$

The higher order derivative $\mathfrak{f}^{(k)}$ of $\mathfrak{f}$ is defined inductively by

$$\mathfrak{f}^{(k)} = [\mathfrak{d}, \mathfrak{f}^{(k-1)}].$$

The operators $\mathfrak{L}_n(\alpha)$ and $\mathfrak{d}$ and the following theorem are due to Lehn [**Leh1**, Theorems 3.3 and 3.10].

THEOREM 3.24. *Let X be a smooth projective complex surface with the canonical class K_X and the Euler class e_X. Let $n, m \in \mathbb{Z}$ and $\alpha, \beta \in H^*(X)$. Then,*

(i) $[\mathfrak{L}_m(\alpha), \mathfrak{a}_n(\beta)] = -n \cdot \mathfrak{a}_{m+n}(\alpha\beta)$.

(ii) $\mathfrak{a}_n'(\alpha) = n \cdot \mathfrak{L}_n(\alpha) - \dfrac{n(|n|-1)}{2} \cdot \mathfrak{a}_n(K_X\alpha)$.

(iii) *The operators $\mathfrak{L}_n(\alpha)$ satisfy the Virasoro algebra commutation relation:*

$$[\mathfrak{L}_m(\alpha), \mathfrak{L}_n(\beta)] = (m-n) \cdot \mathfrak{L}_{m+n}(\alpha\beta) + \frac{m^3 - m}{12} \delta_{m,-n} \int_X (e_X \alpha\beta) \cdot \mathrm{Id}_{\mathbb{H}_X}.$$

PROOF. We refer to **[Leh1]** for the proofs of (ii) and (iii) since the arguments are very technical. To prove (i), assume first that $m \neq 0$. By (3.33) and Lemma 3.12 (i),

$$\begin{aligned}
[\mathfrak{L}_m(\alpha), \mathfrak{a}_n(\beta)] &= -\frac{1}{2} \cdot \sum_{i \in \mathbb{Z}} [\mathfrak{a}_i \mathfrak{a}_{m-i}(\tau_{2*}\alpha), \mathfrak{a}_n(\beta)] \\
&= \frac{1}{2} \cdot \sum_{i \in \mathbb{Z}} \{ i\delta_{i,-n} \mathfrak{a}_{m-i}(\alpha\beta) + (m-i)\delta_{m-i,-n} \mathfrak{a}_i(\alpha\beta) \} \\
&= \frac{1}{2} \cdot \{ -n\mathfrak{a}_{m+n}(\alpha\beta) - n\mathfrak{a}_{m+n}(\alpha\beta) \} \\
&= -n \cdot \mathfrak{a}_{m+n}(\alpha\beta).
\end{aligned}$$

The case when $m = 0$ can be similarly proved. □

Next, we address the issue of how to identify two linear operators in $\mathrm{End}(\mathbb{H}_X)$. We begin with a lemma.

LEMMA 3.25. *Let $n \geq 1$, $\alpha \in H^*(X)$, and $\mathfrak{f} \in \mathrm{End}(\mathbb{H}_X)$ with $\mathfrak{f}' = 0$. Then,*

$$[\mathfrak{f}, \mathfrak{a}_{-(n+1)}(\alpha)] = -\frac{1}{n} \cdot \Big\{ [[\mathfrak{f}, \mathfrak{a}_{-1}(1_X)]', \mathfrak{a}_{-n}(\alpha)] + [\mathfrak{a}'_{-1}(1_X), [\mathfrak{f}, \mathfrak{a}_{-n}(\alpha)]] \Big\}.$$

PROOF. By the Jacobi identity (3.4), we have

$$\begin{aligned}
[\mathfrak{f}, \mathfrak{a}_{-1}(1_X)]' &= [\mathfrak{f}', \mathfrak{a}_{-1}(1_X)] + [\mathfrak{f}, \mathfrak{a}_{-1}(1_X)'] \\
&= [\mathfrak{f}, \mathfrak{a}_{-1}(1_X)'] \\
&= -[\mathfrak{a}_{-1}(1_X)', \mathfrak{f}]
\end{aligned}$$

since $\mathfrak{f}' = 0$. Applying (3.4) one more time, we see that

$$\begin{aligned}
[\mathfrak{a}'_{-1}(1_X), [\mathfrak{f}, \mathfrak{a}_{-n}(\alpha)]] &= [[\mathfrak{a}'_{-1}(1_X), \mathfrak{f}], \mathfrak{a}_{-n}(\alpha)] + [\mathfrak{f}, [\mathfrak{a}'_{-1}(1_X), \mathfrak{a}_{-n}(\alpha)]] \\
&= -[[\mathfrak{f}, \mathfrak{a}_{-1}(1_X)]', \mathfrak{a}_{-n}(\alpha)] + [\mathfrak{f}, [-\mathfrak{L}_{-1}(1_X), \mathfrak{a}_{-n}(\alpha)]] \\
&= -[[\mathfrak{f}, \mathfrak{a}_{-1}(1_X)]', \mathfrak{a}_{-n}(\alpha)] - [\mathfrak{f}, n\mathfrak{a}_{-(n+1)}(\alpha)]
\end{aligned}$$

where Theorem 3.24 (ii) and (i) have been used. So the lemma holds. □

The following proposition allows us to identify two linear operators in $\mathrm{End}(\mathbb{H}_X)$.

PROPOSITION 3.26. *For $i = 1, 2$, let $\mathfrak{f}_i \in \mathrm{End}(\mathbb{H}_X)$ such that $\mathfrak{f}_i \cdot |0\rangle = 0$ and $\mathfrak{f}'_i = 0$. If $[\mathfrak{f}_1, \mathfrak{a}_{-1}(\alpha)] = [\mathfrak{f}_2, \mathfrak{a}_{-1}(\alpha)]$ for every $\alpha \in H^*(X)$, then $\mathfrak{f}_1 = \mathfrak{f}_2$.*

PROOF. Recall that the space $\mathbb{H}_X$ is linearly spanned by the Heisenberg monomial classes (3.10). Therefore, it suffices to show that

$$[\mathfrak{f}_1, \mathfrak{a}_{-n}(\alpha)] = [\mathfrak{f}_2, \mathfrak{a}_{-n}(\alpha)] \tag{3.34}$$

for every $n \geq 1$ and $\alpha \in H^*(X)$. By assumption, (3.34) holds for $n = 1$. So (3.34) follows from Lemma 3.25 and an induction on n. □

We end this section with a lemma which will play an important role in the proof of Lemma 3.29 below. It enables us to use induction on the order of derivatives.

LEMMA 3.27. *Let $k \ge 1$ and $\alpha \in H^*(X)$. Then $(\mathfrak{a}_{n_1} \cdots \mathfrak{a}_{n_k}(\tau_{k*}\alpha))'$ is equal to*

$$-\sum_{i=1}^{k} \frac{n_i}{2} \cdot \sum_{m_1+m_2=n_i} \mathfrak{a}_{n_1} \cdots \mathfrak{a}_{n_{i-1}} : \mathfrak{a}_{m_1}\mathfrak{a}_{m_2} : \mathfrak{a}_{n_{i+1}} \cdots \mathfrak{a}_{n_k}(\tau_{(k+1)*}\alpha)$$
$$-\sum_{i=1}^{k} \frac{n_i(|n_i|-1)}{2} \cdot \mathfrak{a}_{n_1} \cdots \mathfrak{a}_{n_k}(\tau_{k*}(K_X\alpha)).$$

PROOF. Let

$$\tau_{k*}\alpha = \sum_j \alpha_{j,1} \otimes \cdots \otimes \alpha_{j,k}$$

via the Künneth decomposition of $H^*(X^k)$. By (3.5),

$$\begin{aligned}
&(\mathfrak{a}_{n_1} \cdots \mathfrak{a}_{n_k}(\tau_{k*}\alpha))' \\
= &\sum_j \big(\mathfrak{a}_{n_1}(\alpha_{j,1}) \cdots \mathfrak{a}_{n_k}(\alpha_{j,k})\big)' \\
= &\sum_{i=1}^{k} \sum_j \prod_{s=1}^{i-1} \mathfrak{a}_{n_s}(\alpha_{j,s}) \cdot \mathfrak{a}'_{n_i}(\alpha_{j,i}) \cdot \prod_{s=i+1}^{k} \mathfrak{a}_{n_s}(\alpha_{j,s}).
\end{aligned}$$

By Theorem 3.24 (ii),

$$\begin{aligned}
&(\mathfrak{a}_{n_1} \cdots \mathfrak{a}_{n_k}(\tau_{k*}\alpha))' \\
= &-\sum_{i=1}^{k} \frac{n_i}{2} \cdot \sum_{m_1+m_2=n_i} \sum_j \prod_{s=1}^{i-1} \mathfrak{a}_{n_s}(\alpha_{j,s}) \cdot : \mathfrak{a}_{m_1}\mathfrak{a}_{m_2} : (\tau_{2*}\alpha_{j,i}) \cdot \prod_{s=i+1}^{k} \mathfrak{a}_{n_s}(\alpha_{j,s}) \\
&-\sum_{i=1}^{k} \frac{n_i(|n_i|-1)}{2} \cdot \sum_j \prod_{s=1}^{i-1} \mathfrak{a}_{n_s}(\alpha_{j,s}) \cdot \mathfrak{a}_{n_i}(K_X\alpha_{j,i}) \cdot \prod_{s=i+1}^{k} \mathfrak{a}_{n_s}(\alpha_{j,s}) \\
= &-\sum_{i=1}^{k} \frac{n_i}{2} \cdot \sum_{m_1+m_2=n_i} \mathfrak{a}_{n_1} \cdots \mathfrak{a}_{n_{i-1}} : \mathfrak{a}_{m_1}\mathfrak{a}_{m_2} : \mathfrak{a}_{n_{i+1}} \cdots \mathfrak{a}_{n_k}(\tau_{(k+1)*}\alpha) \\
&-\sum_{i=1}^{k} \frac{n_i(|n_i|-1)}{2} \cdot \mathfrak{a}_{n_1} \cdots \mathfrak{a}_{n_k}(\tau_{k*}(K_X\alpha))
\end{aligned}$$

where we have used the third and first formulas in Lemma 3.11 at the last step. □

3.6. Higher order derivatives of Heisenberg operators

The derivative of a linear operator in $\mathrm{End}(\mathbb{H}_X)$ is defined in Definition 3.23. In this section, we will determine the structure of the higher order derivatives of Heisenberg operators. We will prove that $\mathfrak{a}_n^{(k)}(\alpha)$ has a universal expression, depending only on the canonical class K_X and Euler class e_X of the surface X. The result is essential to the understanding of the Chern character operators in Section 4.1.

Recall the generalized partitions from Definition 1.6.

DEFINITION 3.28. Let X be a smooth projective complex surface.

(i) Let $\alpha \in H^*(X)$, and $\lambda = (\cdots(-2)^{m_{-2}}(-1)^{m_{-1}}1^{m_1}2^{m_2}\cdots)$. Define

$$\mathfrak{a}_\lambda(\alpha) = \prod_i \big(\mathfrak{a}_i(\alpha)\big)^{m_i}, \tag{3.35}$$

$$\mathfrak{a}_\lambda(\tau_*\alpha) = \left(\prod_i (\mathfrak{a}_i)^{m_i}\right)(\tau_{\ell(\lambda)*}\alpha) \tag{3.36}$$

where the product $\prod_i(\mathfrak{a}_i)^{m_i}$ is understood to be

$$\cdots \mathfrak{a}_{-2}^{m_{-2}}\mathfrak{a}_{-1}^{m_{-1}}\mathfrak{a}_1^{m_1}\mathfrak{a}_2^{m_2}\cdots.$$

(ii) Define

$$\mathfrak{A}_X = \{\alpha \in H^*(X) | K_X\alpha = 0\}$$

which is an ideal in the cohomology ring $H^*(X)$.

The following lemma determines $\mathfrak{a}_n^{(k)}(\alpha)$ when $\alpha \in \mathfrak{A}_X$. The main idea in its proof is to use induction on k, Lemma 3.27, and Lemma 3.12 (ii).

LEMMA 3.29. *Let $k \ge 0$, $n \in \mathbb{Z}$, and $\alpha \in \mathfrak{A}_X$. Then, $\mathfrak{a}_n^{(k)}(\alpha)$ is equal to*

$$(-n)^k k! \left(\sum_{\ell(\lambda)=k+1, |\lambda|=n} \frac{1}{\lambda!}\mathfrak{a}_\lambda(\tau_*\alpha) - \sum_{\ell(\lambda)=k-1, |\lambda|=n} \frac{s(\lambda)-1}{24\lambda!}\mathfrak{a}_\lambda(\tau_*(e_X\alpha))\right).$$

PROOF. Use induction on k. When $k = 0$ or 1, the lemma is trivially true. Next, assume that the lemma is true for $\mathfrak{a}_n^{(k)}(\alpha)$. Since $\mathfrak{a}_n^{(k+1)}(\alpha) = \big(\mathfrak{a}_n^{(k)}(\alpha)\big)'$ and $K_X\alpha = 0$, we see from Lemma 3.27 and Lemma 3.12 (ii) that $\mathfrak{a}_n^{(k+1)}(\alpha)$ is of the form

$$\sum_{\ell(\lambda)=k+2, |\lambda|=n} \tilde{f}_{1_X}(\lambda)\mathfrak{a}_\lambda(\tau_*\alpha) + \sum_{\ell(\lambda)=k, |\lambda|=n} \tilde{f}_{e_X}(\lambda)\mathfrak{a}_\lambda(\tau_*(e_X\alpha)).$$

Fix a generalized partition $\lambda = (\cdots(-2)^{m_{-2}}(-1)^{m_{-1}}1^{m_1}2^{m_2}\cdots)$ with $|\lambda| = n$. In the following, we compute the coefficients $\tilde{f}_{1_X}(\lambda)$ and $\tilde{f}_{e_X}(\lambda)$ separately.

(i) We start with the computation of $\tilde{f}_{1_X}(\lambda)$. Note that $\ell(\lambda) = k+2$. Going from $\mathfrak{a}_n^{(k)}(\alpha)$ to $\mathfrak{a}_n^{(k+1)}(\alpha) = (\mathfrak{a}_n^{(k)}(\alpha))'$, we see that those terms in $\mathfrak{a}_n^{(k)}(\alpha)$ whose derivatives contain the term $\mathfrak{a}_\lambda(\tau_*\alpha)$ in $\mathfrak{a}_n^{(k+1)}(\alpha)$ are of the form $\mathfrak{a}_{\lambda_{i,j}}(\tau_*\alpha)$ where $i \le j$, $m_i \ge 1$, $m_j \ge 1$, and $\lambda_{i,j}$ stands for the generalized partition obtained from λ by subtracting 1 from the multiplicities of i and j and by adding 1 to the multiplicity of $(i+j)$. For fixed i and j, $\{\mathfrak{a}_{\lambda_{i,j}}(\tau_*\alpha)\}'$ is equal to

$$\left(-\frac{i+j}{2}\right)(m_{i+j}+1)(2-\delta_{i,j})\mathfrak{a}_\lambda(\tau_*\alpha) + \sum_{\substack{\mu\neq\lambda \\ \ell(\mu)=k+2, |\mu|=n}} d_\mu \mathfrak{a}_\mu(\tau_*\alpha) + \sum_{\ell(\mu)=k, |\mu|=n} d_\mu\mathfrak{a}_\mu(\tau_*(e_X\alpha))$$

for some constant d_μ. By induction, the coefficient of $\mathfrak{a}_{\lambda_{i,j}}(\tau_*\alpha)$ in $\mathfrak{a}_n^{(k)}(\alpha)$ is

$$\frac{(-n)^k k!}{(\lambda_{i,j})!} = \frac{(-n)^k k!}{\lambda!}\cdot\frac{m_i(m_j-\delta_{i,j})}{m_{i+j}+1}.$$

Combining all of these, we see that the coefficient of $\mathfrak{a}_\lambda(\tau_*\alpha)$ in $\mathfrak{a}_n^{(k+1)}(\alpha)$ is

$$\begin{aligned}
&\sum_{i\le j} \frac{(-n)^k k!}{\lambda!}\cdot\frac{m_i(m_j-\delta_{i,j})}{m_{i+j}+1}\left(-\frac{i+j}{2}\right)(m_{i+j}+1)(2-\delta_{i,j})\\
=\ & \frac{(-n)^k k!}{\lambda!}\left(-\sum_{i<j} m_i m_j(i+j) - \sum_i m_i(m_i-1)i\right)\\
=\ & \frac{(-n)^k k!}{\lambda!}\left(-\frac{1}{2}\sum_{i,j} m_i m_j(i+j) + \sum_i m_i i\right)\\
=\ & \frac{(-n)^{k+1}(k+1)!}{\lambda!}
\end{aligned}$$

where we have used $\sum_i m_i = \ell(\lambda) = k+2$ and $\sum_i i m_i = |\lambda| = n$ in the last step.

(ii) Before going into the proof, let us use one example to illustrate the idea. Take $k=2$, $n=-3$, and $\lambda = ((-5)2)$. To figure out the coefficient of $\mathfrak{a}_\lambda(\tau_*(e_X\alpha))$ in $\mathfrak{a}_{-3}^{(3)}(\alpha)$, we apply the induction hypothesis to $\mathfrak{a}_{-3}^{(2)}(\alpha)$ and take the derivative one more time. In view of Lemma 3.12 (ii) and Lemma 3.27, there are exactly two ways to get $\mathfrak{a}_\lambda(\tau_*(e_X\alpha))$. The first one is similar to (i) above, and comes from the term $\{\mathfrak{a}_{\lambda_{-5,2}}(\tau_*(e_X\alpha))\}'$ where $\lambda_{-5,2} = ((-3))$, i.e., the generalized partition $\lambda_{-5,2}$ is obtained from λ by subtracting 1 from the multiplicities of (-5) and 2 and by adding 1 to the multiplicity of $(-3) = (-5)+2$. The second source comes from three terms $\{\mathfrak{a}_{\tilde\lambda}(\tau_*\alpha)\}'$ where $\tilde\lambda = ((-5)1^2)$, $((-5+j)(-j)2)$ with $j=1,2$. For instance, when $\tilde\lambda = ((-5+j)(-j)2)$ with $j=1$ or 2, $\{\mathfrak{a}_{\tilde\lambda}(\tau_*(\alpha))\}'$ contains

$$\begin{aligned}
\mathfrak{a}_{-5}\mathfrak{a}_j\mathfrak{a}_{-j}\mathfrak{a}_2(\tau_{4*}\alpha) &= \mathfrak{a}_{-5}\mathfrak{a}_{-j}\mathfrak{a}_j\mathfrak{a}_2(\tau_{4*}\alpha) + (-j)\mathfrak{a}_{-5}\mathfrak{a}_2(\tau_{2*}(e_X\alpha))\\
&= \mathfrak{a}_{-5}\mathfrak{a}_{-j}\mathfrak{a}_j\mathfrak{a}_2(\tau_{4*}\alpha) + (-j)\mathfrak{a}_\lambda(\tau_*(e_X\alpha)).
\end{aligned}$$

Keeping track of the coefficients in these derivatives

$$\{\mathfrak{a}_{\lambda_{-5,2}}(\tau_*(e_X\alpha))\}', \quad \{\mathfrak{a}_{\tilde\lambda}(\tau_*\alpha)\}',$$

we obtain the coefficient of $\mathfrak{a}_\lambda(\tau_*(e_X\alpha))$ in $\mathfrak{a}_{-3}^{(3)}(\alpha)$.

Now we go back to the proof and compute $\tilde f_{e_X}(\lambda)$. In this case, $\ell(\lambda) = k$. We will prove that the coefficient of $\mathfrak{a}_\lambda(\tau_*(e_X\alpha))$ in $\mathfrak{a}_n^{(k+1)}(\alpha)$ is equal to

$$\frac{(-n)^{k+1}(k+1)!}{\lambda!}\cdot c$$

where

$$c = \frac{1-s(\lambda)}{24} = \frac{1-\sum_i i^2 m_i}{24}.$$

In view of Lemma 3.27 and Lemma 3.12 (ii), there are exactly two sources contributing to the term $\mathfrak{a}_\lambda(\tau_*(e_X\alpha))$ in $\mathfrak{a}_n^{(k+1)}(\alpha)$ from the derivatives of the terms in $\mathfrak{a}_n^{(k)}(\alpha)$. In the following, we handle these two sources separately.

The first is similar to (i) above, and comes from the derivatives $\{\mathfrak{a}_{\lambda_{i,j}}(\tau_*(e_X\alpha))\}'$ where $i\le j$, $m_i\ge 1$, $m_j\ge 1$. For fixed i and j, $\{\mathfrak{a}_{\lambda_{i,j}}(\tau_*(e_X\alpha))\}'$ is equal to

$$\left(-\frac{i+j}{2}\right)(m_{i+j}+1)(2-\delta_{i,j})\mathfrak{a}_\lambda(\tau_*(e_X\alpha)) + \sum_{\substack{\mu\ne\lambda\\ \ell(\mu)=k,|\mu|=n}} d_\mu\mathfrak{a}_\mu(\tau_*(e_X\alpha))$$

for some d_μ. By induction, the coefficient of $\mathfrak{a}_{\lambda_{i,j}}(\tau_*(e_X\alpha))$ in $\mathfrak{a}_n^{(k)}(\alpha)$ is

$$\frac{(-n)^k k!}{(\lambda_{i,j})^!}\cdot\frac{1-s(\lambda_{i,j})}{24}=\frac{(-n)^k k!}{\lambda^!}\cdot\frac{m_i(m_j-\delta_{i,j})}{m_{i+j}+1}\cdot\left(c-\frac{ij}{12}\right).$$

Combining these and noting $\sum_i m_i = k$ and $\sum_i im_i = n$, we see that the total contribution of the first source to the coefficient of $\mathfrak{a}_\lambda(\tau_*(e_X\alpha))$ in $\mathfrak{a}_n^{(k+1)}(\alpha)$ is

$$\begin{aligned}(3.37)\quad &\sum_{i\le j}\frac{(-n)^k k!}{\lambda^!}\cdot\frac{m_i(m_j-\delta_{i,j})}{m_{i+j}+1}\cdot\left(c-\frac{ij}{12}\right)\left(-\frac{i+j}{2}\right)(m_{i+j}+1)(2-\delta_{i,j})\\
&=\frac{(-n)^k k!}{\lambda^!}\sum_{i\le j}m_i(m_j-\delta_{i,j})\left(c-\frac{ij}{12}\right)\left(-\frac{i+j}{2}\right)(2-\delta_{i,j})\\
&=\frac{(-n)^k k!}{\lambda^!}\left(-\sum_{i<j}m_im_j\left(c-\frac{ij}{12}\right)(i+j)-\sum_i m_i(m_i-1)\left(c-\frac{i^2}{12}\right)i\right)\\
&=\frac{(-n)^k k!}{\lambda^!}\left(-\frac{1}{2}\sum_{i,j}m_im_j\left(c-\frac{ij}{12}\right)(i+j)+\sum_i m_i\left(c-\frac{i^2}{12}\right)i\right)\\
&=\frac{(-n)^k k!}{\lambda^!}\left(nc-knc+\frac{n}{12}\sum_i i^2m_i-\frac{1}{12}\sum_i i^3m_i\right).\end{aligned}$$

The second source comes from the derivatives $\{\mathfrak{a}_{\tilde\lambda_{i,j}}(\tau_*\alpha)\}'$ where $i\neq 0$, $0<2j\le|i|$, $m_i\ge 1$, and $\tilde\lambda_{i,j}$ is defined as follows. There are two cases, depending on whether $i>0$ or $i<0$. For $i>0$, define $\tilde\lambda_{i,j}$ to be the generalized partition obtained from λ by subtracting 1 from the multiplicity of i and by adding 1 to the multiplicities of j and $(i-j)$. Then, $\{\mathfrak{a}_{\tilde\lambda_{i,j}}(\tau_*\alpha)\}'$ contains a term $(\ldots\mathfrak{a}_j\ldots\mathfrak{a}'_{i-j}\ldots)$ which contains a term $(\ldots\mathfrak{a}_j\ldots\mathfrak{a}_{-j}\mathfrak{a}_i\ldots)$ after expanding the derivative $\mathfrak{a}'_{i-j}$. When we switch $\mathfrak{a}_j$ with $\mathfrak{a}_{-j}$, we get the term $\mathfrak{a}_\lambda(\tau_*(e_X\alpha))$ in $\mathfrak{a}_n^{(k+1)}(\alpha)$. For $i<0$, define $\tilde\lambda_{i,j}$ to be the generalized partition obtained from λ by subtracting 1 from the multiplicity of i and by adding 1 to the multiplicities of $-j$ and $(i+j)$. Then $\{\mathfrak{a}_{\tilde\lambda_{i,j}}(\tau_*\alpha)\}'$ contains a term $(\ldots\mathfrak{a}'_{i+j}\ldots\mathfrak{a}_{-j}\ldots)$ which contains a term $(\ldots\mathfrak{a}_i\mathfrak{a}_j\ldots\mathfrak{a}_{-j}\ldots)$ after expanding $\mathfrak{a}'_{i+j}$. Switching $\mathfrak{a}_j$ with $\mathfrak{a}_{-j}$, we get $\mathfrak{a}_\lambda(\tau_*(e_X\alpha))$ in $\mathfrak{a}_n^{(k+1)}(\alpha)$.

More precisely, for fixed $i>0$ and j, the derivative $\{\mathfrak{a}_{\tilde\lambda_{i,j}}(\tau_*\alpha)\}'$ is

$$\begin{aligned}&j(i-j)\cdot\frac{(m_{i-j}+1)(m_j+1+\delta_{i,2j})}{1+\delta_{i,2j}}\cdot\mathfrak{a}_\lambda(\tau_*(e_X\alpha))+\\
&+\sum_{\substack{\mu\neq\lambda\\ \ell(\mu)=k,|\mu|=n}}d_\mu\mathfrak{a}_\mu(\tau_*(e_X\alpha))+\sum_{\ell(\mu)=k+2,|\mu|=n}d_\mu\mathfrak{a}_\mu(\tau_*\alpha)\end{aligned}$$

for some constant d_μ. By induction, the coefficient of $\mathfrak{a}_{\tilde\lambda_{i,j}}(\tau_*(e\alpha))$ in $\mathfrak{a}_n^{(k)}(\alpha)$ is

$$\frac{(-n)^k k!}{(\tilde\lambda_{i,j})^!}=\frac{(-n)^k k!}{\lambda^!}\cdot\frac{m_i}{(m_{i-j}+1)(m_j+1+\delta_{i,2j})}.$$

It follows that the total contribution of the second source with $i > 0$ to the coefficient of the term $\mathfrak{a}_\lambda(\tau_*(e_X\alpha))$ in $\mathfrak{a}_n^{(k+1)}(\alpha)$ is equal to

$$\begin{aligned}
(3.38) \qquad & \sum_{i>0}\sum_{0<2j\le i} \frac{(-n)^k k!}{\lambda!}\cdot\frac{m_i}{(m_{i-j}+1)(m_j+1+\delta_{i,2j})}\cdot \\
& \qquad \cdot j(i-j)\cdot\frac{(m_{i-j}+1)(m_j+1+\delta_{i,2j})}{1+\delta_{i,2j}} \\
= \; & \frac{(-n)^k k!}{\lambda!}\sum_{i>0} m_i \sum_{0<2j\le i}\frac{j(i-j)}{1+\delta_{i,2j}} \\
= \; & \frac{(-n)^k k!}{\lambda!}\sum_{i>0} m_i \left(\frac{1}{2}\sum_{0<j<i} j(i-j)\right) \\
= \; & \frac{(-n)^k k!}{\lambda!}\sum_{i>0} m_i\cdot\frac{i^3-i}{12} \\
= \; & \frac{(-n)^k k!}{\lambda!}\cdot\frac{1}{12}\sum_{i>0}(i^3m_i - im_i).
\end{aligned}$$

Similarly, we see that the total contribution of the second source with $i < 0$ to the coefficient of $\mathfrak{a}_\lambda(\tau_*(e_X\alpha))$ in $\mathfrak{a}_n^{(k+1)}(\alpha)$ is equal to

$$(3.39) \qquad \frac{(-n)^k k!}{\lambda!}\cdot\frac{1}{12}\sum_{i<0}(i^3m_i - im_i).$$

Next, we see from (3.37), (3.38) and (3.39) that the coefficient of the term $\mathfrak{a}_\lambda(\tau_*(e_X\alpha))$ in the derivative $\mathfrak{a}_n^{(k+1)}(\alpha)$ is equal to

$$\begin{aligned}
& \frac{(-n)^k k!}{\lambda!}\left(nc - knc + \frac{n}{12}\sum_i i^2m_i - \frac{1}{12}\sum_i i^3m_i\right) \\
+ \; & \frac{(-n)^k k!}{\lambda!}\cdot\frac{1}{12}\sum_i(i^3m_i - im_i).
\end{aligned}$$

Since $n = \sum_i im_i$, the coefficient of the term $\mathfrak{a}_\lambda(\tau_*(e_X\alpha))$ in $\mathfrak{a}_n^{(k+1)}(\alpha)$ is equal to

$$\begin{aligned}
& \frac{(-n)^k k!}{\lambda!}\left(nc - knc + \frac{n}{12}\sum_i i^2m_i - \frac{n}{12}\right) \\
= \; & \frac{(-n)^k k!}{\lambda!}(nc - knc - 2nc) \\
= \; & (-n)^{k+1}(k+1)!\,\frac{1-s(\lambda)}{24\lambda!}.
\end{aligned}$$

Combining (i) and (ii), we have proved the lemma for $\mathfrak{a}_n^{(k+1)}(\alpha)$. □

REMARK 3.30. Let $\alpha \in \mathfrak{A}_X$. From the proof of Lemma 3.29, we can read that

$$\left(\sum_{\ell(\lambda)=k+1,|\lambda|=n} \frac{1}{\lambda^!}\mathfrak{a}_\lambda(\tau_*\alpha) - \sum_{\ell(\lambda)=k-1,|\lambda|=n} \frac{s(\lambda)-1}{24\lambda^!}\mathfrak{a}_\lambda(\tau_*(e_X\alpha)) \right)'$$
$$= -n(k+1)\left(\sum_{\ell(\lambda)=k+2,|\lambda|=n} \frac{1}{\lambda^!}\mathfrak{a}_\lambda(\tau_*\alpha) - \sum_{\ell(\lambda)=k,|\lambda|=n} \frac{s(\lambda)-1}{24\lambda^!}\mathfrak{a}_\lambda(\tau_*(e_X\alpha)) \right).$$

Note that when $n = 0$, this formula is not covered by Lemma 3.29. Applying this formula twice in the second equality below, we obtain that for a fixed constant d,

$$\left(\sum_{\ell(\lambda)=k+1,|\lambda|=n} \frac{1}{\lambda^!}\mathfrak{a}_\lambda(\tau_*\alpha) - \sum_{\ell(\lambda)=k-1,|\lambda|=n} \frac{s(\lambda)+d}{24\lambda^!}\mathfrak{a}_\lambda(\tau_*(e_X\alpha)) \right)'$$
$$= \left(\sum_{\ell(\lambda)=k+1,|\lambda|=n} \frac{1}{\lambda^!}\mathfrak{a}_\lambda(\tau_*\alpha) - \sum_{\ell(\lambda)=k-1,|\lambda|=n} \frac{s(\lambda)-1}{24\lambda^!}\mathfrak{a}_\lambda(\tau_*(e_X\alpha)) \right)'$$
$$- \frac{d+1}{24}\left(\sum_{\ell(\lambda)=k-1,|\lambda|=n} \frac{1}{\lambda^!}\mathfrak{a}_\lambda(\tau_*(e_X\alpha)) \right)'$$
$$= -n(k+1)\left(\sum_{\ell(\lambda)=k+2,|\lambda|=n} \frac{1}{\lambda^!}\mathfrak{a}_\lambda(\tau_*\alpha) - \sum_{\ell(\lambda)=k,|\lambda|=n} \frac{s(\lambda)-1}{24\lambda^!}\mathfrak{a}_\lambda(\tau_*(e_X\alpha)) \right)$$
$$+ \frac{n(k-1)(d+1)}{24} \cdot \sum_{\ell(\lambda)=k,|\lambda|=n} \frac{1}{\lambda^!}\mathfrak{a}_\lambda(\tau_*(e_X\alpha)).$$

Therefore, we have

$$\left(\sum_{\ell(\lambda)=k+1,|\lambda|=n} \frac{1}{\lambda^!}\mathfrak{a}_\lambda(\tau_*\alpha) - \sum_{\ell(\lambda)=k-1,|\lambda|=n} \frac{s(\lambda)+d}{24\lambda^!}\mathfrak{a}_\lambda(\tau_*(e_X\alpha)) \right)'$$
$$= -n(k+1)\left(\sum_{\ell(\lambda)=k+2,|\lambda|=n} \frac{1}{\lambda^!}\mathfrak{a}_\lambda(\tau_*\alpha) - \sum_{\ell(\lambda)=k,|\lambda|=n} \frac{s(\lambda)+d}{24\lambda^!}\mathfrak{a}_\lambda(\tau_*(e_X\alpha)) \right)$$
$$- \frac{n(d+1)}{12} \cdot \sum_{\ell(\lambda)=k,|\lambda|=n} \frac{1}{\lambda^!}\mathfrak{a}_\lambda(\tau_*(e_X\alpha)).$$

We will need this formula in the proof of Lemma 4.3.

Now we describe a formula of $\mathfrak{a}_n^{(k)}(\alpha)$ for general $\alpha \in H^*(X)$.

THEOREM 3.31. *Let $k \ge 0$, $n \in \mathbb{Z}$, and $\alpha \in H^*(X)$. Then, $\mathfrak{a}_n^{(k)}(\alpha)$ is equal to*

$$(-n)^k k! \left(\sum_{\ell(\lambda)=k+1,|\lambda|=n} \frac{1}{\lambda^!}\mathfrak{a}_\lambda(\tau_*\alpha) - \sum_{\ell(\lambda)=k-1,|\lambda|=n} \frac{s(\lambda)-1}{24\lambda^!}\mathfrak{a}_\lambda(\tau_*(e_X\alpha)) \right)$$
$$+ \sum_{\epsilon\in\{K_X,K_X^2\}} \sum_{\ell(\lambda)=k+1-|\epsilon|/2,|\lambda|=n} \frac{f_\epsilon(\lambda)}{\lambda^!}\mathfrak{a}_\lambda(\tau_*(\epsilon\alpha))$$

where all the numbers $f_\epsilon(\lambda)$ are independent of X and α.

PROOF. We see from Lemma 3.27 and Lemma 3.12 (ii) that

$$\mathfrak{a}_n^{(k)}(\alpha) = \sum_{\epsilon \in \{1_X, e_X, K_X, K_X^2\}} \sum_{\ell(\lambda)=k+1-|\epsilon|/2, |\lambda|=n} \frac{f_\epsilon(\lambda)}{\lambda^!} \mathfrak{a}_\lambda(\tau_*(\epsilon\alpha)) \tag{3.40}$$

where all the coefficients $f_\epsilon(\lambda)$ are independent of X and α.

To determine $f_{1_X}(\lambda)$ and $f_{e_X}(\lambda)$, let X be a $K3$ surface and $\alpha = 1_X$. Then, $K_X\alpha = 0$ but $e_X\alpha \neq 0$. By Lemma 3.29, $\mathfrak{a}_n^{(k)}(\alpha)$ is equal to

$$(-n)^k k! \left(\sum_{\ell(\lambda)=k+1, |\lambda|=n} \frac{1}{\lambda^!} \mathfrak{a}_\lambda(\tau_*\alpha) - \sum_{\ell(\lambda)=k-1, |\lambda|=n} \frac{s(\lambda)-1}{24\lambda^!} \mathfrak{a}_\lambda(\tau_*(e_X\alpha)) \right).$$

Comparing with (3.40) and noting that the set of all the Heisenberg monomials $\mathfrak{a}_\lambda(\tau_*\alpha), \mathfrak{a}_{\tilde{\lambda}}(\tau_*(e_X\alpha))$ with $\ell(\lambda) = k+1, |\lambda| = n, \ell(\tilde{\lambda}) = k-1, |\tilde{\lambda}| = n$ is linearly independent, we see that $f_{1_X}(\lambda) = (-n)^k k!$ and

$$f_{e_X}(\lambda) = -\frac{(-n)^k k!(s(\lambda)-1)}{24}.$$

This completes the proof of the theorem. □

3.7. The Ext vertex operators of Carlsson and Okounkov

The Nekrasov partition function in supersymmetric quantum gauge theory is mathematically formulated as an equivariant integral over certain moduli spaces of sheaves on a complex surface. Okounkov and Nekrasov [**NO**] showed that this function, together with the related correlation function, may be studied using the representation theory of the infinite wedge representation and vertex operators. Subsequently, Carlsson and Okounkov [**Car1, Car2, CO**] defined a new vertex operator out of the Chern class of a vector bundle on a pair of Hilbert schemes of points on the same surface. When the pair consists of the same Hilbert scheme, the restriction of this vector bundle to the diagonal of the pair coincides with the tangent bundle of the Hilbert scheme. In this section, we will review the construction of Carlsson and Okounkov, and survey the properties of their Ext vertex operators which play important roles in understanding the intersection theory of the Hilbert schemes $X^{[n]}$ when the Chern classes of the tangent bundles $T_{X^{[n]}}$ are involved.

Let X be either a smooth projective complex surface, or a smooth quasi-projective complex surface with a $\mathbb{C}^*$-action whose fixed point set is compact. Let L be a line bundle over the smooth surface X. Let $\mathbb{E}_L$ be the virtual vector bundle on $X^{[k]} \times X^{[\ell]}$ whose fiber at $(\xi, \eta) \in X^{[k]} \times X^{[\ell]}$ is given by

$$\mathbb{E}_L|_{(\xi,\eta)} = \chi(\mathcal{O}, L) - \chi(I_\eta, I_\xi \otimes L) \tag{3.41}$$

where I_ξ and I_η denote the ideal sheaves of the 0-dimensional closed subschemes ξ and η respectively, and we have used the notation

$$\chi(F, G) = \sum_{i=0}^{2} (-1)^i \operatorname{Ext}^i(F, G).$$

By the Riemann-Roch Theorem, the virtual rank of $\mathbb{E}_L$ is

$$\operatorname{rk}(\mathbb{E}_L) = k + \ell.$$

Let $\mathfrak{L}_m$ be the trivial line bundle on X with a scaling action of $\mathbb{C}^*$ of character z^m, and let Δ_n be the diagonal in $X^{[n]} \times X^{[n]}$. Then,

$$\mathbb{E}_{\mathfrak{L}_m}|_{\Delta_n} = T_{X^{[n]},m}, \tag{3.42}$$

the tangent bundle $T_{X^{[n]}}$ with a scaling action of $\mathbb{C}^*$ of character z^m.

DEFINITION 3.32. Let z be a formal variable. The Ext operator

$$\mathbf{W}(L, z): \quad \mathbb{H}_X[[z, z^{-1}]] \to \mathbb{H}_X[[z, z^{-1}]]$$

is the linear operator defined by

$$\langle \mathbf{W}(L, z)\eta, \xi \rangle = \int_{X^{[k]} \times X^{[\ell]}} (\eta \otimes \xi)\, c_{k+\ell}(\mathbb{E}_L) \tag{3.43}$$

for $\eta \in H^*(X^{[k]})$ and $\xi \in H^*(X^{[\ell]})$.

LEMMA 3.33. *Let $\mathfrak{n}$ be the number-of-points operator, i.e., $\mathfrak{n}|_{H^*(X^{[n]})} = n\,\mathrm{Id}$. Let $T_{X^{[n]},m}$ be the tangent bundle $T_{X^{[n]}}$ with a scaling action of $\mathbb{C}^*$ of character z^m. Then,*

$$\sum_{n=0}^{+\infty} q^n \chi(T_{X^{[n]},m}) = \mathrm{Tr}\, q^{\mathfrak{n}}\, \mathbf{W}(\mathfrak{L}_m, z). \tag{3.44}$$

PROOF. We will show that the coefficients of q^n on both sides of (3.44) are equal. Let $\{e_j\}_j$ be a linear basis of $H^*(X^{[n]})$. Then the fundamental class of the diagonal Δ_n in $X^{[n]} \times X^{[n]}$ is given by

$$[\Delta_n] = \sum_j (-1)^{|e_j|} e_j \otimes e_j^* \tag{3.45}$$

where $\{e_j^*\}_j$ is the linear basis of $H^*(X^{[n]})$ dual to $\{e_j\}_j$ in the sense that $\langle e_j, e_{j'}^* \rangle = \delta_{j,j'}$. By the definition of $\mathbf{W}(\mathfrak{L}_m, z)$,

$$\begin{aligned}
\mathrm{Tr}\, q^n\, \mathbf{W}(\mathfrak{L}_m, z) &= q^n \sum_j (-1)^{|e_j|} \langle \mathbf{W}(\mathfrak{L}_m, z) e_j, e_j^* \rangle \\
&= q^n \sum_j (-1)^{|e_j|} \int_{X^{[n]} \times X^{[n]}} (e_j \otimes e_j^*)\, c_{2n}(\mathbb{E}_{\mathfrak{L}_m}) \\
&= q^n \int_{X^{[n]} \times X^{[n]}} [\Delta_n]\, c_{2n}(\mathbb{E}_{\mathfrak{L}_m}) \\
&= q^n \int_{\Delta_n} c_{2n}(\mathbb{E}_{\mathfrak{L}_m}|_{\Delta_n}).
\end{aligned}$$

By (3.42), we have $c_{2n}(\mathbb{E}_{\mathfrak{L}_m})|_{\Delta_n} = c_{2n}(T_{X^{[n]},m})$. Therefore,

$$\mathrm{Tr}\, q^n\, \mathbf{W}(\mathfrak{L}_m, z) = q^n \int_{X^{[n]}} c_{2n}(T_{X^{[n]},m}) = q^n \chi(T_{X^{[n]},m}).$$

This completes the proof of the lemma. □

To describe the operator $\mathbf{W}(L, z)$, we need to define the operators $\Gamma_\pm(L, z)$. By abusing notations, we also use L to denote its first Chern class. Define

$$\Gamma_\pm(L, z) = \exp\left(\sum_{n>0} \frac{z^{\mp n}}{n} \mathfrak{a}_{\pm n}(L) \right). \tag{3.46}$$

REMARK 3.34. There is a sign difference between the Heisenberg commutation relations used in [**Car1**] (see p.3 there) and in (3.9). So for $n > 0$, our Heisenberg operators $\mathfrak{a}_{-n}(L)$ and $\mathfrak{a}_n(-L)$ are equal to the Heisenberg operators $\mathfrak{a}_{-n}(L)$ and $\mathfrak{a}_n(L)$ in [**Car1**] respectively. Accordingly, the operators $\Gamma_-(L,z)$ and $\Gamma_+(-L,z)$ in (3.46) are equal to the operators $\Gamma_-(L,z)$ and $\Gamma_+(L,z)$ in [**Car1**] respectively.

LEMMA 3.35. *The following commutation relations hold*

$$[\Gamma_+(L,x), \Gamma_+(L',y)] = [\Gamma_-(L,x), \Gamma_-(L',y)] = 0, \tag{3.47}$$

$$\Gamma_+(L,x)\Gamma_-(L',y) = (1-y/x)^{\langle L,L'\rangle}\,\Gamma_-(L',y)\Gamma_+(L,x). \tag{3.48}$$

PROOF. Formula (3.47) follows immediately from the Heisenberg commutation relations (3.9). To prove (3.48), we notice from (3.9) again that for $n > 0$,

$$\begin{aligned}
&\exp\left(\frac{x^{-n}}{n}\mathfrak{a}_n(L)\right)\cdot\exp\left(\frac{y^n}{n}\mathfrak{a}_{-n}(L')\right)\\
=\;&\sum_{i,j\ge 0}\frac{1}{i!\cdot j!}\left(\frac{x^{-n}}{n}\mathfrak{a}_n(L)\right)^i\cdot\left(\frac{y^n}{n}\mathfrak{a}_{-n}(L')\right)^j\\
=\;&\sum_{i,j\ge 0}\frac{1}{i!\cdot j!}\sum_{r\le i,j}\left(\frac{y^n}{n}\mathfrak{a}_{-n}(L')\right)^{j-r}\cdot\left(\frac{x^{-n}}{n}\mathfrak{a}_n(L)\right)^{i-r}\\
&\qquad\cdot\binom{i}{r}\binom{j}{r}r!\left[\frac{x^{-n}}{n}\mathfrak{a}_n(L),\frac{y^n}{n}\mathfrak{a}_{-n}(L')\right]^r\\
=\;&\exp\left(\frac{y^n}{n}\mathfrak{a}_{-n}(L')\right)\cdot\exp\left(\frac{x^{-n}}{n}\mathfrak{a}_n(L)\right)\cdot\exp\left(-\frac{1}{n}\left(\frac{y}{x}\right)^n\langle L,L'\rangle\right).
\end{aligned}$$

It follows that

$$\begin{aligned}
&\Gamma_+(L,x)\Gamma_-(L',y)\\
=\;&\prod_{n>0}\exp\left(-\frac{1}{n}\left(\frac{y}{x}\right)^n\langle L,L'\rangle\right)\cdot\Gamma_-(L',y)\Gamma_+(L,x)\\
=\;&(1-y/x)^{\langle L,L'\rangle}\,\Gamma_-(L',y)\Gamma_+(L,x).
\end{aligned}$$

This completes the proof of the lemma. □

The following two results are proved in [**Car1**, **CO**], but we follow the notations in [**Car1**, Theorem 1] adjusted in view of Remark 3.34.

THEOREM 3.36. *Let L be a line bundle over the smooth surface X. Then,*

$$\mathbf{W}(L,z) = \Gamma_-(L-K_X,z)\,\Gamma_+(-L,z). \tag{3.49}$$

We outline the proof of Theorem 3.36, and refer to [**Car1**, **Car2**, **CO**] for further details. The first step is to use certain universality results and reduce to the special case of the equivariant cohomology of a smooth toric surface X. Then for a smooth toric surface X, choose a torus action on X such that the fixed points are isolated. Some decomposition argument with respect to the isolated fixed points further reduces the proof to the cases of $X = \mathbb{C}^2$ and $\mathbb{P}^1\times\mathbb{P}^1$. Finally, the cases of $X = \mathbb{C}^2$ and $\mathbb{P}^1\times\mathbb{P}^1$ can be verified directly by using equivariant localization.

COROLLARY 3.37. *Let $T_{X^{[n]},m}$ and $\mathfrak{L}_m$ be the tangent bundle $T_{X^{[n]}}$ and the trivial line bundle on X respectively with a scaling action of $\mathbb{C}^*$ of character z^m. Let q be a real number with $|q| < 1$. Then,*

$$\sum_{n=0}^{+\infty} q^n \chi(T_{X^{[n]},m}) = (q;q)_\infty^{\langle \mathfrak{L}_m, K_X - \mathfrak{L}_m\rangle - \chi(X)} \tag{3.50}$$

where $(a;q)_n = (1-a)(1-aq)\cdots(1-aq^n)$.

PROOF. By Lemma 3.33, it suffices to prove that

$$\operatorname{Tr} q^{\mathfrak{n}}\, \mathbf{W}(\mathfrak{L}_m, z) = (q;q)_\infty^{\langle \mathfrak{L}_m, K_X - \mathfrak{L}_m\rangle - \chi(X)}. \tag{3.51}$$

By Theorem 3.36, we obtain

$$\begin{aligned} \operatorname{Tr} q^{\mathfrak{n}}\, \mathbf{W}(\mathfrak{L}_m, z) &= \operatorname{Tr} q^{\mathfrak{n}}\, \Gamma_-(\mathfrak{L}_m - K_X, z)\, \Gamma_+(-\mathfrak{L}_m, z) \\ &= \operatorname{Tr} \Gamma_-(\mathfrak{L}_m - K_X, zq)\, q^{\mathfrak{n}}\, \Gamma_+(-\mathfrak{L}_m, z) \\ &= \operatorname{Tr} q^{\mathfrak{n}}\, \Gamma_+(-\mathfrak{L}_m, z)\, \Gamma_-(\mathfrak{L}_m - K_X, zq). \end{aligned}$$

By (3.48) and since $|q| < 1$, we get

$$\begin{aligned} &\operatorname{Tr} q^{\mathfrak{n}}\, \mathbf{W}(\mathfrak{L}_m, z) \\ = \;& (1-q)^{\langle \mathfrak{L}_m, K_X - \mathfrak{L}_m\rangle} \operatorname{Tr} q^{\mathfrak{n}}\, \Gamma_-(\mathfrak{L}_m - K_X, zq)\, \Gamma_+(-\mathfrak{L}_m, z) \\ = \;& (q;q)_\infty^{\langle \mathfrak{L}_m, K_X - \mathfrak{L}_m\rangle} \operatorname{Tr} q^{\mathfrak{n}}\, \Gamma_-(\mathfrak{L}_m - K_X, 0)\, \Gamma_+(-\mathfrak{L}_m, z) \\ = \;& (q;q)_\infty^{\langle \mathfrak{L}_m, K_X - \mathfrak{L}_m\rangle} \operatorname{Tr} q^{\mathfrak{n}} \\ = \;& (q;q)_\infty^{\langle \mathfrak{L}_m, K_X - \mathfrak{L}_m\rangle - \chi(X)} \end{aligned}$$

where we have used (1.22) in the last step. □

CHAPTER 4

Chern character operators

Understanding intersection theory over moduli spaces of sheaves is a fascinating aspect of research in both mathematics and theoretical physics, with topics including Donaldson theory, Donaldson-Thomas theory, S-duality, and the Nekrasov partition function, to name a few. The Hilbert schemes of points on surfaces can be viewed as the basic models for moduli spaces of sheaves. To study intersection over the Hilbert schemes of points on surfaces, one key step is to obtain meaningful ring generators for their cohomology rings, which can be achieved (Section 7.1) via Chern character operators. This chapter is devoted to these Chern character operators which were introduced in [**LQW1**] and studied further in [**LQW4**]. We will define these operators, and prove a universal formula expressing them in terms of the Heisenberg operators, the canonical class K_X, and the Euler class e_X. As applications, we will obtain certain universality of the intersections of Chern characters on $X^{[n]}$, and describe a connection between the $\mathcal{W}$ algebras and the Hilbert scheme $X^{[n]}$. Moreover, we will address the characteristic classes of the tautological bundles over $X^{[n]}$. Theorem 4.2, Theorem 4.7, Theorem 4.15 and their proofs are from [**LQW1**, Section 5], [**LQW4**, Subsection 4.2], [**LQW3**, Section 4] respectively. The proof of Theorem 4.25 is outlined, and the details can be found in [**LQW4**, Section 6].

4.1. Chern character operators

In this section, we will define the Chern character operators $\mathfrak{G}_k(\gamma)$ which are the generalization of Lehn's boundary operator $\mathfrak{d}$ introduced in Definition 3.23 (see also Definition 4.17 below, which is from [**Leh1**]). Then we will compute the commutator $[\mathfrak{G}_k(\gamma), \mathfrak{a}_{-1}(\alpha)]$, and determine the universal structure of $\mathfrak{G}_k(\gamma)$ in terms of the Heisenberg operators.

DEFINITION 4.1. Let X be a smooth projective complex surface.

(i) For $\gamma \in H^*(X)$, we define $G(\gamma, 0) = 0$.
(ii) For $\gamma \in H^*(X)$ and $n > 0$, define

$$G(\gamma, n) = p_{1*}\big(\mathrm{ch}(\mathcal{O}_{\mathcal{Z}_n}) \cdot p_2^*\mathrm{td}(X) \cdot p_2^*\gamma\big) \in H^*(X^{[n]}) \tag{4.1}$$

where $\mathrm{ch}(\mathcal{O}_{\mathcal{Z}_n})$ is the Chern character of the sheaf $\mathcal{O}_{\mathcal{Z}_n}$, $\mathrm{td}(X)$ is the Todd class of X, and p_1, p_2 are the natural projections of $X^{[n]} \times X$ to $X^{[n]}, X$ respectively. Define the *Chern character operator* $\mathfrak{G}(\gamma) \in \mathrm{End}(\mathbb{H}_X)$ by

$$\mathfrak{G}(\gamma) = \bigoplus_{n=0}^{+\infty} \big(G(\gamma, n) \cup\big)$$

where $G(\gamma, n)$ acts on $\mathbb{H}_X^n = H^*(X^{[n]})$ by the cup product.

(iii) For $i \in \mathbb{Z}$ and $\gamma \in H^s(X)$, define $G_i(\gamma, n)$ to be the component of $G(\gamma, n)$ in $H^{s+2i}(X^{[n]})$. Define the *i-th Chern character operator* $\mathfrak{G}_i(\gamma) \in \operatorname{End}(\mathbb{H}_X)$ by

$$\mathfrak{G}_i(\gamma) = \bigoplus_{n=0}^{+\infty} \big(G_i(\gamma, n) \cup \big)$$

where again $G_i(\gamma, n)$ acts on $\mathbb{H}_X^n = H^*(X^{[n]})$ by the cup product.

Notice that $\mathfrak{G}_i(\gamma) \in \operatorname{End}(\mathbb{H}_X)$ is homogeneous of bi-degree $(0, |\gamma| + 2i)$, and

$$\mathfrak{G}_i(\gamma)' = 0 \qquad \text{and} \qquad \mathfrak{G}_i(\gamma)^\dagger = \mathfrak{G}_i(\gamma). \tag{4.2}$$

Also,

$$G(\gamma, n) = \sum_{i \in \mathbb{Z}} G_i(\gamma, n).$$

Therefore, by the definitions of the operators $\mathfrak{G}(\gamma)$ and $\mathfrak{G}_i(\gamma)$, we get

$$\mathfrak{G}(\gamma) = \sum_{i \in \mathbb{Z}} \mathfrak{G}_i(\gamma).$$

When $\gamma = 1_X$, we have

$$G(1_X, n) = p_{1*}(\operatorname{ch}(\mathcal{O}_{\mathcal{Z}_n}) \cdot p_2^* \operatorname{td}(X)).$$

So we see from the Grothendieck-Riemann-Roch Theorem [**Hart**, Theorem 5.3 in Appendix A] that

$$G(1_X, n) = \operatorname{ch}(p_{1*}\mathcal{O}_{\mathcal{Z}_n}) = \operatorname{ch}\big((\mathcal{O}_X)^{[n]}\big). \tag{4.3}$$

Recall that $(\mathcal{O}_X)^{[n]}$ is a locally free sheaf of rank-n on $X^{[n]}$. By Theorem 1.24 (i), we have $c_1\big((\mathcal{O}_X)^{[n]}\big) = -B_n/2$. It follows that

$$\mathfrak{G}_0(1_X) = \bigoplus_n \big(G_0(1_X, n) \cup \big) = \bigoplus_n (n \cdot \operatorname{Id}_{X^{[n]}}) = \mathfrak{L}_0(1_X), \tag{4.4}$$

$$\mathfrak{G}_1(1_X) = \bigoplus_n \big(G_1(1_X, n) \cup \big) = \bigoplus_n \big(c_1(p_{1*}\mathcal{O}_{\mathcal{Z}_n}) \cup \big) = \mathfrak{d}. \tag{4.5}$$

THEOREM 4.2. *Let $\gamma, \alpha \in H^*(X)$. Then, we have*

$$[\mathfrak{G}(\gamma), \mathfrak{a}_{-1}(\alpha)] = \exp(\operatorname{ad}(\mathfrak{d}))(\mathfrak{a}_{-1}(\gamma\alpha)), \tag{4.6}$$

or equivalently, using the component $\mathfrak{G}_k(\gamma)$ of $\mathfrak{G}(\gamma)$, we obtain

$$[\mathfrak{G}_k(\gamma), \mathfrak{a}_{-1}(\alpha)] = \frac{1}{k!} \cdot \mathfrak{a}_{-1}^{(k)}(\gamma\alpha). \tag{4.7}$$

PROOF. Let notations be as in Section 1.5. By the exact sequence (1.38),

$$g_X^* \operatorname{ch}(\mathcal{O}_{\mathcal{Z}_n}) - f_X^* \operatorname{ch}(\mathcal{O}_{\mathcal{Z}_{n-1}}) = \rho_X^* \operatorname{ch}(\mathcal{O}_{\Delta_X}) \cdot p_1^* \exp(-[E_n]) \tag{4.8}$$

where we have used the fact that

$$\begin{aligned} \operatorname{ch}(\rho_X^*\mathcal{O}_{\Delta_X} \otimes p_1^*\mathcal{O}_{X^{[n,n-1]}}(-E_n)) &= \rho_X^* \operatorname{ch}(\mathcal{O}_{\Delta_X}) \cdot p_1^* \operatorname{ch}(\mathcal{O}_{X^{[n,n-1]}}(-E_n)) \\ &= \rho_X^* \operatorname{ch}(\mathcal{O}_{\Delta_X}) \cdot p_1^* \exp(-[E_n]). \end{aligned}$$

Claim. $g_n^* G(\gamma, n) = f_n^* G(\gamma, n-1) + \rho^*\gamma \cdot \exp(-[E_n])$.

PROOF. For $i \geq 1$, let $p_{i,1}$ and $p_{i,2}$ be the projections of $X^{[i]} \times X$ to $X^{[i]}$ and X respectively. From the definition of $G(\gamma, n)$, we see that

$$\begin{aligned} g_n^* G(\gamma, n) &= g_n^*(p_{n,1})_*(\mathrm{ch}(\mathcal{O}_{\mathcal{Z}_n}) \cdot (p_{n,2})^* \mathrm{td}(X) \cdot (p_{n,2})^*\gamma) \\ &= p_{1*} g_X^*(\mathrm{ch}(\mathcal{O}_{\mathcal{Z}_n}) \cdot (p_{n,2})^* \mathrm{td}(X) \cdot (p_{n,2})^*\gamma) \\ &= p_{1*}(g_X^* \mathrm{ch}(\mathcal{O}_{\mathcal{Z}_n}) \cdot p_2^* \mathrm{td}(X) \cdot p_2^*\gamma) \end{aligned}$$

where p_2 is the projection of $X^{[n,n-1]} \times X$ to X. Similarly,

$$f_n^* G(\gamma, n-1) = p_{1*}(f_X^* \mathrm{ch}(\mathcal{O}_{\mathcal{Z}_{n-1}}) \cdot p_2^* \mathrm{td}(X) \cdot p_2^*\gamma).$$

So combining these with (4.8) above, we conclude that

$$\begin{aligned} & g_n^* G(\gamma, n) - f_n^* G(\gamma, n-1) \\ = \;& p_{1*}((g_X^* \mathrm{ch}(\mathcal{O}_{\mathcal{Z}_n}) - f_X^* \mathrm{ch}(\mathcal{O}_{\mathcal{Z}_{n-1}})) \cdot p_2^* \mathrm{td}(X) \cdot p_2^*\gamma) \\ = \;& p_{1*}(\rho_X^* \mathrm{ch}(\mathcal{O}_{\Delta_X}) \cdot p_1^* \exp(-[E_n]) \cdot p_2^* \mathrm{td}(X) \cdot p_2^*\gamma) \\ = \;& p_{1*}(\rho_X^* \mathrm{ch}(\mathcal{O}_{\Delta_X}) \cdot p_2^* \mathrm{td}(X) \cdot p_2^*\gamma) \cdot \exp(-[E_n]) \\ = \;& p_{1*}\rho_X^*(\mathrm{ch}(\mathcal{O}_{\Delta_X}) \cdot (p_{1,2})^* \mathrm{td}(X) \cdot (p_{1,2})^*\gamma) \cdot \exp(-[E_n]) \\ = \;& \rho^*(p_{1,1})_*(\mathrm{ch}(\mathcal{O}_{\Delta_X}) \cdot (p_{1,2})^* \mathrm{td}(X) \cdot (p_{1,2})^*\gamma) \cdot \exp(-[E_n]). \end{aligned}$$

Applying the Grothendieck-Riemann-Roch Theorem to the diagonal embedding $\tau_2 : X \to X \times X$, we get

$$\mathrm{ch}(\mathcal{O}_{\Delta_X}) \cdot (p_{1,2})^* \mathrm{td}(X) = \tau_{2*}[X].$$

So

$$(p_{1,1})_*(\mathrm{ch}(\mathcal{O}_{\Delta_X}) \cdot (p_{1,2})^* \mathrm{td}(X) \cdot (p_{1,2})^*\gamma) = (p_{1,1})_*(\tau_{2*}[X] \cdot (p_{1,2})^*\gamma) = \gamma.$$

It follows that

$$g_n^* G(\gamma, n) - f_n^* G(\gamma, n-1) = \rho^*\gamma \cdot \exp(-[E_n]).$$

This completes the proof of the claim. $\square$

We continue the proof of the theorem. By Definition 3.3, for $a \in H^*(X^{[n-1]})$,

$$\mathfrak{a}_{-1}(\alpha)(a) = \tilde{p}_{1*}([Q^{[n,n-1]}] \cdot \tilde{\rho}^*\alpha \cdot \tilde{p}_2^* a).$$

Define $\iota : X^{[n,n-1]} \to X^{[n]} \times X \times X^{[n-1]}$ by putting $\iota(\xi, \eta) = (\xi, \rho(\xi, \eta), \eta)$. Then, ι is an embedding. Moreover, $\iota_*[X^{[n,n-1]}] = [Q^{[n,n-1]}]$. Thus, we have

$$\begin{aligned} \mathfrak{a}_{-1}(\alpha)(a) &= \tilde{p}_{1*}(\iota_*[X^{[n,n-1]}] \cdot \tilde{\rho}^*\alpha \cdot \tilde{p}_2^* a) \\ &= \tilde{p}_{1*}\iota_*([X^{[n,n-1]}] \cdot (\iota^* \circ \tilde{\rho}^*)\alpha \cdot (\iota^* \circ \tilde{p}_2^*)a) \\ &= g_{n*}([X^{[n,n-1]}] \cdot \rho^*\alpha \cdot f_n^* a). \end{aligned}$$

Combining this with the above Claim, we conclude that

$$\begin{aligned} \mathfrak{G}(\gamma)\mathfrak{a}_{-1}(\alpha)(a) &= G(\gamma, n) \cdot g_{n*}([X^{[n,n-1]}] \cdot \rho^*\alpha \cdot f_n^* a) \\ &= g_{n*}(g_n^* G(\gamma, n) \cdot [X^{[n,n-1]}] \cdot \rho^*\alpha \cdot f_n^* a) \\ &= g_{n*}(f_n^* G(\gamma, n-1) \cdot [X^{[n,n-1]}] \cdot \rho^*\alpha \cdot f_n^* a) \\ &\quad + g_{n*}(\rho^*\gamma \cdot \exp(-[E_n]) \cdot [X^{[n,n-1]}] \cdot \rho^*\alpha \cdot f_n^* a) \\ &= g_{n*}([X^{[n,n-1]}] \cdot f_n^* G(\gamma, n-1)\rho^*\alpha \cdot f_n^* a) \\ &\quad + g_{n*}((\exp(-[E_n]) \cdot [X^{[n,n-1]}]) \cdot \rho^*(\gamma\alpha) \cdot f_n^* a). \end{aligned}$$

Thus,

$$[\mathfrak{G}(\gamma), \mathfrak{a}_{-1}(\alpha)](a) = g_{n*}((\exp(-[E_n]) \cdot [X^{[n,n-1]}]) \cdot \rho^*(\gamma\alpha) \cdot f_n^* a)$$

which is equal to $\exp(\mathrm{ad}(\mathfrak{d}))(\mathfrak{a}_{-1}(\gamma\alpha))(a)$ by the Lemma 3.9 in [**Leh1**]. □

Recall from Definition 3.28 (ii) that

$$\mathfrak{A}_X = \{\alpha \in H^*(X) | K_X \alpha = 0\}.$$

The next lemma determines $\mathfrak{G}_k(\alpha)$ when $\alpha \in \mathfrak{A}_X$.

LEMMA 4.3. *Let $k \geq 0$, and $\alpha \in \mathfrak{A}_X$. Then, $\mathfrak{G}_k(\alpha)$ is equal to*

$$- \sum_{\ell(\lambda)=k+2, |\lambda|=0} \frac{1}{\lambda^!} \mathfrak{a}_\lambda(\tau_* \alpha) + \sum_{\ell(\lambda)=k, |\lambda|=0} \frac{s(\lambda) - 2}{24\lambda^!} \mathfrak{a}_\lambda(\tau_*(e_X \alpha)). \tag{4.9}$$

PROOF. Denote the expression (4.9) by $\mathfrak{f}_2$. We will apply Proposition 3.26 to the two operators $\mathfrak{f}_1 = \mathfrak{G}_k(\alpha)$ and $\mathfrak{f}_2$. Note that $\mathfrak{G}_k(\alpha)|0\rangle = 0$ and $\mathfrak{G}_k(\alpha)' = 0$. From Theorem 4.2 and Lemma 3.29, we see that

$$\begin{aligned} & [\mathfrak{G}_k(\alpha), \mathfrak{a}_{-1}(\beta)] \\ = & \frac{1}{k!} \cdot \mathfrak{a}_{-1}^{(k)}(\alpha\beta) \\ = & \sum_{\ell(\lambda)=k+1, |\lambda|=-1} \frac{1}{\lambda^!} \mathfrak{a}_\lambda(\tau_*(\alpha\beta)) - \sum_{\ell(\lambda)=k-1, |\lambda|=-1} \frac{s(\lambda) - 1}{24\lambda^!} \mathfrak{a}_\lambda(\tau_*(e_X \alpha\beta)). \end{aligned}$$

Next we need to check the corresponding properties for $\mathfrak{f}_2$. Indeed, $\mathfrak{f}_2 \cdot |0\rangle = 0$ from the definition of $\mathfrak{f}_2$. Also, $\mathfrak{f}_2' = 0$ by Remark 3.30. By Lemma 3.12 (i),

$$\begin{aligned} & [\mathfrak{f}_2, \mathfrak{a}_{-1}(\beta)] \\ = & \sum_{\ell(\lambda)=k+1, |\lambda|=-1} \frac{1}{\lambda^!} \mathfrak{a}_\lambda(\tau_*(\alpha\beta)) - \sum_{\ell(\lambda)=k, |\lambda|=0} \frac{s(\lambda) - 2}{24(\lambda_1)^!} \mathfrak{a}_{\lambda_1}(\tau_*(e_X \alpha\beta)) \\ = & \sum_{\ell(\lambda)=k+1, |\lambda|=-1} \frac{1}{\lambda^!} \mathfrak{a}_\lambda(\tau_*(\alpha\beta)) - \sum_{\ell(\lambda_1)=k-1, |\lambda_1|=-1} \frac{s(\lambda_1) - 1}{24(\lambda_1)^!} \mathfrak{a}_{\lambda_1}(\tau_*(e_X \alpha\beta)) \\ = & [\mathfrak{G}_k(\alpha), \mathfrak{a}_{-1}(\beta)] \end{aligned}$$

where for a generalized partition λ with $\ell(\lambda) = k$, the generalized partition λ_1 is obtained from λ by subtracting 1 from the multiplicity of 1.

Thus, we conclude from Proposition 3.26 that $\mathfrak{G}_k(\alpha) = \mathfrak{f}_2$. □

To understand the operator $\mathfrak{G}_k(\alpha)$ for a general $\alpha \in H^*(X)$, we need the notion of universal linear combinations.

DEFINITION 4.4. Let X be a projective surface, $s \geq 1$, and $\alpha_1, \ldots, \alpha_s \in H^*(X)$. Let $k_1, \ldots, k_s \geq 0$, and $n_{i,j} \in \mathbb{Z}$ with $1 \leq i \leq s$ and $1 \leq j \leq k_i$. A *universal linear combination* of $\mathfrak{a}_{n_{i,1}} \cdots \mathfrak{a}_{n_{i,k_i}}(\tau_{k_i *}(\alpha_i))$, $1 \leq i \leq s$ is a linear combination of the form

$$\sum_{i=1}^{s} f_i(k_i, n_{i,1}, \ldots, n_{i,k_i}) \, \mathfrak{a}_{n_{i,1}} \cdots \mathfrak{a}_{n_{i,k_i}}(\tau_{k_i *}(\alpha_i))$$

where the coefficients $f_i(k_i, n_{i,1}, \ldots, n_{i,k_i})$ are independent of $X, \alpha_1, \ldots, \alpha_s$. A *universal linear combination* of $\mathfrak{a}_{n_{i,1}} \cdots \mathfrak{a}_{n_{i,k_i}}(\tau_{k_i *}(\alpha_i))|0\rangle$, $1 \leq i \leq s$ is defined in a similar way.

LEMMA 4.5. *Let $k \ge 0$, $n \in \mathbb{Z}$, and $\alpha, \beta \in H^*(X)$. Then, $[\mathfrak{G}_k(\alpha), \mathfrak{a}_n(\beta)]$ is a universal linear combination of expressions of the form*

$$\mathfrak{a}_{m_1} \cdots \mathfrak{a}_{m_{k+1-r}}(\tau_{(k+1-r)*}(K_X^r \alpha\beta))$$

where $0 \le r \le 2$ and $m_1 + \ldots + m_{k+1-r} = n$.

PROOF. First of all, note that $\mathfrak{a}_0(\beta) = 0$. So we may let $n \ne 0$. Since $\mathfrak{G}_k(\alpha)^\dagger = \mathfrak{G}_k(\alpha)$ and $\mathfrak{a}_n(\beta)^\dagger = (-1)^n \mathfrak{a}_{-n}(\beta)$, we see from (3.3) that

$$[\mathfrak{G}_k(\alpha), \mathfrak{a}_n(\beta)]^\dagger = -[\mathfrak{G}_k(\alpha)^\dagger, \mathfrak{a}_n(\beta)^\dagger] = (-1)^{1+n}[\mathfrak{G}_k(\alpha), \mathfrak{a}_{-n}(\beta)].$$

By (3.2), $(\mathfrak{a}_{m_1} \cdots \mathfrak{a}_{m_{k+s-r}}(\tau_{(k+s-r)*}(K_X^r \alpha\beta)))^\dagger$ is equal to

$$(-1)^{m_1+\ldots+m_{k+s-r}} \cdot \mathfrak{a}_{-m_{k+s-r}} \cdots \mathfrak{a}_{-m_1}(\tau_{(k+s-r)*}(K_X^r \alpha\beta)).$$

Thus it suffices to prove the lemma for $n \le -1$. When $n = -1$,

$$[\mathfrak{G}_k(\alpha), \mathfrak{a}_n(\beta)] = \frac{1}{k!} \cdot \mathfrak{a}_{-1}^{(k)}(\alpha\beta)$$

by Lemma 4.2. Therefore, the lemma for $n = -1$ follows by repeatedly applying Lemma 3.27. When $n \le -2$, we see from Lemma 3.25 that

$$\begin{aligned}
& [\mathfrak{G}_k(\alpha), \mathfrak{a}_n(\beta)] \\
= \; & \frac{1}{n+1} \cdot \big\{[[\mathfrak{G}_k(\alpha), \mathfrak{a}_{-1}(1_X)]', \mathfrak{a}_{n+1}(\beta)] + [\mathfrak{a}'_{-1}(1_X), [\mathfrak{G}_k(\alpha), \mathfrak{a}_{n+1}(\beta)]]\big\} \\
= \; & \frac{1}{n+1} \cdot \big\{[[\mathfrak{G}_k(\alpha), \mathfrak{a}_{-1}(1_X)]', \mathfrak{a}_{n+1}(\beta)] \\
& + [\mathfrak{a}_{-1}(1_X), [\mathfrak{G}_k(\alpha), \mathfrak{a}_{n+1}(\beta)]]' - [\mathfrak{a}_{-1}(1_X), [\mathfrak{G}_k(\alpha), \mathfrak{a}_{n+1}(\beta)]']\big\}.
\end{aligned}$$

Therefore, the lemma for $n \le -2$ follows from induction, Lemma 3.12 (i), and Lemma 3.27. □

The formulas in the following proposition are referred to as the *transfer properties*.

PROPOSITION 4.6. (The Transfer Properties) *Let $\alpha, \beta \in H^*(X)$.*

(i) *Let $k, s \ge 1$, $n_1, \ldots, n_k, m_1, \ldots, m_s \in \mathbb{Z}$. Then,*

$$\begin{aligned}
& [\mathfrak{a}_{n_1} \cdots \mathfrak{a}_{n_k}(\tau_{k*}\alpha), \mathfrak{a}_{m_1} \cdots \mathfrak{a}_{m_s}(\tau_{s*}\beta)] \\
= \; & [\mathfrak{a}_{n_1} \cdots \mathfrak{a}_{n_k}(\tau_{k*}1_X), \mathfrak{a}_{m_1} \cdots \mathfrak{a}_{m_s}(\tau_{s*}(\alpha\beta))];
\end{aligned}$$

(ii) *Let $k \ge 0$ and $n \in \mathbb{Z}$. Then,*

$$[\mathfrak{G}_k(\alpha), \mathfrak{a}_n(\beta)] = [\mathfrak{G}_k(1_X), \mathfrak{a}_n(\alpha\beta)] = [\mathfrak{G}_k(\alpha\beta), \mathfrak{a}_n(1_X)].$$

PROOF. The formulas in (i) and (ii) follow from Lemma 3.12 (i) and Lemma 4.5 respectively. □

Now we present a formula of $\mathfrak{G}_k(\alpha)$ for a general $\alpha \in H^*(X)$.

THEOREM 4.7. *Let $k \ge 0$ and $\alpha \in H^*(X)$. Then, $\mathfrak{G}_k(\alpha)$ is equal to*

$$\begin{aligned}
& - \sum_{\ell(\lambda)=k+2, |\lambda|=0} \frac{1}{\lambda!} \mathfrak{a}_\lambda(\tau_*\alpha) + \sum_{\ell(\lambda)=k, |\lambda|=0} \frac{s(\lambda)-2}{24\lambda!} \mathfrak{a}_\lambda(\tau_*(e_X\alpha)) \\
& + \sum_{\epsilon \in \{K_X, K_X^2\}} \sum_{\ell(\lambda)=k+2-|\epsilon|/2, |\lambda|=0} \frac{g_\epsilon(\lambda)}{\lambda!} \mathfrak{a}_\lambda(\tau_*(\epsilon\alpha))
\end{aligned}$$

where all the numbers $g_\epsilon(\lambda)$ are independent of X and α.

PROOF. Clearly we may assume $\alpha \neq 0$. By (4.4) and the definition of $\mathfrak{L}_0(1_X)$,

$$\mathfrak{G}_0(1_X) = \mathfrak{L}_0(1_X) = -\sum_{\ell(\lambda)=2, |\lambda|=0} \frac{1}{\lambda!}\mathfrak{a}_\lambda(\tau_* 1_X).$$

So our theorem is true for $k = |\alpha| = 0$. In the following, assume $k + |\alpha| > 0$.

Next, we write $\mathfrak{G}_k(\alpha) \in \mathrm{End}(\mathbb{H}_X)$ as a linear combination of Heisenberg monomials $\prod_i (\mathfrak{a}_i(\alpha_{i,1}) \cdots \mathfrak{a}_i(\alpha_{i,m_i}))$. For degree reasons, $\sum_i i m_i = 0$ and

$$-2\sum_i m_i + \sum_{i,j} |\alpha_{i,j}| = 2k + |\alpha|. \tag{4.10}$$

In other words, we regard $\mathfrak{G}_k(\alpha)$ as an element in the completion of the universal enveloping algebra of the Heisenberg algebra. We can always do so because $\mathbb{H}_X$ is irreducible as a representation of the Heisenberg algebra. Now rewrite $\mathfrak{G}_k(\alpha)$ as

$$\mathfrak{G}_k(\alpha) = \sum_{|\lambda|=0} \mathfrak{g}_\lambda \tag{4.11}$$

where for each fixed generalized partition $\lambda = (\cdots(-2)^{m_{-2}}(-1)^{m_{-1}}1^{m_1}2^{m_2}\cdots)$ with $|\lambda| = 0$, the operator $\mathfrak{g}_\lambda$ stands for the component in $\mathfrak{G}_k(\alpha)$ containing all the expressions of the form $\prod_i (\mathfrak{a}_i(\alpha_{i,1}) \cdots \mathfrak{a}_i(\alpha_{i,m_i}))$.

Fix $n \in \mathbb{Z}$ and $\beta \in H^*(X)$. By Lemma 4.5 and Lemma 3.12 (ii),

$$[\mathfrak{G}_k(\alpha), \mathfrak{a}_n(\beta)] = \sum_{\substack{\epsilon \in \{1_X, K_X, K_X^2, e_X\} \\ \ell(\mu)=k+1-|\epsilon|/2, |\mu|=n}} d(\epsilon, \mu)\mathfrak{a}_\mu(\tau_*(\epsilon\alpha\beta)) \tag{4.12}$$

where all the coefficients $d(\epsilon, \mu) \in \mathbb{Q}$ are independent of the surface X and the cohomology classes $\alpha, \beta \in H^*(X)$. It follows that the generalized partitions λ in (4.11) must satisfy $k \le \ell(\lambda) \le (k+2)$. Moreover, for a fixed generalized partition λ with $|\lambda| = 0$ and $k \le \ell(\lambda) \le (k+2)$, and for every i with $m_{-i} > 0$, we have

$$[\mathfrak{g}_\lambda, \mathfrak{a}_i(\beta)] = \begin{cases} d(1_X, \lambda_{-i})\mathfrak{a}_{\lambda_{-i}}(\tau_*(\alpha\beta)), & \text{if } \ell(\lambda) = k+2 \\ d(K_X, \lambda_{-i})\mathfrak{a}_{\lambda_{-i}}(\tau_*(K_X\alpha\beta)), & \text{if } \ell(\lambda) = k+1 \\ \mathfrak{a}_{\lambda_{-i}}(\tau_*\omega), & \text{if } \ell(\lambda) = k \end{cases} \tag{4.13}$$

where λ_{-i} is the generalized partition obtained from λ by subtracting 1 from the multiplicity of $(-i)$, and $\omega = \big(d(e_X, \lambda_{-i})e_X + d(K_X^2, \lambda_{-i})K_X^2\big)\alpha\beta$. Note that $d(\epsilon, \lambda_{-i})$ is independent of X, α and β.

By choosing special surfaces X and $\alpha, \beta \in H^*(X)$, we now claim that

$$\frac{d(1_X, \lambda_{-i})}{im_{-i}}, \quad \frac{d(K_X, \lambda_{-i})}{im_{-i}}, \quad \frac{d(e_X, \lambda_{-i})}{im_{-i}}e_X + \frac{d(K_X^2, \lambda_{-i})}{im_{-i}}K_X^2 \tag{4.14}$$

are independent of i as long as $m_{-i} > 0$. Indeed, to prove that the 3rd item in (4.14) is independent of i as long as $m_{-i} > 0$, we may assume that $\mathfrak{g}_\lambda \neq 0$ by (4.13). Note that $|\lambda| = 0$ and $\ell(\lambda) = k$. By (4.10), $\sum_{i,j} |\alpha_{i,j}| = 4k + |\alpha|$ for every Heisenberg monomial $\prod_i (\mathfrak{a}_i(\alpha_{i,1}) \cdots \mathfrak{a}_i(\alpha_{i,m_i}))$ contained in $\mathfrak{g}_\lambda$. On the other hand, $|\alpha_{i,j}| \le 4$ and thus $\sum_{i,j} |\alpha_{i,j}| \le 4\ell(\lambda) = 4k$. It follows that $|\alpha| = 0$ and $|\alpha_{i,j}| = 4$ for every i and j. So $\mathfrak{g}_\lambda = b \cdot \mathfrak{a}_\lambda(\tau_*[x])$where b is a nonzero number, and $[x] \in H^4(X)$ is the cohomology class corresponding to a point $x \in X$. Since $\alpha \neq 0$ and $|\alpha| = 0$, $\alpha = \tilde{b} \cdot 1_X$ for some nonzero number $\tilde{b}$. Choosing $\beta = 1_X$, we see from (4.13) that

$$\begin{aligned} m_{-i} ib \cdot \mathfrak{a}_{\lambda_{-i}}(\tau_*[x]) &= [\mathfrak{g}_\lambda, \mathfrak{a}_i(1_X)] \\ &= \tilde{b} \cdot \mathfrak{a}_{\lambda_{-i}}(\tau_*(d(e_X, \lambda_{-i})e_X + d(K_X^2, \lambda_{-i})K_X^2)). \end{aligned}$$

So $b[x] = \tilde{b} \cdot (d(e_X, \lambda_{-i})e_X + d(K_X^2, \lambda_{-i})K_X^2)/(im_{-i})$. Thus

$$\frac{d(e_X, \lambda_{-i})}{im_{-i}} e_X + \frac{d(K_X^2, \lambda_{-i})}{im_{-i}} K_X^2$$

is independent of i as long as $m_{-i} > 0$. Similarly, by choosing $\alpha \in H^2(X)$ with $\langle K_X, \alpha \rangle \neq 0$ (respectively, $\alpha \in H^4(X) - \{0\}$), we see that $d(K_X, \lambda_{-i})/(im_{-i})$ (respectively, $d(1_X, \lambda_{-i})/(im_{-i})$) is independent of i as long as $m_{-i} > 0$. This proves (4.14).

Now, for an arbitrary X and $\alpha \in H^*(X)$, put

$$\mathfrak{f} = \sum_{\epsilon \in \{1_X, K_X, K_X^2, e_X\}} \sum_{\ell(\lambda) = k+2-|\epsilon|/2, |\lambda|=0} \frac{d(\epsilon, \lambda_{-i})}{im_{-i}} \mathfrak{a}_\lambda(\tau_*(\epsilon\alpha)).$$

We claim that $\mathfrak{G}_k(\alpha) = \mathfrak{f}$. Indeed, for every $n \in \mathbb{Z} - \{0\}$ and $\beta \in H^*(X)$, we have

$$\begin{aligned} & [\mathfrak{f}, \mathfrak{a}_n(\beta)] \\ = & \sum_{\epsilon \in \{1_X, K_X, K_X^2, e_X\}} \sum_{\ell(\lambda) = k+2-|\epsilon|/2, |\lambda|=0} nm_{-n} \frac{d(\epsilon, \lambda_{-i})}{im_{-i}} \cdot \mathfrak{a}_{\lambda_{-n}}(\tau_*(\epsilon\alpha\beta)) \\ = & \sum_{\epsilon \in \{1_X, K_X, K_X^2, e_X\}} \sum_{\ell(\lambda) = k+2-|\epsilon|/2, |\lambda|=0} d(\epsilon, \lambda_{-n}) \cdot \mathfrak{a}_{\lambda_{-n}}(\tau_*(\epsilon\alpha\beta)) \\ = & [\mathfrak{G}_k(\alpha), \mathfrak{a}_n(\beta)] \end{aligned}$$

where we have used (4.14) in the second equality, and (4.11) and (4.13) in the last equality. Since $\mathbb{H}_X$ is irreducible, we see from Schur's lemma that $(\mathfrak{G}_k(\alpha) - \mathfrak{f})$ must be a scalar multiple of the identity operator. Recall that the bi-degree of $(\mathfrak{G}_k(\alpha) - \mathfrak{f})$ is $(0, 2k + |\alpha|)$ which is nontrivial since $k + |\alpha| > 0$. Thus, $(\mathfrak{G}_k(\alpha) - \mathfrak{f}) = 0$ and

$$\mathfrak{G}_k(\alpha) = \mathfrak{f} = \sum_{\epsilon \in \{1_X, K_X, K_X^2, e_X\}} \sum_{\ell(\lambda) = k+2-|\epsilon|/2, |\lambda|=0} \frac{d(\epsilon, \lambda_{-i})}{im_{-i}} \mathfrak{a}_\lambda(\tau_*(\epsilon\alpha)).$$

Finally, since the numbers $d(\epsilon, \lambda_{-i})/(im_{-i})$ are independent of X and $\alpha \in H^*(X)$, by using Lemma 4.3 and an argument similar to that in the proof of Theorem 3.31, we conclude that $d(1_X, \lambda_{-i})/(im_{-i}) = -1/\lambda^!$ and

$$\frac{d(e_X, \lambda_{-i})}{im_{-i}} = \frac{s(\lambda) - 2}{24\lambda^!}.$$

In particular, $d(e_X, \lambda_{-i})/(im_{-i})$ is independent of i whenever $m_{-i} > 0$. So by (4.14), $d(K^2, \lambda_{-i})/(im_{-i})$ is also independent of i whenever $m_{-i} > 0$. Now denoting $d(\epsilon, \lambda_{-i})/(im_{-i})$ by $g_\epsilon(\lambda)/\lambda^!$ for $\epsilon = K_X$ and K_X^2, we have proved the theorem. $\square$

It turns out that the operator $\mathfrak{G}_1(\alpha)$ can be determined completely. To state the result, we need the notation $: \mathfrak{a}^3 :_0 (\tau_{3*}\alpha)$. So fix $\alpha \in H^*(X)$. Define a vertex operator (i.e. a field) $a(\alpha)(z)$ by putting

$$a(\alpha)(z) = \sum_{n \in \mathbb{Z}} \mathfrak{a}_n(\alpha) z^{-n-1}. \tag{4.15}$$

If

$$\tau_{p*}(\alpha) = \sum_i \alpha_{i,1} \otimes \alpha_{i,1} \otimes \ldots \otimes \alpha_{i,p} \in H^*(X)^{\otimes p},$$

then we define

$$: a(z)^p : (\tau_* \alpha) = \sum_i : a(\alpha_{i,1})(z) a(\alpha_{i,2})(z) \cdots a(\alpha_{i,p})(z) : .$$

Note that $: a(z)^p : (\tau_* \alpha)$ is of conformal weight p. So we can rewrite

$$: a(z)^p : (\tau_* \alpha) = \sum_m : a^p :_m (\tau_* \alpha)\, z^{-m-p}, \tag{4.16}$$

where the operator $: a^p :_m (\tau_* \alpha) \in \mathrm{End}(\mathbb{H}_X)$ is the coefficient of z^{-m-p} (i.e. the m-th Fourier component of the field $: a(z)^p : (\tau_* \alpha)$), and maps $H^*(X^{[n]})$ to $H^*(X^{[n+m]})$. We have

$$: \mathfrak{a}^3 :_0 (\tau_{3*} \alpha) = \sum_{k, \ell > 0} \big(\mathfrak{a}_{-k} \mathfrak{a}_{-\ell} \mathfrak{a}_{k+\ell} + \mathfrak{a}_{-k-\ell} \mathfrak{a}_k \mathfrak{a}_\ell \big)(\tau_{3*} \alpha). \tag{4.17}$$

PROPOSITION 4.8. *Let $\alpha \in H^*(X)$. Then,*

$$\mathfrak{G}_1(\alpha) = -\frac{1}{6} : \mathfrak{a}^3 :_0 (\tau_{3*} \alpha) - \sum_{n>0} \frac{n-1}{2} : \mathfrak{a}_n \mathfrak{a}_{-n} : (\tau_{2*}(K_X \alpha)). \tag{4.18}$$

PROOF. Observe that both sides of (4.18) annihilate the vacuum vector $|0\rangle$. To prove (4.18), it suffices to show that the commutators of both sides of (4.18) with the operators $\mathfrak{a}_n(\beta)$, $n \in \mathbb{Z}, \beta \in H^*(X)$, coincide. By Lemma 4.5 and Theorem 3.24 (ii),

$$\begin{aligned} [\mathfrak{G}_1(\alpha), \mathfrak{a}_n(\beta)] &= [\mathfrak{G}_1(1_X), \mathfrak{a}_n(\alpha\beta)] \\ &= \mathfrak{a}'_n(\alpha\beta) \\ &= n \cdot \mathfrak{L}_n(\alpha\beta) - \frac{n(|n|-1)}{2} \cdot \mathfrak{a}_n(K_X \alpha\beta). \end{aligned}$$

Using (4.17) and Lemma 3.12 (i), we can easily compute the commutator of the right hand side of (4.18) with $\mathfrak{a}_n(\beta)$. These two commutators indeed coincide. □

In particular, setting $\alpha = 1_X$, we obtain from (4.18) that

$$\mathfrak{d} = \mathfrak{G}_1(1_X) = -\frac{1}{6} : \mathfrak{a}^3 :_0 (\tau_{3*} 1_X) - \sum_{n>0} \frac{n-1}{2} : \mathfrak{a}_n \mathfrak{a}_{-n} : (\tau_{2*} K_X). \tag{4.19}$$

This is equivalent to Theorem 3.24 (ii) of Lehn.

By the definition of Chern character operators (Definition 4.1 (iii)), we have $G_k(\alpha, n) = \mathfrak{G}_k(\alpha) 1_{X^{[n]}}$ where $1_{X^{[n]}}$ denotes the fundamental class of the Hilbert scheme $X^{[n]}$. It follows that the cohomology class $G_k(\alpha, n)$ can be understood via Theorem 4.7. This leads us to the next section.

4.2. Chern characters

In this section, using the results from the previous section, we will study the structure of the cohomology class $G_k(\alpha, n)$ on the Hilbert scheme $X^{[n]}$. A universal expression of $G_k(\alpha, n)$ in terms of the Heisenberg monomial classes will be derived. This further leads to the cup product and intersection number among the classes $G_k(\alpha, n)$.

We begin with an elementary observation.

LEMMA 4.9. *Fix a, b with $1 \le a \le b$. Let $\mathfrak{g} \in \mathrm{End}(\mathbb{H}_X)$ be of bi-degree (ℓ, s), and $A = \mathfrak{a}_{-n_1}(\beta_1) \cdots \mathfrak{a}_{-n_b}(\beta_b)|0\rangle$. Then, $\mathfrak{g}(A)$ is equal to the sum of the following two terms:*

$$\sum_{i=0}^{a-1} \sum_{\sigma_i} \pm \prod_{\ell \in \sigma_i^0} \mathfrak{a}_{-n_\ell}(\beta_\ell)[[\cdots[\mathfrak{g}, \mathfrak{a}_{-n_{\sigma_i(1)}}(\beta_{\sigma_i(1)})], \cdots], \mathfrak{a}_{-n_{\sigma_i(i)}}(\beta_{\sigma_i(i)})]|0\rangle \tag{4.20}$$

$$+ \sum_{\sigma_a} (-1)^{\sum_{k=0}^{a-1}(s+\sum_{\ell=1}^{k}|\beta_{\sigma_a(\ell)}|)\sum_{\sigma_a(k)<j<\sigma_a(k+1)}|\beta_j|} \prod_{\ell \in \sigma_a^1} \mathfrak{a}_{-n_\ell}(\beta_\ell) \cdot \tag{4.21}$$
$$\cdot [[\cdots[\mathfrak{g}, \mathfrak{a}_{-n_{\sigma_a(1)}}(\beta_{\sigma_a(1)})], \cdots], \mathfrak{a}_{-n_{\sigma_a(a)}}(\beta_{\sigma_a(a)})] \prod_{\ell \in \sigma_a^2} \mathfrak{a}_{-n_\ell}(\beta_\ell)|0\rangle$$

where for each fixed i with $0 \le i \le a$, σ_i runs over all the maps

$$\{1, \ldots, i\} \to \{1, \ldots, b\}$$

satisfying $\sigma_i(1) < \cdots < \sigma_i(i)$. Moreover,

$$\begin{aligned} \sigma_i^0 &= \{\ell \,|\, 1 \le \ell \le b, \ell \ne \sigma_i(1), \ldots, \sigma_i(i)\}, \\ \sigma_a^1 &= \{\ell \,|\, 1 \le \ell < \sigma_a(a), \ell \ne \sigma_a(1), \ldots, \sigma_a(a)\}, \\ \sigma_a^2 &= \{\ell \,|\, \sigma_a(a) < \ell \le b\}. \end{aligned}$$

PROOF. Note that for all i with $0 \le i < a$ and for all the above σ_i, we move

$$[\cdots[\mathfrak{g}, \mathfrak{a}_{-n_{\sigma_i(1)}}(\beta_{\sigma_i(1)})], \cdots], \mathfrak{a}_{-n_{\sigma_i(i)}}(\beta_{\sigma_i(i)})]$$

all the way to the right. This produces (4.20). In doing so, we obtain (4.21) by repeatedly applying the fact that

$$\mathfrak{g}_1\mathfrak{g}_2 = [\mathfrak{g}_1, \mathfrak{g}_2] + (-1)^{s_1 s_2}\mathfrak{g}_2\mathfrak{g}_1$$

for two operators $\mathfrak{g}_1, \mathfrak{g}_2 \in \mathrm{End}(\mathbb{H}_X)$ of bi-degrees (ℓ_1, s_1), (ℓ_2, s_2) respectively. □

THEOREM 4.10. *Let $n \ge 1$, $k \ge 0$, and $\alpha \in H^*(X)$. Then, $G_k(\alpha, n)$ is equal to*

$$\sum_{0 \le j \le k} \sum_{\substack{\lambda \vdash (j+1) \\ \ell(\lambda) = k-j+1}} \frac{(-1)^{|\lambda|-1}}{\lambda^! \cdot |\lambda|!} \cdot \mathbf{1}_{-(n-j-1)}\mathfrak{a}_{-\lambda}(\tau_*\alpha)|0\rangle$$
$$+ \sum_{0 \le j \le k} \sum_{\substack{\lambda \vdash (j+1) \\ \ell(\lambda) = k-j-1}} \frac{(-1)^{|\lambda|}}{\lambda^! \cdot |\lambda|!} \cdot \frac{|\lambda| + s(\lambda) - 2}{24} \cdot \mathbf{1}_{-(n-j-1)}\mathfrak{a}_{-\lambda}(\tau_*(e_X\alpha))|0\rangle$$
$$+ \sum_{\substack{\epsilon \in \{K_X, K_X^2\} \\ 0 \le j \le k}} \sum_{\substack{\lambda \vdash (j+1) \\ \ell(\lambda) = k-j+1-|\epsilon|/2}} \frac{(-1)^{|\lambda|} g_\epsilon(\lambda + (1^{j+1}))}{\lambda^! \cdot |\lambda|!} \cdot \mathbf{1}_{-(n-j-1)}\mathfrak{a}_{-\lambda}(\tau_*(\epsilon\alpha))|0\rangle$$

where the universal function g_ϵ is from Theorem 4.7, and $\lambda + (1^{j+1})$ is the partition obtained from λ by adding $(j+1)$ to the multiplicity of 1.

PROOF. By (3.21),

$$1_{X^{[n]}} = \frac{1}{n!} \cdot \mathfrak{a}_{-1}(1_X)^n|0\rangle.$$

From the definition of Chern character operators, we see that $\mathfrak{G}_k(\alpha)|0\rangle = 0$ and $G_k(\alpha, n) = \mathfrak{G}_k(\alpha)1_{X^{[n]}}$. Also, by Theorem 4.7 and Lemma 3.12 (i), we have

$$[\ldots[\mathfrak{G}_k(\alpha), \mathfrak{a}_{n_1}(\alpha_1)], \ldots], \mathfrak{a}_{n_{k+2}}(\alpha_{k+2})] = 0$$

whenever $n_1, \ldots, n_{k+2} < 0$. Combining all of these observations, we obtain

$$\begin{aligned} & G_k(\alpha, n) \\ = & \frac{1}{n!} \cdot \mathfrak{G}_k(\alpha)\mathfrak{a}_{-1}(1_X)^n|0\rangle \\ = & \frac{1}{n!} \cdot \sum_{u=1}^{k+1} \binom{n}{u} \mathfrak{a}_{-1}(1_X)^{n-u}[\cdots[\mathfrak{G}_k(\alpha), \underbrace{\mathfrak{a}_{-1}(1_X)], \cdots], \mathfrak{a}_{-1}(1_X)]}_{u \text{ times}}|0\rangle \\ = & \sum_{j=0}^{k} \frac{1}{(j+1)!} \cdot \mathbf{1}_{-(n-j-1)}[\cdots[\mathfrak{G}_k(\alpha), \underbrace{\mathfrak{a}_{-1}(1_X)], \cdots], \mathfrak{a}_{-1}(1_X)]}_{j+1 \text{ times}}|0\rangle. \end{aligned}$$

Now our theorem follows immediately from Theorem 4.7. □

REMARK 4.11. Fix a cohomology class $\gamma \in H^*(X)$. Then we have

$$G_0(\gamma, n) = \mathbf{1}_{-(n-1)}\mathfrak{a}_{-1}(\gamma)|0\rangle \tag{4.22}$$

by Theorem 4.10. Using Proposition 4.8, we obtain

$$\begin{aligned} G_1(\gamma, n) &= \mathfrak{G}_1(\gamma)1_{X^{[n]}} \\ &= \mathfrak{G}_1(\gamma)\frac{1}{n!}\mathfrak{a}_{-1}(1_X)^n|0\rangle \\ &= -\frac{1}{2} \cdot \mathbf{1}_{-(n-2)}\mathfrak{a}_{-2}(\gamma)|0\rangle. \end{aligned} \tag{4.23}$$

Next, we will study the cup product of the cohomology classes $G_k(\alpha, n)$. Our goal is to express the cup product

$$\prod_{i=1}^{s} G_{k_i}(\alpha_i, n)$$

as a universal linear combination of the Heisenberg monomial classes.

DEFINITION 4.12. Let $s \geq 1$ and $\alpha_1, \ldots, \alpha_s \in H^*(X)$ be homogeneous. For a partition $\pi = \{\pi_1, \ldots, \pi_j\}$ of the set $\{1, \ldots, s\}$, we fix the orders of the elements listed in each subset π_i $(1 \leq i \leq j)$ once and for all, and define $\ell(\pi) = j$, $\alpha_{\pi_i} = \prod_{m \in \pi_i} \alpha_m$, and $\text{sign}(\alpha, \pi)$ by

$$\prod_{i=1}^{j} \alpha_{\pi_i} = \text{sign}(\alpha, \pi) \cdot \prod_{i=1}^{s} \alpha_i.$$

LEMMA 4.13. *Let $n \geq 1, s \geq 0$, $k_1, \ldots, k_s \geq 0$, and $\alpha_1, \ldots, \alpha_s \in H^*(X)$ be homogeneous. Then, $\prod_{i=1}^{s} G_{k_i}(\alpha_i, n)$ is a finite linear combination of expressions:*

$$\text{sign}(\alpha, \pi) \cdot \mathbf{1}_{-\left(n - \sum\limits_{i=1}^{\ell(\pi)} \sum\limits_{j=1}^{m_i - r_i} n_{i,j}\right)} \prod_{\substack{1 \leq i \leq \ell(\pi) \\ 1 \leq j \leq m_i - r_i}} \mathfrak{a}_{-n_{i,j}}(\tau_{(m_i - r_i)*}(\epsilon_i \alpha_{\pi_i})) \cdot |0\rangle \tag{4.24}$$

whose coefficients are independent of X, $\alpha_1, \ldots, \alpha_s$, and the integer n. Here π runs over all partitions of $\{1, \ldots, s\}$, $\epsilon_i \in \{1_X, K_X, K_X^2, e_X\}$,

$$r_i = |\epsilon_i|/2 \leq m_i \leq 2 + \sum_{j \in \pi_i} k_j,$$

$0 < n_{i,1} \le \ldots \le n_{i,m_i - r_i}$, $\sum_{j=1}^{m_i - r_i} n_{i,j} \le \sum_{j \in \pi_i} (k_j + 1)$ *for each* i, *and*

$$\sum_{i=1}^{\ell(\pi)} \left(m_i - 2 + \sum_{j=1}^{m_i - r_i} n_{i,j} \right) = \sum_{i=1}^{s} k_i. \tag{4.25}$$

PROOF. Use induction on s. When $s = 1$, the statement is true by Theorem 4.10. Next, let $s \ge 2$. By induction, $\prod_{i=2}^{s} G_{k_i}(\alpha_i, n)$ is a linear combination of expressions:

$$\begin{aligned} & \operatorname{sign}(\alpha, \sigma) \cdot \mathbf{1}_{-(n-\tilde{n})} \left(\prod_{i=1}^{\ell(\sigma)} \left(\prod_{j=1}^{m_i - r_i} \mathfrak{a}_{-n_{i,j}} \right) (\tau_{(m_i - r_i)*}(\epsilon_i \alpha_{\sigma_i})) \right) |0\rangle \\ = \; & \frac{\operatorname{sign}(\alpha, \sigma)}{(n - \tilde{n})!} \cdot \mathfrak{a}_{-1}(1_X)^{n-\tilde{n}} \left(\prod_{i=1}^{\ell(\sigma)} \left(\prod_{j=1}^{m_i - r_i} \mathfrak{a}_{-n_{i,j}} \right) (\tau_{(m_i - r_i)*}(\epsilon_i \alpha_{\sigma_i})) \right) |0\rangle \end{aligned}$$

where σ runs over all partitions of $\{2, \ldots, s\}$, $\epsilon_i \in \{1_X, K_X, K_X^2, e_X\}$,

$$r_i = |\epsilon_i|/2 \le m_i \le 2 + \sum_{j \in \sigma_i} k_j,$$

$0 < n_{i,1} \le \ldots \le n_{i,m_i - r_i}$, $\sum_{j=1}^{m_i - r_i} n_{i,j} \le \sum_{j \in \sigma_i} (k_j + 1)$, and $\tilde{n} = \sum_{i=1}^{\ell(\sigma)} \sum_{j=1}^{m_i - r_i} n_{i,j}$. Moreover, the coefficients in the linear combination are independent of $X, \alpha_2, \ldots, \alpha_s$ and n. Now apply $\mathfrak{G}_{k_1}(\alpha_1)$ to $\prod_{i=2}^{s} G_{k_i}(\alpha_i, n)$, and move $\mathfrak{G}_{k_1}(\alpha_1)$ to the right by using Lemma 4.9. Note that $|\tau_{(m_i - r_i)*}(\epsilon_i \alpha_{\sigma_i})| \equiv |\alpha_{\sigma_i}| \pmod 2$. By Theorem 4.7,

$$[\ldots [\mathfrak{G}_{k_1}(\alpha_1), \mathfrak{a}_{\ell_1}(\beta_1)], \ldots], \mathfrak{a}_{\ell_{k_1+2}}(\beta_{k_1+2})] = 0$$

when $\ell_1, \ldots, \ell_{k_1+2} < 0$. Since $\mathfrak{G}_{k_1}(\alpha_1)|0\rangle = 0$ and

$$\prod_{i=1}^{s} G_{k_i}(\alpha_i, n) = \mathfrak{G}_{k_1}(\alpha_1) \left(\prod_{i=2}^{s} G_{k_i}(\alpha_i, n) \right),$$

we see from Lemma 4.9 that $\prod_{i=1}^{s} G_{k_i}(\alpha_i, n)$ is a universal linear combination of expressions:

$$\begin{aligned}(4.26)\qquad &\frac{\mathrm{sign}(\alpha,\sigma)}{(n-\tilde{n})!}\binom{n-\tilde{n}}{t}\cdot(-1)^{|\alpha_1|\sum\limits_{1\le i\le \ell(\sigma), i\notin U}|\alpha_{\sigma_i}|+\sum\limits_{v=1}^{u}\sum\limits_{w>i_v, w\notin U}|\alpha_{\sigma_{i_v}}||\alpha_{\sigma_w}|}\cdot\\ &\cdot\mathfrak{a}_{-1}(1_X)^{n-\tilde{n}-t}\left(\prod_{\substack{1\le i\le \ell(\sigma)\\ i\notin U}}\left(\prod_{j=1}^{m_i-r_i}\mathfrak{a}_{-n_{i,j}}\right)(\tau_{(m_i-r_i)*}(\epsilon_i\alpha_{\sigma_i}))\right)\cdot\\ &\cdot\Big[\cdots[\underbrace{\mathfrak{G}_{k_1}(\alpha_1),\mathfrak{a}_{-1}(1_X)],\cdots],\mathfrak{a}_{-1}(1_X)]}_{t\text{ times}},\prod_{j=1}^{m_{i_1}-r_{i_1}}\mathfrak{a}_{-n_{i_1,j}}(\tau_{(m_{i_1}-r_{i_1})*}(\epsilon_{i_1}\alpha_{\sigma_{i_1}}))],\\ &\cdots],\left(\prod_{j=1}^{m_{i_u}-r_{i_u}}\mathfrak{a}_{-n_{i_u,j}}\right)(\tau_{(m_{i_u}-r_{i_u})*}(\epsilon_{i_u}\alpha_{\sigma_{i_u}}))\Big]\ |0\rangle\end{aligned}$$

where $0\le t\le (k_1+1), u\ge 0$, $(t+u)\ge 1$,

$$U=\{i_1,\ldots,i_u\}\subset\{1,\ldots,\ell(\sigma)\}$$

with $i_1<\ldots<i_u$. Let π be the partition of $\{1,\ldots,s\}$ consisting of all the σ_i with $1\le i\le \ell(\sigma)$ and $i\notin U$, and $\{1\}\coprod\left(\coprod_{i\in U}\sigma_i\right)$. Then, $\alpha_{\pi_{\ell(\pi)}}=\alpha_1\alpha_{\sigma_{i_1}}\cdots\alpha_{\sigma_{i_u}}$ and

$$(4.27)\qquad \mathrm{sign}(\alpha,\pi)=\mathrm{sign}(\alpha,\sigma)\cdot(-1)^{|\alpha_1|\sum\limits_{1\le i\le \ell(\sigma), i\notin U}|\alpha_{\sigma_i}|+\sum\limits_{v=1}^{u}\sum\limits_{w>i_v, w\notin U}|\alpha_{\sigma_{i_v}}||\alpha_{\sigma_w}|}.$$

In view of (4.26) and (4.27), $\prod_{i=1}^{s} G_{k_i}(\alpha_i, n)$ is a linear combination of

$$\begin{aligned}(4.28)\qquad &\mathrm{sign}(\alpha,\pi)\cdot\mathbf{1}_{-(n-\tilde{n}-t)}\left(\prod_{1\le i<\ell(\pi)}\left(\prod_{j=1}^{m_i-r_i}\mathfrak{a}_{-n_{i,j}}\right)(\tau_{(m_i-r_i)*}(\epsilon_i\alpha_{\pi_i}))\right)\cdot\\ &\cdot\Big[\cdots[\underbrace{\mathfrak{G}_{k_1}(\alpha_1),\mathfrak{a}_{-1}(1_X)],\cdots],\mathfrak{a}_{-1}(1_X)]}_{t\text{ times}},\prod_{j=1}^{m_{i_1}-r_{i_1}}\mathfrak{a}_{-n_{i_1,j}}(\tau_{(m_{i_1}-r_{i_1})*}(\epsilon_{i_1}\alpha_{\sigma_{i_1}}))],\\ &\cdots],\left(\prod_{j=1}^{m_{i_u}-r_{i_u}}\mathfrak{a}_{-n_{i_u,j}}\right)(\tau_{(m_{i_u}-r_{i_u})*}(\epsilon_{i_u}\alpha_{\sigma_{i_u}}))\Big]\ |0\rangle.\end{aligned}$$

with all the coefficients being independent of $X,\alpha_1,\ldots,\alpha_s$ and n. Also notice that in the expression (4.28), the only factor depending on n is $\mathbf{1}_{-(n-\tilde{n}-t)}$.

By Theorem 4.7 and Lemma 3.12, (4.28) is a universal linear combination of

$$\begin{aligned}
&\mathrm{sign}(\alpha,\pi)\cdot \mathbf{1}_{-(n-\tilde{n}-t)}\left(\prod_{1\le i<\ell(\pi)}\left(\prod_{j=1}^{m_i-r_i}\mathfrak{a}_{-n_{i,j}}\right)(\tau_{(m_i-r_i)*}(\epsilon_i\alpha_{\pi_i}))\right)\cdot\\
&\cdot\mathfrak{a}_{-n_1}\cdots\mathfrak{a}_{-n_{m-r}}(\tau_{(m-r)*}(\epsilon\alpha_1\alpha_{\sigma_{i_1}}\cdots\alpha_{\sigma_{i_u}}))|0\rangle\\
= \; &\mathrm{sign}(\alpha,\pi)\cdot \mathbf{1}_{-(n-\tilde{n}-t)}\left(\prod_{1\le i<\ell(\pi)}\left(\prod_{j=1}^{m_i-r_i}\mathfrak{a}_{-n_{i,j}}\right)(\tau_{(m_i-r_i)*}(\epsilon_i\alpha_{\pi_i}))\right)\cdot\\
&\cdot\mathfrak{a}_{-n_1}\cdots\mathfrak{a}_{-n_{m-r}}(\tau_{(m-r)*}(\epsilon\alpha_{\pi_{\ell(\pi)}}))|0\rangle
\end{aligned}$$

which is of the form (4.24). Here

$$m = (k_1 - t + 2) + m_{i_1} + \ldots + m_{i_u} - 2u \le 2 + \sum_{j\in\pi_{\ell(\pi)}} k_j,$$

$\epsilon \in \{1_X, K_X, K_X^2, e_X\}$, $r = |\epsilon|/2 \le m$, $0 < n_1 \le \ldots \le n_{m-r}$, and

$$n_1 + \ldots + n_{m-r} = t + \sum_{i\in U}\sum_{j=1}^{m_i-r_i} n_{i,j} \le \sum_{j\in\pi_{\ell(\pi)}} (k_j+1)$$

since $t \le (k_1+1)$.

Finally, the cohomology degree of $\prod_{i=1}^{s} G_{k_i}(\alpha_i, n)$ is equal to $\sum_{i=1}^{s}(2k_i + |\alpha_i|)$. Comparing this with the cohomology degree of (4.24), we obtain (4.25). □

The following lemma determines the leading term in the cup product

$$\prod_{i=1}^{s} G_{k_i}(\alpha_i, n).$$

LEMMA 4.14. *Let notations be the same as in Lemma 4.13, and* $n_0 = \sum_{i=1}^{s}(k_i+1)$.

(i) *If an expression from (4.24) satisfies* $\sum_{i=1}^{\ell(\pi)}\sum_{j=1}^{m_i-r_i} n_{i,j} = n_0$, *then it is equal to*

$$\mathbf{1}_{-(n-n_0)}\prod_{i=1}^{s}\mathfrak{a}_{-(k_i+1)}(\alpha_i)\cdot|0\rangle.$$

(ii) *The coefficient of* $\mathbf{1}_{-(n-n_0)}\prod_{i=1}^{s}\mathfrak{a}_{-(k_i+1)}(\alpha_i)\cdot|0\rangle$ *in* $\prod_{i=1}^{s} G_{k_i}(\alpha_i,n)$ *is*

$$\prod_{i=1}^{s}\frac{(-1)^{k_i}}{(k_i+1)!}.$$

PROOF. (i) We may let $s \ge 1$. Since

$$\sum_{i=1}^{\ell(\pi)}\sum_{j=1}^{m_i-r_i} n_{i,j} = \sum_{i=1}^{s}(k_i+1),$$

we see from Lemma 4.13 that

$$\sum_{j=1}^{m_i-r_i} n_{i,j} = \sum_{j\in\pi_i}(k_j+1)$$

for every i. By (4.25), we obtain

$$\sum_{i=1}^{\ell(\pi)}(m_i - 2 + |\pi_i|) = 0 \tag{4.29}$$

where $|\pi_i|$ stands for the number of elements in the subset π_i. Note that for every i with $1 \le i \le \ell(\pi)$, we have $(m_i - r_i) \ge 1$ since

$$\sum_{j=1}^{m_i-r_i} n_{i,j} = \sum_{j\in\pi_i}(k_j+1) \ge 1.$$

So for every i, $m_i \ge 1$ and $r_i = 0$ if $m_i = 1$. By (4.29), $m_i = |\pi_i| = 1$ for $1 \le i \le \ell(\pi)$. Thus for $1 \le i \le \ell(\pi)$, we have $r_i = 0$, $\epsilon_i = 1_X$, and $n_{i,1} = (k_j+1)$ if $\pi_i = \{j\}$.

Now let $\pi_i = \{t_i\}$ for $1 \le i \le \ell(\pi) = s$. Then the expression (4.24) is

$$\begin{aligned} &\operatorname{sign}(\alpha,\pi)\cdot \mathbf{1}_{-(n-n_0)}\left(\prod_{i=1}^{s}\mathfrak{a}_{-(k_{t_i}+1)}(\alpha_{t_i})\right)|0\rangle \\ = \ &\mathbf{1}_{-(n-n_0)}\left(\prod_{i=1}^{s}\mathfrak{a}_{-(k_i+1)}(\alpha_i)\right)|0\rangle. \end{aligned}$$

(ii) The idea is to use induction on s and track the proof of Lemma 4.13 more carefully. When $s = 1$, the statement is true by Theorem 4.10. Next, let $s \ge 2$ and $\tilde{n}_0 = \sum_{j=2}^{s}(k_j+1)$. Assume that the coefficient of $\mathbf{1}_{-(n-\tilde{n}_0)}\left(\prod_{i=2}^{s}\mathfrak{a}_{-n_i}(\alpha_i)\right)|0\rangle$ in the cup product $\prod_{i=2}^{s} G_{k_i}(\alpha_i,n)$ is equal to $\prod_{i=2}^{s}\frac{(-1)^{k_i}}{(k_i+1)!}$. Tracking the proof of Lemma 4.13 and applying Theorem 4.7, we conclude that the coefficient of $\mathbf{1}_{-(n-n_0)}\left(\prod_{i=1}^{s}\mathfrak{a}_{-(k_i+1)}(\alpha_i)\right)|0\rangle$ in $\prod_{i=1}^{s} G_{k_i}(\alpha_i,n)$ is equal to $\prod_{i=1}^{s}\frac{(-1)^{k_i}}{(k_i+1)!}$. □

We now apply Lemma 4.13 to study certain intersection numbers in the Hilbert scheme $X^{[n]}$. For this purpose, we establish the notation

$$\langle \omega_1,\ldots,\omega_k\rangle = \int_Y \omega_1\cdots\omega_k. \tag{4.30}$$

where $\omega_1,\ldots,\omega_k \in H^*(Y)$ and Y stands for a smooth projective variety.

THEOREM 4.15. *Let $n, s \ge 1$, $k_1,\ldots,k_s \ge 0$, and let $\alpha_1,\ldots,\alpha_s \in H^*(X)$ be homogeneous cohomology classes. Assume that $\sum_{i=1}^{s}(2k_i + |\alpha_i|) = 4n$. Then,*

$$\left\langle \prod_{i=1}^{s} G_{k_i}(\alpha_i,n)\right\rangle$$

is a finite linear combination of $\operatorname{sign}(\alpha,\pi)\cdot\prod_{i=1}^{\ell(\pi)}\langle\epsilon_i\alpha_{\pi_i}\rangle$ *where* π *runs over all partitions of* $\{1,\ldots,s\}$, $\epsilon_i\in\{1_X,K_X,K_X^2,e_X\}$. *Moreover, all the coefficients in this linear combination are independent of* X, $\alpha_1,\ldots,\alpha_s$ *and* n.

PROOF. Note that $\mathfrak{a}_{-1}(x)^n|0\rangle$ is a generator of $H^{4n}(X^{[n]})\cong\mathbb{C}$. So in view of Lemma 4.13, an expression (4.24) nontrivially contributing to $\left\langle\prod_{i=1}^{s}G_{k_i}(\alpha_i,n)\right\rangle$ must satisfy:

(i) $n_{i,j}=1$ for all $1\le i\le\ell(\pi)$ and $1\le j\le m_i-r_i$,
(ii) $\epsilon_i\alpha_{\pi_i}=\langle\epsilon_i\alpha_{\pi_i}\rangle\cdot x$ for all $1\le i\le\ell(\pi)$, and
(iii) $n-\sum_{i=1}^{\ell(\pi)}(m_i-r_i)=0$.

Since

$$\tau_{k*}(x)=\underbrace{x\otimes\cdots\otimes x}_{k\,\text{times}}$$

for all $k\ge 0$, our conclusion follows immediately from (4.24). □

REMARK 4.16. Assume that $\sum_{i=1}^{s}(k_i+2)=2n$. By Theorem 4.2,

$$[\mathfrak{G}_{k_i}(x),\mathfrak{a}_{-1}(x)]=0$$

for all $1\le i\le s$. Using Theorem 4.7 and Theorem 4.10 repeatedly, we see that

$$\left\langle\prod_{i=1}^{s}G_{k_i}(x,n)\right\rangle=\sum_{\substack{0\le j_1\le k_1,\ldots,0\le j_s\le k_s\\(j_1+1)+\ldots+(j_s+1)=n}}\prod_{i=1}^{s}\sum_{\substack{\lambda_i\vdash(j_i+1)\\ \ell(\lambda_i)=k_i-j_i+1}}\frac{(-1)^{|\lambda_i|-1}}{(\lambda_i)^!|\lambda_i|!}.$$

Note that this number is independent of the projective surface X.

4.3. Characteristic classes of tautological bundles

A basic and important class of sheaves over the Hilbert scheme $X^{[n]}$ consists of the tautological bundles defined by (1.25). This section is devoted to the characteristic classes of these tautological bundles. We will state Lehn's formula for the generating series of the total Chern class of the rank-n tautological bundle $L^{[n]}$ where L denotes a line bundle on X. Moreover, we will prove its generalization by inserting a formal variable $\hbar$.

To begin with, for a smooth quasi-projective complex variety Y, let $K(Y)$ be the Grothendieck group generated (over $\mathbb{C}$) by locally free sheaves, or equivalently by arbitrary coherent sheaves.

Let X be a smooth projective complex surface. Recall from (1.25) that if F is a rank-r locally free sheaf on X, then $F^{[n]}$ is a rank-rn locally free sheaf on $X^{[n]}$. If

$$0\to F_1\to F_2\to F_3\to 0$$

is an exact sequence of locally free sheaves on X, then we have an exact sequence

$$0\to(F_1)^{[n]}\to(F_2)^{[n]}\to(F_3)^{[n]}\to 0$$

of locally free sheaves on $X^{[n]}$. Therefore, we obtain a well-defined group homomorphism:

$$-^{[n]}: \quad K(X) \to K(X^{[n]}).$$

DEFINITION 4.17. Let $u \in K(X)$. Define $\mathfrak{c}(u)$ and $\mathfrak{ch}(u) \in \mathrm{End}(\mathbb{H}_X)$ to be the linear operators which act on the n-th component $H^*(X^{[n]})$ by cup product with the total Chern class $c(u^{[n]})$ and the Chern character $\mathrm{ch}(u^{[n]})$ respectively. Define $\mathfrak{c}_i(u)$ and $\mathfrak{ch}_i(u) \in \mathrm{End}(\mathbb{H}_X)$ to be the i-th components of $\mathfrak{c}(u)$ and $\mathfrak{ch}(u)$ corresponding to the cup products by $c_i(u^{[n]})$ and $\mathrm{ch}_i(u^{[n]})$ respectively.

The definition above and theorem below are from [**Leh1**, Section 4]. Definition 4.1 and Theorem 4.2 are the generalizations of Definition 4.17 and (4.31) respectively.

THEOREM 4.18.

(i) *Let $u \in K(X)$ be a class of rank r, and $\alpha \in H^*(X)$. Then,*

$$[\mathfrak{ch}(u), \mathfrak{a}_{-1}(\alpha)] = \exp(\mathrm{ad}(\mathfrak{d}))(\mathfrak{a}_{-1}(\mathrm{ch}(u)\alpha)), \tag{4.31}$$

$$\mathfrak{c}(u) \cdot \mathfrak{a}_{-1}(\alpha) \cdot \mathfrak{c}(u)^{-1} = \sum_{v,k \ge 0} \binom{r-k}{v} \mathfrak{a}_{-1}^{(v)}(c_k(u)\alpha). \tag{4.32}$$

(ii) *Let $\mathfrak{C}(u)$ be the right-hand-side of (4.32) with $\alpha = 1_X$. Then,*

$$\sum_{n=0}^{+\infty} c(u^{[n]}) = \exp(\mathfrak{C}(u)) \cdot |0\rangle. \tag{4.33}$$

In particular, for a line bundle L on X, we have

$$\sum_{n=0}^{+\infty} c(L^{[n]}) = \exp\left(\sum_{r\ge 1} \frac{(-1)^{r-1}}{r} \mathfrak{a}_{-r}(c(L))\right) \cdot |0\rangle. \tag{4.34}$$

PROOF. By the Grothendieck-Riemann-Roch Theorem and (4.1),

$$G(\mathrm{ch}(u), n) = \mathrm{ch}(u^{[n]}).$$

So $\mathfrak{G}(\mathrm{ch}(u)) = \mathfrak{ch}(u)$, and (4.31) follows from (4.6). The proof of (4.32) is similar to the proof of Theorem 4.2. Next, recall that $1_{X^{[n]}} = 1/n! \cdot \mathfrak{a}_{-1}(1_X)^n|0\rangle$. Thus,

$$\begin{aligned}
\sum_{n=0}^{+\infty} c(u^{[n]}) &= \mathfrak{c}(u) \sum_{n=0}^{+\infty} 1_{X^{[n]}} \\
&= \mathfrak{c}(u) \exp\big(\mathfrak{a}_{-1}(1_X)\big) \cdot |0\rangle \\
&= \mathfrak{c}(u) \exp\big(\mathfrak{a}_{-1}(1_X)\big)\mathfrak{c}(u)^{-1} \cdot |0\rangle \\
&= \exp\big(\mathfrak{c}(u)\mathfrak{a}_{-1}(1_X)\mathfrak{c}(u)^{-1}\big) \cdot |0\rangle \\
&= \exp(\mathfrak{C}(u)) \cdot |0\rangle
\end{aligned}$$

where we have used (4.32) in the last step. This proves (4.33). Formula (4.34) and its proof will be generalized in Proposition 4.19 and its proof below. □

We remark that (4.34) was conjectured by Göttsche. Assume that C is a smooth curve in the surface X. Then, $C^{[n]}$ is an n-dimensional closed subscheme of $X^{[n]}$. Let $L = \mathcal{O}_X(C)$. By (1.28),

$$[C^{[n]}] = c_n(L^{[n]}).$$

Reading the top Chern classes $c_n(L^{[n]})$ from (4.34), we recover the following formula of Nakajima [**Nak5**, Formula (9.16)]:

$$\sum_{n=0}^{+\infty} [C^{[n]}] = \exp\left(\sum_{r\geq 1} \frac{(-1)^{r-1}}{r} \mathfrak{a}_{-r}([C])\right) \cdot |0\rangle. \tag{4.35}$$

PROPOSITION 4.19. *Let $\hbar$ and w be two formal variables, and let*

$$c_\hbar(L^{[n]}) = \sum_{i=0}^{n} c_i(L^{[n]})\hbar^i$$

where L denotes a line bundle on the surface X. Then,

$$\sum_{n=0}^{+\infty} c_\hbar(L^{[n]})w^n = \exp\left(\sum_{r\geq 1} \frac{(-\hbar)^{r-1}}{r} \mathfrak{a}_{-r}(c_\hbar(L))w^r\right) \cdot |0\rangle. \tag{4.36}$$

PROOF. We modify the proof of (4.34) in [**Leh1**] by inserting the formal variable $\hbar$ in suitable positions. First of all, for a class $u \in K(X)$, put

$$\mathfrak{c}_\hbar(u) = \sum_i \mathfrak{c}_i(u)\hbar^i.$$

Then the modification of (4.32) is

$$\mathfrak{c}_\hbar(u) \cdot \mathfrak{a}_{-1}(\alpha) \cdot \mathfrak{c}_\hbar(u)^{-1} = \sum_{v,k\geq 0} \binom{r-k}{v} \mathfrak{a}_{-1}^{(v)}(c_k(u)\alpha)\hbar^{v+k}. \tag{4.37}$$

where r denotes the rank of u. In particular,

$$\mathfrak{c}_\hbar(L) \cdot \mathfrak{a}_{-1}(1_X) \cdot \mathfrak{c}_\hbar(L)^{-1} = \mathfrak{C}_\hbar(L) \tag{4.38}$$

where

$$\mathfrak{C}_\hbar(L) = \mathfrak{a}_{-1}(c_\hbar(L)) + \mathfrak{a}'_{-1}(1_X)\hbar.$$

As in the proof of (4.33), we obtain

$$\sum_{n=0}^{+\infty} c_\hbar(L^{[n]})w^n = \mathfrak{c}_\hbar(L)\exp\left(\mathfrak{a}_{-1}(1_X)w\right) \cdot |0\rangle. \tag{4.39}$$

Since $\mathfrak{c}_\hbar(L) \cdot \mathfrak{a}_{-1}(1_X) = \mathfrak{C}_\hbar(L) \cdot \mathfrak{c}_\hbar(L)$, we have

$$\begin{aligned}
&\frac{\mathrm{d}}{\mathrm{d}w}\left(\sum_{n=0}^{+\infty} c_\hbar(L^{[n]})w^n\right) \\
= \; & \frac{\mathrm{d}}{\mathrm{d}w}\left(\mathfrak{c}_\hbar(L)\exp\left(\mathfrak{a}_{-1}(1_X)w\right) \cdot |0\rangle\right) \\
= \; & \mathfrak{c}_\hbar(L) \cdot \mathfrak{a}_{-1}(1_X) \cdot \exp\left(\mathfrak{a}_{-1}(1_X)w\right) \cdot |0\rangle \\
= \; & \mathfrak{C}_\hbar(L) \cdot \mathfrak{c}_\hbar(L) \cdot \exp\left(\mathfrak{a}_{-1}(1_X)w\right) \cdot |0\rangle \\
= \; & \mathfrak{C}_\hbar(L) \cdot \sum_{n=0}^{+\infty} c_\hbar(L^{[n]})w^n.
\end{aligned}$$

So (4.39) is the solution to the initial-value-problem:

$$\frac{\mathrm{d}}{\mathrm{d}w}\mathfrak{X} = \mathfrak{C}_\hbar(L) \cdot \mathfrak{X}, \quad \mathfrak{X}(0) = |0\rangle. \tag{4.40}$$

Next, define two operators:

$$\begin{aligned} \mathfrak{R}_\hbar(w) &= \sum_{r\geq 1} \frac{(-\hbar)^{r-1}}{r}\mathfrak{a}_{-r}(c_\hbar(L))w^r, \\ \mathfrak{S}_\hbar(w) &= \exp\left(\sum_{r\geq 1} \frac{(-\hbar)^{r-1}}{r}\mathfrak{a}_{-r}(c_\hbar(L))w^r\right) = \exp\left(\mathfrak{R}_\hbar(w)\right). \end{aligned}$$

Then, $\mathfrak{S}_\hbar(0)|0\rangle = |0\rangle$, and

$$\frac{\mathrm{d}}{\mathrm{d}w}\mathfrak{S}_\hbar(w)|0\rangle = \mathfrak{S}_\hbar(w)\cdot\sum_{r\geq 0}(-\hbar)^r\mathfrak{a}_{-r-1}(c_\hbar(L))w^r\cdot|0\rangle. \tag{4.41}$$

Moreover, by Theorem 3.24 (i) and (ii),

$$\begin{aligned} [\mathfrak{a}'_{-1}(1_X)\hbar, \mathfrak{R}_\hbar(w)] &= -\hbar[\mathfrak{L}_{-1}(1_X), \mathfrak{R}_\hbar(w)] \\ &= \sum_{r\geq 1}(-\hbar)^r\mathfrak{a}_{-r-1}(c_\hbar(L))w^r. \end{aligned} \tag{4.42}$$

In particular, $[\mathfrak{a}'_{-1}(1_X)\hbar, \mathfrak{R}_\hbar(w)]$ commutes with $\mathfrak{R}_\hbar(w)$. It follows that

$$[\mathfrak{a}'_{-1}(1_X)\hbar, \mathfrak{R}_\hbar(w)^m] = m\mathfrak{R}_\hbar(w)^{m-1}[\mathfrak{a}'_{-1}(1_X)\hbar, \mathfrak{R}_\hbar(w)].$$

Therefore, we have

$$\begin{aligned} [\mathfrak{C}_\hbar(L), \mathfrak{S}_\hbar(w)] &= [\mathfrak{a}'_{-1}(1_X)\hbar, \exp\left(\mathfrak{R}_\hbar(w)\right)] \\ &= \exp\left(\mathfrak{R}_\hbar(w)\right)\cdot[\mathfrak{a}'_{-1}(1_X)\hbar, \mathfrak{R}_\hbar(w)] \\ &= \mathfrak{S}_\hbar(w)\cdot\sum_{r\geq 1}(-\hbar)^r\mathfrak{a}_{-r-1}(c_\hbar(L))w^r \end{aligned}$$

where we have used (4.42) in the last step. Combining with $\mathfrak{a}'_{-1}(1_X)|0\rangle = 0$, we obtain

$$\begin{aligned} &\mathfrak{C}_\hbar(L)\mathfrak{S}_\hbar(w)|0\rangle \\ =\ &\mathfrak{S}_\hbar(w)\mathfrak{C}_\hbar(L)|0\rangle + \mathfrak{S}_\hbar(w)\cdot\sum_{r\geq 1}(-\hbar)^r\mathfrak{a}_{-r-1}(c_\hbar(L))w^r\cdot|0\rangle \\ =\ &\mathfrak{S}_\hbar(w)\mathfrak{a}_{-1}(c_\hbar(L))|0\rangle + \mathfrak{S}_\hbar(w)\cdot\sum_{r\geq 1}(-\hbar)^r\mathfrak{a}_{-r-1}(c_\hbar(L))w^r\cdot|0\rangle \\ =\ &\mathfrak{S}_\hbar(w)\cdot\sum_{r\geq 0}(-\hbar)^r\mathfrak{a}_{-r-1}(c_\hbar(L))w^r\cdot|0\rangle \\ =\ &\frac{\mathrm{d}}{\mathrm{d}w}\mathfrak{S}_\hbar(w)|0\rangle \end{aligned}$$

where we have used (4.41) in the last step. Hence, $\mathfrak{S}_\hbar(w)|0\rangle$ is also a solution to (4.40), and is equal to (4.39). □

COROLLARY 4.20. *Let L be a line bundle on the surface X, and let $L^{[n]*}$ be the dual of $L^{[n]}$. Then,*

$$\sum_{n=0}^{+\infty} c(L^{[n]*})w^n = \exp\left(\sum_{r\geq 1}\frac{1}{r}\mathfrak{a}_{-r}(c(L^*))w^r\right)\cdot|0\rangle. \tag{4.43}$$

PROOF. This follows from (4.36) by setting $\hbar = -1$. □

The generating series

$$\sum_{n=0}^{+\infty} c(u^{[n]}), \quad \sum_{n=0}^{+\infty} \mathrm{ch}(u^{[n]})$$

and similar generating series for the tangent bundles $T_{X^{[n]}}$ of the Hilbert schemes $X^{[n]}$ have been further investigated in [**Boi1**, **BN**, **QY**].

4.4. $\mathcal{W}$ algebras and Hilbert schemes

The geometric construction of Lehn's Virasoro algebras acting on the cohomology of the Hilbert schemes of points has been presented in Section 3.5. The Chern character operators constructed in Section 4.1 are vertex operators of higher conformal weights. Another distinguished class of vertex algebras, which are called $\mathcal{W}$ algebras and are higher-spin generalizations of Virasoro algebras, often appears in connection with conformal field theories and with representation theory of affine Kac-Moody Lie algebras [**FB**]. Among $\mathcal{W}$ algebras, a well-known example is given by the so-called $\mathcal{W}_{1+\infty}$ algebra which is the central extension of the Lie algebra of differential operators on the circle [**FKRW**, **Kac**]. In this section, we will study the interplay between $\mathcal{W}$ algebras and the Hilbert schemes $X^{[n]}$. By introducing the linear operators $\mathfrak{J}^p_n(\alpha)$, we will construct geometrically a $\mathcal{W}$ algebra action on the Fock space $\mathbb{H}_X$. The proof of the main result (Theorem 4.25) will be sketched.

We begin with the definition of $\mathcal{W}$ algebras. Let

$$A = \bigoplus_{n=0}^{d} A_n$$

(where $A_d \cong \mathbb{C}$) be a finite dimensional graded complex vector space. We write $|\alpha| = n$ if $\alpha \in A_n$, and define a super structure (i.e. the $\mathbb{Z}_2$-grading) on A by letting

$$A_{\bar{0}} = \bigoplus_{n \in 2\mathbb{Z}_+} A_n,$$
$$A_{\bar{1}} = \bigoplus_{n \in 1+2\mathbb{Z}_+} A_n$$

(here $\mathbb{Z}_+$ stands for the set of all nonnegative integers). We assume that A affords an algebra structure which is super commutative and compatible with the $\mathbb{Z}_2$-grading. We further assume that there exists a linear operator $\mathrm{Tr} : A \to \mathbb{C}$ which is zero on the graded subspace A_n unless n is the top degree. This defines a supercommutative bilinear form $\langle \cdot, \cdot \rangle : A \times A \to \mathbb{C}$ by

$$\langle \alpha, \beta \rangle = \mathrm{Tr}(\alpha\beta).$$

Let t be an indeterminate and let $\partial_t = \frac{d}{dt}$. Let $\mathcal{W}_{as}$ be the associative algebra of regular differential operators on the circle S^1. Denote by $\mathcal{W}(A)_{as}$ the associative superalgebra $\mathcal{W}_{as} \otimes A$. It has a linear basis given by

$$J^p_k(\alpha) = -t^{p+k}(\partial_t)^p \otimes \alpha, \quad p \in \mathbb{Z}_+, k \in \mathbb{Z}, \alpha \in A.$$

A different linear basis of $\mathcal{W}(A)_{as}$ is given by

$$L^p_k(\alpha) = -t^k D^p \otimes \alpha, \quad p \in \mathbb{Z}_+, k \in \mathbb{Z}, \alpha \in A,$$

where $D = t\partial_t$. Note that $f(D)t = tf(D+1)$ for every $f(w) \in \mathbb{C}[w]$. Hence,

$$J_k^p(\alpha) = -t^k D(D-1)\cdots(D-p+1) \otimes \alpha.$$

Let $\mathcal{W}(A)$ denote the Lie superalgebra obtained from the associative superalgebra $\mathcal{W}(A)_{as}$ by taking the usual super bracket of operators:

$$[X \otimes \alpha, Y \otimes \beta] = (XY - YX) \otimes (\alpha\beta).$$

Assuming that α and β are of homogeneous degree, we see that

$$[X \otimes \alpha, Y \otimes \beta] = (-1)^{1+|\alpha|\cdot|\beta|}[Y \otimes \beta, X \otimes \alpha].$$

The commutation relation in $\mathcal{W}(A)$ is given by

$$\begin{aligned} & [t^r f(D) \otimes \alpha, t^s g(D) \otimes \beta] \\ = \; & t^{r+s}\left(f(D+s)g(D) - f(D)g(D+r)\right) \otimes (\alpha\beta). \end{aligned} \tag{4.44}$$

When $A = \mathbb{C}$ (hence $d = 0$), we will simply write the Lie superalgebra $\mathcal{W}(A)$ as $\mathcal{W}$, which is the usual Lie algebra of differential operator on the circle [**FKRW**, **Kac**]. The algebra $\mathcal{W}$ affords a universal central extension which is usually referred to as the $\mathcal{W}_{1+\infty}$ algebra. In $\mathcal{W}_{1+\infty}$, the operators L_k^0 generate a Heisenberg algebra, while the operators L_k^1 generate a Virasoro algebra.

The algebra $\mathcal{W}(A)$ has a natural *weight* filtration

$$\mathcal{W}(A)^0 \subset \mathcal{W}(A)^1 \subset \mathcal{W}(A)^2 \subset \ldots \subset \mathcal{W}(A)$$

by letting the weight of $L_k^p(\alpha)$ be p. We have

$$[\mathcal{W}(A)^p, \mathcal{W}(A)^q] \subset \mathcal{W}(A)^{p+q-1}.$$

Denote the associated graded algebra by $\mathcal{GW}(A)$. The leading term (according to the weight filtration) of $J_k^p(\alpha)$ is just $L_k^p(\alpha)$, and so $J_k^p(\alpha)$ and $L_k^p(\alpha)$ give rise to the same element in the graded algebra $\mathcal{GW}(A)$ which is denoted by $\mathfrak{L}_k^p(\alpha)$. We derive the following commutation relation from (4.44):

$$[\mathfrak{L}_m^p(\alpha), \mathfrak{L}_n^q(\beta)] \;=\; (qm - pn)\cdot \mathfrak{L}_{m+n}^{p+q-1}(\alpha\beta).$$

The $\mathcal{W}$ (super)algebra $\widehat{\mathcal{W}}(A)$ is a central extension of the Lie superalgebra $\mathcal{GW}(A)$ by a one-dimensional center with a specified generator C:

$$0 \longrightarrow \mathbb{C}C \longrightarrow \widehat{\mathcal{W}}(A) \longrightarrow \mathcal{GW}(A) \longrightarrow 0, \tag{4.45}$$

such that the commutators in $\widehat{\mathcal{W}}(A) = \mathcal{GW}(A) + \mathbb{C}C$ are given by:

$$[\mathfrak{L}_m^p(\alpha), \mathfrak{L}_n^q(\beta)] \;=\; \begin{cases} m\delta_{m,-n}\operatorname{Tr}(\alpha\beta)\cdot C, & \text{if } p = q = 0, \\ (qm-pn)\cdot \mathfrak{L}_{m+n}^{p+q-1}(\alpha\beta), & \text{otherwise.} \end{cases} \tag{4.46}$$

In this section, we will be mainly interested in the $\mathcal{W}$ (super)algebra $\widehat{\mathcal{W}}(A)$ when A is (a subring of) the cohomology ring $H^*(X)$ of a projective surface X, with the trace defined by

$$\operatorname{Tr}(\alpha) = -\int_X \alpha$$

for $\alpha \in H^*(X)$. First of all, we introduce the following definitions.

DEFINITION 4.21. Let X be a smooth projective complex surface.

(i) For $p \geq 0$, $n \in \mathbb{Z}$ and $\alpha \in H^*(X)$, define $\mathfrak{J}_n^p(\alpha) \in \mathrm{End}(\mathbb{H}_X)$ to be

$$p! \cdot \left(- \sum_{\ell(\lambda)=p+1, |\lambda|=n} \frac{1}{\lambda^!} \mathfrak{a}_\lambda(\tau_* \alpha) + \sum_{\ell(\lambda)=p-1, |\lambda|=n} \frac{s(\lambda)+n^2-2}{24\lambda^!} \mathfrak{a}_\lambda(\tau_*(e_X \alpha)) \right);$$

(ii) We define $\mathcal{W}_X$ to be the linear span of the identity operator $\mathrm{Id}_{\mathbb{H}_X}$ and the operators $\mathfrak{J}_n^p(\alpha)$ in $\mathrm{End}(\mathbb{H}_X)$, where $p \geq 0, n \in \mathbb{Z}$ and $\alpha \in H^*(X)$.

Some of the operators $\mathfrak{J}_n^p(\alpha)$ can be identified with the familiar ones, by using the above definition, Theorem 4.7 and Theorem 3.31. For example, for $\alpha \in H^*(X)$, we see that

$$\begin{aligned} \mathfrak{J}_n^0(\alpha) &= -\mathfrak{a}_n(\alpha), \\ \mathfrak{J}_n^1(\alpha) &= \mathfrak{L}_n(\alpha). \end{aligned}$$

For $\alpha \in \mathfrak{A}_X$, we obtain

$$\begin{aligned} \mathfrak{J}_0^p(\alpha) &= p! \cdot \mathfrak{G}_{p-1}(\alpha), \\ \mathfrak{J}_{-1}^p(\alpha) &= -\mathfrak{a}_{-1}^{(p)}(\alpha). \end{aligned}$$

In general, we have the following.

LEMMA 4.22. *For $p \geq 0$, $\alpha \in \mathfrak{A}_X$ and $\beta \in H^*(X)$, the following identity holds:*

$$[\mathfrak{G}_p(\alpha), \mathfrak{a}_n(\beta)] = \frac{n}{p!} \cdot \mathfrak{J}_n^p(\alpha\beta).$$

PROOF. Follows from Definition 4.21 (i), Theorem 4.7, and Lemma 3.12 (i). $\square$

By Lemma 4.22, we may regard the linear operators $\mathfrak{J}_n^p(\alpha) \in \mathrm{End}(\mathbb{H}_X)$ as being geometric since both the Chern character operators $\mathfrak{G}_p(\alpha)$ and the Heisenberg operators $\mathfrak{a}_n(\beta)$ are geometric by their constructions in Section 4.1 and Section 3.2 respectively.

We are interested in the commutation relation among the operators $\mathfrak{J}_m^p(\alpha)$. To this end, it is useful to adopt the vertex algebra language [**Bor**, **FB**, **Kac**] for our present setup. Our convention for *fields* is to write them in a form

$$\phi(z) = \sum_n \phi_n z^{-n-\Delta}$$

where Δ is the conformal weight of the field $\phi(z)$. Define

$$\partial\phi(z) = \sum_n (-n-\Delta)\phi_n z^{-n-\Delta-1}$$

which is called the *derivative field* of $\phi(z)$. We set

$$\begin{aligned} \phi_-(z) &= \sum_{n \geq 0} \phi_n z^{-n-\Delta}, \\ \phi_+(z) &= \sum_{n<0} \phi_n z^{-n-\Delta}. \end{aligned}$$

If $\psi(z)$ is another vertex operator, we define a new vertex operator, which is called the *normally ordered product* of $\phi(z)$ and $\psi(z)$, to be:

$$:\phi(z)\psi(z): \; = \; \phi_+(z)\psi(z) + (-1)^{\phi\psi}\psi(z)\phi_-(z)$$

where $(-1)^{\phi\psi}$ is -1 if both $\phi(z)$ and $\psi(z)$ are odd (i.e. fermionic) fields and 1 otherwise. Inductively we can define the normally ordered product of k vertex operators $\phi_1(z), \phi_2(z), \ldots, \phi_k(z)$ from right to left by

$$:\phi_1(z)\phi_2(z)\cdots\phi_k(z): \;=\; :\phi_1(z)(:\phi_2(z)\cdots\phi_k(z):):.$$

For $\alpha \in H^*(X)$, recall the field $a(\alpha)(z)$ from (4.15), and the notations $:a(z)^p:(\tau_*\alpha)$ and $:a^p:_m(\tau_*\alpha)$ from (4.16). For $r \geq 1$, we can similarly define the field $:(\partial^r a(z))a(z)^{p-1}:(\tau_*\alpha)$, and define the operator $:(\partial^r a)a^{p-1}:_m(\tau_*\alpha)$ as the coefficient of z^{-m-r-p} in $:(\partial^r a(z))a(z)^{p-1}:(\tau_*\alpha)$.

We remark that when the variable z or the cohomology class α is clear or irrelevant in the context, we will drop z or α from the notations of fields. For instance, we will sometimes use $:a^p:(\tau_*\alpha)$ or $:a^p:$, $:(\partial^r a)a^p:(\tau_*\alpha)$ or $:(\partial^r a)a^p:$, etc., to stand for $:a(z)^p:(\tau_*\alpha)$, $:(\partial^r a(z))a(z)^p:(\tau_*\alpha)$, etc., respectively.

LEMMA 4.23. *In terms of fields, the operator $\mathfrak{J}^p_m(\alpha)$ can be rewritten as:*

$$\begin{aligned}
(4.47)\qquad & -\frac{1}{(p+1)}:a^{p+1}:_m(\tau_*\alpha) \\
&+ \frac{1}{24}p(m^2-3m-2p):a^{p-1}:_m(\tau_*(e_X\alpha)) \\
&+ \frac{1}{24}p(p-1):(\partial^2 a)\,a^{p-2}:_m(\tau_*(e_X\alpha)).
\end{aligned}$$

PROOF. First of all, by the definition of $:a^{p+1}:_m(\tau_*\alpha)$, we have

$$\begin{aligned}
& p!\cdot \sum_{\ell(\lambda)=p+1,|\lambda|=m} \frac{1}{\lambda!}\mathfrak{a}_\lambda(\tau_*\alpha) \\
=\;& \frac{1}{(p+1)} \sum_{i_1+\ldots+i_{p+1}=m} :\mathfrak{a}_{i_1}\cdots\mathfrak{a}_{i_{p+1}}:(\tau_*\alpha) \\
=\;& \frac{1}{(p+1)}:a^{p+1}:_m(\tau_*\alpha).
\end{aligned}$$

Next, recalling the definition of $s(\lambda)$, we obtain

$$\begin{aligned}
& p!\cdot \sum_{\ell(\lambda)=p-1,|\lambda|=m} \frac{s(\lambda)+m^2-2}{\lambda!}\mathfrak{a}_\lambda(\tau_*(e_X\alpha)) \\
=\;& p\cdot \sum_{i_1+\ldots+i_{p-1}=m} \left(\sum_{b=1}^{p-1} i_b^2 + m^2 - 2\right) :\mathfrak{a}_{i_1}\cdots\mathfrak{a}_{i_{p-1}}:(\tau_*(e_X\alpha)) \\
=\;& p\cdot \sum_{i_1+\ldots+i_{p-1}=m} \left(\sum_{b=1}^{p-1}(i_b+1)(i_b+2) + m^2 - 3m - 2p\right)\cdot \\
& \qquad :\mathfrak{a}_{i_1}\cdots\mathfrak{a}_{i_{p-1}}:(\tau_*(e_X\alpha)) \\
=\;& p(p-1):(\partial^2 a)\,a^{p-2}:_m(\tau_*(e_X\alpha)) + p(m^2-3m-2p):a^{p-1}:_m(\tau_*(e_X\alpha)).
\end{aligned}$$

Now the lemma follows from the definition of the operator $\mathfrak{J}^p_m(\alpha)$. □

REMARK 4.24. More explicitly, we can rewrite the operator $\mathfrak{J}^p_m(\alpha)$ as the m-th Fourier component of a vertex operator as follows:

$$\begin{aligned} & \mathfrak{J}^p_m(\alpha) \\ = \; & -\frac{1}{(p+1)} : a^{p+1} :_m (\tau_*\alpha) + \frac{p}{24}(\partial^2 : a^{p-1} :)_m(\tau_*(e_X\alpha)) \\ & + \frac{(p+1)p}{12}(\partial : a^{p-1} :)_m(\tau_*(e_X\alpha)) + \frac{p(p^2-p-2)}{24} : a^{p-1} :_m (\tau_*(e_X\alpha)) \\ & + \frac{1}{24}p(p-1) : (\partial^2 a)a^{p-2} :_m (\tau_*(e_X\alpha)). \end{aligned}$$

In practice, (4.47) suffices for the purpose of computations below. Also, since

$$\mathfrak{J}^{p+1}_0(\alpha) = (p+1)! \cdot \mathfrak{G}_p(\alpha)$$

for $\alpha \in \mathfrak{A}_X$, we conclude that the Chern character operator $\mathfrak{G}_p(\alpha)$ with $\alpha \in \mathfrak{A}_X$ is the zero-mode of a vertex operator.

Next, define an integer $\Omega^{p;q}_{m,n}$ for $m, n, p, q \in \mathbb{Z}$ as follows:

$$\begin{aligned} \Omega^{p;q}_{m,n} \;=\; & mp^3n^2 + 3mp^2n^2q - p^2nq + p^2qn^3 - 3mp^2n^2 + pnq \\ & + 3m^2pnq - 3mpn^2q - m^3q^2p - pqn^3 - mpq + m^3pq \\ & + mpq^2 + 2mpn^2 - 3m^2pnq^2 - 2m^2nq + 3m^2nq^2 - m^2nq^3. \end{aligned} \tag{4.48}$$

THEOREM 4.25. *The space $\mathcal{W}_X$ is closed under the Lie bracket. More explicitly, for $m, n \in \mathbb{Z}$ and $\alpha, \beta \in H^*(X)$, the following commutation relation holds:*

$$[\mathfrak{J}^p_m(\alpha), \mathfrak{J}^q_n(\beta)] = (qm - pn) \cdot \mathfrak{J}^{p+q-1}_{m+n}(\alpha\beta) - \frac{\Omega^{p;q}_{m,n}}{12} \cdot \mathfrak{J}^{p+q-3}_{m+n}(e_X\alpha\beta)$$

where $(p, q) \in \mathbb{Z}^2_+$ except for the unordered *pairs $(0,0), (1,0), (2,0)$ and $(1,1)$. In addition, for these four exceptional cases, the following identities hold:*

$$\begin{aligned} [\mathfrak{J}^0_m(\alpha), \mathfrak{J}^0_n(\beta)] &= -m\delta_{m,-n} \int_X (\alpha\beta) \cdot \mathrm{Id}_{\mathbb{H}_X}, \\ [\mathfrak{J}^1_m(\alpha), \mathfrak{J}^0_n(\beta)] &= -n \cdot \mathfrak{J}^0_{m+n}(\alpha\beta), \\ [\mathfrak{J}^2_m(\alpha), \mathfrak{J}^0_n(\beta)] &= -2n \cdot \mathfrak{J}^1_{m+n}(\alpha\beta) + \frac{m^3-m}{6}\delta_{m,-n} \int_X (e_X\alpha\beta) \cdot \mathrm{Id}_{\mathbb{H}_X}, \\ [\mathfrak{J}^1_m(\alpha), \mathfrak{J}^1_n(\beta)] &= (m-n) \cdot \mathfrak{J}^1_{m+n}(\alpha\beta) + \frac{m^3-m}{12}\delta_{m,-n} \int_X (e_X\alpha\beta) \cdot \mathrm{Id}_{\mathbb{H}_X}. \end{aligned}$$

Note that our algebra $\mathcal{W}_X$ contains as subalgebras the Heisenberg algebra (Section 3.2) of Nakajima and Grojnowski generated by the operators

$$\mathfrak{J}^0_m(\alpha) = -\mathfrak{a}_m(\alpha),$$

and the Virasoro algebra (Section 3.5) of Lehn generated by the operators

$$\mathfrak{J}^1_m(\alpha) = \mathfrak{L}_m(\alpha).$$

We denote

$$\mathfrak{B}_X = \{\alpha \in H^*(X) | e_X\alpha = K_X\alpha = 0\}$$

which is an ideal in the ring $H^*(X)$. Obviously, $\mathfrak{B}_X \subset \mathfrak{A}_X$. Denote by

$$\mathcal{W}^{\mathfrak{B}}_X$$

the linear span of $\mathfrak{J}_n^p(\alpha)$, where $p \geq 0, n \in \mathbb{Z}$ and $\alpha \in \mathfrak{B}_X$. From the commutation relation in Theorem 4.25 we observe that $\mathcal{W}_X^{\mathfrak{B}}$ is a Lie subalgebra of the Lie (super)algebra $\mathcal{W}_X$. Recall the (super)algebra $\widehat{\mathcal{W}}(A)$ introduced in (4.45) for a general ring A. The next theorem follows from comparing the commutator (4.46) and the ones given in Theorem 4.25.

THEOREM 4.26. *Let X be a smooth projective complex surface. Then, the Lie (super)algebra $\widehat{\mathcal{W}}(\mathfrak{B}_X)$ is isomorphic to the Lie (super)algebra $\mathcal{W}_X^{\mathfrak{B}}$ by sending $C \mapsto \mathrm{Id}_{\mathbb{H}_X}$ and $\mathfrak{L}_n^p(\alpha) \mapsto \mathfrak{J}_n^p(\alpha)$, where $p \geq 0$, $n \in \mathbb{Z}$, and $\alpha \in \mathfrak{B}_X$.*

REMARK 4.27. Let X be a smooth projective complex surface.

(i) If the canonical class K_X and Euler class e_X are numerically trivial, then the ideal $\mathfrak{B}_X$ is the entire cohomology ring $H^*(X)$, and thus the algebra $\mathcal{W}_X^{\mathfrak{B}}$ coincides with $\mathcal{W}_X$. In addition, Theorem 4.25 implies that the field
$$\mathfrak{J}^p(\alpha)(z) = \sum_{n \in \mathbb{Z}} \mathfrak{J}_n^p(\alpha) z^{-n-p-1}$$
is a primary field of conformal weight $(p+1)$ with respect to the Virasoro field.

(ii) In general, we note that the leading term in the commutators of the algebra $\mathcal{W}_X$ given in Theorem 4.25 is precisely the $\mathcal{W}$ algebra associated to the ring $H^*(X)$. Therefore, we are justified to regard the $\mathcal{W}$ algebras $\mathcal{W}_X$ and $\mathcal{W}_X^{\mathfrak{B}}$ in general as certain topological deformations of the $\mathcal{W}_{1+\infty}$ algebra in the framework of Hilbert schemes.

The proof of Theorem 4.25 is very technical and uses the operator product expansion (OPE) technique. In the rest of this section, we will outline the proof, and refer to [**LQW4**, Section 6] for further details. Denote the three terms in (4.47) by ${}^1\mathfrak{J}_m^p(\alpha), {}^2\mathfrak{J}_m^p(\alpha)$ and ${}^3\mathfrak{J}_m^p(\alpha)$ respectively, that is,
$$\mathfrak{J}_m^p(\alpha) = {}^1\mathfrak{J}_m^p(\alpha) + {}^2\mathfrak{J}_m^p(\alpha) + {}^3\mathfrak{J}_m^p(\alpha).$$
Note that $[{}^i\mathfrak{J}_m^p(\alpha), {}^j\mathfrak{J}_n^q(\beta)] = 0$ $(i, j = 2, 3)$ by Lemma 3.12 (i) and $e_X^2 = 0$. Thus,
$$\begin{aligned} (4.49) \qquad & [\mathfrak{J}_m^p(\alpha), \mathfrak{J}_n^q(\beta)] \\ = \; & [{}^1\mathfrak{J}_m^p(\alpha), {}^1\mathfrak{J}_n^q(\beta)] + [{}^1\mathfrak{J}_m^p(\alpha), {}^2\mathfrak{J}_n^q(\beta)] + [{}^1\mathfrak{J}_m^p(\alpha), {}^3\mathfrak{J}_n^q(\beta)] \\ & + [{}^2\mathfrak{J}_m^p(\alpha), {}^1\mathfrak{J}_n^q(\beta)] + [{}^3\mathfrak{J}_m^p(\alpha), {}^1\mathfrak{J}_n^q(\beta)]. \end{aligned}$$

Following [**Kac**, **FB**], we see that the commutation relations of the Heisenberg generators can be recast equivalently in terms of the operator product expansion (OPE) as:
$$(4.50) \qquad a(\alpha)(z)\, a(\beta)(w) \sim \frac{-\int_X(\alpha\beta)}{(z-w)^2}.$$

Here and below, $\sim$ means that the regular terms with respect to $(z-w)$ on the right hand side of the OPEs are omitted; it is well known that the meromorphic terms carry all the information about the corresponding commutation relations. We derive from (4.50) the following two OPEs:
$$(4.51) \qquad a(\alpha)(z)\, \partial a(\beta)(w) \sim \frac{-2\int_X(\alpha\beta)}{(z-w)^3}, \quad a(\alpha)(z)\, \partial^2 a(\beta)(w) \sim \frac{-6\int_X(\alpha\beta)}{(z-w)^4}.$$

Our goal is to compute the commutator $[\mathfrak{J}^p_m(\alpha), \mathfrak{J}^q_n(\beta)]$ by using the OPE technique. To this end, we need to compute the following OPEs

$$(: a(z)^{p+1} : (\tau_*\alpha))(: a(w)^{q+1} : (\tau_*\beta)), \tag{4.52}$$

$$(: a(z)^{p+1} : (\tau_*\alpha))(: a(w)^{q-1} : (\tau_*(e_X\beta))), \tag{4.53}$$

$$(: a(z)^{p+1} : (\tau_*\alpha))(: \partial^2 a(w)a(w)^{q-2} : (\tau_*(e_X\beta))), \tag{4.54}$$

which correspond to the first three commutators on the right-hand-side of (4.49) respectively. Then we will be able to rewrite the terms appearing in these OPEs as combinations of (the derivative fields of) the four fields

$$\begin{gathered} : a(z)^{p+q} : (\tau_*(\alpha\beta)), \\ : a(z)^{p+q-2} : (\tau_*(e_X\alpha\beta)), \\ : \partial^2(a(z))a(z)^{p+q-3} : (\tau_*(e_X\alpha\beta)), \\ : \partial^3(a(z))a(z)^{p+q-3} : (\tau_*(e_X\alpha\beta)), \end{gathered}$$

by using the following simple lemma.

LEMMA 4.28. *As in the previous section, write the fields $a^r(z), \partial^s a(z)$, etc., simply as $a^r, \partial^s a$, etc., respectively. Then for $N \geq 0$, we have the following:*

(i) $\partial^2 : a^N := N : (\partial^2 a)a^{N-1} : +N(N-1) : (\partial a)^2 a^{N-2} :$

(ii) $\partial^3 : a^N :$ *is equal to*

$$N : (\partial^3 a)a^{N-1} : +6\binom{N}{2} : (\partial a)(\partial^2 a)a^{N-2} : +6\binom{N}{3} : (\partial a)^3 a^{N-3} :$$

(iii) $\partial : a^N \partial^2 a :=: (\partial^3 a)a^N : +N : (\partial a)(\partial^2 a)a^{N-1} :$

(iv) $\partial : (\partial a)^2 a^N := 2 : (\partial a)(\partial^2 a)a^N : +N : (\partial a)^3 a^{N-1} :.$

PROOF. Using the Leibniz rule, we obtain

$$\begin{aligned} \partial^2 : a^N : &= \partial(N : (\partial a)a^{N-1} :) \\ &= N : (\partial^2 a)a^{N-1} : +N(N-1) : (\partial a)^2 a^{N-2} : . \end{aligned}$$

This proves part (i). The proofs of the remaining formulas are similar. □

Another ingredient in calculating various OPEs is the Wick Theorem [**Kac**, Theorem 3.3]. The Wick Theorem in the vertex algebra literature is a very powerful tool for calculating the OPE of two normally ordered products of free fields.

THEOREM 4.29. (Wick Theorem) *Let $a^1(z), \ldots, a^M(z)$ and $b^1(z), \ldots, b^N(z)$ be two collections of free fields such that the following properties hold:*

(i) $\big[[a^i(z)_-, b^j(w)], c^k(z)\big] = 0$ *for all i, j, k, and $c = a$ or b,*

(ii) $[a^i(z)_\pm, b^j(w)_\pm] = 0$ *for all i and j.*

Let $[a^i b^j] = [a^i(z)_-, b^j(w)]$ be the contraction of $a^i(z)$ and $b^j(w)$. The one has the following OPE in the domain $|z| > |w|$:

$$\begin{aligned} &: a^1(z) \cdots a^M(z) :: b^1(w) \cdots b^N(w) : \\ = &\sum_{s=0}^{\min(M,N)} \sum_{\substack{i_1 < \cdots < i_s \\ j_1 \neq \cdots \neq j_s}} \Big(\pm [a^{i_1} b^{j_1}] \cdots [a^{i_s} b^{j_s}] \\ &\qquad : a^1(z) \cdots a^M(z) b^1(w) \cdots b^N(w) :_{(i_1, \ldots, i_s; j_1, \ldots, j_s)} \Big) \end{aligned}$$

where the subscript $(i_1, \ldots, i_s; j_1, \ldots, j_s)$ *means that the fields*

$$a^{i_1}(z), \ldots, a^{i_s}(z), b^{j_1}(w), \ldots, b^{j_s}(w)$$

are removed, and the sign $\pm$ *is obtained by the usual super rule: each permutation of the adjacent odd fields changes the sign.*

The Wick Theorem has to be modified in order to apply to our setup. First of all, we naturally incorporate the transfer property, Lemma 3.12 (i). Next, by Lemma 3.12 (ii), the contraction of two pairs of free fields will give rise to a term depending on $e_X \alpha\beta$. So the contraction of more than two pairs of fields will be zero thanks to $e_X^2 = 0$.

CHAPTER 5

Multiple q-zeta values and Hilbert schemes

As applications of the results in Chapter 3 and Chapter 4, in this chapter we will compute the generating series for the intersection pairings between the total Chern classes of the tangent bundles of the Hilbert schemes of points on a smooth projective complex surface and the Chern characters of tautological bundles over these Hilbert schemes. As pointed out in [**Car1**], these generating series may be regarded as the generalization of the Nekrasov partition function and the related correlation function from the equivariant setting to the non-equivariant setting. Modulo the lower weight term, we will verify Okounkov's conjecture [**Ok2**, Conjecture 2] which connects the Hilbert schemes of points on surfaces and multiple q-zeta values. In addition, this conjecture is completely proved when the surface is abelian. Okounkov's conjecture is another example of the interplay of algebraic geometry, representation theory and analytic number theory. The main approach is to use the Ext vertex operators of Carlsson and Okounkov discussed in Section 3.7 and the structure of the Chern character operators proved in Section 4.1. This chapter follows the presentation of [**QY**], and we refer to [**QY**, Section 6] for further applications which determine some universal constants in the sense of Boissière and Nieper-Wisskirchen [**Boi1**, **BN**] regarding the total Chern classes of the tangent bundles of these Hilbert schemes.

5.1. Okounkov's conjecture

In this section, we will review Okounkov's conjecture from [**Ok2**]. For integers $k_1, \ldots, k_N \geq 0$ and cohomology classes $\alpha_1, \ldots, \alpha_N \in H^*(X)$ on a smooth projective complex surface X, we will define the generating series $F^{\alpha_1,\ldots,\alpha_N}_{k_1,\ldots,k_N}(q)$ and express it in terms of the Ext vertex operators of Carlsson and Okounkov discussed in Section 3.7.

In the region $\mathrm{Re}\, s > 1$, the Riemann zeta function is defined by

$$\zeta(s) = \sum_{n=1}^{+\infty} \frac{1}{n^s}.$$

The integers $s > 1$ give rise to a sequence of special values of the Riemann zeta function. Multiple zeta values are series of the form

$$\zeta(s_1, \ldots, s_k) = \sum_{n_1 > \cdots > n_k > 0} \frac{1}{n_1^{s_1} \cdots n_k^{s_k}}$$

where $n_1, \ldots, n_k$ denote positive integers, and $s_1, \ldots, s_k$ are positive integers with $s_1 > 1$. Their $\mathbb{Q}$-linear span is a subalgebra of $\mathbb{R}$ and denoted by **MZV**. Multiple q-zeta values $\zeta_q(s_1, \ldots, s_k)$ are q-deformations of $\zeta(s_1, \ldots, s_k)$, which may take different forms as indicated by [**Bra1**, **Bra2**, **OT**, **Zhao**, **Zud**]. In this chapter, we adopt the convention that $\zeta_q(s_1, \ldots, s_k)$ is a multiple q-zeta value if

$$\lim_{q\to 1}(1-q)^{s_1+\ldots+s_k}\zeta_q(s_1,\ldots,s_k) \quad \in \quad \mathbf{MZV}.$$

The *weight* of $\zeta_q(s_1, \ldots, s_k)$ is defined to be $s_1 + \ldots + s_k$.

Next, we switch to the setting of Hilbert schemes of points. Let X be a smooth projective complex surface, and let L be a line bundle on X. Let $\mathrm{ch}_k(L^{[n]})$ be the k-th Chern character of the tautological rank-n bundle $L^{[n]}$ over the Hilbert scheme $X^{[n]}$. We define the generating series $\langle \mathrm{ch}_{k_1}^{L_1} \cdots \mathrm{ch}_{k_N}^{L_N} \rangle$ by

$$\langle \mathrm{ch}_{k_1}^{L_1} \cdots \mathrm{ch}_{k_N}^{L_N} \rangle = \sum_{n\geq 0} q^n \int_{X^{[n]}} \mathrm{ch}_{k_1}(L_1^{[n]}) \cdots \mathrm{ch}_{k_N}(L_N^{[n]}) \cdot c(T_{X^{[n]}}) \tag{5.1}$$

where $0 < q < 1$, and $c(T_{X^{[n]}})$ is the total Chern class of the tangent bundle $T_{X^{[n]}}$. By (1.22),

$$\langle 1 \rangle = (q;q)_\infty^{-\chi(X)}$$

where

$$(a;q)_n = \prod_{i=0}^{n}(1-aq^i).$$

The reduced series $\langle \mathrm{ch}_{k_1}^{L} \cdots \mathrm{ch}_{k_N}^{L} \rangle'$ is defined by

$$\begin{aligned} \langle \mathrm{ch}_{k_1}^{L_1} \cdots \mathrm{ch}_{k_N}^{L_N} \rangle' &= \langle \mathrm{ch}_{k_1}^{L_1} \cdots \mathrm{ch}_{k_N}^{L_N} \rangle / \langle 1 \rangle \\ &= (q;q)_\infty^{\chi(X)} \cdot \langle \mathrm{ch}_{k_1}^{L_1} \cdots \mathrm{ch}_{k_N}^{L_N} \rangle. \end{aligned} \tag{5.2}$$

In [**Car1**, **Car2**], for $X = \mathbb{C}^2$ with a suitable $\mathbb{C}^*$-action and $L = \mathcal{O}_X$, the series $\langle \mathrm{ch}_{k_1}^{L} \cdots \mathrm{ch}_{k_\ell}^{L} \rangle'$ in the equivariant setting has been studied, and proved to be quasi-modular in q [**Car2**, Theorem 2 on p.2904]. Using this as evidence, Okounkov [**Ok2**, Conjecture 2] proposed the following interesting conjecture.

CONJECTURE 5.1. *The reduced series* $\langle \mathrm{ch}_{k_1}^{L} \cdots \mathrm{ch}_{k_N}^{L} \rangle'$ *is a multiple q-zeta value of weight* $\sum_{i=1}^{N}(k_i+2)$.

A motivation for Conjecture 5.1 is Okounkov's observation that the series $\langle \mathrm{ch}_{k_1}^{L_1} \cdots \mathrm{ch}_{k_N}^{L_N} \rangle$ can be written as the trace over the Fock space of a product of a certain vertex operator (that is, the Ext operator) and a certain integral of motion of the second quantized trigonometric Calogero-Sutherland system. This will be carried out via (5.41) and Lemma 5.2 below in terms of the Chern character operators (the relation with Calogero-Sutherland systems is the subject of Section 7.4). The structure (Theorem 4.7) of these Chern character operators will help with understanding $\langle \mathrm{ch}_{k_1}^{L_1} \cdots \mathrm{ch}_{k_N}^{L_N} \rangle$ and hence the reduced series $\langle \mathrm{ch}_{k_1}^{L} \cdots \mathrm{ch}_{k_N}^{L} \rangle'$. Conjecture 5.1 may lead to further understanding of the graded algebra of multiple q-zeta values, and provides another example of the interplay of algebraic geometry, representation theory, and analytic number theory.

We will partially verify Conjecture 5.1. To achieve this, for cohomology classes $\alpha_1, \ldots, \alpha_N \in H^*(X)$ and integers $k_1, \ldots, k_N \geq 0$, we define the series

$$F_{k_1,\ldots,k_N}^{\alpha_1,\ldots,\alpha_N}(q) = \sum_n q^n \int_{X^{[n]}} \left(\prod_{i=1}^{N} G_{k_i}(\alpha_i, n)\right) c(T_{X^{[n]}}) \tag{5.3}$$

where the classes $G_{k_i}(\alpha_i, n)$ are defined in Definition 4.1 (iii). Note that

$$F(q) = \langle 1 \rangle = \sum_n q^n \int_{X^{[n]}} c(T_{X^{[n]}}) = \sum_n \chi(X^{[n]}) q^n = (q;q)_\infty^{-\chi(X)}$$

by (1.22). Recall the line bundle $\mathfrak{L}_m$ and Ext operator $\mathbf{W}(L, z)$ from Section 3.7. The following lemma is similar to Lemma 3.33.

LEMMA 5.2. *Let* $\mathfrak{n}$ *be the number-of-points operator, i.e.,* $\mathfrak{n}|_{H^*(X^{[n]})} = n\,\mathrm{Id}$. *Then,*

$$F^{\alpha_1,\ldots,\alpha_N}_{k_1,\ldots,k_N}(q) = \mathrm{Tr}\, q^{\mathfrak{n}}\, \mathbf{W}(\mathfrak{L}_1, z) \prod_{i=1}^N \mathfrak{G}_{k_i}(\alpha_i). \tag{5.4}$$

PROOF. We use the same arguments as in the proof of Lemma 3.33. We will show that the coefficients of q^n on both sides of (5.4) are equal. Let $\{e_j\}_j$ be a linear basis of $H^*(X^{[n]})$. By (3.45),

$$[\Delta_n] = \sum_j (-1)^{|e_j|} e_j \otimes e_j^*$$

where $\{e_j^*\}_j$ is the linear basis of $H^*(X^{[n]})$ dual to $\{e_j\}_j$ in the sense that $\langle e_j, e_{j'}^* \rangle = \delta_{j,j'}$. By the definitions of $\mathbf{W}(L, z)$ and $\mathfrak{G}_k(\alpha)$, we have

$$\begin{aligned}
&\mathrm{Tr}\, q^n \mathbf{W}(\mathfrak{L}_1, z) \prod_{i=1}^N \mathfrak{G}_{k_i}(\alpha_i) \\
= \;& q^n \sum_j (-1)^{|e_j|} \left\langle \mathbf{W}(\mathfrak{L}_1, z) \left(\prod_{i=1}^N \mathfrak{G}_{k_i}(\alpha_i) \right) e_j, e_j^* \right\rangle \\
= \;& q^n \sum_j (-1)^{|e_j|} \int_{X^{[n]} \times X^{[n]}} \left(\left(\prod_{i=1}^N \mathfrak{G}_{k_i}(\alpha_i) \right) e_j \otimes e_j^* \right) c_{2n}(\mathbb{E}_{\mathfrak{L}_1}) \\
= \;& q^n \sum_j (-1)^{|e_j|} \int_{X^{[n]} \times X^{[n]}} \left(\left(\prod_{i=1}^N G_{k_i}(\alpha_i, n) \right) e_j \otimes e_j^* \right) c_{2n}(\mathbb{E}_{\mathfrak{L}_1}) \\
= \;& q^n \sum_j (-1)^{|e_j|} \int_{X^{[n]} \times X^{[n]}} (e_j \otimes e_j^*)\, p_1^* \left(\prod_{i=1}^N G_{k_i}(\alpha_i, n) \right) c_{2n}(\mathbb{E}_{\mathfrak{L}_1}) \\
= \;& q^n \int_{X^{[n]} \times X^{[n]}} [\Delta_n]\, p_1^* \left(\prod_{i=1}^N G_{k_i}(\alpha_i, n) \right) c_{2n}(\mathbb{E}_{\mathfrak{L}_1})
\end{aligned}$$

where $p_1 : X^{[n]} \times X^{[n]} \to X^{[n]}$ denotes the first projection. By (3.42), we have $c_{2n}(\mathbb{E}_{\mathfrak{L}_1})|_{\Delta_n} = c(T_{X^{[n]}})$. Therefore,

$$\mathrm{Tr}\, q^n \mathbf{W}(\mathfrak{L}_1, z) \prod_{i=1}^N \mathfrak{G}_{k_i}(\alpha_i) = q^n \int_{X^{[n]}} \left(\prod_{i=1}^N G_{k_i}(\alpha_i, n) \right) c(T_{X^{[n]}}).$$

This completes the proof of the lemma. □

5.2. The series $F^{\alpha_1,\ldots,\alpha_N}_{k_1,\ldots,k_N}(q)$

In this section, we will first determine the structure of the trace

$$\mathrm{Tr}\, q^{\mathfrak{n}}\, \mathbf{W}(\mathfrak{L}_1, z) \prod_{i=1}^{N} \frac{\mathfrak{a}_{\lambda^{(i)}}(\tau_*\alpha_i)}{\lambda^{(i)}!} \tag{5.5}$$

where $\lambda^{(1)}, \ldots, \lambda^{(N)}$ are generalized partitions, and the notation $\mathfrak{a}_{\lambda^{(i)}}(\tau_*\alpha_i)$ is defined in (3.36). Then, the structure of the generating series $F^{\alpha_1,\ldots,\alpha_N}_{k_1,\ldots,k_N}(q)$ will follow from Lemma 5.2, Theorem 4.7 and the structure of the trace (5.5).

We begin with four technical lemmas. To explain the ideas behind these lemmas, note from (3.49) that the trace (5.5) is equal to

$$\mathrm{Tr}\, q^{\mathfrak{n}}\, \Gamma_-(\mathfrak{L}_1 - K_X, z)\Gamma_+(-\mathfrak{L}_1, z) \prod_{i=1}^{N} \frac{\mathfrak{a}_{\lambda^{(i)}}(\tau_*\alpha_i)}{\lambda^{(i)}!}. \tag{5.6}$$

Lemma 5.3 deals with the commutator between $\dfrac{\mathfrak{a}_\lambda(\tau_*\alpha)}{\lambda!}$ and $\exp\left(\dfrac{z^n}{n}\mathfrak{a}_{-n}(\gamma)\right)$. It enables us to eliminate $\Gamma_-(\mathfrak{L}_1 - K_X, z)$ from (5.6) in Lemma 5.4, and allows us to eliminate $\Gamma_+(-\mathfrak{L}_1, z)$ from (5.6) in Lemma 5.5. Lemma 5.7 determines the structure of the trace

$$\mathrm{Tr}\, q^{\mathfrak{n}} \prod_{i=1}^{N} \frac{\mathfrak{a}_{\lambda^{(i)}}(\tau_*\alpha_i)}{\lambda^{(i)}!}.$$

The proofs of these lemmas are standard but lengthy.

Recall from Definition 1.6 that $\widetilde{\mathcal{P}}$ denotes the set of generalized partitions. If $\lambda = \left(\cdots(-2)^{s_{-2}}(-1)^{s_{-1}}1^{s_1}2^{s_2}\cdots\right)$ and $\mu = \left(\cdots(-2)^{t_{-2}}(-1)^{t_{-1}}1^{t_1}2^{t_2}\cdots\right)$, let

$$\lambda - \mu = \left(\cdots(-2)^{s_{-2}-t_{-2}}(-1)^{s_{-1}-t_{-1}}1^{s_1-t_1}2^{s_2-t_2}\cdots\right)$$

with the convention that $\lambda - \mu = \emptyset$ if $s_i < t_i$ for some i.

LEMMA 5.3. *Let $\lambda \in \widetilde{\mathcal{P}}$. Assume that $\gamma \in H^{\mathrm{even}}(X)$. Then,*

$$\begin{aligned} &\frac{\mathfrak{a}_\lambda(\tau_*\alpha)}{\lambda!}\exp\left(\frac{z^n}{n}\mathfrak{a}_{-n}(\gamma)\right) \\ &= \exp\left(\frac{z^n}{n}\mathfrak{a}_{-n}(\gamma)\right)\cdot \sum_{i\ge 0}\frac{(-z^n)^i}{i!}\frac{\mathfrak{a}_{\lambda-(n^i)}(\tau_*(\gamma^i\alpha))}{(\lambda-(n^i))!}, \end{aligned} \tag{5.7}$$

$$\begin{aligned} &\exp\left(\frac{z^n}{n}\mathfrak{a}_{n}(\gamma)\right)\cdot\frac{\mathfrak{a}_\lambda(\tau_*\alpha)}{\lambda!} \\ &= \sum_{i\ge 0}\frac{(-z^n)^i}{i!}\frac{\mathfrak{a}_{\lambda-((-n)^i)}(\tau_*(\gamma^i\alpha))}{(\lambda-((-n)^i))!}\exp\left(\frac{z^n}{n}\mathfrak{a}_{n}(\gamma)\right). \end{aligned} \tag{5.8}$$

PROOF. Note that the adjoint of $\mathfrak{a}_m(\beta)$ is equal to $(-1)^m\mathfrak{a}_{-m}(\beta)$. So (5.8) follows from (5.7) by taking the adjoint on both sides of (5.7) and by making suitable adjustments. To prove (5.7), put

$$A = \frac{\mathfrak{a}_\lambda(\tau_*\alpha)}{\lambda!}\exp\left(\frac{z^n}{n}\mathfrak{a}_{-n}(\gamma)\right).$$

Then,

$$\begin{aligned} A &= \frac{\mathfrak{a}_\lambda(\tau_*\alpha)}{\lambda^!}\sum_{t\geq 0}\frac{1}{t!}\left(\frac{z^n}{n}\mathfrak{a}_{-n}(\gamma)\right)^t \\ &= \frac{1}{\lambda^!}\sum_{t\geq 0}\frac{1}{t!}\left(\frac{z^n}{n}\right)^t\sum_{i=0}^t\binom{t}{i}\big(\mathfrak{a}_{-n}(\gamma)\big)^{t-i}\Big[\cdots[\mathfrak{a}_\lambda(\tau_*\alpha),\underbrace{\mathfrak{a}_{-n}(\gamma)],\dots,\mathfrak{a}_{-n}(\gamma)}_{i \text{ times}}\Big]. \end{aligned}$$

Let $\lambda = \big(\cdots(-2)^{s_{-2}}(-1)^{s_{-1}}1^{s_1}2^{s_2}\cdots\big)$. We conclude from Lemma 3.12 (i) that the commutator $\Big[\cdots[\mathfrak{a}_\lambda(\tau_*\alpha),\underbrace{\mathfrak{a}_{-n}(\gamma)],\dots,\mathfrak{a}_{-n}(\gamma)}_{i \text{ times}}\Big]$ is equal to

$$s_n(s_n-1)\cdots(s_n+1-i)\,(-n)^i\,\mathfrak{a}_{\lambda-(n^i)}(\tau_*(\gamma^i\alpha))$$

where by our convention, $\lambda-(n^i)=\emptyset$ if $s_n < i$. So A is equal to

$$\begin{aligned} &\frac{1}{\lambda^!}\sum_{t\geq 0}\frac{1}{t!}\left(\frac{z^n}{n}\right)^t\sum_{i=0}^t\binom{t}{i}\big(\mathfrak{a}_{-n}(\gamma)\big)^{t-i}\cdot \\ &\qquad\cdot s_n(s_n-1)\cdots(s_n+1-i)\,(-n)^i\,\mathfrak{a}_{\lambda-(n^i)}(\tau_*(\gamma^i\alpha)). \end{aligned}$$

Simplifying this, we complete the proof of our formula (5.7). □

Let $\widetilde{\mathcal{P}}_+ = \mathcal{P}$ be the subset of $\widetilde{\mathcal{P}}$ consisting of the usual partitions, and $\widetilde{\mathcal{P}}_-$ be the subset of $\widetilde{\mathcal{P}}$ consisting of generalized partitions of the form $\big(\cdots(-2)^{s_{-2}}(-1)^{s_{-1}}\big)$.

LEMMA 5.4. *Let $\lambda^{(1)},\dots,\lambda^{(N)}\in\widetilde{\mathcal{P}}$ be generalized partitions, and $\alpha_1,\dots,\alpha_N\in H^*(X)$. Then, the trace*

$$\mathrm{Tr}\, q^{\mathfrak{n}}\,\mathbf{W}(\mathfrak{L}_1,z)\prod_{i=1}^N\frac{\mathfrak{a}_{\lambda^{(i)}}(\tau_*\alpha_i)}{\lambda^{(i)!}} \tag{5.9}$$

is equal to

$$\sum_{\substack{\mu^{(i,s)}\in\widetilde{\mathcal{P}}_+\\ 1\leq i\leq N, s\geq 1}}\ \prod_{\substack{1\leq i\leq N\\ s,n\geq 1}}\frac{(-(zq^s)^n)^{m_n^{(i,s)}}}{m_n^{(i,s)}!}. \tag{5.10}$$

$$\cdot\mathrm{Tr}\, q^{\mathfrak{n}}\,\Gamma_+(-\mathfrak{L}_1,z)\prod_{i=1}^N\frac{\mathfrak{a}_{\lambda^{(i)}-\sum_{s\geq 1}\mu^{(i,s)}}\big(\tau_*((1_X-K_X)^{\sum_{s,n\geq 1}m_n^{(i,s)}}\alpha_i)\big)}{\big(\lambda^{(i)}-\sum_{s\geq 1}\mu^{(i,s)}\big)^!}$$

where $\mu^{(i,s)}=\big(1^{m_1^{(i,s)}}\cdots n^{m_n^{(i,s)}}\cdots\big)\in\widetilde{\mathcal{P}}_+$ for $1\leq i\leq N$ and $s\geq 1$.

PROOF. For simplicity, denote the trace (5.9) by Q_1. By (3.49),

$$\begin{aligned} Q_1 &= \mathrm{Tr}\, q^{\mathfrak{n}}\,\Gamma_-(\mathfrak{L}_1-K_X,z)\,\Gamma_+(-\mathfrak{L}_1,z)\prod_{i=1}^N\frac{\mathfrak{a}_{\lambda^{(i)}}(\tau_*\alpha_i)}{\lambda^{(i)!}} \\ &= \mathrm{Tr}\,\Gamma_-(\mathfrak{L}_1-K_X,zq)\,q^{\mathfrak{n}}\,\Gamma_+(-\mathfrak{L}_1,z)\prod_{i=1}^N\frac{\mathfrak{a}_{\lambda^{(i)}}(\tau_*\alpha_i)}{\lambda^{(i)!}} \qquad (5.11) \\ &= \mathrm{Tr}\, q^{\mathfrak{n}}\,\Gamma_+(-\mathfrak{L}_1,z)\prod_{i=1}^N\frac{\mathfrak{a}_{\lambda^{(i)}}(\tau_*\alpha_i)}{\lambda^{(i)!}}\cdot\Gamma_-(\mathfrak{L}_1-K_X,zq). \end{aligned}$$

By (3.46) and applying (5.7) repeatedly, we obtain

$$\begin{aligned}
&\prod_{i=1}^{N} \frac{\mathfrak{a}_{\lambda^{(i)}}(\tau_* \alpha_i)}{\lambda^{(i)}!} \cdot \Gamma_-(\mathfrak{L}_1 - K_X, zq) \\
= &\prod_{i=1}^{N} \frac{\mathfrak{a}_{\lambda^{(i)}}(\tau_* \alpha_i)}{\lambda^{(i)}!} \cdot \exp\left(\sum_{n>0} \frac{(zq)^n}{n} \mathfrak{a}_{-n}(\mathfrak{L}_1 - K_X)\right) \\
= &\Gamma_-(\mathfrak{L}_1 - K_X, zq) \sum_{\substack{\mu^{(i,1)} \in \widetilde{\mathcal{P}}_+ \\ 1 \le i \le N}} \prod_{\substack{1 \le i \le N \\ n \ge 1}} \frac{(-(zq)^n)^{m_n^{(i,1)}}}{m_n^{(i,1)}!} \cdot \\
&\cdot \prod_{i=1}^{N} \frac{\mathfrak{a}_{\lambda^{(i)} - \mu^{(i,1)}}\big(\tau_*((1_X - K_X)^{\sum_{n \ge 1} m_n^{(i,1)}} \alpha_i)\big)}{\big(\lambda^{(i)} - \mu^{(i,1)}\big)!}.
\end{aligned}$$

where $\mu^{(i,1)} = \big(1^{m_1^{(i,1)}} \cdots n^{m_n^{(i,1)}} \cdots\big)$. Therefore, Q_1 is equal to

$$\operatorname{Tr} q^{\mathfrak{n}}\, \Gamma_+(-\mathfrak{L}_1, z)\Gamma_-(\mathfrak{L}_1 - K_X, zq) \sum_{\substack{\mu^{(i,1)} \in \widetilde{\mathcal{P}}_+ \\ 1 \le i \le N}} \prod_{\substack{1 \le i \le N \\ n \ge 1}} \frac{(-(zq)^n)^{m_n^{(i,1)}}}{m_n^{(i,1)}!} \cdot$$

$$\cdot \prod_{i=1}^{N} \frac{\mathfrak{a}_{\lambda^{(i)} - \mu^{(i,1)}}\big(\tau_*((1_X - K_X)^{\sum_{n \ge 1} m_n^{(i,1)}} \alpha_i)\big)}{\big(\lambda^{(i)} - \mu^{(i,1)}\big)!}.$$

Since $\langle \mathfrak{L}_1, \mathfrak{L}_1 - K_X \rangle = 0$, we see from (3.48) that Q_1 is equal to

$$\operatorname{Tr} q^{\mathfrak{n}}\, \Gamma_-(\mathfrak{L}_1 - K_X, zq)\Gamma_+(-\mathfrak{L}_1, z) \cdot \sum_{\substack{\mu^{(i,1)} \in \widetilde{\mathcal{P}}_+ \\ 1 \le i \le N}} \prod_{\substack{1 \le i \le N \\ n \ge 1}} \frac{(-(zq)^n)^{m_n^{(i,1)}}}{m_n^{(i,1)}!} \cdot$$

$$\cdot \prod_{i=1}^{N} \frac{\mathfrak{a}_{\lambda^{(i)} - \mu^{(i,1)}}\big(\tau_*((1_X - K_X)^{\sum_{n \ge 1} m_n^{(i,1)}} \alpha_i)\big)}{\big(\lambda^{(i)} - \mu^{(i,1)}\big)!}.$$

Repeat the above process beginning at line (5.11) s times. Then, Q_1 is equals to

$$\operatorname{Tr} q^{\mathfrak{n}}\, \Gamma_-(\mathfrak{L}_1 - K_X, zq^s)\Gamma_+(-\mathfrak{L}_1, z) \cdot \sum_{\substack{\mu^{(i,r)} \in \widetilde{\mathcal{P}}_+ \\ 1 \le i \le N, 1 \le r \le s}} \prod_{\substack{1 \le i \le N \\ 1 \le r \le s, n \ge 1}} \frac{(-(zq^r)^n)^{m_n^{(i,r)}}}{m_n^{(i,r)}!} \cdot$$

$$\cdot \prod_{i=1}^{N} \frac{\mathfrak{a}_{\lambda^{(i)} - \sum_{r=1}^{s} \mu^{(i,r)}}\big(\tau_*((1_X - K_X)^{\sum_{r=1}^{s} \sum_{n \ge 1} m_n^{(i,r)}} \alpha_i)\big)}{\big(\lambda^{(i)} - \sum_{r=1}^{s} \mu^{(i,r)}\big)!}$$

where $\mu^{(i,r)} = \big(1^{m_1^{(i,r)}} \cdots n^{m_n^{(i,r)}} \cdots\big)$. Letting $s \to +\infty$ proves our lemma. $\square$

LEMMA 5.5. *Let $\tilde{\lambda}^{(1)}, \ldots, \tilde{\lambda}^{(N)} \in \widetilde{\mathcal{P}}$ be generalized partitions, and $\tilde{\alpha}_1, \ldots, \tilde{\alpha}_N \in H^*(X)$. Then, the trace*

$$\operatorname{Tr} q^{\mathfrak{n}}\, \Gamma_+(-\mathfrak{L}_1, z) \prod_{i=1}^{N} \frac{\mathfrak{a}_{\tilde{\lambda}^{(i)}}(\tau_* \tilde{\alpha}_i)}{\tilde{\lambda}^{(i)}!} \tag{5.12}$$

is equal to

$$\sum_{\sum_{i=1}^N(|\tilde\lambda^{(i)}|-\sum_{t\ge1}|\tilde\mu^{(i,t)}|)=0}\ \prod_{\substack{1\le i\le N\\ t,n\ge1}} \frac{(z^{-1}q^{t-1})^{n\tilde m_n^{(i,t)}}}{\tilde m_n^{(i,t)}!}$$
$$\cdot \operatorname{Tr} q^{\mathfrak{n}} \prod_{i=1}^N \frac{\mathfrak{a}_{\tilde\lambda^{(i)}-\sum_{t\ge1}\tilde\mu^{(i,t)}}(\tau_*\tilde\alpha_i)}{\left(\tilde\lambda^{(i)}-\sum_{t\ge1}\tilde\mu^{(i,t)}\right)^!}$$

where $\tilde\mu^{(i,t)} = \left(\cdots(-n)^{\tilde m_n^{(i,t)}}\cdots(-1)^{\tilde m_1^{(i,t)}}\right)\in\widetilde{\mathcal{P}}_-$ *for* $1\le i\le N$ *and* $t\ge1$.

PROOF. For simplicity, denote the trace (5.12) by Q_2. By (3.46),

$$Q_2 = \operatorname{Tr} q^{\mathfrak{n}} \exp\left(\sum_{n>0}\frac{z^{-n}}{n}\mathfrak{a}_n(-\mathfrak{L}_1)\right)\prod_{i=1}^N \frac{\mathfrak{a}_{\tilde\lambda^{(i)}}(\tau_*\tilde\alpha_i)}{\tilde\lambda^{(i)}!}.$$

Applying (5.8) repeatedly, we see that Q_2 is equal to

$$\sum_{\substack{\tilde\mu^{(i,1)}\in\widetilde{\mathcal{P}}_-\\ 1\le i\le N}}\ \prod_{\substack{1\le i\le N\\ n\ge1}} \frac{z^{-n\tilde m_n^{(i,1)}}}{\tilde m_n^{(i,1)}!}\cdot \operatorname{Tr} q^{\mathfrak{n}} \prod_{i=1}^N \frac{\mathfrak{a}_{\tilde\lambda^{(i)}-\tilde\mu^{(i,1)}}(\tau_*\tilde\alpha_i)}{\left(\tilde\lambda^{(i)}-\tilde\mu^{(i,1)}\right)^!}\cdot\Gamma_+(-\mathfrak{L}_1,z)$$

where $\tilde\mu^{(i,1)} = \left(\cdots(-n)^{\tilde m_n^{(i,1)}}\cdots(-1)^{\tilde m_1^{(i,1)}}\right)\in\widetilde{\mathcal{P}}_-$. Now

$$\begin{aligned}
&Q_2\\
&= \sum_{\substack{\tilde\mu^{(i,1)}\in\widetilde{\mathcal{P}}_-\\ 1\le i\le N}}\ \prod_{\substack{1\le i\le N\\ n\ge1}} \frac{z^{-n\tilde m_n^{(i,1)}}}{\tilde m_n^{(i,1)}!}\cdot q^{\sum_{i=1}^N(|\tilde\mu^{(i,1)}|-|\tilde\lambda^{(i)}|)}\\
&\qquad\cdot \operatorname{Tr}\prod_{i=1}^N \frac{\mathfrak{a}_{\tilde\lambda^{(i)}-\tilde\mu^{(i,1)}}(\tau_*\tilde\alpha_i)}{\left(\tilde\lambda^{(i)}-\tilde\mu^{(i,1)}\right)^!}\cdot q^{\mathfrak{n}}\Gamma_+(-\mathfrak{L}_1,z)\\
&= \sum_{\substack{\tilde\mu^{(i,1)}\in\widetilde{\mathcal{P}}_-\\ 1\le i\le N}}\ \prod_{\substack{1\le i\le N\\ n\ge1}} \frac{z^{-n\tilde m_n^{(i,1)}}}{\tilde m_n^{(i,1)}!}\cdot q^{\sum_{i=1}^N(|\tilde\mu^{(i,1)}|-|\tilde\lambda^{(i)}|)}\\
&\qquad\cdot \operatorname{Tr} q^{\mathfrak{n}}\Gamma_+(-\mathfrak{L}_1,z)\prod_{i=1}^N \frac{\mathfrak{a}_{\tilde\lambda^{(i)}-\tilde\mu^{(i,1)}}(\tau_*\tilde\alpha_i)}{\left(\tilde\lambda^{(i)}-\tilde\mu^{(i,1)}\right)^!}.
\end{aligned}$$

By degree reasons, $\operatorname{Tr} q^{\mathfrak{n}}\Gamma_+(-\mathfrak{L}_1,z)\mathfrak{a}_\mu(\beta)=0$ if $|\mu|>0$. If $|\mu|=0$, then we have

$$\operatorname{Tr} q^{\mathfrak{n}}\Gamma_+(-\mathfrak{L}_1,z)\mathfrak{a}_\mu(\beta) = \operatorname{Tr} q^{\mathfrak{n}}\mathfrak{a}_\mu(\beta).$$

So Q_2 is equal to

$$\sum_{\sum_{i=1}^N(|\tilde\lambda^{(i)}|-|\tilde\mu^{(i,1)}|)<0}\ \prod_{\substack{1\le i\le N\\ n\ge1}} \frac{z^{-n\tilde m_n^{(i,1)}}}{\tilde m_n^{(i,1)}!}\cdot q^{\sum_{i=1}^N(|\tilde\mu^{(i,1)}|-|\tilde\lambda^{(i)}|)}$$
$$\cdot \operatorname{Tr} q^{\mathfrak{n}}\Gamma_+(-\mathfrak{L}_1,z)\prod_{i=1}^N \frac{\mathfrak{a}_{\tilde\lambda^{(i)}-\tilde\mu^{(i,1)}}(\tau_*\tilde\alpha_i)}{\left(\tilde\lambda^{(i)}-\tilde\mu^{(i,1)}\right)^!}$$
$$+\sum_{\sum_{i=1}^N(|\tilde\lambda^{(i)}|-|\tilde\mu^{(i,1)}|)=0}\ \prod_{\substack{1\le i\le N\\ n\ge1}} \frac{z^{-n\tilde m_n^{(i,1)}}}{\tilde m_n^{(i,1)}!}\cdot \operatorname{Tr} q^{\mathfrak{n}} \prod_{i=1}^N \frac{\mathfrak{a}_{\tilde\lambda^{(i)}-\tilde\mu^{(i,1)}}(\tau_*\tilde\alpha_i)}{\left(\tilde\lambda^{(i)}-\tilde\mu^{(i,1)}\right)^!}.$$

Repeating the process in the previous paragraph t times, we conclude that

$$Q_2 = U(t) - V(t)$$

where $U(t)$ is given by

$$\sum_{\sum_{i=1}^N (|\tilde\lambda^{(i)}| - \sum_{r=1}^t |\tilde\mu^{(i,r)}|) < 0} \prod_{r=1}^t \left(\prod_{\substack{1 \le i \le N \\ n \ge 1}} \frac{z^{-n\tilde m_n^{(i,r)}}}{\tilde m_n^{(i,r)}!} \cdot q^{\sum_{i=1}^N (\sum_{\ell=1}^r |\tilde\mu^{(i,\ell)}| - |\tilde\lambda^{(i)}|)} \right) \cdot \tag{5.13}$$

$$\cdot \mathrm{Tr}\, q^{\mathfrak{n}} \Gamma_+(-\mathfrak{L}_1, z) \prod_{i=1}^N \frac{\mathfrak{a}_{\tilde\lambda^{(i)} - \sum_{r=1}^t \tilde\mu^{(i,r)}}(\tau_* \tilde\alpha_i)}{\left(\tilde\lambda^{(i)} - \sum_{r=1}^t \tilde\mu^{(i,r)}\right)^!} \tag{5.14}$$

with $\tilde\mu^{(i,r)} = \left(\cdots (-n)^{\tilde m_n^{(i,r)}} \cdots (-1)^{\tilde m_1^{(i,r)}}\right) \in \widetilde{\mathcal{P}}_-$, and $V(t)$ is given by

$$\sum_{\sum_{i=1}^N (|\tilde\lambda^{(i)}| - \sum_{r=1}^t |\tilde\mu^{(i,r)}|) = 0} \prod_{r=1}^t \left(\prod_{\substack{1 \le i \le N \\ n \ge 1}} \frac{z^{-n\tilde m_n^{(i,r)}}}{\tilde m_n^{(i,r)}!} \cdot q^{\sum_{i=1}^N (\sum_{\ell=1}^r |\tilde\mu^{(i,\ell)}| - |\tilde\lambda^{(i)}|)} \right) \cdot$$

$$\cdot \mathrm{Tr}\, q^{\mathfrak{n}} \prod_{i=1}^N \frac{\mathfrak{a}_{\tilde\lambda^{(i)} - \sum_{r=1}^t \tilde\mu^{(i,r)}}(\tau_* \tilde\alpha_i)}{\left(\tilde\lambda^{(i)} - \sum_{r=1}^t \tilde\mu^{(i,r)}\right)^!}.$$

Denote line (5.13) by $\widetilde{U}(t)$. Since

$$\sum_{i=1}^N (|\tilde\lambda^{(i)}| - \sum_{r=1}^t |\tilde\mu^{(i,r)}|) < 0$$

and $|\tilde\mu^{(i,r)}| < 0$, $\widetilde{U}(t)$ is a polynomial in q with coefficients being bounded in terms of $-\sum_{i=1}^N |\tilde\lambda^{(i)}|$. Moreover, $q^t | \widetilde{U}(t)$. Line (5.14) is bounded in terms of the generalized partitions $\tilde\lambda^{(i)}$. Since $0 < q < 1$, $U(t) \to 0$ as $t \to +\infty$. Letting $t \to +\infty$, we see that Q_2 equals

$$\sum_{\sum_{i=1}^N (|\tilde\lambda^{(i)}| - \sum_{t \ge 1} |\tilde\mu^{(i,t)}|) = 0} \prod_{t \ge 1} \left(\prod_{\substack{1 \le i \le N \\ n \ge 1}} \frac{z^{-n\tilde m_n^{(i,t)}}}{\tilde m_n^{(i,t)}!} \cdot q^{\sum_{i=1}^N (\sum_{\ell=1}^t |\tilde\mu^{(i,\ell)}| - |\tilde\lambda^{(i)}|)} \right) \cdot$$

$$\cdot \mathrm{Tr}\, q^{\mathfrak{n}} \prod_{i=1}^N \frac{\mathfrak{a}_{\tilde\lambda^{(i)} - \sum_{t \ge 1} \tilde\mu^{(i,t)}}(\tau_* \tilde\alpha_i)}{\left(\tilde\lambda^{(i)} - \sum_{t \ge 1} \tilde\mu^{(i,t)}\right)^!}.$$

Replacing $q^{\sum_{i=1}^N (\sum_{\ell=1}^t |\tilde\mu^{(i,\ell)}| - |\tilde\lambda^{(i)}|)}$ by $q^{-\sum_{i=1}^N \sum_{\ell \ge t+1} |\tilde\mu^{(i,\ell)}|}$ proves our lemma. □

We remark that the assumption $0 < q < 1$ was used in the above proof. However, for most parts of this chapter (e.g. in (5.4)), q can also be viewed as a formal variable.

Let

$$A_N = \mathrm{Tr}\, q^{\mathfrak{n}} \prod_{i=1}^N \frac{\mathfrak{a}_{\lambda^{(i)}}(\tau_* \alpha_i)}{\lambda^{(i)}!} \tag{5.15}$$

where $\lambda^{(1)}, \ldots, \lambda^{(N)} \in \widetilde{\mathcal{P}}$ are generalized partitions, and $\alpha_1, \ldots, \alpha_N \in H^*(X)$ are homogeneous. Our Lemma 5.7 below deals with the trace A_N. To illustrate its proof, let us consider the following example.

EXAMPLE 5.6. Let $N = 3, \alpha_1 = \alpha_3 = 1_X, \alpha_2 = x, \lambda^{(1)} = (-4), \lambda^{(2)} = (4)$, and $\lambda^{(3)} = ((-5)5)$ in (5.15). Then,

$$A_3 = \operatorname{Tr} q^{\mathfrak{n}}\, \mathfrak{a}_{-4}(1_X)\mathfrak{a}_4(x)\mathfrak{a}_{((-5)5)}(\tau_{2*}1_X).$$

To evaluate A_3, we first apply Case 2 in the proof of Lemma 5.7 to eliminate $\mathfrak{a}_{-4}(1_X)\mathfrak{a}_4(x)$. By (5.21) with $i_0 = 1$ and $n_0 = 4$,

$$\begin{aligned} A_3 &= \frac{q^4}{1-q^4} \operatorname{Tr} q^{\mathfrak{n}}\, [\mathfrak{a}_4(x), \mathfrak{a}_{-4}(1_X)]\mathfrak{a}_{((-5)5)}(\tau_{2*}1_X) \\ &= \frac{(-4)q^4}{1-q^4} \cdot \langle x, 1_X\rangle \cdot \operatorname{Tr} q^{\mathfrak{n}}\, \mathfrak{a}_{((-5)5)}(\tau_{2*}1_X). \end{aligned}$$

Then we apply Case 1 in the proof of Lemma 5.7 to eliminate $\mathfrak{a}_{((-5)5)}(\tau_{2*}1_X)$. By (5.20) with $n_1 = 5$,

$$\begin{aligned} A_3 &= \frac{(-4)q^4}{1-q^4} \cdot \langle x, 1_X\rangle \cdot (q;q)_\infty^{-\chi(X)} \cdot \langle e_X, 1_X\rangle \cdot \frac{(-5)q^5}{1-q^5} \\ &= (q;q)_\infty^{-\chi(X)} \cdot \langle x, 1_X\rangle \cdot \langle e_X, 1_X\rangle \cdot \frac{4q^4}{1-q^4} \cdot \frac{5q^5}{1-q^5}. \end{aligned}$$

LEMMA 5.7. *Let $\lambda^{(1)}, \ldots, \lambda^{(N)} \in \widetilde{\mathcal{P}}$ be generalized partitions, and $\alpha_1, \ldots, \alpha_N \in H^*(X)$ be homogeneous. Then, the trace (5.15) can be computed by an induction on N, and is a linear combination of expressions of the form:*

$$(q;q)_\infty^{-\chi(X)} \cdot \operatorname{Sign}(\pi) \cdot \prod_{i=1}^{u} \left\langle e_X^{m_i}, \prod_{j\in\pi_i} \alpha_j \right\rangle \cdot \prod_{i=1}^{v} \frac{n_i^{k_i} q^{n_i}}{1-q^{n_i}} \tag{5.16}$$

where $0 \le v \le \sum_{i=1}^N \ell(\lambda^{(i)})/2$, $m_i \ge 0$, $n_i > 0$, all the integers involved and the partition $\{\pi_1, \ldots, \pi_u\}$ of $\{1, \ldots, N\}$ depend only on $\lambda^{(1)}, \ldots, \lambda^{(N)}$, and $\operatorname{Sign}(\pi)$ *depends on the parity of the degrees of $\alpha_1, \ldots, \alpha_N$ and is the sign defined by*

$$\prod_{i=1}^{u} \prod_{j\in\pi_i} \alpha_j = \operatorname{Sign}(\pi) \cdot \alpha_1 \cdots \alpha_N. \tag{5.17}$$

Moreover, the coefficients of this linear combination are independent of q, α_i, n_i, X.

PROOF. Note that the bi-degree of $\mathfrak{a}_{\lambda^{(i)}}(\tau_*\alpha_i)$ is equal to

$$\Big(|\lambda^{(i)}|, 2(\ell(\lambda^{(i)}) - 2) + |\alpha_i|\Big). \tag{5.18}$$

It follows that $A_N = 0$ unless

$$\sum_{i=1}^{N} |\lambda^{(i)}| = 0.$$

In the rest of the proof, we will assume $\sum_{i=1}^N |\lambda^{(i)}| = 0$. We will divide the proof into two cases.

Case 1: $|\lambda^{(i)}| = 0$ for every $1 \le i \le N$. Then, $\ell(\lambda^{(i)}) \ge 2$ for every i. Since the bi-degree of $\mathfrak{a}_{\lambda^{(i)}}(\tau_*\alpha_i)$ is given by (5.18), $A_N = 0$ unless $\ell(\lambda^{(i)}) = 2$ and $|\alpha_i| = 0$ for all $1 \le i \le N$. Assume that $\ell(\lambda^{(i)}) = 2$ and $|\alpha_i| = 0$ for all $1 \le i \le N$. Then for every $1 \le i \le N$, we have $\lambda^{(i)} = ((-n_i)n_i)$ for some $n_i > 0$. We further assume that $n_1 = \ldots = n_r$ for some $1 \le r \le N$ and $n_i \ne n_1$ if $r < i \le N$. Let $\alpha_1 = a 1_X$ and

$$\tau_{2*} 1_X = \sum_j (-1)^{|\beta_j|} \beta_j \otimes \gamma_j$$

with $\langle \beta_j, \gamma_{j'} \rangle = \delta_{j,j'}$. Then, A_N is equal to

$$\begin{aligned} & a \sum_j (-1)^{|\beta_j|} \mathrm{Tr}\, q^{\mathfrak{n}}\, \mathfrak{a}_{-n_1}(\beta_j)\, \mathfrak{a}_{n_1}(\gamma_j) \prod_{i=2}^N \mathfrak{a}_{\lambda^{(i)}}(\tau_* \alpha_i) \\ = \quad & a q^{n_1} \sum_j (-1)^{|\beta_j|} \mathrm{Tr}\, \mathfrak{a}_{-n_1}(\beta_j)\, q^{\mathfrak{n}}\, \mathfrak{a}_{n_1}(\gamma_j) \prod_{i=2}^N \mathfrak{a}_{\lambda^{(i)}}(\tau_* \alpha_i) \\ = \quad & a q^{n_1} \sum_j \mathrm{Tr}\, q^{\mathfrak{n}}\, \mathfrak{a}_{n_1}(\gamma_j) \prod_{i=2}^N \mathfrak{a}_{\lambda^{(i)}}(\tau_* \alpha_i) \cdot \mathfrak{a}_{-n_1}(\beta_j). \end{aligned}$$

Moving the operator $\mathfrak{a}_{-n_1}(\beta_j)$ to the middle, we see that A_N is equal to

$$\begin{aligned} & a q^{n_1} \sum_j \mathrm{Tr}\, q^{\mathfrak{n}}\, \mathfrak{a}_{n_1}(\gamma_j) \mathfrak{a}_{-n_1}(\beta_j) \prod_{i=2}^N \mathfrak{a}_{\lambda^{(i)}}(\tau_* \alpha_i) \\ + \quad & a q^{n_1} \sum_j \sum_{i=2}^r \mathrm{Tr}\, q^{\mathfrak{n}}\, \mathfrak{a}_{n_1}(\gamma_j) \prod_{k=2}^{i-1} \mathfrak{a}_{\lambda^{(k)}}(\tau_* \alpha_k) \\ & \cdot [\mathfrak{a}_{\lambda^{(i)}}(\alpha_i), \mathfrak{a}_{-n_1}(\beta_j)] \cdot \prod_{k=i+1}^N \mathfrak{a}_{\lambda^{(k)}}(\tau_* \alpha_k). \end{aligned}$$

By Lemma 3.12 (i), A_N is equal to the sum of the expressions

$$\left\langle e_X, \alpha_1 \prod_{i=1}^{k_1} \alpha_{j_i} \right\rangle \cdot \frac{(-n_1)^{k_1} q^{n_1}}{1 - q^{n_1}} \cdot \mathrm{Tr}\, q^{\mathfrak{n}} \prod_{i \in \{2,\ldots,N\} - \{j_1,\ldots,j_{k_1}\}} \mathfrak{a}_{\lambda^{(i)}}(\tau_* \alpha_i) \tag{5.19}$$

where $0 \le k_1 \le r-1$, $\{j_1, \ldots, j_{k_1}\} \subset \{2, \ldots, r\}$, every factor in $(-n_1)^{k_1}$ comes from a commutator of type $[\mathfrak{a}_{n_1}(\cdot), \mathfrak{a}_{-n_1}(\cdot)]$, and the coefficients of this linear combination depend only on $\lambda^{(1)}, \ldots, \lambda^{(N)}$. In particular, we have

$$A_1 = \mathrm{Tr}\, q^{\mathfrak{n}}\, \mathfrak{a}_{\lambda^{(1)}}(\alpha_1) = (q;q)_\infty^{-\chi(X)} \cdot \langle e_X, \alpha_1 \rangle \cdot \frac{(-n_1) q^{n_1}}{1 - q^{n_1}}. \tag{5.20}$$

Combining with (5.19), we see that our lemma holds in this case.

Case 2: $\sum_{i=1}^N |\lambda^{(i)}| = 0$ but $|\lambda^{(i_0)}| \ne 0$ for some i_0. Then $N \ge 2$, and we may assume that $|\lambda^{(i_0)}| < 0$. To simplify the expressions, we further assume that every α_i has an even degree. Note that A_N can be rewritten as

$$\begin{aligned} & \mathrm{Tr}\, q^{\mathfrak{n}} \frac{\mathfrak{a}_{\lambda^{(i_0)}}(\tau_* \alpha_{i_0})}{\lambda^{(i_0)}!} \prod_{\substack{1 \le i \le N \\ i \ne i_0}} \frac{\mathfrak{a}_{\lambda^{(i)}}(\tau_* \alpha_i)}{\lambda^{(i)}!} \\ + \quad & \sum_{r=1}^{i_0-1} \mathrm{Tr}\, q^{\mathfrak{n}} \prod_{i=1}^{r-1} \frac{\mathfrak{a}_{\lambda^{(i)}}(\tau_* \alpha_i)}{\lambda^{(i)}!} \cdot \left[\frac{\mathfrak{a}_{\lambda^{(r)}}(\tau_* \alpha_r)}{\lambda^{(r)}!}, \frac{\mathfrak{a}_{\lambda^{(i_0)}}(\tau_* \alpha_{i_0})}{\lambda^{(i_0)}!} \right] \cdot \prod_{\substack{r+1 \le i \le N \\ i \ne i_0}} \frac{\mathfrak{a}_{\lambda^{(i)}}(\tau_* \alpha_i)}{\lambda^{(i)}!}. \end{aligned}$$

Since

$$q^{\mathfrak{n}}\mathfrak{a}_{\lambda^{(i_0)}}(\tau_*\alpha_{i_0}) = q^{-|\lambda^{(i_0)}|}\mathfrak{a}_{\lambda^{(i_0)}}(\tau_*\alpha_{i_0})q^{\mathfrak{n}},$$

we see that A_N is equal to

$$q^{-|\lambda^{(i_0)}|}\,\mathrm{Tr}\,\frac{\mathfrak{a}_{\lambda^{(i_0)}}(\tau_*\alpha_{i_0})}{\lambda^{(i_0)}!}q^{\mathfrak{n}}\prod_{\substack{1\le i\le N\\ i\ne i_0}}\frac{\mathfrak{a}_{\lambda^{(i)}}(\tau_*\alpha_i)}{\lambda^{(i)}!}$$

$$+\sum_{r=1}^{i_0-1}\mathrm{Tr}\,q^{\mathfrak{n}}\prod_{i=1}^{r-1}\frac{\mathfrak{a}_{\lambda^{(i)}}(\tau_*\alpha_i)}{\lambda^{(i)}!}\cdot\left[\frac{\mathfrak{a}_{\lambda^{(r)}}(\tau_*\alpha_r)}{\lambda^{(r)}!},\frac{\mathfrak{a}_{\lambda^{(i_0)}}(\tau_*\alpha_{i_0})}{\lambda^{(i_0)}!}\right]\cdot\prod_{\substack{r+1\le i\le N\\ i\ne i_0}}\frac{\mathfrak{a}_{\lambda^{(i)}}(\tau_*\alpha_i)}{\lambda^{(i)}!}.$$

Since

$$\mathrm{Tr}\,\frac{\mathfrak{a}_{\lambda^{(i_0)}}(\tau_*\alpha_{i_0})}{\lambda^{(i_0)}!}q^{\mathfrak{n}}\prod_{\substack{1\le i\le N\\ i\ne i_0}}\frac{\mathfrak{a}_{\lambda^{(i)}}(\tau_*\alpha_i)}{\lambda^{(i)}!} = \mathrm{Tr}\,q^{\mathfrak{n}}\prod_{\substack{1\le i\le N\\ i\ne i_0}}\frac{\mathfrak{a}_{\lambda^{(i)}}(\tau_*\alpha_i)}{\lambda^{(i)}!}\cdot\frac{\mathfrak{a}_{\lambda^{(i_0)}}(\tau_*\alpha_{i_0})}{\lambda^{(i_0)}!},$$

A_N is equal to

$$q^{-|\lambda^{(i_0)}|}\,\mathrm{Tr}\,q^{\mathfrak{n}}\prod_{\substack{1\le i\le N\\ i\ne i_0}}\frac{\mathfrak{a}_{\lambda^{(i)}}(\tau_*\alpha_i)}{\lambda^{(i)}!}\cdot\frac{\mathfrak{a}_{\lambda^{(i_0)}}(\tau_*\alpha_{i_0})}{\lambda^{(i_0)}!}$$

$$+\sum_{r=1}^{i_0-1}\mathrm{Tr}\,q^{\mathfrak{n}}\prod_{i=1}^{r-1}\frac{\mathfrak{a}_{\lambda^{(i)}}(\tau_*\alpha_i)}{\lambda^{(i)}!}\cdot\left[\frac{\mathfrak{a}_{\lambda^{(r)}}(\tau_*\alpha_r)}{\lambda^{(r)}!},\frac{\mathfrak{a}_{\lambda^{(i_0)}}(\tau_*\alpha_{i_0})}{\lambda^{(i_0)}!}\right]\cdot\prod_{\substack{r+1\le i\le N\\ i\ne i_0}}\frac{\mathfrak{a}_{\lambda^{(i)}}(\tau_*\alpha_i)}{\lambda^{(i)}!}.$$

Note that

$$\mathrm{Tr}\,q^{\mathfrak{n}}\prod_{\substack{1\le i\le N\\ i\ne i_0}}\frac{\mathfrak{a}_{\lambda^{(i)}}(\tau_*\alpha_i)}{\lambda^{(i)}!}\cdot\frac{\mathfrak{a}_{\lambda^{(i_0)}}(\tau_*\alpha_{i_0})}{\lambda^{(i_0)}!}$$

is equal to

$$A_N$$

$$+\sum_{r=i_0+1}^{N}\mathrm{Tr}\,q^{\mathfrak{n}}\prod_{\substack{1\le i\le r-1\\ i\ne i_0}}\frac{\mathfrak{a}_{\lambda^{(i)}}(\tau_*\alpha_i)}{\lambda^{(i)}!}\cdot\left[\frac{\mathfrak{a}_{\lambda^{(r)}}(\tau_*\alpha_r)}{\lambda^{(r)}!},\frac{\mathfrak{a}_{\lambda^{(i_0)}}(\tau_*\alpha_{i_0})}{\lambda^{(i_0)}!}\right]\cdot\prod_{i=r+1}^{N}\frac{\mathfrak{a}_{\lambda^{(i)}}(\tau_*\alpha_i)}{\lambda^{(i)}!}.$$

Therefore, we conclude that $(1-q^{-|\lambda^{(i_0)}|})A_N$ is equal to

$$q^{-|\lambda^{(i_0)}|}\sum_{r=i_0+1}^{N}\mathrm{Tr}\,q^{\mathfrak{n}}\prod_{1\le i\le r-1,i\ne i_0}\frac{\mathfrak{a}_{\lambda^{(i)}}(\tau_*\alpha_i)}{\lambda^{(i)}!}$$

$$\cdot\left[\frac{\mathfrak{a}_{\lambda^{(r)}}(\tau_*\alpha_r)}{\lambda^{(r)}!},\frac{\mathfrak{a}_{\lambda^{(i_0)}}(\tau_*\alpha_{i_0})}{\lambda^{(i_0)}!}\right]\cdot\prod_{i=r+1}^{N}\frac{\mathfrak{a}_{\lambda^{(i)}}(\tau_*\alpha_i)}{\lambda^{(i)}!}$$

$$+\sum_{r=1}^{i_0-1}\mathrm{Tr}\,q^{\mathfrak{n}}\prod_{i=1}^{r-1}\frac{\mathfrak{a}_{\lambda^{(i)}}(\tau_*\alpha_i)}{\lambda^{(i)}!}\cdot\left[\frac{\mathfrak{a}_{\lambda^{(r)}}(\tau_*\alpha_r)}{\lambda^{(r)}!},\frac{\mathfrak{a}_{\lambda^{(i_0)}}(\tau_*\alpha_{i_0})}{\lambda^{(i_0)}!}\right]\cdot\prod_{\substack{r+1\le i\le N\\ i\ne i_0}}\frac{\mathfrak{a}_{\lambda^{(i)}}(\tau_*\alpha_i)}{\lambda^{(i)}!}.$$

Put $n_0 = -|\lambda^{(i_0)}| > 0$. It follows that A_N is equal to

$$
\begin{aligned}
&\frac{q^{n_0}}{1-q^{n_0}} \sum_{r=i_0+1}^{N} \operatorname{Tr} q^{\mathfrak{n}} \prod_{1\le i\le r-1, i\ne i_0} \frac{\mathfrak{a}_{\lambda^{(i)}}(\tau_*\alpha_i)}{\lambda^{(i)}!} \\
&\cdot \left[\frac{\mathfrak{a}_{\lambda^{(r)}}(\tau_*\alpha_r)}{\lambda^{(r)}!}, \frac{\mathfrak{a}_{\lambda^{(i_0)}}(\tau_*\alpha_{i_0})}{\lambda^{(i_0)}!}\right] \cdot \prod_{i=r+1}^{N} \frac{\mathfrak{a}_{\lambda^{(i)}}(\tau_*\alpha_i)}{\lambda^{(i)}!} \\
&+ \frac{1}{1-q^{n_0}} \sum_{r=1}^{i_0-1} \operatorname{Tr} q^{\mathfrak{n}} \prod_{i=1}^{r-1} \frac{\mathfrak{a}_{\lambda^{(i)}}(\tau_*\alpha_i)}{\lambda^{(i)}!} \cdot \left[\frac{\mathfrak{a}_{\lambda^{(r)}}(\tau_*\alpha_r)}{\lambda^{(r)}!}, \frac{\mathfrak{a}_{\lambda^{(i_0)}}(\tau_*\alpha_{i_0})}{\lambda^{(i_0)}!}\right] \\
&\cdot \prod_{r+1\le i\le N, i\ne i_0} \frac{\mathfrak{a}_{\lambda^{(i)}}(\tau_*\alpha_i)}{\lambda^{(i)}!}.
\end{aligned} \tag{5.21}
$$

By Lemma 3.12 (i) and (ii), our lemma holds in this case as well. □

The following theorem provides the structure of the trace

$$\operatorname{Tr} q^{\mathfrak{n}}\, \mathbf{W}(\mathfrak{L}_1, z) \prod_{i=1}^{N} \frac{\mathfrak{a}_{\lambda^{(i)}}(\tau_*\alpha_i)}{\lambda^{(i)}!}.$$

THEOREM 5.8. *For $1 \le i \le N$, let*

$$\lambda^{(i)} = \big(\cdots(-n)^{\tilde{m}_n^{(i)}} \cdots (-1)^{\tilde{m}_1^{(i)}} 1^{m_1^{(i)}} \cdots n^{m_n^{(i)}} \cdots\big)$$

and $\alpha_i \in H^*(X)$ be homogeneous. Then, the trace

$$\operatorname{Tr} q^{\mathfrak{n}}\, \mathbf{W}(\mathfrak{L}_1, z) \prod_{i=1}^{N} \frac{\mathfrak{a}_{\lambda^{(i)}}(\tau_*\alpha_i)}{\lambda^{(i)}!} \tag{5.22}$$

is equal to

$$
\begin{aligned}
&z^{\sum_{i=1}^N |\lambda^{(i)}|} \cdot (q;q)_\infty^{-\chi(X)} \cdot \prod_{i=1}^{N} \big\langle (1_X - K_X)^{\sum_{n\ge 1} m_n^{(i)}}, \alpha_i \big\rangle \\
&\cdot \prod_{1\le i\le N, n\ge 1} \left(\frac{(-1)^{m_n^{(i)}}}{m_n^{(i)}!} \frac{q^{n m_n^{(i)}}}{(1-q^n)^{m_n^{(i)}}} \frac{1}{\tilde{m}_n^{(i)}!} \frac{1}{(1-q^n)^{\tilde{m}_n^{(i)}}} \right) + \widetilde{W},
\end{aligned}
$$

and the lower weight term $\widetilde{W}$ is a linear combination of expressions of the form:

$$
\begin{aligned}
&z^{\sum_{i=1}^N |\lambda^{(i)}|} \cdot (q;q)_\infty^{-\chi(X)} \cdot \operatorname{Sign}(\pi) \cdot \prod_{i=1}^{u} \Big\langle K_X^{r_i} e_X^{r_i'}, \prod_{j\in\pi_i} \alpha_j \Big\rangle \\
&\cdot \prod_{i=1}^{v} \frac{q^{n_i w_i p_i}}{(1-q^{n_i})^{w_i}}
\end{aligned} \tag{5.23}
$$

where $\sum_{i=1}^{v} w_i < \sum_{i=1}^{N} \ell(\lambda^{(i)})$, the integers $u, v, r_i, r_i' \ge 0, n_i > 0, w_i > 0, p_i \in \{0,1\}$ and the partition $\pi = \{\pi_1, \ldots, \pi_u\}$ of $\{1, \ldots, N\}$ depend only on the generalized partitions $\lambda^{(1)}, \ldots, \lambda^{(N)}$, and $\operatorname{Sign}(\pi)$ is from (5.17). Moreover, the coefficients of this linear combination are independent of $q, \alpha_1, \ldots, \alpha_N$ and X.

PROOF. For simplicity, denote the trace (5.22) by Tr_λ. Combining Lemma 5.4 and Lemma 5.5, we conclude that Tr_λ is equal to

$$\sum_{\substack{\sum_{i=1}^N (|\lambda^{(i)}|-\sum_{s\ge 1}|\mu^{(i,s)}|-\sum_{t\ge 1}|\tilde\mu^{(i,t)}|)=0\\ \mu^{(i,s)}\in\widetilde{\mathcal{P}}_+,\,\tilde\mu^{(i,t)}\in\widetilde{\mathcal{P}}_-}} \prod_{\substack{1\le i\le N\\ s,n\ge 1}} \frac{(-(zq^s)^n)^{m_n^{(i,s)}}}{m_n^{(i,s)}!}\cdot \prod_{\substack{1\le i\le N\\ t,n\ge 1}} \frac{(z^{-1}q^{t-1})^{n\tilde m_n^{(i,t)}}}{\tilde m_n^{(i,t)}!}$$

$$\cdot \mathrm{Tr}\, q^{\mathfrak{n}} \prod_{i=1}^N \frac{\mathfrak{a}_{\lambda^{(i)}-\sum_{s\ge 1}\mu^{(i,s)}-\sum_{t\ge 1}\tilde\mu^{(i,t)}}\big(\tau_*((1_X-K_X)^{\sum_{s,n\ge 1} m_n^{(i,s)}}\alpha_i)\big)}{\big(\lambda^{(i)}-\sum_{s\ge 1}\mu^{(i,s)}-\sum_{t\ge 1}\tilde\mu^{(i,t)}\big)^!}$$

where $\mu^{(i,s)} = \big(1^{m_1^{(i,s)}}\cdots n^{m_n^{(i,s)}}\cdots\big)$ and $\tilde\mu^{(i,t)} = \big(\cdots(-n)^{\tilde m_n^{(i,t)}}\cdots(-1)^{\tilde m_1^{(i,t)}}\big)$. The sum of all the exponents of z is $\sum_{i=1}^N |\lambda^{(i)}|$. So Tr_λ is equal to

$$z^{\sum_{i=1}^N |\lambda^{(i)}|}\cdot \sum_{\substack{\sum_{i=1}^N (|\lambda^{(i)}|-\sum_{s\ge 1}|\mu^{(i,s)}|-\sum_{t\ge 1}|\tilde\mu^{(i,t)}|)=0\\ \mu^{(i,s)}\in\widetilde{\mathcal{P}}_+,\,\tilde\mu^{(i,t)}\in\widetilde{\mathcal{P}}_-}}$$

$$\prod_{\substack{1\le i\le N\\ s,n\ge 1}} \frac{(-q^{sn})^{m_n^{(i,s)}}}{m_n^{(i,s)}!}\cdot \prod_{\substack{1\le i\le N\\ t,n\ge 1}} \frac{q^{(t-1)n\tilde m_n^{(i,t)}}}{\tilde m_n^{(i,t)}!}\cdot$$

$$\cdot \mathrm{Tr}\, q^{\mathfrak{n}} \prod_{i=1}^N \frac{\mathfrak{a}_{\lambda^{(i)}-\sum_{s\ge 1}\mu^{(i,s)}-\sum_{t\ge 1}\tilde\mu^{(i,t)}}\big(\tau_*((1_X-K_X)^{\sum_{s,n\ge 1} m_n^{(i,s)}}\alpha_i)\big)}{\big(\lambda^{(i)}-\sum_{s\ge 1}\mu^{(i,s)}-\sum_{t\ge 1}\tilde\mu^{(i,t)}\big)^!}. \tag{5.24}$$

By our convention,

$$\sum_{s\ge 1}\mu^{(i,s)} + \sum_{t\ge 1}\tilde\mu^{(i,t)} \le \lambda^{(i)}$$

for every $1\le i\le N$. In the rest of the proof, we group the possible terms in the above summand into two cases: Case A and Case B.

Case A: $\sum_{s\ge 1}\mu^{(i,s)} + \sum_{t\ge 1}\tilde\mu^{(i,t)} = \lambda^{(i)}$ for every $1\le i\le N$. Then line (5.24) is equal to

$$\begin{aligned} &\mathrm{Tr}\, q^{\mathfrak{n}}\cdot \prod_{i=1}^N \big\langle (1_X-K_X)^{\sum_{s,n\ge 1} m_n^{(i,s)}}, \alpha_i\big\rangle \\ =\ & (q;q)_\infty^{-\chi(X)} \prod_{i=1}^N \big\langle (1_X-K_X)^{\sum_{s,n\ge 1} m_n^{(i,s)}}, \alpha_i\big\rangle. \end{aligned}$$

Therefore, the contribution C_1 of this case to Tr_λ is equal to

$$z^{\sum_{i=1}^N |\lambda^{(i)}|}\cdot (q;q)_\infty^{-\chi(X)} \prod_{i=1}^N \big\langle (1_X-K_X)^{\sum_{n\ge 1} m_n^{(i)}}, \alpha_i\big\rangle\cdot$$

$$\cdot \sum_{\substack{\sum_{s\ge 1} m_n^{(i,s)} = m_n^{(i)}\\ 1\le i\le N,\, n\ge 1}} \prod_{\substack{1\le i\le N\\ s,n\ge 1}} \frac{(-q^{sn})^{m_n^{(i,s)}}}{m_n^{(i,s)}!}\cdot \sum_{\substack{\sum_{t\ge 1}\tilde m_n^{(i,t)} = \tilde m_n^{(i)}\\ 1\le i\le N,\, n\ge 1}} \prod_{\substack{1\le i\le N\\ t,n\ge 1}} \frac{q^{(t-1)n\tilde m_n^{(i,t)}}}{\tilde m_n^{(i,t)}!}.$$

Rewrite q^{sn} as $q^{(s-1)n}q^n$. Then the contribution C_1 is equal to

$$z^{\sum_{i=1}^N |\lambda^{(i)}|} \cdot (q;q)_\infty^{-\chi(X)} \prod_{i=1}^N \langle (1_X - K_X)^{\sum_{n\geq 1} m_n^{(i)}}, \alpha_i \rangle \cdot \prod_{1\leq i\leq N, n\geq 1} (-q^n)^{m_n^{(i)}} \cdot$$

$$\cdot \sum_{\substack{\sum_{s\geq 1} m_n^{(i,s)} = m_n^{(i)} \\ 1\leq i\leq N,\, n\geq 1}} \prod_{\substack{1\leq i\leq N \\ s,n\geq 1}} \frac{q^{(s-1)nm_n^{(i,s)}}}{m_n^{(i,s)}!} \cdot \sum_{\substack{\sum_{t\geq 1} \tilde{m}_n^{(i,t)} = \tilde{m}_n^{(i)} \\ 1\leq i\leq N,\, n\geq 1}} \prod_{\substack{1\leq i\leq N \\ t,n\geq 1}} \frac{q^{(t-1)n\tilde{m}_n^{(i,t)}}}{\tilde{m}_n^{(i,t)}!}.$$

Since

$$\sum_{\sum_{s\geq 1} i_{s,n} = i_n,\, n\geq 1} \prod_{s,n\geq 1} \frac{(q^{(s-1)n})^{i_{s,n}}}{i_{s,n}!} = \prod_{n\geq 1} \left(\frac{1}{i_n!} \frac{1}{(1-q^n)^{i_n}} \right),$$

C_1 is equal to

$$z^{\sum_{i=1}^N |\lambda^{(i)}|} \cdot (q;q)_\infty^{-\chi(X)} \prod_{i=1}^N \langle (1_X - K_X)^{\sum_{n\geq 1} m_n^{(i)}}, \alpha_i \rangle \cdot \tag{5.25}$$

$$\cdot \prod_{1\leq i\leq N, n\geq 1} \left(\frac{(-1)^{m_n^{(i)}}}{m_n^{(i)}!} \frac{q^{nm_n^{(i)}}}{(1-q^n)^{m_n^{(i)}}} \right) \cdot \prod_{1\leq i\leq N, n\geq 1} \left(\frac{1}{\tilde{m}_n^{(i)}!} \frac{1}{(1-q^n)^{\tilde{m}_n^{(i)}}} \right).$$

Case B: $\sum_{s\geq 1} \mu^{(i,s)} + \sum_{t\geq 1} \tilde{\mu}^{(i,t)} < \lambda^{(i)}$ for some $1 \leq i \leq N$. Without loss of generality, we may assume that

$$\sum_{s\geq 1} \mu^{(i,s)} + \sum_{t\geq 1} \tilde{\mu}^{(i,t)} = \lambda^{(i)}$$

for every $1 \leq i \leq N_1$ where $N_1 < N$, and

$$\sum_{s\geq 1} \mu^{(i,s)} + \sum_{t\geq 1} \tilde{\mu}^{(i,t)} < \lambda^{(i)}$$

for every $N_1 + 1 \leq i \leq N$. For $N_1 + 1 \leq i \leq N$, put

$$\sum_{s\geq 1} \mu^{(i,s)} + \sum_{t\geq 1} \tilde{\mu}^{(i,t)} = \tilde{\lambda}^{(i)} = (\cdots (-n)^{\tilde{p}_n^{(i)}} \cdots (-1)^{\tilde{p}_1^{(i)}} 1^{p_1^{(i)}} \cdots n^{p_n^{(i)}} \cdots).$$

An argument similar to that in the previous paragraph shows that for the fixed generalized partitions $\tilde{\lambda}^{(i)}$ with $N_1 + 1 \leq i \leq N$, the contribution C_2 of this case to Tr_λ is equal to

$$z^{\sum_{i=1}^N |\lambda^{(i)}|} \cdot \prod_{i=1}^{N_1} \langle (1_X - K_X)^{\sum_{n\geq 1} m_n^{(i)}}, \alpha_i \rangle \tag{5.26}$$

$$\cdot \prod_{\substack{1\leq i\leq N_1 \\ n\geq 1}} \left(\frac{(-1)^{m_n^{(i)}}}{m_n^{(i)}!} \frac{q^{nm_n^{(i)}}}{(1-q^n)^{m_n^{(i)}}} \frac{1}{\tilde{m}_n^{(i)}!} \frac{1}{(1-q^n)^{\tilde{m}_n^{(i)}}} \right)$$

$$\cdot \prod_{\substack{N_1+1\leq i\leq N \\ n\geq 1}} \left(\frac{(-1)^{p_n^{(i)}}}{p_n^{(i)}!} \frac{q^{np_n^{(i)}}}{(1-q^n)^{p_n^{(i)}}} \frac{1}{\tilde{p}_n^{(i)}!} \frac{1}{(1-q^n)^{\tilde{p}_n^{(i)}}} \right)$$

$$\cdot \mathrm{Tr}\, q^{\mathfrak{n}} \prod_{i=N_1+1}^N \frac{\mathfrak{a}_{\lambda^{(i)} - \tilde{\lambda}^{(i)}} \left(\tau_* ((1_X - K_X)^{\sum_{n\geq 1} p_n^{(i)}} \alpha_i) \right)}{\left(\lambda^{(i)} - \tilde{\lambda}^{(i)} \right)^!}.$$

By Lemma 5.7, C_2 is a linear combination of expressions of the form:

$$z^{\sum_{i=1}^N |\lambda^{(i)}|} \cdot \prod_{i=1}^{N_1} \langle (1_X - K_X)^{\sum_{n\geq 1} m_n^{(i)}}, \alpha_i\rangle$$

$$\cdot \prod_{\substack{1\leq i\leq N_1\\ n\geq 1}} \left(\frac{(-1)^{m_n^{(i)}}}{m_n^{(i)}!} \frac{q^{nm_n^{(i)}}}{(1-q^n)^{m_n^{(i)}}} \frac{1}{\tilde{m}_n^{(i)}!} \frac{1}{(1-q^n)^{\tilde{m}_n^{(i)}}} \right)$$

$$\cdot \prod_{\substack{N_1+1\leq i\leq N\\ n\geq 1}} \left(\frac{(-1)^{p_n^{(i)}}}{p_n^{(i)}!} \frac{q^{np_n^{(i)}}}{(1-q^n)^{p_n^{(i)}}} \frac{1}{\tilde{p}_n^{(i)}!} \frac{1}{(1-q^n)^{\tilde{p}_n^{(i)}}} \right).$$

$$(q;q)_\infty^{-\chi(X)} \cdot \mathrm{Sign}(\pi) \cdot \prod_{i=1}^{u} \left\langle e_X^{m_i}, \prod_{j\in\pi_i} ((1_X - K_X)^{\sum_{n\geq 1} p_n^{(j)}} \alpha_j) \right\rangle \cdot \prod_{i=1}^{v} \frac{q^{n_i}}{1-q^{n_i}}$$

where $v < \sum_{i=N_1+1}^N \ell(\lambda^{(i)} - \tilde{\lambda}^{(i)})$, $n_i > 0$, $m_i \geq 0$, $\{\pi_1, \dots, \pi_u\}$ is a partition of $\{N_1+1, \dots, N\}$, and $\mathrm{Sign}(\pi)$ compensates the formal difference between $\prod_{i=1}^u \prod_{j\in\pi_i} \alpha_j$ and $\alpha_{N_1+1}\cdots\alpha_N$. The coefficients of this linear combination are independent of $q, \alpha_1, \dots, \alpha_N$ and X, and depend only on the partitions $\lambda^{(i)} - \tilde{\lambda}^{(i)}$. Note that for nonnegative integers a and b, the pairing

$$\langle e_X^a, (1_X - K_X)^b \beta\rangle = \langle e_X^a (1_X - K_X)^b, \beta\rangle$$

is a linear combination of $\langle e_X^a K_X^c, \beta\rangle, 0 \leq c \leq b$. In addition, we have

$$\sum_{1\leq i\leq N_1, n\geq 1} (m_n^{(i)} + \tilde{m}_n^{(i)}) + \sum_{N_1+1\leq i\leq N, n\geq 1} (p_n^{(i)} + \tilde{p}_n^{(i)}) + v < \sum_{i=1}^N \ell(\lambda^{(i)})$$

regarding the weights in C_2. It follows that C_2 is a linear combination of the expressions (5.23). Combining with (5.25) completes the proof of our theorem. □

REMARK 5.9. When $N = 1$, we can work out the lower weight term $\widetilde{W}$ in Theorem 5.8 by examining its proof more carefully and by using (5.20). To state the result, let $\lambda = (\cdots(-n)^{\tilde{m}_n}\cdots(-1)^{\tilde{m}_1} 1^{m_1}\cdots n^{m_n}\cdots) \in \widetilde{\mathcal{P}}$. For $n_1 \geq 1$ with $m_{n_1} \cdot \tilde{m}_{n_1} \geq 1$, define $m_{n_1}(n_1) = m_{n_1} - 1$, $\tilde{m}_{n_1}(n_1) = \tilde{m}_{n_1} - 1$, and $m_n(n_1) = m_n$ and $\tilde{m}_n(n_1) = \tilde{m}_n$ if $n \neq n_1$. Then,

$$\mathrm{Tr}\, q^{\mathfrak{n}}\, \mathbf{W}(\mathfrak{L}_1, z)\, \frac{\mathfrak{a}_\lambda(\tau_*\alpha)}{\lambda!}$$

is equal to the sum

$$z^{|\lambda|} \cdot (q;q)_\infty^{-\chi(X)} \cdot \langle (1_X - K_X)^{\sum_{n\geq 1} m_n}, \alpha\rangle \cdot \prod_{n\geq 1} \left(\frac{(-1)^{m_n}}{m_n!} \frac{q^{nm_n}}{(1-q^n)^{m_n}} \frac{1}{\tilde{m}_n!} \frac{1}{(1-q^n)^{\tilde{m}_n}} \right)$$
$$+ z^{|\lambda|} \cdot (q;q)_\infty^{-\chi(X)} \cdot \langle e_X, \alpha\rangle \cdot \sum_{n_1\geq 1 \text{ with } m_{n_1}\cdot\tilde{m}_{n_1}\geq 1} \frac{n_1 q^{n_1}}{1-q^{n_1}} \cdot \prod_{n\geq 1} \left(\frac{(-1)^{m_n(n_1)}}{m_n(n_1)!} \frac{q^{nm_n(n_1)}}{(1-q^n)^{m_n(n_1)}} \frac{1}{\tilde{m}_n(n_1)!} \frac{1}{(1-q^n)^{\tilde{m}_n(n_1)}} \right) \tag{5.27}$$

where the first two lines come from Case A in the proof of Theorem 5.8, and the last two lines come from Case B in the proof of Theorem 5.8.

EXAMPLE 5.10. Let $\lambda = ((-5)(-4)5)$. Then, we see from (5.27) with $n_1 = 5$ that $\mathrm{Tr}\, q^{\mathfrak{n}}\, \mathbf{W}(\mathfrak{L}_1, z)\, \mathfrak{a}_{((-5)(-4)5)}(\tau_{3*}\alpha)$ is equal to

$$\begin{aligned} & z^{-4} \cdot (q;q)_\infty^{-\chi(X)} \cdot \langle 1_X - K_X, \alpha\rangle \cdot \frac{q^4}{1-q^4} \cdot \frac{-q^5}{1-q^5} \frac{q^5}{1-q^5} \\ + \quad & z^{-4} \cdot (q;q)_\infty^{-\chi(X)} \cdot \langle e_X, \alpha\rangle \cdot \frac{5q^5}{1-q^5} \cdot \frac{1}{1-q^4} \end{aligned}$$

where the first line is the leading term and the second line is the lower weight term.

The next lemma is used to organize the leading term in Theorem 5.8.

LEMMA 5.11. *For $\alpha \in H^*(X)$ and $k \geq 0$, define $\Theta_k^\alpha(q)$ to be*

$$(5.28) \qquad - \sum_{\ell(\lambda)=k+2, |\lambda|=0} \left\langle (1_X - K_X)^{\sum_{n\geq 1} i_n}, \alpha \right\rangle \cdot \prod_{n \geq 1} \left(\frac{(-1)^{i_n}}{i_n!} \frac{q^{n i_n}}{(1-q^n)^{i_n}} \frac{1}{\tilde{i}_n!} \frac{1}{(1-q^n)^{\tilde{i}_n}} \right)$$

where $\lambda = (\cdots (-n)^{\tilde{i}_n} \cdots (-1)^{\tilde{i}_1} 1^{i_1} \cdots n^{i_n} \cdots)$. Then,

$$\Theta_k^\alpha(q) = \mathrm{Coeff}_{z^0} \Theta_k^\alpha(q, z)$$

which denotes the coefficient of z^0 in $\Theta_k^\alpha(q, z)$ defined by

$$(5.29) \qquad - \sum_{\substack{a, s_1, \ldots, s_a, b, t_1, \ldots, t_b \geq 1 \\ \sum_{i=1}^a s_i + \sum_{j=1}^b t_j = k+2}} \left\langle (1_X - K_X)^{\sum_{i=1}^a s_i}, \alpha \right\rangle \prod_{i=1}^a \frac{(-1)^{s_i}}{s_i!} \cdot \prod_{j=1}^b \frac{1}{t_j!} \cdot \sum_{n_1 > \cdots > n_a > 0} \prod_{i=1}^a \frac{(qz)^{n_i s_i}}{(1-q^{n_i})^{s_i}} \cdot \sum_{m_1 > \cdots > m_b > 0} \prod_{j=1}^b \frac{z^{-m_j t_j}}{(1-q^{m_j})^{t_j}}.$$

PROOF. Put

$$A = \left\langle (1_X - K_X)^{\sum_{n\geq 1} i_n}, \alpha \right\rangle$$

which implicitly depends on $\sum_{n\geq 1} i_n$. Rewrite $|\lambda|$ and $\ell(\lambda)$ in terms of the integers i_n and $\tilde{i}_n$. Then, $\Theta_k^\alpha(q)$ is equal to

$$(5.30) \qquad - \sum_{\substack{\sum_{n\geq 1} i_n + \sum_{n\geq 1} \tilde{i}_n = k+2 \\ \sum_{n\geq 1} n i_n = \sum_{n \geq 1} n \tilde{i}_n > 0}} A \prod_{n\geq 1} \left(\frac{(-1)^{i_n}}{i_n!} \frac{q^{n i_n}}{(1-q^n)^{i_n}} \frac{1}{\tilde{i}_n!} \frac{1}{(1-q^n)^{\tilde{i}_n}} \right).$$

Denote the *positive* integers in the *ordered* list $\{i_1, \ldots, i_n, \ldots\}$ by $s_a, \ldots, s_1$ respectively (e.g., if the ordered list $\{i_1, \ldots, i_n, \ldots\}$ is $\{2, 0, 5, 4, 0, \ldots\}$, then $a = 3$ with $s_3 = 2, s_2 = 5, s_1 = 4$). We have $a \geq 1$. Similarly, denote the *positive* integers in the *ordered* list $\{\tilde{i}_1, \ldots, \tilde{i}_n, \ldots\}$ by $t_b, \ldots, t_1$ respectively. Then $b \geq 1$. Since $\sum_{n\geq 1} i_n = \sum_{i=1}^a s_i$, we get

$$A = \left\langle (1_X - K_X)^{\sum_{i=1}^a s_i}, \alpha \right\rangle.$$

Rewriting (5.30) in terms of $s_a, \ldots, s_1$ and $t_b, \ldots, t_1$, we see that

$$\Theta_k^\alpha(q) = \mathrm{Coeff}_{z^0} \Theta_k^\alpha(q, z).$$

This completes the proof of the lemma. ∎

We remark that the multiple q-zeta value $\Theta^{\alpha}_{k}(q,z)$ has weight $(k+2)$.

THEOREM 5.12. *For $1 \le i \le N$, let $k_i \ge 0$ and $\alpha_i \in H^*(X)$ be homogeneous. Then,*

$$F^{\alpha_1,\ldots,\alpha_N}_{k_1,\ldots,k_N}(q) = (q;q)_\infty^{-\chi(X)} \cdot \mathrm{Coeff}_{z_1^0\cdots z_N^0}\left(\prod_{i=1}^N \Theta^{\alpha_i}_{k_i}(q,z_i)\right) + W_1, \tag{5.31}$$

and the lower weight term W_1 is an infinite linear combination of the expressions:

$$(q;q)_\infty^{-\chi(X)} \cdot \mathrm{Sign}(\pi) \cdot \prod_{i=1}^u \left\langle K_X^{r_i} e_X^{r_i'}, \prod_{j\in\pi_i} \alpha_j \right\rangle \cdot \prod_{i=1}^v \frac{q^{n_i w_i p_i}}{(1-q^{n_i})^{w_i}} \tag{5.32}$$

where $\sum_{i=1}^v w_i < \sum_{i=1}^N (k_i+2)$, and the integers $u, v, r_i, r_i' \ge 0, n_i > 0, w_i > 0, p_i \in \{0,1\}$ and the partition $\pi = \{\pi_1, \ldots, \pi_u\}$ of $\{1,\ldots,N\}$ depend only on the integers k_i. Moreover, the coefficients of this linear combination are independent of q, α_i, X.

PROOF. By Lemma 5.2,

$$F^{\alpha_1,\ldots,\alpha_N}_{k_1,\ldots,k_N}(q) = \mathrm{Tr}\, q^{\mathfrak{n}}\, \mathbf{W}(\mathfrak{L}_1, z) \prod_{i=1}^N \mathfrak{G}_{k_i}(\alpha_i).$$

Combining with Theorem 4.7 and Theorem 5.8, we conclude that

$$F^{\alpha_1,\ldots,\alpha_N}_{k_1,\ldots,k_N}(q) = \widetilde{F}^{\alpha_1,\ldots,\alpha_N}_{k_1,\ldots,k_N}(q) + W_{1,1} \tag{5.33}$$

where $W_{1,1}$ is an infinite linear combination of the expressions (5.32), and the term $\widetilde{F}^{\alpha_1,\ldots,\alpha_N}_{k_1,\ldots,k_N}(q)$ is defined to be

$$(-1)^N \cdot \sum_{\substack{\ell(\lambda^{(i)})=k_i+2, |\lambda^{(i)}|=0\\ 1\le i\le N}} \mathrm{Tr}\, q^{\mathfrak{n}}\, \mathbf{W}(\mathfrak{L}_1, z) \prod_{i=1}^N \frac{\mathfrak{a}_{\lambda^{(i)}}(\tau_*\alpha_i)}{\lambda^{(i)}!}. \tag{5.34}$$

Applying Theorem 5.8 again, we see that $\widetilde{F}^{\alpha_1,\ldots,\alpha_N}_{k_1,\ldots,k_N}(q)$ is equal to

$$\begin{aligned}(-1)^N (q;q)_\infty^{-\chi(X)} \cdot \sum_{\substack{\ell(\lambda^{(i)})=k_i+2, |\lambda^{(i)}|=0\\ 1\le i\le N}} \prod_{i=1}^N \langle (1_X - K_X)^{\sum_{n\ge1} m_n^{(i)}}, \alpha_i\rangle \cdot \\ \cdot \prod_{1\le i\le N, n\ge 1} \left(\frac{(-1)^{m_n^{(i)}}}{m_n^{(i)}!} \frac{q^{nm_n^{(i)}}}{(1-q^n)^{m_n^{(i)}}} \frac{1}{\tilde{m}_n^{(i)}!} \frac{1}{(1-q^n)^{\tilde{m}_n^{(i)}}} \right) + W_{1,2}\end{aligned} \tag{5.35}$$

where the lower weight term $W_{1,2}$ is an infinite linear combination of the expressions (5.32), and we have put $\lambda^{(i)} = \left(\cdots (-n)^{\tilde{m}_n^{(i)}} \cdots (-1)^{\tilde{m}_1^{(i)}} 1^{m_1^{(i)}} \cdots n^{m_n^{(i)}} \cdots\right)$. So

$$\begin{aligned} & \widetilde{F}^{\alpha_1,\ldots,\alpha_N}_{k_1,\ldots,k_N}(q) \\ = \;& (q;q)_\infty^{-\chi(X)} \cdot \prod_{i=1}^N \Theta^{\alpha_i}_{k_i}(q) + W_{1,2} \\ = \;& (q;q)_\infty^{-\chi(X)} \cdot \mathrm{Coeff}_{z_1^0\cdots z_N^0}\left(\prod_{i=1}^N \Theta^{\alpha_i}_{k_i}(q,z_i)\right) + W_{1,2}\end{aligned} \tag{5.36}$$

by Lemma 5.11. Putting $W_1 = W_{1,1} + W_{1,2}$ completes the proof of (5.31). $\square$

In order to relate the lower weight term $W_{1,2}$ in (5.35) and (5.36) to multiple q-zeta values (with additional variables $z_1, \ldots, z_N$ inserted), we will assume $e_X \alpha_i = 0$ for all $1 \le i \le N$. The following lemma strengthens Lemma 5.7, and follows from the same arguments.

LEMMA 5.13. *Let $\lambda^{(1)}, \ldots, \lambda^{(N)} \in \widetilde{\mathcal{P}}$, and $\alpha_1, \ldots, \alpha_N \in H^*(X)$ be homogeneous. Assume that $e_X \alpha_i = 0$ for every $1 \le i \le N$, and $\sum_{i=1}^N |\lambda^{(i)}| = 0$. Put*

$$A_N = \mathrm{Tr}\, q^{\mathfrak{n}} \prod_{i=1}^N \frac{\mathfrak{a}_{\lambda^{(i)}}(\tau_* \alpha_i)}{\lambda^{(i)}!}.$$

(i) *If $\ell(\lambda^{(i)}) \ge 2$ for every $1 \le i \le N$, then $A_N = 0$.*
(ii) *If $A_N \ne 0$, then A_N is a linear combination of the expressions:*

$$\begin{aligned} &(q;q)_\infty^{-\chi(X)} \cdot \mathrm{Sign}(\pi) \cdot \prod_{i=1}^u \left\langle 1_X, \prod_{j \in \pi_i} \alpha_j \right\rangle \cdot \prod_{i=1}^{\tilde{\ell}} \frac{(-\tilde{n}_i) q^{\tilde{n}_i \tilde{p}_i}}{1 - q^{\tilde{n}_i}} \\ = \; &(q;q)_\infty^{-\chi(X)} \cdot \mathrm{Sign}(\pi) \cdot \prod_{i=1}^u \left\langle 1_X, \prod_{j \in \pi_i} \alpha_j \right\rangle \cdot \prod_{i=1}^{\ell} \frac{(-n_i')^{w_i} q^{n_i' p_i}}{(1 - q^{n_i'})^{w_i}} \end{aligned}$$

where

$$\tilde{\ell} = \sum_{i=1}^N \ell(\lambda^{(i)})/2 = \sum_{i=1}^{\ell} w_i,$$

$\tilde{p}_i \in \{0, 1\}$, $0 \le p_i \le w_i$, the partition $\pi = \{\pi_1, \ldots, \pi_u\}$ of $\{1, \ldots, N\}$ depend only on $\lambda^{(1)}, \ldots, \lambda^{(N)}$, the integers $\tilde{n}_1, \ldots, \tilde{n}_{\tilde{\ell}}$ are the positive parts (repeated with multiplicities) in $\lambda^{(1)}, \ldots, \lambda^{(N)}$, the integers $n_1', \ldots, n_\ell'$ denote the different integers in $\tilde{n}_1, \ldots, \tilde{n}_{\tilde{\ell}}$, and each n_i' appears w_i times in $\tilde{n}_1, \ldots, \tilde{n}_{\tilde{\ell}}$.

THEOREM 5.14. *For $1 \le i \le N$, let $k_i \ge 0$ and $\alpha_i \in H^*(X)$ be homogeneous. Assume that $e_X \alpha_i = 0$ for every $1 \le i \le N$. Then,*

$$\widetilde{F}^{\alpha_1, \ldots, \alpha_N}_{k_1, \ldots, k_N}(q) = (q;q)_\infty^{-\chi(X)} \cdot \mathrm{Coeff}_{z_1^0 \cdots z_N^0} \left(\prod_{i=1}^N \Theta^{\alpha_i}_{k_i}(q, z_i) \right) + W_{1,2}, \tag{5.37}$$

and $(q;q)_\infty^{\chi(X)} \cdot W_{1,2}$ is a linear combination of the coefficients of $z_1^0 \cdots z_N^0$ in some multiple q-zeta values (with variables $z_1, \ldots, z_N$ inserted) of weights $< \sum_{i=1}^N (k_i+2)$. Moreover, the coefficients in this linear combination are independent of q.

We now sketch the proof of Theorem 5.14, and refer to the proof of [**QY**, Theorem 4.10] for further details. Recall that $\widetilde{F}^{\alpha_1, \ldots, \alpha_N}_{k_1, \ldots, k_N}(q)$ is defined in (5.34), and that (5.37) is just (5.36). From the proofs of (5.36) and Theorem 5.8, we see that the lower weight term $W_{1,2}$ in (5.37) is the contribution of Case B in the proof of Theorem 5.8 to the right-hand-side of (5.34). A careful analysis of these contributions together with Lemma 5.13 completes the proof.

We will end this section with three propositions about $F^\alpha_k(q)$, which provide some insight into the lower weight term W_1 in Theorem 5.12. Proposition 5.15 deals with $F^\alpha_0(q)$ for an arbitrary $\alpha \in H^*(X)$. Proposition 5.17 calculates $F^\alpha_1(q)$ by assuming $e_X \alpha = 0$. Proposition 5.18 computes $F^\alpha_k(q), k \ge 2$ by assuming $e_X \alpha = K_X \alpha = 0$.

PROPOSITION 5.15. *The generating series $F^{\alpha}_0(q)$ is equal to*

$$(q;q)_\infty^{-\chi(X)} \cdot \langle 1_X - K_X, \alpha\rangle \cdot \sum_{n>0} \frac{q^n}{(1-q^n)^2} \tag{5.38}$$
$$+ \quad (q;q)_\infty^{-\chi(X)} \cdot \langle e_X, \alpha\rangle \cdot \sum_{n>0} \frac{nq^n}{1-q^n}.$$

PROOF. By Lemma 5.2,

$$F^{\alpha}_k(q) = \mathrm{Tr}\, q^{\mathfrak{n}}\, \mathbf{W}(\mathfrak{L}_1, z)\, \mathfrak{G}_k(\alpha).$$

By Theorem 4.7, we have

$$\mathfrak{G}_0(\alpha) = -\sum_{n>0} (\mathfrak{a}_{-n}\mathfrak{a}_n)(\tau_{2*}\alpha).$$

Now (5.38) follows from Remark 5.9. □

REMARK 5.16. By (5.38),

$$F^{1_X}_0(q) = (q;q)_\infty^{-\chi(X)} \cdot \chi(X) \cdot \sum_{n>0} \frac{nq^n}{1-q^n} = q\frac{\mathrm{d}}{\mathrm{d}q}(q;q)_\infty^{-\chi(X)}.$$

PROPOSITION 5.17. *Let $\alpha \in H^*(X)$ be a homogeneous class satisfying $e_X\alpha = 0$. Then, the generating series $F^{\alpha}_1(q)$ is the coefficient of z^0 in*

$$(q;q)_\infty^{-\chi(X)} \cdot \frac{\langle K_X - K_X^2, \alpha\rangle}{2} \cdot \left(\sum_{n>0} \frac{(n-1)q^n}{(1-q^n)^2}\right.$$
$$\left. + \sum_{n>0} \frac{(qz)^n}{1-q^n} \cdot \left(\sum_{m>0} \frac{z^{-2m}}{(1-q^m)^2} + 2 \sum_{m_1>m_2>0} \frac{z^{-m_1}}{1-q^{m_1}} \frac{z^{-m_2}}{1-q^{m_2}}\right)\right).$$

PROOF. We have

$$F^{\alpha}_1(q) = \mathrm{Tr}\, q^{\mathfrak{n}}\, \mathbf{W}(\mathfrak{L}_1, z)\, \mathfrak{G}_1(\alpha).$$

By (4.18),

$$\mathfrak{G}_1(\alpha) = -\sum_{\ell(\lambda)=3, |\lambda|=0} \frac{\mathfrak{a}_\lambda(\tau_{3*}\alpha)}{\lambda^!} - \sum_{n>0} \frac{n-1}{2} (\mathfrak{a}_{-n}\mathfrak{a}_n)(\tau_{2*}(K_X\alpha)).$$

Applying Remark 5.9 to

$$-\sum_{n>0} \frac{n-1}{2} \mathrm{Tr}\, q^{\mathfrak{n}}\, \mathbf{W}(\mathfrak{L}_1, z)\, (\mathfrak{a}_{-n}\mathfrak{a}_n)(\tau_{2*}(K_X\alpha))$$

yields the weight-2 terms in our proposition. Again by Remark 5.9, the trace

$$\mathrm{Tr}\, q^{\mathfrak{n}}\, \mathbf{W}(\mathfrak{L}_1, z)\, \frac{\mathfrak{a}_\lambda(\tau_{3*}\alpha)}{\lambda^!}$$

with $\ell(\lambda) = 3$ and $|\lambda| = 0$ contains only weight-3 terms (i.e., does not contain lower weight terms). So the proof of Theorem 5.12 shows that

$$-\sum_{\ell(\lambda)=3, |\lambda|=0} \mathrm{Tr}\, q^{\mathfrak{n}}\, \mathbf{W}(\mathfrak{L}_1, z)\, \frac{\mathfrak{a}_\lambda(\tau_{3*}\alpha)}{\lambda^!} = (q;q)_\infty^{-\chi(X)} \cdot \mathrm{Coeff}_{z^0} \Theta^{\alpha}_1(q,z).$$

Expanding $\mathrm{Coeff}_{z^0} \Theta^{\alpha}_1(q,z)$ yields the weight-3 terms in our proposition. □

PROPOSITION 5.18. *Let $\alpha \in H^*(X)$ be homogeneous satisfying $K_X\alpha = e_X\alpha = 0$.*

(i) *If $|\alpha| < 4$, then $F_k^\alpha(q) = 0$ for every $k \ge 0$;*
(ii) *Let $|\alpha| = 4$ and $k \ge 0$. Then, $F_k^\alpha(q)$ is the coefficient of z^0 in*

$$-(q;q)_\infty^{-\chi(X)} \cdot \langle 1_X, \alpha\rangle \cdot \sum_{\substack{a,s_1,\ldots,s_a,b,t_1,\ldots,t_b \ge 1 \\ \sum_{i=1}^a s_i + \sum_{j=1}^b t_j = k+2}} \prod_{i=1}^a \frac{(-1)^{s_i}}{s_i!} \cdot \prod_{j=1}^b \frac{1}{t_j!}$$

$$\cdot \sum_{n_1 > \cdots > n_a > 0} \prod_{i=1}^a \frac{(qz)^{n_i s_i}}{(1-q^{n_i})^{s_i}} \cdot \sum_{m_1 > \cdots > m_b > 0} \prod_{j=1}^b \frac{z^{-m_j t_j}}{(1-q^{m_j})^{t_j}}. \tag{5.39}$$

In particular, if $2 \nmid k$, then $F_k^\alpha(q) = 0$.

PROOF. Since $K_X\alpha = e_X\alpha = 0$, we conclude from Theorem 4.7 that

$$\mathfrak{G}_k(\alpha) = -\sum_{\ell(\lambda)=k+2, |\lambda|=0} \frac{\mathfrak{a}_\lambda(\tau_*\alpha)}{\lambda!}.$$

As in the proof of Proposition 5.17, Remark 5.9 and the proof of Theorem 5.12 yield

$$F_k^\alpha(q) = (q;q)_\infty^{-\chi(X)} \cdot \mathrm{Coeff}_{z^0}\Theta_k^\alpha(q,z).$$

By the definition of $\Theta_k^\alpha(q,z)$ in (5.29), we see that (i) holds and that our formula for $F_k^\alpha(q)$ with $|\alpha| = 4$ and $k \ge 0$ holds. Note that line (5.39) can be rewritten as

$$\sum_{n_1 > \cdots > n_a > 0} \prod_{i=1}^a \frac{(qz^2)^{n_i s_i/2}}{(1-q^{n_i})^{s_i}} \cdot \sum_{m_1 > \cdots > m_b > 0} \prod_{j=1}^b \frac{(qz^{-2})^{m_j t_j/2}}{(1-q^{m_j})^{t_j}}.$$

Therefore, if $|\alpha| = 4$ and $2 \nmid k$, then the roles of $a, s_1, \ldots, s_a$ and $b, t_1, \ldots, t_b$ in the above formula of $F_k^\alpha(q)$ are anti-symmetric; so $F_k^\alpha(q) = 0$. $\square$

5.3. The reduced series $\langle \mathbf{ch}_{k_1}^{L_1} \cdots \mathbf{ch}_{k_N}^{L_N} \rangle'$

In this section, using the calculations from the previous section, we will prove Conjecture 5.1 modulo the lower weight term. Moreover, for abelian surfaces, we will verify Conjecture 5.1. The main idea is to express the Chern character of the tautological bundle $L^{[n]}$ in terms of the classes $G_i(\alpha, n)$.

Let L be a line bundle on the smooth projective surface X. We see from (4.1) and the Grothendieck-Riemann-Roch Theorem that

$$\begin{aligned} \mathrm{ch}(L^{[n]}) &= p_{1*}(\mathrm{ch}(\mathcal{O}_{\mathcal{Z}_n}) \cdot p_2^*\mathrm{ch}(L) \cdot p_2^*\mathrm{td}(X)) \\ &= p_{1*}(\mathrm{ch}(\mathcal{O}_{\mathcal{Z}_n}) \cdot p_2^*(1_X + L + L^2/2) \cdot p_2^*\mathrm{td}(X)) \\ &= G(1_X, n) + G(L, n) + G(L^2/2, n). \end{aligned} \tag{5.40}$$

Since the cohomology degree of $G_i(\alpha, n)$ is $2i + |\alpha|$, we have

$$\mathrm{ch}_k(L^{[n]}) = G_k(1_X, n) + G_{k-1}(L, n) + G_{k-2}(L^2/2, n). \tag{5.41}$$

Next, we recall the generating series $\langle \mathrm{ch}_{k_1}^{L_1} \cdots \mathrm{ch}_{k_N}^{L_N} \rangle$ and its reduced version $\langle \mathrm{ch}_{k_1}^{L_1} \cdots \mathrm{ch}_{k_N}^{L_N} \rangle'$ from (5.1) and (5.2) respectively.

THEOREM 5.19. *Let $L_1, \ldots, L_N$ be line bundles over X, and $k_1, \ldots, k_N \geq 0$. Then,*

$$\langle\, \mathrm{ch}_{k_1}^{L_1} \cdots \mathrm{ch}_{k_N}^{L_N} \,\rangle' = \mathrm{Coeff}_{z_1^0 \cdots z_N^0} \left(\prod_{i=1}^{N} \Theta_{k_i}^{1_X}(q, z_i) \right) + W, \tag{5.42}$$

and the lower weight term W is an infinite linear combination of the expressions:

$$\prod_{i=1}^{u} \left\langle K_X^{r_i} e_X^{r_i'}, L_1^{\ell_{i,1}} \cdots L_N^{\ell_{i,N}} \right\rangle \cdot \prod_{i=1}^{v} \frac{q^{n_i w_i p_i}}{(1-q^{n_i})^{w_i}}$$

where $\sum_{i=1}^{v} w_i < \sum_{i=1}^{N} (k_i + 2)$, and the integers $u, v,\ r_i, r_i', \ell_{i,j} \geq 0, n_i > 0, w_i > 0, p_i \in \{0, 1\}$ depend only on $k_1, \ldots, k_N$. Furthermore, all the coefficients of this linear combination are independent of $q, L_1, \ldots, L_N$ and X.

PROOF. We conclude from (5.1), (5.2), (5.41) and (5.3) that

$$\langle\, \mathrm{ch}_{k_1}^{L} \cdots \mathrm{ch}_{k_N}^{L} \,\rangle' = (q;q)_\infty^{\chi(X)} \cdot F_{k_1, \ldots, k_N}^{1_X, \ldots, 1_X}(q) + (q;q)_\infty^{\chi(X)} \cdot A$$

where A is the sum of the series $F_{k_1', \ldots, k_N'}^{\alpha_1, \ldots, \alpha_N}$ such that for every $1 \leq i \leq N$,

$$(\alpha_i, k_i') \in \{(1_X, k_i), (L_i, k_i - 1), (L_i^2/2, k_i - 2)\},$$

and $\sum_{i=1}^{N} k_i' < \sum_{i=1}^{N} k_i$. Now our result follows from Theorem 5.12. □

Theorem 5.19 proves Conjecture 5.1, modulo the lower weight term W. Note that the leading term

$$\mathrm{Coeff}_{z_1^0 \cdots z_N^0} \left(\prod_{i=1}^{N} \Theta_{k_i}^{1_X}(q, z_i) \right)$$

in $\langle\, \mathrm{ch}_{k_1}^{L_1} \cdots \mathrm{ch}_{k_N}^{L_N} \,\rangle'$ has weight

$$\sum_{i=1}^{N} (k_i + 2),$$

and is a multiple of $\langle K_X, K_X \rangle^N$ whose coefficient depends only on $k_1, \ldots, k_N$ and is independent of the line bundles $L_1, \ldots, L_N$ and the surface X.

In general, it is unclear how to organize the lower weight term W in Theorem 5.19 into multiple q-zeta values. On the other hand, we have the following result which, together with Theorem 5.19, verifies Conjecture 5.1 when X is an abelian surface.

THEOREM 5.20. *Let $L_1, \ldots, L_N$ be line bundles over an abelian surface X, and $k_1, \ldots, k_N \geq 0$. Then, the lower weight term W in (5.42) is a linear combination of the coefficients of $z_1^0 \cdots z_N^0$ in some multiple q-zeta values (with additional variables $z_1, \ldots, z_N$ inserted) of weights $< \sum_{i=1}^{N} (k_i + 2)$. Moreover, the coefficients in this linear combination are independent of q.*

PROOF. Since $e_X = K_X = 0$,

$$F_{\tilde{k}_1, \ldots, \tilde{k}_N}^{\tilde{\alpha}_1, \ldots, \tilde{\alpha}_N} = \widetilde{F}_{\tilde{k}_1, \ldots, \tilde{k}_N}^{\tilde{\alpha}_1, \ldots, \tilde{\alpha}_N}$$

by Lemma 5.2, Theorem 4.7 and (5.34). By Theorem 5.14 and the proof of Theorem 5.19, our theorem follows. □

We remark that some of the multiple q-zeta values mentioned in Theorem 5.20 are in the generalized sense, i.e., in the following form:

$$\sum_{n_1>\cdots>n_\ell>0}\prod_{i=1}^{\ell}\frac{(-n_i)^{w_i}q^{n_ip_i}f_i(z_1,\ldots,z_N)^{n_i}}{(1-q^{n_i})^{w_i}}$$

where $0\le p_i\le w_i$, and each $f_i(z_1,\ldots,z_N)$ is a monomial of $z_1^{\pm 1},\ldots,z_N^{\pm 1}$. As indicated in [**Ok2**, Section 2], the factors $(-n_i)^{w_i}$ in the above expression may be related to the operator $q\dfrac{\mathrm{d}}{\mathrm{d}q}$.

Our next two propositions compute the series $\langle\,\mathrm{ch}_k^L\,\rangle$ completely, and should offer some insight into the lower weight term W in Theorem 5.19. Proposition 5.21 calculates $\langle\,\mathrm{ch}_1^L\,\rangle$ by assuming $e_X=0$, while Proposition 5.22 deals with the series $\langle\,\mathrm{ch}_k^L\,\rangle, k\ge 2$ by assuming $e_X=K_X=0$ (i.e., by assuming that X is an abelian surface). Note from (5.41) that when $\chi(X)=0$, we have

$$\langle\,\mathrm{ch}_k^L\,\rangle'=\langle\,\mathrm{ch}_k^L\,\rangle=F_k^{1_X}(q)+F_{k-1}^L(q)+\frac{1}{2}\cdot F_{k-2}^{L^2}(q). \tag{5.43}$$

PROPOSITION 5.21. *Let L be a line bundle over a smooth projective surface X with $e_X=0$. Then, the series $\langle\,\mathrm{ch}_1^L\,\rangle'$ is the coefficient of z^0 in*

$$-\langle K_X,L\rangle\cdot\sum_{n>0}\frac{q^n}{(1-q^n)^2}-\frac{\langle K_X,K_X\rangle}{2}\cdot\sum_{n>0}\frac{(n-1)q^n}{(1-q^n)^2}$$

$$-\frac{\langle K_X,K_X\rangle}{2}\cdot\sum_{n>0}\frac{(qz)^n}{1-q^n}\cdot\left(\sum_{m>0}\frac{z^{-2m}}{(1-q^m)^2}+2\sum_{m_1>m_2>0}\frac{z^{-m_1}}{1-q^{m_1}}\frac{z^{-m_2}}{1-q^{m_2}}\right).$$

PROOF. Our formula follows from (5.43), (5.38) and Proposition 5.17. □

PROPOSITION 5.22. *Let L be a line bundle over an abelian surface X. If $2\nmid k$, then $\langle\,\mathrm{ch}_k^L\,\rangle'=0$. If $2|k$, the generating series $\langle\,\mathrm{ch}_k^L\,\rangle'$ is the coefficient of z^0 in*

$$-\frac{\langle L,L\rangle}{2}\cdot\sum_{\substack{a,s_1,\ldots,s_a,b,t_1,\ldots,t_b\ge 1\\ \sum_{i=1}^a s_i+\sum_{j=1}^b t_j=k}}\prod_{i=1}^{a}\frac{(-1)^{s_i}}{s_i!}\cdot\prod_{j=1}^{b}\frac{1}{t_j!}$$

$$\cdot\sum_{n_1>\cdots>n_a>0}\prod_{i=1}^{a}\frac{(qz)^{n_is_i}}{(1-q^{n_i})^{s_i}}\cdot\sum_{m_1>\cdots>m_b>0}\prod_{j=1}^{b}\frac{z^{-m_jt_j}}{(1-q^{m_j})^{t_j}}.$$

PROOF. Follows immediately from (5.43) and Proposition 5.18. □

CHAPTER 6

Lie algebras and incidence Hilbert schemes

Let X be a smooth projective complex surface. It is known [**Che2**, Theorem on p.43] that besides the Hilbert scheme $X^{[n]}$, the incidence Hilbert scheme $X^{[n,n+1]}$ is the only class of (generalized or nested) Hilbert schemes of points on smooth varieties of dimension bigger than one which are smooth for all n. We refer to Section 1.5 for some basic properties of $X^{[n,n+1]}$. Inspired by the generating series (1.41) of the Poincaré polynomial of $X^{[n,n+1]}$ and the Heisenberg algebra actions (in Section 3.2) on the cohomology of the Hilbert schemes $X^{[n]}$, it is natural to consider the sum of the cohomology group of $X^{[n,n+1]}$ over all $n \geq 0$:

$$\widetilde{\mathbb{H}}_X = \bigoplus_{n=0}^{+\infty} H^*\big(X^{[n,n+1]}\big). \tag{6.1}$$

The space $\widetilde{\mathbb{H}}_X$ has richer structures than the Fock space

$$\mathbb{H}_X = \bigoplus_{n=0}^{+\infty} H^*\big(X^{[n]}\big).$$

For instance, $\widetilde{\mathbb{H}}_X$ is an $H^*(X)$-module which is induced by the natural morphisms $X^{[n,n+1]} \to X$, and explains the factor $P_t(X)$ in (1.41). In addition to the Heisenberg algebra action, $\widetilde{\mathbb{H}}_X$ admits the action of a translation operator $\tilde{\mathfrak{t}}$, which explains the factor $1/(1-t^2q)$ in (1.41). This chapter follows the presentation of [**LQ2**]. Furthermore, we refer to [**LQ2**, Section 6] for potential applications to Donaldson-Thomas theory, where the infinite dimensional Lie algebra actions on the cohomology of the moduli space of ideal sheaves on certain 3-folds were constructed. We also refer to [**LQ4**] for the equivariant cohomology of incidence Hilbert schemes $(\mathbb{C}^2)^{[n,n+1]}$ and its relations with infinite dimensional Lie algebras.

6.1. Heisenberg algebra actions for incidence Hilbert schemes

In this section, we will define the Heisenberg operators acting on the space $\widetilde{\mathbb{H}}_X$. We will prove that these operators satisfy the Heisenberg commutation relations, and are $H^*(X)$-linear. Moreover, via the natural morphisms $X^{[n,n+1]} \to X^{[n]}$ and $X^{[n,n+1]} \to X^{[n+1]}$, we will compare these Heisenberg operators with the Heisenberg operators acting on the Fock space $\mathbb{H}_X$.

We begin with the definition of the Heisenberg operators. For $m \geq 0$ and $n > 0$, let $\widetilde{Q}^{[m,m]} = \emptyset$ and define $\widetilde{Q}^{[m+n,m]}$ to be the following closed subset of $X^{[m+n,m+n+1]} \times X \times X^{[m,m+1]}$:

$$\widetilde{Q}^{[m+n,m]} = \Big\{((\xi,\xi'),x,(\eta,\eta'))\,|\quad \xi \supset \eta, \xi' \supset \eta', \quad \mathrm{Supp}(I_\eta/I_\xi) = \{x\}, \\ \mathrm{Supp}(I_\xi/I_{\xi'}) = \mathrm{Supp}(I_\eta/I_{\eta'})\Big\}.$$

The dimension of the subset $\widetilde{Q}^{[m+n,m]}$ is given by the following lemma.

LEMMA 6.1. *For* $m \geq 0$ *and* $n > 0$, $\dim \widetilde{Q}^{[m+n,m]} = 2m + n + 3$.

PROOF. Take an element $\big((\xi, \xi'), x, (\eta, \eta')\big)$ in $\widetilde{Q}^{[m+n,m]}$. Let

$$\{t\} = \mathrm{Supp}(I_\xi / I_{\xi'}) = \mathrm{Supp}(I_\eta / I_{\eta'})$$

for some $t \in X$. First of all, assume that $x \neq t$. Then η can be decomposed as

$$\eta = \eta_0 + \eta_x + \eta_t$$

where $\eta_x \in M_i(x)$ for some $i \geq 0$, $\eta_t \in M_j(t)$ for some $j \geq 0$, $\eta_0 \in X^{[m-i-j]}$, and $x, t \notin \mathrm{Supp}(\eta_0)$. Then, η', ξ and ξ' can be written as

$$\begin{aligned} \eta' &= \eta_0 + \eta_x + \eta_t', \\ \xi &= \eta_0 + \xi_x + \eta_t, \\ \xi' &= \eta_0 + \xi_x + \eta_t', \end{aligned}$$

with

$$(\eta_t, \eta_t') \in M_{j,j+1}(t), \quad (\eta_x, \xi_x) \in M_{i,i+n}(x).$$

When $i = j = 0$, we conclude from (1.17) that the number of moduli of these triples $\big((\xi, \xi'), x, (\eta, \eta')\big) \in \widetilde{Q}^{[m+n,m]}$ is equal to

$$\begin{aligned} (6.2) \qquad & \#(\text{moduli of } \eta_0) + \#(\text{moduli of } \xi_x) + \#(\text{moduli of } t) \\ = \; & 2m + (n-1) + 4 \\ = \; & 2m + n + 3. \end{aligned}$$

In general, when $i \geq 1$ or $j \geq 1$, we see from (1.17) again that the number of moduli of these triples $\big((\xi, \xi'), x, (\eta, \eta')\big) \in \widetilde{Q}^{[m+n,m]}$ is at most

$$\begin{aligned} (6.3) \qquad & \#(\text{moduli of } \eta_0) + \#(\text{moduli of } \eta_x) + \\ & \quad \#(\text{moduli of } \eta_t \subset \eta_t') + \#(\text{moduli of } \xi_x) \\ = \; & 2(m-i-j) + \max(i-1, 0) + j + (i+n-1) + 4 \\ < \; & 2m + n + 3. \end{aligned}$$

Next, let $x = t$. This time we decompose the 0-cycle η into

$$\eta = \eta_0 + \eta_x$$

where $\eta_x \in M_i(x)$ for some $i \geq 0$, $\eta_0 \in X^{[m-i]}$, and $x \notin \mathrm{Supp}(\eta_0)$. Then,

$$\begin{aligned} \eta' &= \eta_0 + \eta_x', \\ \xi &= \eta_0 + \xi_x, \\ \xi' &= \eta_0 + \xi_x', \end{aligned}$$

with

$$\begin{aligned} (\eta_x, \eta_x') \in M_{i,i+1}(x), &\quad (\eta_x, \xi_x) \in M_{i,i+n}(x), \\ (\xi_x, \xi_x') \in M_{i+n,i+n+1}(x), &\quad (\eta_x', \xi_x') \in M_{i+1,i+n+1}(x). \end{aligned}$$

So the number of moduli of these triples $\big((\xi, \xi'), x, (\eta, \eta')\big) \in \widetilde{Q}^{[m+n,m]}$ is at most

$$\begin{aligned} (6.4) \qquad & \#(\text{moduli of } \eta_0) + \#(\text{moduli of } \eta_x \subset \eta_x') + \#(\text{moduli of } \xi_x \subset \xi_x') \\ = \; & 2(m-i) + i + (i+n) + 2 \\ < \; & 2m + n + 3. \end{aligned}$$

Combining (6.2), (6.3) and (6.4), we get $\dim \widetilde{Q}^{[m+n,m]} = 2m+n+3$. $\square$

Recall the notation $\widetilde{\mathbb{H}}_X$ from (6.1). Now we introduce the operators on $\widetilde{\mathbb{H}}_X$ induced by the correspondence of $\widetilde{Q}^{[m+n,m]}$.

DEFINITION 6.2. (i) Let $\alpha \in H^*(X)$. Define $\tilde{\mathfrak{a}}_0(\alpha) = 0$ in $\mathrm{End}\,(\widetilde{\mathbb{H}}_X)$.

(ii) Let $n > 0$ and $\alpha \in H^*(X)$. Define $\tilde{\mathfrak{a}}_{-n}(\alpha) \in \mathrm{End}\,(\widetilde{\mathbb{H}}_X)$ by

$$\tilde{\mathfrak{a}}_{-n}(\alpha)(\widetilde{A}) = \tilde{p}_{1*}\left([\widetilde{Q}^{[m+n,m]}] \cdot \tilde{\rho}^*\alpha \cdot \tilde{p}_2^*\widetilde{A}\right) \tag{6.5}$$

for $\widetilde{A} \in H^*(X^{[m,m+1]})$, where $\tilde{p}_1, \tilde{\rho}, \tilde{p}_2$ are the projections of

$$X^{[m+n,m+n+1]} \times X \times X^{[m,m+1]}$$

to $X^{[m+n,m+n+1]}, X, X^{[m,m+1]}$ respectively.

(iii) Define $\tilde{\mathfrak{a}}_n(\alpha) \in \mathrm{End}\,(\widetilde{\mathbb{H}}_X)$ to be $(-1)^n$ times the operator obtained from the definition of $\tilde{\mathfrak{a}}_{-n}(\alpha)$ by switching the roles of $\tilde{p}_1$ and $\tilde{p}_2$.

A non-degenerate super-symmetric bilinear form $\langle \cdot, \cdot \rangle$ on $\widetilde{\mathbb{H}}_X$ is induced from the standard one on $H^*(X^{[n,n+1]})$ defined by

$$\langle \widetilde{A}, \widetilde{B} \rangle = \int_{X^{[n,n+1]}} \widetilde{A}\widetilde{B}, \qquad \widetilde{A}, \widetilde{B} \in H^*(X^{[n,n+1]}).$$

This allows us to define the *adjoint* $\tilde{\mathfrak{f}}^\dagger \in \mathrm{End}\,(\widetilde{\mathbb{H}}_X)$ for $\tilde{\mathfrak{f}} \in \mathrm{End}\,(\widetilde{\mathbb{H}}_X)$. As in Lemma 3.7, we have

$$\tilde{\mathfrak{a}}_n(\alpha) = (-1)^n \cdot \tilde{\mathfrak{a}}_{-n}(\alpha)^\dagger. \tag{6.6}$$

Let $n \neq 0$. By (6.5) and Lemma 6.1, $\tilde{\mathfrak{a}}_{-n}(\alpha)$ has bi-degree $(n, 2n-2+|\alpha|)$, i.e.,

$$\tilde{\mathfrak{a}}_{-n}(\alpha) : H^r\big(X^{[m,m+1]}\big) \to H^{r+2n-2+|\alpha|}\big(X^{[m+n,m+n+1]}\big).$$

Next, notice that $\widetilde{\mathbb{H}}_X$ is an $H^*(X)$-module. The module structure is induced by the natural morphisms $\tilde{\rho}_n : X^{[n,n+1]} \to X$.

LEMMA 6.3. *The map $\tilde{\mathfrak{a}}_{-n}(\alpha) : \widetilde{\mathbb{H}}_X \to \widetilde{\mathbb{H}}_X$ is an $H^*(X)$-module homomorphism.*

PROOF. We will only prove the lemma for $n > 0$ since the proof for $n < 0$ is similar. Let $n > 0$, $\beta \in H^*(X)$, and $\widetilde{A} \in H^*\big(X^{[m,m+1]}\big)$. We need to show that

$$\tilde{\mathfrak{a}}_{-n}(\alpha)\big(\tilde{\rho}_m^*\beta \cdot \widetilde{A}\big) = (-1)^{|\alpha||\beta|}\ \tilde{\rho}_{m+n}^*\beta \cdot \tilde{\mathfrak{a}}_{-n}(\alpha)(\widetilde{A}). \tag{6.7}$$

By definition,

$$\begin{aligned}
\tilde{\mathfrak{a}}_{-n}(\alpha)\big(\tilde{\rho}_m^*\beta \cdot \widetilde{A}\big) &= \tilde{p}_{1*}\big([\widetilde{Q}^{[m+n,m]}] \cdot \tilde{\rho}^*\alpha \cdot \tilde{p}_2^*(\tilde{\rho}_m^*\beta \cdot \widetilde{A})\big) \\
&= (-1)^{|\alpha||\beta|}\ \tilde{p}_{1*}\big([\widetilde{Q}^{[m+n,m]}] \cdot \tilde{p}_2^*\tilde{\rho}_m^*\beta \cdot \tilde{\rho}^*\alpha \cdot \tilde{p}_2^*\widetilde{A}\big) \\
&= (-1)^{|\alpha||\beta|}\ \tilde{p}_{1*}\big(\tau_*((\tilde{\rho}_m \circ \tilde{p}_2 \circ \tau)^*\beta) \cdot \tilde{\rho}^*\alpha \cdot \tilde{p}_2^*\widetilde{A}\big)
\end{aligned}$$

where $\tau : \widetilde{Q}^{[m+n,m]} \to X^{[m+n,m+n+1]} \times X \times X^{[m,m+1]}$ is the inclusion map.

From the proof of Lemma 6.1, we see that for $\big((\xi,\xi'), x, (\eta,\eta')\big) \in \widetilde{Q}^{[m+n,m]}$,

$$\mathrm{Supp}(I_{\eta'}/I_{\xi'}) = \mathrm{Supp}(I_\eta/I_\xi) = \{x\}.$$

Therefore, $\tilde{\rho}_m \circ \tilde{p}_2 \circ \tau = \rho_{m+n} \circ \tilde{p}_1 \circ \tau$. Hence, we obtain

$$\begin{aligned}
\tilde{\mathfrak{a}}_{-n}(\alpha)(\tilde{\rho}_m^*\beta \cdot \widetilde{A}) &= (-1)^{|\alpha||\beta|}\, \tilde{p}_{1*}\big(\tau_*((\tilde{\rho}_{m+n} \circ \tilde{p}_1 \circ \tau)^*\beta) \cdot \tilde{\rho}^*\alpha \cdot \tilde{p}_2^*\widetilde{A}\big) \\
&= (-1)^{|\alpha||\beta|}\, \tilde{p}_{1*}\big([\widetilde{Q}^{[m+n,m]}] \cdot \tilde{p}_1^*\tilde{\rho}_{m+n}^*\beta \cdot \tilde{\rho}^*\alpha \cdot \tilde{p}_2^*\widetilde{A}\big) \\
&= (-1)^{|\alpha||\beta|}\, \tilde{\rho}_{m+n}^*\beta \cdot \tilde{p}_{1*}\big([\widetilde{Q}^{[m+n,m]}] \cdot \cdot \tilde{\rho}^*\alpha \cdot \tilde{p}_2^*\widetilde{A}\big)
\end{aligned}$$

by the projection formula. It follows immediately that (6.7) holds. □

There are two natural morphisms from $X^{[m,m+1]}$ to $X^{[m]}$ and $X^{[m+1]}$ respectively:

$$\begin{array}{ccc}
X^{[m,m+1]} & \stackrel{g_{m+1}}{\longrightarrow} & X^{[m+1]} \\
\downarrow f_{m+1} & & \\
X^{[m]}. & &
\end{array}$$

Since there are Heisenberg operators $\mathfrak{a}_{-n}(\alpha)$ on $\mathbb{H}_X$, it is natural to ask whether the following diagram is commutative or not:

$$(6.8)\qquad \begin{array}{ccc}
H^*(X^{[m]}) & \xrightarrow{\mathfrak{a}_{-n}(\alpha)} & H^*(X^{[m+n]}) \\
\big\downarrow f_{m+1}^* & & \big\downarrow f_{m+n+1}^* \\
H^*(X^{[m,m+1]}) & \xrightarrow{\tilde{\mathfrak{a}}_{-n}(\alpha)} & H^*(X^{[m+n,m+n+1]}).
\end{array}$$

Lemma 6.4. *The diagram (6.8) is commutative.*

This lemma not only relates different Heisenberg operators on different spaces, but also is used in the proof of Proposition 6.5 to determine a constant.

Proof. Again, we will only prove the lemma for $n > 0$.

Let $n > 0$ and $A \in H^*(X^{[m]})$. By Definition 3.3 and (6.5), we obtain

$$f_{m+n+1}^*\mathfrak{a}_{-n}(\alpha)(A) = f_{m+n+1}^* p_{1*}([Q^{[m+n,m]}] \cdot \rho^*\alpha \cdot p_2^*A), \tag{6.9}$$

$$\tilde{\mathfrak{a}}_{-n}(\alpha) f_{m+1}^*(A) = \tilde{p}_{1*}([\widetilde{Q}^{[m+n,m]}] \cdot \tilde{\rho}^*\alpha \cdot \tilde{p}_2^* f_{m+1}^*(A)). \tag{6.10}$$

Let $g = \mathrm{Id}_{X^{[m+n,m+n+1]} \times X} \times f_{m+1}$ and $h = f_{m+n+1} \times \mathrm{Id}_{X \times X^{[m]}}$:

$$\begin{aligned}
g:&\quad X^{[m+n,m+n+1]} \times X \times X^{[m,m+1]} \to X^{[m+n,m+n+1]} \times X \times X^{[m]}, \\
h:&\quad X^{[m+n,m+n+1]} \times X \times X^{[m]} \to X^{[m+n]} \times X \times X^{[m]},
\end{aligned}$$

and let

$$\tilde{p}_1' : X^{[m+n,m+n+1]} \times X \times X^{[m]} \to X^{[m+n,m+n+1]}$$

be the first projection. By (6.9) and base change, we conclude that

$$\begin{aligned}
&f_{m+n+1}^*\mathfrak{a}_{-n}(\alpha)(A) \\
=\;& (\tilde{p}_1')_* h^*([Q^{[m+n,m]}] \cdot \rho^*\alpha \cdot p_2^*A) \\
=\;& (\tilde{p}_1')_*\big(h^*[Q^{[m+n,m]}] \cdot (\rho \circ h)^*\alpha \cdot (p_2 \circ h)^*A\big).
\end{aligned} \tag{6.11}$$

Similarly, using (6.10), $\tilde{p}_1 = \tilde{p}_1' \circ g$, and the projection formula, we obtain

$$\begin{aligned}
&\tilde{\mathfrak{a}}_{-n}(\alpha) f_{m+1}^*(A) \\
=\;& (\tilde{p}_1')_* g_*\Big([\widetilde{Q}^{[m+n,m]}] \cdot g^*\big((\rho \circ h)^*\alpha \cdot (p_2 \circ h)^*A\big)\Big) \\
=\;& (\tilde{p}_1')_*\big(g_*[\widetilde{Q}^{[m+n,m]}] \cdot (\rho \circ h)^*\alpha \cdot (p_2 \circ h)^*A\big).
\end{aligned} \tag{6.12}$$

Next, we compare the two cycles $h^*[Q^{[m+n,m]}]$ and $g_*[\widetilde{Q}^{[m+n,m]}]$. Define the subset

$$W = \{((\xi,\xi'),x,\eta) \mid \xi \supset \eta \text{ and } \mathrm{Supp}(I_\eta/I_\xi) = \{x\}\}$$

of $X^{[m+n,m+n+1]} \times X \times X^{[m]}$. Note that both $h^{-1}(Q^{[m+n,m]})$ and $g(\widetilde{Q}^{[m+n,m]})$ are equal to W. A typical element $(\xi,x,\eta) \in Q^{[m+n,m]}$ is of the form:

$$(x_1 + \ldots + x_m + \xi_x,\ x,\ x_1 + \ldots + x_m)$$

where $x_1,\ldots,x_m,x \in X$ are distinct, and $\xi_x \in M_n(x)$. Hence,

$$h^*[Q^{[m+n,m]}] = [W]. \tag{6.13}$$

On the other hand, a typical element $((\xi,\xi'),x,(\eta,\eta')) \in \widetilde{Q}^{[m+n,m]}$ is of the form:

$$((\xi_0 + \xi_x,\ \xi_0 + x_{m+1} + \xi_x),\ x,\ (\xi_0,\ \xi_0 + x_{m+1}))$$

where $x_1,\ldots,x_{m+1},x \in X$ are distinct, $\xi_0 = x_1 + \ldots + x_m$, and $\xi_x \in M_n(x)$. Hence,

$$g_*[\widetilde{Q}^{[m+n,m]}] = [W]. \tag{6.14}$$

The combinations of (6.11), (6.12), (6.13) and (6.14) prove the lemma. □

We remark that in the next section, another comparison of the Heisenberg operators $\mathfrak{a}_{-n}(\alpha)$ and $\tilde{\mathfrak{a}}_{-n}(\alpha)$ will be made by using the pullbacks g_{m+1}^*.

In the rest of this section, we will determine the commutation relations among the operators

$$\tilde{\mathfrak{a}}_n(\alpha), n \in \mathbb{Z}, \alpha \in H^*(X).$$

The following proposition is the analogue of (3.9). Its proof is a bit long, and will be divided into many cases. The main idea in the proof is to find the geometric representation (with the correct dimension) of the composition of two Heisenberg operators $\tilde{\mathfrak{a}}_{-n}(\alpha)$ and $\tilde{\mathfrak{a}}_{-k}(\beta)$, and then compare the geometric representation of $\tilde{\mathfrak{a}}_{-n}(\alpha)\tilde{\mathfrak{a}}_{-k}(\beta)$ with that of $\tilde{\mathfrak{a}}_{-k}(\beta)\tilde{\mathfrak{a}}_{-n}(\alpha)$.

PROPOSITION 6.5. *The operators $\tilde{\mathfrak{a}}_n(\alpha), n \in \mathbb{Z}, \alpha \in H^*(X)$ satisfy the following Heisenberg algebra commutation relation:*

$$[\tilde{\mathfrak{a}}_n(\alpha), \tilde{\mathfrak{a}}_k(\beta)] = -n\,\delta_{n,-k} \int_X (\alpha\beta) \cdot \mathrm{Id}_{\widetilde{\mathbb{H}}_X}. \tag{6.15}$$

PROOF. In view of (6.6), formula (6.15) is equivalent to the formulas:

$$[\tilde{\mathfrak{a}}_{-n}(\alpha), \tilde{\mathfrak{a}}_{-k}(\beta)] = 0, \tag{6.16}$$

$$[\tilde{\mathfrak{a}}_{-n}(\alpha), \tilde{\mathfrak{a}}_k(\beta)] = n\,\delta_{n,k} \int_X (\alpha\beta) \cdot \mathrm{Id}_{\widetilde{\mathbb{H}}_X} \tag{6.17}$$

where $n, k > 0$. We will prove these two formulas separately.

Proof of (6.16):

Let $n, k > 0$. Then the operator $\tilde{\mathfrak{a}}_{-n}(\alpha)\tilde{\mathfrak{a}}_{-k}(\beta)$ is induced by the class

$$w = \pi_{1245*}\left(\pi_{123}^*[\widetilde{Q}^{[m+n+k,m+k]}] \cdot \pi_{345}^*[\widetilde{Q}^{[m+k,m]}]\right) \tag{6.18}$$

in the Chow group $A_{2m+k+n+4}(W')$, and $W' = \pi_{1245}(W)$ with

$$W = \pi_{123}^{-1}(\widetilde{Q}^{[m+n+k,m+k]}) \cap \pi_{345}^{-1}(\widetilde{Q}^{[m+k,m]}). \tag{6.19}$$

Note that W is a closed subset of the ambient space

$$X^{[m+n+k,m+n+k+1]} \times X \times X^{[m+k,m+k+1]} \times X \times X^{[m,m+1]}.$$

The map $\pi_{i_1,\ldots,i_k}$ in (6.18) denotes the projection of this ambient space to the product of its i_1-th, ..., i_k-th factors.

In the sequel, the ambient spaces in similar situations will not be explicitly presented since they can be written out easily from the context.

From Lemma 6.1, we know that the expected dimension of the intersection W should be $2m+n+k+4$. The subset W may have many irreducible components. Those that are mapped by π_{1245} to subsets of dimension less than $2m+n+k+4$ will not contribute to the cycle w. The aim of the computations below is to pick out those components with dimension no less than the expected dimension.

Any element $((\xi,\xi'),x,(\eta,\eta'),t,(\zeta,\zeta'))$ in W must satisfy the conditions:

$$\zeta\subset\zeta',\quad \eta\subset\eta',\quad \xi\subset\xi',\quad \zeta\subset\eta\subset\xi,\quad \zeta'\subset\eta'\subset\xi',$$
$$\mathrm{Supp}(I_\eta/I_\xi)=\{x\},\quad \mathrm{Supp}(I_\zeta/I_\eta)=\{t\},$$
$$\mathrm{Supp}(I_\zeta/I_{\zeta'})=\mathrm{Supp}(I_\eta/I_{\eta'})=\mathrm{Supp}(I_\xi/I_{\xi'})=\{p\}$$

for some $p\in X$. In the following, we consider four different cases separately. We use W_i to denote the subset of W consisting of all the points satisfying Case i, and put

$$\mathfrak{w}_i=\dim\,\pi_{1245}(W_i)$$

(similar notations such as $U_i,\mathfrak{u}_i,V_i,\mathfrak{v}_i$ will be used throughout the rest of this chapter).

Case 1: x,t,p are distinct. We have the following decompositions:

$$\begin{aligned}\zeta&=\zeta_0+\zeta_x+\zeta_t+\zeta_p, & \zeta'&=\zeta_0+\zeta_x+\zeta_t+\zeta'_p,\\ \eta&=\zeta_0+\zeta_x+\eta_t+\zeta_p, & \eta'&=\zeta_0+\zeta_x+\eta_t+\zeta'_p,\\ \xi&=\zeta_0+\xi_x+\eta_t+\zeta_p, & \xi'&=\zeta_0+\xi_x+\eta_t+\zeta'_p,\end{aligned}$$

where $\mathrm{Supp}(\zeta_0)\cap\{x,t,p\}=\emptyset$, $\ell(\zeta_x)=i$, $\ell(\zeta_t)=j$, $\ell(\zeta_p)=\ell$, and $\zeta_x,\zeta_t,\ldots$ are supported at $\{x\},\{t\},\ldots$ respectively. Then,

$$\begin{aligned}\mathfrak{w}_1 &= \#(\text{moduli of }\zeta_0)+\#(\text{moduli of }\zeta_x)+\#(\text{moduli of }\zeta_t)+\\ &\quad+\#(\text{moduli of }\eta_t)+\#(\text{moduli of }\xi_x)+\#(\text{moduli of }\zeta_p\subset\zeta'_p)\\ &= 2(m-i-j-\ell)+\max(i-1,0)+2+\max(j-1,0)+2+\\ &\quad+(j+k-1)+(i+n-1)+\ell+2\\ &= (2m+k+n+4)-i+\max(i-1,0)-j+\max(j-1,0)-\ell.\end{aligned}$$

Case 1.1: If $i=j=\ell=0$, then we have $\mathfrak{w}_1=(2m+k+n+4)$.

Case 1.2: If one of the integers i,j,ℓ is positive, then $\mathfrak{w}_1<(2m+k+n+4)$.

The three remaining cases are listed by:

Case 2: $x=t\neq p$;
Case 3: $x=p\neq t$ (by symmetry, this also covers the case $t=p\neq x$);
Case 4: $x=p=t$.

For these three cases, we skip the arguments which are in the same style as in Case 1. They all have dimension less than the expected dimension $(2m+n+k+4)$. Therefore, only Case 1.1 has contribution to the cohomological operation. In this subcase, it is not difficult to show that the intersection (6.19) along W_1 is transversal. Moreover, $\pi_{1245}(W_1)$ consists of all the points of the form:

$$((\zeta_0+\xi_x+\eta_t,\zeta_0+\xi_x+\eta_t+p),x,t,(\zeta_0,\zeta_0+p))$$

in $X^{[m+n+k,m+n+k+1]} \times X \times X \times X^{[m,m+1]}$. So the contribution of this subcase to the operator $\tilde{\mathfrak{a}}_{-n}(\alpha)\tilde{\mathfrak{a}}_{-k}(\beta)$ coincides with the corresponding contribution for the operator $(-1)^{|\alpha||\beta|} \cdot \tilde{\mathfrak{a}}_{-k}(\beta)\tilde{\mathfrak{a}}_{-n}(\alpha)$. In other words, we obtain the identity:

$$\tilde{\mathfrak{a}}_{-n}(\alpha)\tilde{\mathfrak{a}}_{-k}(\beta) = (-1)^{|\alpha||\beta|} \cdot \tilde{\mathfrak{a}}_{-k}(\beta)\tilde{\mathfrak{a}}_{-n}(\alpha).$$

This completes the proof of the commutation relation (6.16).

Proof of (6.17):
Let $n, k > 0$. Let

$$\tau_{m,m+k} : X^{[m+k,m+k+1]} \times X \times X^{[m,m+1]} \to X^{[m,m+1]} \times X \times X^{[m+k,m+k+1]}$$

be the isomorphism switching $X^{[m,m+1]}$ and $X^{[m+k,m+k+1]}$. Then the operator $\tilde{\mathfrak{a}}_{-n}(\alpha)\tilde{\mathfrak{a}}_{k}(\beta)$ is induced by the class

$$u = \pi_{1245*}\Big(\pi_{123}^*\big[\widetilde{Q}^{[m+n,m]}\big] \cdot \pi_{345}^*\big[\tau_{m,m+k}(\widetilde{Q}^{[m+k,m]})\big]\Big) \tag{6.20}$$

in the Chow group $A_{2m+k+n+4}(U')$, and $U' = \pi_{1245}(U)$ with

$$U = \pi_{123}^{-1}(\widetilde{Q}^{[m+n,m]}) \cap \pi_{345}^{-1}(\tau_{m,m+k}(\widetilde{Q}^{[m+k,m]})). \tag{6.21}$$

Any element $((\xi, \xi'), x, (\eta, \eta'), t, (\zeta, \zeta'))$ in U must satisfy the conditions

$$\begin{gathered}\zeta \subset \zeta', \quad \eta \subset \eta', \quad \xi \subset \xi', \quad \xi \supset \eta \subset \zeta, \quad \xi' \supset \eta' \subset \zeta', \\ \operatorname{Supp}(I_\eta/I_\xi) = \{x\}, \quad \operatorname{Supp}(I_\eta/I_\zeta) = \{t\}, \\ \operatorname{Supp}(I_\zeta/I_{\zeta'}) = \operatorname{Supp}(I_\eta/I_{\eta'}) = \operatorname{Supp}(I_\xi/I_{\xi'}) = \{p\}\end{gathered}$$

for some $p \in X$. In the following, we consider four different cases separately.

Case 1: x, t, p are distinct. We have the following decompositions:

$$\begin{aligned} \eta &= \eta_0 + \eta_x + \eta_t + \eta_p, & \eta' &= \eta_0 + \eta_x + \eta_t + \eta_p', \\ \xi &= \eta_0 + \xi_x + \eta_t + \eta_p, & \xi' &= \eta_0 + \xi_x + \eta_t + \eta_p', \\ \zeta &= \eta_0 + \eta_x + \zeta_t + \eta_p, & \zeta' &= \eta_0 + \eta_x + \zeta_t + \eta_p' \end{aligned}$$

where $\operatorname{Supp}(\eta_0) \cap \{x, t, p\} = \emptyset$, $\ell(\eta_x) = i$, $\ell(\eta_t) = j$, and $\ell(\eta_p) = \ell$. Then,

$$\begin{aligned} \mathfrak{u}_1 &= \#(\text{moduli of } \eta_0) + \#(\text{moduli of } \xi_x) + \#(\text{moduli of } \zeta_t) + \\ &\quad + \#(\text{moduli of } \eta_x) + \#(\text{moduli of } \eta_t) + \#(\text{moduli of } \eta_p \subset \eta_p') \\ &= 2(m - i - j - \ell) + (i + n - 1) + (j + k - 1) + \\ &\quad + \max(i - 1, 0) + \max(j - 1, 0) + \ell + 6 \\ &= (2m + k + n + 4) - i + \max(i - 1, 0) - j + \max(j - 1, 0) - \ell. \end{aligned}$$

Case 1.1: If $i = j = \ell = 0$, then we have $\mathfrak{u}_1 = (2m + k + n + 4)$.

Case 1.2: If one of the integers i, j, ℓ is positive, then $\mathfrak{u}_1 < (2m + k + n + 4)$.

The three remaining cases are listed by:

Case 2: $x = t \neq p$;
Case 3: $x = p \neq t$ (by symmetry, this also covers the case $t = p \neq x$);
Case 4: $x = p = t$.

In these three cases, all the dimensions are smaller than the expected dimension $(2m+n+k+4)$. So only Case 1.1 contributes to the class u in (6.20).

Next we consider the operator $\tilde{\mathfrak{a}}_k(\beta)\tilde{\mathfrak{a}}_{-n}(\alpha)$. This is the only case where there are two components for V below with the expected dimension. One of the components will cancel out with the one from $\tilde{\mathfrak{a}}_{-n}(\alpha)\tilde{\mathfrak{a}}_k(\beta)$, and the other is a non-transversal intersection which carries a multiplicity . Using Lemma 6.4 which compares $\tilde{\mathfrak{a}}_n(\alpha)$ with the Heisenberg operators $\mathfrak{a}_n(\alpha)$ on $\mathbb{H}_X$, we determine the multiplicity.

More precisely, the operator $\tilde{\mathfrak{a}}_k(\beta)\tilde{\mathfrak{a}}_{-n}(\alpha)$ is induced by the class:

(6.22)
$$v = \pi_{1245*}\Big(\pi_{123}^*\big[\tau_{m+n,m+k+n}(\widetilde{Q}^{[m+k+n,m+n]})\big]\cdot\pi_{345}^*\big[\widetilde{Q}^{[m+k+n,m+k]}\big]\Big)$$

in $A_{2m+k+n+4}(V')$, where $V'=\pi_{1245}(V)$ and V is given by

$$V=\pi_{123}^{-1}\big(\tau_{m+n,m+k+n}(\widetilde{Q}^{[m+k+n,m+n]})\big)\cap\pi_{345}^{-1}\big(\widetilde{Q}^{[m+k+n,m+k]}\big). \tag{6.23}$$

Any element $((\xi,\xi'),x,(\eta,\eta'),t,(\zeta,\zeta'))$ in (6.23) must satisfy the conditions:

$$\begin{gathered}\zeta\subset\zeta',\quad \eta\subset\eta',\quad \xi\subset\xi',\quad \xi\subset\eta\supset\zeta,\quad \xi'\subset\eta'\supset\zeta',\\ \mathrm{Supp}(I_\xi/I_\eta)=\{x\},\quad \mathrm{Supp}(I_\zeta/I_\eta)=\{t\},\\ \mathrm{Supp}(I_\zeta/I_{\zeta'})=\mathrm{Supp}(I_\eta/I_{\eta'})=\mathrm{Supp}(I_\xi/I_{\xi'})=\{p\}\end{gathered}$$

for some $p\in X$. In the following, we consider four different cases separately.

Case 1$'$: x,t,p are distinct. We have the following decompositions:

$$\begin{aligned}\zeta&=\zeta_0+\zeta_x+\zeta_t+\zeta_p, & \zeta'&=\zeta_0+\zeta_x+\zeta_t+\zeta_p',\\ \eta&=\zeta_0+\zeta_x+\eta_t+\zeta_p, & \eta'&=\zeta_0+\zeta_x+\eta_t+\zeta_p',\\ \xi&=\zeta_0+\xi_x+\eta_t+\zeta_p, & \xi'&=\zeta_0+\xi_x+\eta_t+\zeta_p',\end{aligned}$$

where $\mathrm{Supp}(\zeta_0)\cap\{x,t,p\}=\emptyset$, $\ell(\xi_x)=i$, $\ell(\zeta_t)=j$, and $\ell(\zeta_p)=\ell$. Then,

$$\begin{aligned}\mathfrak{v}_1&=\#(\text{moduli of }\zeta_0)+\#(\text{moduli of }\zeta_x)+\#(\text{moduli of }\zeta_t)+\\ &\quad+\#(\text{moduli of }\eta_t)+\#(\text{moduli of }\xi_x)+\#(\text{moduli of }\zeta_p\subset\zeta_p')\\ &=2(m-i-j-\ell)+(i+k-1)+\max(j-1,0)+\\ &\quad+(j+n-1)+\max(i-1,0)+\ell+6\\ &=(2m+k+n+4)-i+\max(i-1,0)-j+\max(j-1,0)-\ell.\end{aligned}$$

Case 1$'$.1: If $i=j=\ell=0$, then we have $\mathfrak{v}_1=(2m+k+n+4)$.

Case 1$'$.2: If one of the integers i,j and ℓ is positive, then $\mathfrak{v}_1<(2m+k+n+4)$.

Case 2$'$: $x=t\neq p$. We have the following decompositions:

$$\begin{aligned}\zeta&=\zeta_0+\zeta_x+\zeta_p, & \zeta'&=\zeta_0+\zeta_x+\zeta_p',\\ \eta&=\zeta_0+\eta_x+\zeta_p, & \eta'&=\zeta_0+\eta_x+\zeta_p',\\ \xi&=\zeta_0+\xi_x+\zeta_p, & \xi'&=\zeta_0+\xi_x+\zeta_p',\end{aligned}$$

where $\mathrm{Supp}(\zeta_0)\cap\{x,p\}=\emptyset$, $\ell(\xi_x)=i$, $\ell(\zeta_x)=j$, and $\ell(\zeta_p)=\ell$. Note that

$$k+i=\ell(\eta_x)=n+j. \tag{6.24}$$

Case 2$'$.1: $i=j=\ell=0$. By (6.24), we have $k=n$. Thus,

$$\mathfrak{v}_2=2(m+k)+4=2m+k+n+4.$$

Case 2′.2: If one of the integers i, j and ℓ is positive, then $\mathfrak{v}_2 < (2m+k+n+4)$.

The two remaining cases are listed by:

Case 3′: $x = p \neq t$ (by symmetry, this also covers the case $t = p \neq x$);
Case 4′: $x = p = t$.

In these cases, the dimensions are smaller than $(2m+k+n+4)$. Therefore these two cases have no contributions to the class v in (6.22).

Finally, note that the contribution of Case 1.1 to the class u in (6.20) and the contribution of Case 1′.1 to the class v in (6.22) cancel out in the commutation:

$$[\tilde{\mathfrak{a}}_{-n}(\alpha), \tilde{\mathfrak{a}}_k(\beta)]. \tag{6.25}$$

So $[\tilde{\mathfrak{a}}_{-n}(\alpha), \tilde{\mathfrak{a}}_k(\beta)] = 0$ when $n \neq k$. This proves (6.17) when $n \neq k$.

When $n = k$, Case 2′.1 also contributes to the operator $\tilde{\mathfrak{a}}_k(\beta)\tilde{\mathfrak{a}}_{-n}(\alpha)$, and hence to (6.25). Note from Case 2′.1 that $\pi_{1245}(V_2)$ consists of all the points of the form:

$$\big((\zeta_0, \zeta_0 + p), x, x, (\zeta_0, \zeta_0 + p)\big)$$

in $X^{[m+n,m+n+1]} \times X \times X \times X^{[m+n,m+n+1]}$, where $x \neq p$ and $p \notin \mathrm{Supp}(\zeta_0)$. Thus,

$$[\tilde{\mathfrak{a}}_{-n}(\alpha), \tilde{\mathfrak{a}}_n(\beta)]|_{H^*(X^{[m+n,m+n+1]})} = c \cdot \int_X (\alpha\beta) \cdot \mathrm{Id}_{H^*(X^{[m+n,m+n+1]})}$$

for some constant c. Now we conclude from Lemma 6.4 and (3.9) that $c = n$. This completes the proof of the commutation relation (6.17) when $n = k$. □

6.2. A translation operator for incidence Hilbert schemes

In this section, we will introduce a new operator $\tilde{\mathfrak{t}}$ on $\widetilde{\mathbb{H}}_X$, called the translation operator. The operator $\tilde{\mathfrak{t}}$ is constructed via a correspondence and has many nice properties. Indeed, we will show that it is an $H^*(X)$-module homomorphism, commutes with the Heisenberg operators, and has a left inverse. These properties imply that it is responsible for the factor $1/(1 - t^2 q)$ in (1.41).

From Lemma 6.4, we may tend to infer that the Heisenberg operators $\tilde{\mathfrak{a}}_n(\alpha)$ are no different from the pullback of the Heisenberg operators $\mathfrak{a}_n(\alpha)$ of Nakajima and Grojnowski. This is almost the case when we compare these two types of operators via the map

$$f_{m+1} \colon X^{[m,m+1]} \to X^{[m]}.$$

However, we will see that when we compare them via the map

$$g_{m+1} \colon X^{[m,m+1]} \to X^{[m+1]},$$

they are far from being the same. The difference is measured by the new translation operator. Once we are presented with the fact that the space $\widetilde{\mathbb{H}}_X$ is a highest weight module of the algebra generated by the operators $\tilde{\mathfrak{a}}_n(\alpha)$, the cohomology $H^*(X)$ and the translation operator $\tilde{\mathfrak{t}}$, we will realize that the naive choice of the pullback of Heisenberg algebras on Hilbert schemes either by the map f_{m+1} or g_{m+1} won't provide the right algebra. In this regard, the new algebra we are going to construct is subtler and richer than the Heisenberg algebra on the Hilbert scheme $X^{[m]}$. The fundamental difference between these two algebras is the translation operator $\tilde{\mathfrak{t}}$.

To define the translation operator, let $\widetilde{Q}_m$ be the closed subset:

$$\begin{aligned} \widetilde{Q}_m &= \big\{((\xi', \xi''), (\xi, \xi')) \,|\, \mathrm{Supp}(I_\xi/I_{\xi'}) = \mathrm{Supp}(I_{\xi'}/I_{\xi''})\big\} \\ &\subset X^{[m+1,m+2]} \times X^{[m,m+1]}. \end{aligned}$$

As in Lemma 6.1, we can verify that

$$\dim \widetilde{Q}_m = 2m + 3.$$

DEFINITION 6.6. The *translation operator* is the linear operator $\tilde{\mathfrak{t}} \in \mathrm{End}\left(\widetilde{\mathbb{H}}_X\right)$ defined by

$$\tilde{\mathfrak{t}}(\widetilde{A}) = \tilde{p}_{1*}\left([\widetilde{Q}_m] \cdot \tilde{p}_2^*\widetilde{A}\right) \tag{6.26}$$

for $\widetilde{A} \in H^*(X^{[m,m+1]})$, where $\tilde{p}_1, \tilde{p}_2$ are the two projections of $X^{[m+1,m+2]} \times X^{[m,m+1]}$.

The bi-degrees of $\tilde{\mathfrak{t}}$ and its adjoint $\tilde{\mathfrak{t}}^\dagger$ are $(1, 2)$ and $(-1, -2)$ respectively.

LEMMA 6.7. (i) *The operator* $(-\tilde{\mathfrak{t}}^\dagger)$ *is the left inverse of* $\tilde{\mathfrak{t}}$*;*
(ii) *The maps* $\tilde{\mathfrak{t}}, \tilde{\mathfrak{t}}^\dagger : \widetilde{\mathbb{H}}_X \to \widetilde{\mathbb{H}}_X$ *are* $H^*(X)$*-module homomorphisms.*

PROOF. We skip the proof of (ii) since it is similar to the proof of Lemma 6.3. In the following, we prove (i). Note that the operator $\tilde{\mathfrak{t}}^\dagger \tilde{\mathfrak{t}}$ is induced by the class:

$$w = \pi_{13*}\left(\pi_{12}^*\big[\tau(\widetilde{Q}_m)\big] \cdot \pi_{23}^*\big[\widetilde{Q}_m\big]\right) \in A_{2m+2}(W') \tag{6.27}$$

where $\tau : X^{[m+1,m+2]} \times X^{[m,m+1]} \to X^{[m,m+1]} \times X^{[m+1,m+2]}$ is the isomorphism switching the two factors, $W' = \pi_{13}(W)$, and W is given by

$$W = \pi_{12}^{-1}\big(\tau(\widetilde{Q}_m)\big) \cap \pi_{23}^{-1}\big(\widetilde{Q}_m\big). \tag{6.28}$$

The expected dimension of the intersection W is $2m + 2$.

Any element $\big((\eta, \xi'), (\xi', \xi''), (\xi, \xi')\big)$ in W must satisfy the conditions:

$$\eta, \xi \subset \xi' \subset \xi'',$$
$$\mathrm{Supp}(I_\eta/I_{\xi'}) = \mathrm{Supp}(I_{\xi'}/I_{\xi''}) = \mathrm{Supp}(I_\xi/I_{\xi'}) = \{p\}$$

for some $p \in X$. We have the following decompositions:

$$\xi = \xi_0 + \xi_p, \qquad \xi' = \xi_0 + \xi_p',$$
$$\xi'' = \xi_0 + \xi_p'', \qquad \eta = \xi_0 + \eta_p,$$

where $p \notin \mathrm{Supp}(\xi_0)$ and $\ell(\xi_p) = i$. Note that $\ell(\eta_p) = i$. Fix the integer i.

If $i > 0$, then the projection of the subset of W, consisting of all the elements

$$\big((\eta, \xi'), (\xi', \xi''), (\xi, \xi')\big),$$

to $X^{[m,m+1]} \times X^{[m,m+1]}$ has dimension at most equal to:

$$\begin{aligned} & \#(\text{moduli of } \xi_0) + \#(\text{moduli of } \xi_p \subset \xi_p') + \#(\text{moduli of } \eta_p) \\ = \ & 2(m - i) + i + (i - 1) + 2 \\ < \ & (2m + 2). \end{aligned}$$

So the case $i > 0$ does not contribute to the class w in (6.27).

If $i = 0$, then $\xi_p' = p$, $\xi_p'' \in M_2(p)$, and $\xi_p = \eta_p = \emptyset$. So $\eta = \xi = \xi_0$. The projection of this part of W to $X^{[m,m+1]} \times X^{[m,m+1]}$ consists of elements of the form:

$$\big\{\big((\xi_0, \xi_0 + p), (\xi_0, \xi_0 + p)\big)\big\}.$$

It follows that $w = c[\Delta]$ for some constant c, where Δ denotes the diagonal in $X^{[m,m+1]} \times X^{[m,m+1]}$. Hence, we get

$$\tilde{\mathfrak{t}}^\dagger \, \tilde{\mathfrak{t}} = c \, \mathrm{Id}_{H^*(X^{[m,m+1]})}.$$

To determine c, note that we can split off the factor ξ_0 from our consideration, i.e., we can simply consider the case $m = 0$. Then we have the morphism:

$$\pi_{13} : X \times X^{[1,2]} \times X \to X \times X.$$

Fix a point $p \in X$. Since $[\Delta] \cdot [\{p\} \times X] = 1$, we see from (6.27) that

$$\begin{aligned} c &= w \cdot [\{p\} \times X] \\ &= \pi_{13*}\left(\pi_{12}^*\big[\tau(\widetilde{Q}_0)\big] \cdot \pi_{23}^*\big[\widetilde{Q}_0\big]\right) \cdot [\{p\} \times X] \\ &= \pi_{12}^*\big[\tau(\widetilde{Q}_0)\big] \cdot \pi_{23}^*\big[\widetilde{Q}_0\big] \cdot [\{p\} \times X^{[1,2]} \times X] \\ &= [\{p\} \times U_p \times X] \cdot [\{p\} \times \widetilde{Q}_0] \end{aligned}$$

where $U_p = \big\{(p, \xi_p'') |\, \xi_p'' \in M_2(p)\big\}$. It follows that

$$c = [U_p \times X] \cdot [\widetilde{Q}_0] = [U_p] \cdot \phi_{1*}[\widetilde{Q}_0] = [U_p] \cdot [\phi_1(\widetilde{Q}_0)] \tag{6.29}$$

where $\phi_1 : X^{[1,2]} \times X \to X^{[1,2]}$ is the projection. We have

$$\phi_1(\widetilde{Q}_0) = \{(x, \xi_x'') |\, x \in X \text{ and } \xi_x'' \in M_2(x)\}.$$

Recall the natural morphism $g_2 : X^{[1,2]} \to X^{[2]}$. By (1.40),

$$[\phi_1(\widetilde{Q}_0)] = \frac{1}{2}\, g_2^*[M_2(X)]$$

where $M_2(X) = \cup_{x \in X} M_2(x)$. Hence, we see from (6.29) that

$$\begin{aligned} c &= [U_p] \cdot \frac{1}{2}\, g_2^*[M_2(X)] \\ &= \frac{1}{2}\, g_{2*}[U_p] \cdot [M_2(X)] \\ &= \frac{1}{2}\, [g_2(U_p)] \cdot [M_2(X)] \\ &= \frac{1}{2}\, [M_2(p)] \cdot [M_2(X)] \\ &= -1 \end{aligned}$$

where we have used the fact that $[M_2(p)] \cdot [M_2(X)] = -2$ from (1.35). □

The translation operator $\tilde{\mathfrak{t}}$ may look similar to the creation operators at the first glimpse. However, we now show that it differs from the creation operators in the essential way in that it commutes with all the annihilation operators.

PROPOSITION 6.8. *The translation operator* $\tilde{\mathfrak{t}}$ *and its adjoint* $\tilde{\mathfrak{t}}^\dagger$ *commute with the Heisenberg operators* $\tilde{\mathfrak{a}}_{-n}(\alpha)$ *for all* n *and* α, *i.e.,*

$$[\tilde{\mathfrak{t}}, \tilde{\mathfrak{a}}_{-n}(\alpha)] = [\tilde{\mathfrak{t}}^\dagger, \tilde{\mathfrak{a}}_{-n}(\alpha)] = 0.$$

PROOF. Let $n > 0$. Then Proposition 6.8 is decomposed into two parts:

$$[\tilde{\mathfrak{t}}, \tilde{\mathfrak{a}}_{-n}(\alpha)] = 0, \tag{6.30}$$

$$[\tilde{\mathfrak{t}}, \tilde{\mathfrak{a}}_{n}(\alpha)] = 0. \tag{6.31}$$

We prove them separately, and will compare the proof with that of Proposition 6.5.

Proof of (6.30):

Let $n > 0$. This part is similar to the proof of (6.16).

The operator $\tilde{\mathfrak{a}}_{-n}(\alpha)\tilde{\mathfrak{t}}$ is induced by the class:

$$w = \pi_{124*}\left(\pi_{123}^*\left[\widetilde{Q}^{[m+n+1,m+1]}\right] \cdot \pi_{34}^*\left[\widetilde{Q}_m\right]\right) \in A_{2m+n+4}(W') \tag{6.32}$$

where $W' = \pi_{124}(W)$ and the subset W is defined by

$$W = \pi_{123}^{-1}(\widetilde{Q}^{[m+n+1,m+1]}) \cap \pi_{34}^{-1}(\widetilde{Q}_m). \tag{6.33}$$

The expect dimension of the intersection W is $2m + n + 4$.

Any element $((\eta, \eta'), x, (\xi', \xi''), (\xi, \xi'))$ in (6.33) must satisfy the conditions

$$\eta \subset \eta', \quad \xi \subset \xi' \subset \xi'', \quad \xi' \subset \eta, \quad \xi'' \subset \eta',$$
$$\mathrm{Supp}(I_{\xi'}/I_\eta) = \{x\},$$
$$\mathrm{Supp}(I_\eta/I_{\eta'}) = \mathrm{Supp}(I_{\xi'}/I_{\xi''}) = \mathrm{Supp}(I_\xi/I_{\xi'}) = \{p\}$$

for some $p \in X$. In the following, we consider two cases separately.

Case 1: $x \neq p$. We have the following decompositions:

$$\xi = \xi_0 + \xi_x + \xi_p, \qquad \xi' = \xi_0 + \xi_x + \xi_p', \quad \xi'' = \xi_0 + \xi_x + \xi_p'',$$
$$\eta = \xi_0 + \eta_x + \xi_p', \qquad \eta' = \xi_0 + \eta_x + \xi_p'',$$

where $\mathrm{Supp}(\xi_0) \cap \{x, p\} = \emptyset$, $\ell(\xi_x) = i$, and $\ell(\xi_p) = j$.

Case 1.1: If $i = j = 0$, then we have $\mathfrak{w}_1 = (2m + n + 4)$.

Case 1.2: If $i > 0$ or $j > 0$, then we obtain $\mathfrak{w}_1 < (2m + n + 4)$.

Case 2: $x = p$. We have $\mathfrak{w}_2 < (2m + n + 4)$.

In summary, only Case 1.1 contributes to the class w. In this subcase, the intersection (6.33) along W_1 is transversal, and $\pi_{1245}(W_1)$ consists of all the elements:

$$((\xi_0 + \eta_x + p, \xi_0 + \eta_x + \xi_p''), x, (\xi_0, \xi_0 + p))$$

in $X^{[m+n+1,m+n+2]} \times X \times X^{[m,m+1]}$. So the contribution of this subcase to the operator $\tilde{\mathfrak{a}}_{-n}(\alpha)\tilde{\mathfrak{t}}$ coincides with the corresponding contribution for the operator $\tilde{\mathfrak{t}}\tilde{\mathfrak{a}}_{-n}(\alpha)$. This completes the proof of the commutation relation (6.30).

Proof of (6.31):

This part looks similar to the proof of (6.17). However, it is fundamentally different. Recall that the subset V in (6.23) whose correspondence defines $\tilde{\mathfrak{a}}_n(\alpha)\tilde{\mathfrak{a}}_{-n}(\beta)$ has one more component with the expected dimension than the subset for the operator $\tilde{\mathfrak{a}}_{-n}(\alpha)\tilde{\mathfrak{a}}_n(\beta)$. That doesn't occur here. If there were two components, they would come from the operator $\tilde{\mathfrak{a}}_n(\alpha)\tilde{\mathfrak{t}}$. So pay a special attention to Case 2 below and compare it with Case $2'$.1 in the proof of Proposition 6.5. In order to illustrate this subtle difference, we present the full content of the proof.

Let $n > 0$. Then the operator $\tilde{\mathfrak{a}}_n(\alpha)\tilde{\mathfrak{t}}$ is induced by the class:

$$u = \pi_{124*}\left(\pi_{123}^*\left[\tau_{m+1,m+n+1}(\widetilde{Q}^{[m+n+1,m+1]})\right] \cdot \pi_{34}^*\left[\widetilde{Q}_{m+n}\right]\right) \tag{6.34}$$

in $A_{2m+n+4}(U')$, where $U' = \pi_{124}(U)$ and the subset U is defined by

$$U = \pi_{123}^{-1}(\tau_{m+1,m+n+1}(\widetilde{Q}^{[m+n+1,m+1]})) \cap \pi_{34}^{-1}(\widetilde{Q}_{m+n}). \tag{6.35}$$

Any element $\big((\eta,\eta'),x,(\xi',\xi''),(\xi,\xi')\big)$ in (6.35) must satisfy the conditions

$$\begin{gathered}\eta\subset\eta',\quad \xi\subset\xi'\subset\xi'',\quad \eta\subset\xi',\quad \eta'\subset\xi'',\\ \mathrm{Supp}(I_\eta/I_{\xi'})=\{x\},\\ \mathrm{Supp}(I_\eta/I_{\eta'})=\mathrm{Supp}(I_{\xi'}/I_{\xi''})=\mathrm{Supp}(I_\xi/I_{\xi'})=\{p\}\end{gathered}$$

for some $p\in X$. In the following, we consider two cases separately.

Case 1: $x\neq p$. We have the following decompositions:

$$\begin{array}{lll}\xi=\xi_0+\xi_x+\xi_p, & \xi'=\xi_0+\xi_x+\xi'_p, & \xi''=\xi_0+\xi_x+\xi''_p,\\ \eta=\xi_0+\eta_x+\xi'_p, & \eta'=\xi_0+\eta_x+\xi''_p, & \end{array}$$

where $\mathrm{Supp}(\xi_0)\cap\{x,p\}=\emptyset$, $\ell(\eta_x)=i$, and $\ell(\xi_p)=j$.

Case 1.1: If $i=j=0$, then we have $\mathfrak{u}_1=(2m+n+4)$.

Case 1.2: If $i>0$ or $j>0$, then we obtain

$$\begin{aligned}\mathfrak{u}_1 &\le \#(\text{moduli of }\xi_0)+\#(\text{moduli of }\xi_x)+\#(\text{moduli of }\eta_x)+\\ &\quad +\#(\text{moduli of }\xi_p)+\#(\text{moduli of }\xi'_p\subset\xi''_p)\\ &= 2(m-i-j)+(i+n-1)+\max(i-1,0)+\\ &\quad +\max(j-1,0)+(j+1)+4\\ &< (2m+n+4).\end{aligned}$$

Case 2: $x=p$. We have the following decompositions:

$$\begin{array}{lll}\xi=\xi_0+\xi_x, & \xi'=\xi_0+\xi'_x, & \xi''=\xi_0+\xi''_x,\\ \eta=\xi_0+\eta_x, & \eta'=\xi_0+\eta'_x, & \end{array}$$

where $x\notin\mathrm{Supp}(\xi_0)$ and $\ell(\eta_x)=i$. In this case, we see that

$$\begin{aligned}\mathfrak{u}_2 &\le \#(\text{moduli of }\xi_0)+\#(\text{moduli of }\xi_x\subset\xi'_x)+\#(\text{moduli of }\eta_x\subset\eta'_x)\\ &\le 2(m+1-i)+(i+n-1)+i+2\\ &< (2m+n+4).\end{aligned}$$

In summary, only Case 1.1 contributes to the class u. The contribution of this subcase to the operator $\tilde{\mathfrak{a}}_n(\alpha)\,\tilde{\mathfrak{t}}$ coincides with that of the operator $\tilde{\mathfrak{t}}\,\tilde{\mathfrak{a}}_n(\alpha)$. This completes the proof of the commutation relation (6.31). □

Next, we compare the Heisenberg operators $\tilde{\mathfrak{a}}_n(\alpha)$ with the pull-back of the Heisenberg operators $\mathfrak{a}_n(\alpha)$ via the morphism

$$g_{m+1}\colon X^{[m,m+1]}\to X^{[m+1]}.$$

The comparison of annihilation operators is similar to Lemma 6.4. The proof of the following lemma is omitted since it is similar to the proof of Lemma 6.4.

LEMMA 6.9. *Let $n>0$ and $\alpha\in H^*(X)$. Then, we have a commutative diagram:*

$$\begin{array}{ccc}H^*\big(X^{[m+1]}\big) & \xleftarrow{\mathfrak{a}_n(\alpha)} & H^*\big(X^{[m+n+1]}\big)\\ \big\downarrow{\scriptstyle g^*_{m+1}} & & \big\downarrow{\scriptstyle g^*_{m+n+1}}\\ H^*\big(X^{[m,m+1]}\big) & \xleftarrow{\tilde{\mathfrak{a}}_n(\alpha)} & H^*\big(X^{[m+n,m+n+1]}\big).\end{array}$$

Lemma 6.9 will not hold for the creation operators. However, Proposition 6.11 below provides an explicit formula relating the creation operators. In order to prove Proposition 6.11, we begin with a technical lemma.

Let $m \geq 0$. Define $\widetilde{Q}_{m,0}$ to be the diagonal of $X^{[m,m+1]} \times X^{[m,m+1]}$. For $n \geq 1$, define $\widetilde{Q}_{m,n}$ to be the closed subset of $X^{[m+n,m+n+1]} \times X^{[m,m+1]}$:

$$\widetilde{Q}_{m,n} = \Big\{ \big((\xi,\xi'),(\eta,\eta')\big) \mid \quad \xi' \supset \xi \supset \eta' \supset \eta, \\ \mathrm{Supp}(I_\xi/I_{\xi'}) = \mathrm{Supp}(I_{\eta'}/I_\xi) = \mathrm{Supp}(I_\eta/I_{\eta'}) \Big\}.$$

Note that $\widetilde{Q}_{m,n}$ has dimension $(2m+n+2)$, and contains exactly one irreducible component of dimension $(2m+n+2)$ whose generic elements are of the form:

$$((\eta+\xi_x, \eta+\xi'_x),(\eta,\eta+x)), \qquad x \notin \mathrm{Supp}(\eta). \tag{6.36}$$

LEMMA 6.10. *The restriction of $\tilde{\mathfrak{t}}^n$ to $H^*(X^{[m,m+1]})$ is given by the cycle $[\widetilde{Q}_{m,n}]$.*

PROOF. Use induction on n. Note that $\widetilde{Q}_{m,1} = \widetilde{Q}_m$ by definition. So the Lemma is trivially true when $n = 0, 1$. Next, assume that $n \geq 2$ and the restriction of $\tilde{\mathfrak{t}}^{n-1}$ to $H^*(X^{[m,m+1]})$ is given by the cycle $[\widetilde{Q}_{m,n-1}]$. Then we see that the restriction of $\tilde{\mathfrak{t}}^n = \tilde{\mathfrak{t}}\tilde{\mathfrak{t}}^{n-1}$ to $H^*(X^{[m,m+1]})$ is given by the cycle

$$w = \pi_{13*}\Big(\pi_{12}^*[\widetilde{Q}_{m+n-1}] \cdot \pi_{23}^*[\widetilde{Q}_{m,n-1}]\Big) \in A_{2m+n+2}(W') \tag{6.37}$$

where $W' = \pi_{13}(W)$ and the subset W is given by

$$W = \pi_{12}^{-1}(\widetilde{Q}_{m+n-1}) \cap \pi_{23}^{-1}(\widetilde{Q}_{m,n-1}). \tag{6.38}$$

Any element $\big((\xi,\xi'),(\theta,\theta'),(\eta,\eta')\big)$ in (6.38) must satisfy:

$$\xi' \supset \xi = \theta' \supset \theta \supset \eta' \supset \eta,$$
$$\mathrm{Supp}(I_\xi/I_{\xi'}) = \mathrm{Supp}(I_\theta/I_{\theta'}) = \mathrm{Supp}(I_{\eta'}/I_\theta) = \mathrm{Supp}(I_\eta/I_{\eta'}) = \{x\}$$

for some point $x \in X$. We have the following decompositions:

$$\eta = \eta_0 + \eta_x, \qquad \eta' = \eta_0 + \eta'_x, \qquad \theta = \eta_0 + \theta_x,$$
$$\theta' = \xi = \eta_0 + \xi_x, \qquad \xi' = \eta_0 + \xi'_x,$$

where $x \notin \mathrm{Supp}(\eta_0)$ and $\eta_x \subset \eta'_x \subset \theta_x \subset \xi_x \subset \xi'_x$. Put $\ell = \ell(\eta_x)$. When $\ell \geq 1$, the relation $\eta'_x \subset \xi_x$ imposes a nontrivial condition on the points

$$\big((\eta_x,\eta'_x),(\xi_x,\xi'_x)\big) \in M_{\ell,\ell+1}(x) \times M_{\ell+n,\ell+n+1}(x).$$

Hence the subset consisting of the images $\big((\xi,\xi'),(\eta,\eta')\big)$ of those points

$$\big((\xi,\xi'),(\theta,\theta'),(\eta,\eta')\big)$$

in (6.38) with $\ell = \ell(\eta_x) \geq 1$ has dimension at most

$$2(m-\ell) + 2 + [\ell + (n+\ell) - 1] = (2m+n+1)$$

which is less than the expected dimension $(2m+n+2)$. So the case $\ell \geq 1$ does not contribute to the cycle w in (6.37). When $\ell = 0$, we have

$$\eta = \eta_0, \qquad \eta' = \eta_0 + x, \qquad \theta = \eta_0 + \theta_x,$$
$$\theta' = \xi = \eta_0 + \xi_x, \qquad \xi' = \eta_0 + \xi'_x$$

where $x \notin \mathrm{Supp}(\eta_0)$ and $\theta_x \subset \xi_x \subset \xi'_x$. The images $\big((\eta_0 + \xi_x, \eta_0 + \xi'_x), (\eta_0, \eta_0 + x)\big)$ form a subset of dimension $(2m + n + 2)$ in $X^{[m+n,m+n+1]} \times X^{[m,m+1]}$. From the description (6.36) of the generic points in $\widetilde{Q}_{m,n}$, we conclude that

$$w = c \cdot [\widetilde{Q}_{m,n}] \tag{6.39}$$

where c is the intersection multiplicity of (6.38) along the generic points.

To determine c, choose a primitive integral class $\widetilde{A} \in H^*(X^{[m,m+1]})$, i.e.,

$$\widetilde{A} \in H^*(X^{[m,m+1]}; \mathbb{Z})/\mathrm{Tor} \subset H^*(X^{[m,m+1]}).$$

By (6.39), $\tilde{\mathfrak{t}}^n(\widetilde{A}) = c\widetilde{B}$ for some integral class $\widetilde{B}$. By Lemma 6.7 (i), we see that

$$\widetilde{A} = (-\tilde{\mathfrak{t}}^\dagger)^n \tilde{\mathfrak{t}}^n(\widetilde{A}) = c(-\tilde{\mathfrak{t}}^\dagger)^n(\widetilde{B}).$$

Since $\widetilde{A}$ is primitive and $(\tilde{\mathfrak{t}}^\dagger)^n(\widetilde{B})$ is an integral class, we must have $c = 1$. □

PROPOSITION 6.11. *Let $n > 0$, $\alpha \in H^*(X)$, and $A \in H^*\big(X^{[m+1]}\big)$. Then,*
(i) $g_n^* \mathfrak{a}_{-n}(\alpha)|0\rangle = n \cdot \tilde{\mathfrak{t}}^{n-1} \tilde{\rho}_0^*(\alpha)$;
(ii) $g_{m+n+1}^* \mathfrak{a}_{-n}(\alpha)(A) = \tilde{\mathfrak{a}}_{-n}(\alpha)(g_{m+1}^* A) + n \cdot \tilde{\mathfrak{t}}^{n-1}\big(\tilde{\rho}_{m+1}^*(\alpha) \cdot f_{m+2}^*(A)\big)$.

PROOF. (i) It suffices to prove the formula for $\alpha = 1_X$. The cohomology class $\mathfrak{a}_{-n}(1_X)|0\rangle$ is represented by the subscheme $M_n(X)$. Recall $M_{n-1,n}(X)$ from (1.37). By Theorem 1.31 (iii),

$$\frac{1}{n} g_n^*[M_n(X)] = [M_{n-1,n}(X)].$$

By Lemma 6.10, $\tilde{\mathfrak{t}}^{n-1}(1_{X^{[0,1]}})$ is also represented by $M_{n-1,n}(X)$. Therefore,

$$g_n^* \mathfrak{a}_{-n}(1_X)|0\rangle = n \cdot \tilde{\mathfrak{t}}^{n-1} \tilde{\rho}_0^*(1_X).$$

(ii) By Definition 3.3 and (6.5), we obtain

$$g_{m+n+1}^* \mathfrak{a}_{-n}(\alpha)(A) = g_{m+n+1}^* p_{1*}([Q^{[m+n+1,m+1]}] \cdot \rho^*\alpha \cdot p_2^* A), \tag{6.40}$$

$$\tilde{\mathfrak{a}}_{-n}(\alpha) g_{m+1}^*(A) = \tilde{p}_{1*}([\widetilde{Q}^{[m+n,m]}] \cdot \tilde{\rho}^*\alpha \cdot \tilde{p}_2^* g_{m+1}^*(A)). \tag{6.41}$$

Let $g = \mathrm{Id}_{X^{[m+n,m+n+1]} \times X} \times g_{m+1}$ and $h = g_{m+n+1} \times \mathrm{Id}_{X \times X^{[m+1]}}$:

$$\begin{aligned} g: &\quad X^{[m+n,m+n+1]} \times X \times X^{[m,m+1]} \to X^{[m+n,m+n+1]} \times X \times X^{[m+1]}, \\ h: &\quad X^{[m+n,m+n+1]} \times X \times X^{[m+1]} \to X^{[m+n+1]} \times X \times X^{[m+1]}. \end{aligned}$$

Let

$$\tilde{p}_1' : X^{[m+n,m+n+1]} \times X \times X^{[m+1]} \to X^{[m+n,m+n+1]}$$

be the first projection. By (6.40) and base change, we conclude that

$$\begin{aligned} &g_{m+n+1}^* \mathfrak{a}_{-n}(\alpha)(A) \\ = \;& (\tilde{p}_1')_* h^*([Q^{[m+n+1,m+1]}] \cdot \rho^*\alpha \cdot p_2^* A) \\ = \;& (\tilde{p}_1')_*\big(h^*[Q^{[m+n+1,m+1]}] \cdot (\rho \circ h)^*\alpha \cdot (p_2 \circ h)^* A\big). \end{aligned} \tag{6.42}$$

Similarly, using (6.41), $\tilde{p}_{1*} = \tilde{p}_1' \circ g$ and the projection formula, we obtain

$$\begin{aligned} &\tilde{\mathfrak{a}}_{-n}(\alpha) g_{m+1}^*(A) \\ = \;& (\tilde{p}_1')_* g_* \Big([\widetilde{Q}^{[m+n,m]}] \cdot g^*\big((\rho \circ h)^*\alpha \cdot (p_2 \circ h)^* A\big)\Big) \\ = \;& (\tilde{p}_1')_*\big(g_*[\widetilde{Q}^{[m+n,m]}] \cdot (\rho \circ h)^*\alpha \cdot (p_2 \circ h)^* A\big). \end{aligned} \tag{6.43}$$

Next, we compare $h^*[Q^{[m+n+1,m+1]}]$ and $g_*[\widetilde{Q}^{[m+n,m]}]$. Let U_1 be the closure of the subset of $X^{[m+n,m+n+1]} \times X \times X^{[m+1]}$ consisting of all the elements of the form:

$$\left(\left(\sum_{i=1}^{m} x_i + \xi'_x, \sum_{i=1}^{m+1} x_i + \xi'_x\right), x, \sum_{i=1}^{m+1} x_i\right)$$

where $x_1, \ldots, x_{m+1}, x \in X$ are distinct, and $\xi'_x \in M_n(x)$ is curvi-linear. Let U_2 be the closure of the subset of $X^{[m+n,m+n+1]} \times X \times X^{[m+1]}$ consisting of all the elements:

$$\left(\left(\sum_{i=1}^{m+1} x_i + \xi_x, \sum_{i=1}^{m+1} x_i + \xi'_x\right), x, \sum_{i=1}^{m+1} x_i\right) \tag{6.44}$$

where $x_1, \ldots, x_{m+1}, x \in X$ are distinct, $\xi'_x \in M_n(x)$ is curvi-linear, and $\xi_x \subset \xi'_x$. Note that general elements in $M_n(x)$ are curvi-linear. Moreover, since $\xi'_x \in M_n(x)$ is curvi-linear, ξ'_x uniquely determines ξ_x with $\xi_x \subset \xi'_x$.

A typical element $(\xi', x, \eta') \in Q^{[m+n+1,m+1]}$ is:

$$(x_1 + \ldots + x_{m+1} + \xi'_x, \; x, \; x_1 + \ldots + x_{m+1})$$

where $x_1, \ldots, x_{m+1}, x \in X$ are distinct, and $\xi'_x \in M_n(x)$ is curvi-linear. Hence,

$$h^*[Q^{[m+n+1,m+1]}] = [U_1] + n[U_2]. \tag{6.45}$$

On the other hand, a typical element $((\xi, \xi'), x, (\eta, \eta')) \in \widetilde{Q}^{[m+n,m]}$ is of the form:

$$\left(\left(\sum_{i=1}^{m} x_i + \xi'_x, \sum_{i=1}^{m+1} x_i + \xi'_x\right), x, \left(\sum_{i=1}^{m} x_i, \sum_{i=1}^{m+1} x_i\right)\right)$$

where $x_1, \ldots, x_{m+1}, x \in X$ are distinct, and $\xi'_x \in M_n(x)$ is curvi-linear. Hence,

$$g_*[\widetilde{Q}^{[m+n,m]}] = [U_1]. \tag{6.46}$$

Combining (6.42), (6.43), (6.45) and (6.46), we obtain

$$\begin{aligned} & g_{m+n+1}^* \mathfrak{a}_{-n}(\alpha)(A) - \tilde{\mathfrak{a}}_{-n}(\alpha) g_{m+1}^*(A) \\ = \; & n \cdot (\tilde{p}'_1)_* \big([U_2] \cdot (\rho \circ h)^* \alpha \cdot (p_2 \circ h)^* A\big). \end{aligned} \tag{6.47}$$

Let $\phi = \mathrm{Id}_{X^{[m+n,m+n+1]}} \times \tilde{\rho}_{m+1} \times f_{m+2}$ be the morphism:

$$X^{[m+n,m+n+1]} \times X^{[m+1,m+2]} \to X^{[m+n,m+n+1]} \times X \times X^{[m+1]}.$$

From the descriptions (6.36) and (6.44), we conclude that

$$[U_2] = \phi_*[\widetilde{Q}_{m+1,n-1}].$$

By (6.47), the projection formula and Lemma 6.10, we obtain

$$\begin{aligned} & g_{m+n+1}^* \mathfrak{a}_{-n}(\alpha)(A) - \tilde{\mathfrak{a}}_{-n}(\alpha) g_{m+1}^*(A) \\ = \; & n \cdot (\tilde{p}'_1 \circ \phi)_* \Big([\widetilde{Q}_{m+1,n-1}] \cdot (\rho \circ h \circ \phi)^* \alpha \cdot (p_2 \circ h \circ \phi)^* A\Big) \\ = \; & n \cdot \pi_{1*} \Big([\widetilde{Q}_{m+1,n-1}] \cdot \pi_2^* \big(\tilde{\rho}_{m+1}^*(\alpha) \cdot f_{m+2}^*(A)\big)\Big) \\ = \; & n \cdot \mathfrak{t}^{n-1} \big(\tilde{\rho}_{m+1}^*(\alpha) \cdot f_{m+2}^*(A)\big) \end{aligned}$$

where π_1 and π_2 are the two projections of $X^{[m+n,m+n+1]} \times X^{[m+1,m+2]}$. □

6.3. Lie algebras and incidence Hilbert schemes

With all the results obtained in previous sections, we are ready to formulate and prove the main theorem.

DEFINITION 6.12. (i) Define $\tilde{\mathfrak{h}}_X$ to be the Heisenberg algebra generated by the operators $\tilde{\mathfrak{a}}_n(\alpha)$, $n \in \mathbb{Z}$, $\alpha \in H^*(X)$ and the identity operator $\mathrm{Id}_{\widetilde{\mathbb{H}}_X}$;

(ii) Define a Lie algebra structure on

$$\tilde{\mathfrak{H}}_X = \tilde{\mathfrak{h}}_X \oplus H^*(X) \oplus \mathbb{C}\tilde{\mathfrak{t}}$$

by declaring

$$[\tilde{\mathfrak{a}}_n(\alpha), \beta] = 0, \quad [\beta, \gamma] = 0, \quad [\tilde{\mathfrak{t}}, \tilde{\mathfrak{a}}_n(\alpha)] = 0, \quad [\tilde{\mathfrak{t}}, \beta] = 0 \tag{6.48}$$

for operators $\tilde{\mathfrak{a}}_n(\alpha) \in \tilde{\mathfrak{h}}_X$ and cohomology classes $\beta, \gamma \in H^*(X)$.

THEOREM 6.13. *The space $\widetilde{\mathbb{H}}_X$ is a representation of the Lie algebra $\tilde{\mathfrak{H}}_X$ with a highest weight vector being the vacuum vector*

$$|0\rangle = 1_X \in H^0(X) = H^0(X^{[0,1]})$$

where 1_X denotes the fundamental cohomology class of X.

PROOF. By Lemma 6.3, Proposition 6.5, Lemma 6.7 (ii) and Proposition 6.8, we see that the space $\widetilde{\mathbb{H}}_X$ is a representation of the Lie algebra $\tilde{\mathfrak{H}}_X$.

Next, for each $0 \le i \le 4$, we fix a linear basis

$$\{\alpha_{i,1}, \ldots, \alpha_{i,b_i(X)}\}$$

of $H^i(X)$. By Proposition 6.5 and the existence of the left inverse $\tilde{\mathfrak{t}}^\dagger$ of the operator $\tilde{\mathfrak{t}}$ in Lemma 6.7 (i), the following cohomology classes:

$$\begin{aligned} &\tilde{\mathfrak{t}}^m \cdot \prod_{i \text{ even}} \tilde{\mathfrak{a}}_{-n_{i,j_i}}(\alpha_{i,j_i})^{m_{i,j_i}} \cdot \prod_{i \text{ odd}} \tilde{\mathfrak{a}}_{-n_{i,j_i}}(\alpha_{i,j_i})^{m_{i,j_i}} \cdot \alpha_{k,j_k}|0\rangle \\ = \;&\tilde{\mathfrak{t}}^m \cdot \prod_{i \text{ even}} \tilde{\mathfrak{a}}_{-n_{i,j_i}}(\alpha_{i,j_i})^{m_{i,j_i}} \cdot \prod_{i \text{ odd}} \tilde{\mathfrak{a}}_{-n_{i,j_i}}(\alpha_{i,j_i})^{m_{i,j_i}} \cdot \alpha_{k,j_k} \end{aligned} \tag{6.49}$$

are linearly independent, where $0 \le i, k \le 4$, $0 \le j_i \le b_i(X)$, $0 \le j_k \le b_k(X)$, $n_{i,j_i} > 0$, $m \ge 0$, $m_{i,j_i} \ge 0$ for all the i and j_i, and $m_{i,j_i} = 0$ or 1 for odd i. Since the bi-degrees of the operators $\tilde{\mathfrak{t}}$ and $\tilde{\mathfrak{a}}_{-n}(\alpha)$ are $(1, 2)$ and $(n, 2n - 2 + |\alpha|)$ respectively, we conclude from (1.41) that cohomology classes in (6.49) form a linear basis of $\widetilde{\mathbb{H}}_X$. Therefore, the representation is the highest weight representation. $\square$

Part 3

Cohomology rings of Hilbert schemes of points

CHAPTER 7

The cohomology rings of Hilbert schemes of points on surfaces

Let X be a smooth projective complex surface. One way to understand the intersection theory on the Hilbert scheme $X^{[n]}$ is to construct a meaningful set of generators for the cohomology ring of $X^{[n]}$. In this chapter, we will construct two sets of generators for the cohomology ring of the Hilbert scheme $X^{[n]}$, which are related to the Mumford principle. For a moduli space of sheaves, the conjectural *Mumford principle* asserts that the Künneth components of the Chern character of the universal sheaf form a set of generators for the cohomology ring of the moduli space. Regarding the Hilbert schemes $X^{[n]}$ as the moduli spaces parametrizing the ideal sheaves, the results in Section 7.1 imply that the Mumford principle holds for $X^{[n]}$. Moreover, using the stability theorem (Theorem 7.7), we will define a Hilbert ring $\mathfrak{H}_X$ associated to the projective surface X. The Hilbert ring $\mathfrak{H}_X$ encodes all the information about the cohomology ring of $X^{[n]}$ for every $n \ge 1$. However, these descriptions of the cohomology ring of the Hilbert scheme $X^{[n]}$ are implicit in terms of the multiplication operators or explicit generators with implicitly given relations. The approaches of Lehn-Sorger [**LS2**] and Costello-Grojnowski [**CoG**] provide explicit descriptions of the cohomology ring of the Hilbert scheme $X^{[n]}$ via the cohomogy ring of the surface X. Both approaches are based on the observation [**Leh1**,**LQW1**] that the ring structure of $H^*(X^{[n]}), n \ge 1$ is uniquely determined by the theorem of Nakajima and Grojnowski (Theorem 3.8), Lehn's boundary operator $\mathfrak{d}$ (Theorem 3.24 (ii) and (4.19)), and Theorem 4.2. The approach of Lehn and Sorger works only when the canonical divisor of X is numerically trivial, while the approach of Costello and Grojnowski uses the Calogero-Sutherland operators and the Dunkl-Cherednik operators. We will survey both approaches in this chapter. Theorem 7.5 is from [**LQW1**,**LQW2**], while Theorem 7.7, Theorem 7.11 and their proofs are from [**LQW3**].

7.1. Two sets of ring generators for the cohomology

In this section, we will construct two sets of generators for the cohomology ring of the Hilbert scheme $X^{[n]}$. The first set consists of the cohomology classes $G_i(\alpha, n)$ where $0 \le i < n$ and α runs over a fixed linear basis of $H^*(X)$. The cohomology classes $G_i(\alpha, n)$ are closely related to the Grothendieck-Riemann-Roch Theorem [**Hart**, Theorem 5.3 in Appendix A], and are thus easier to handle. The second set of generators consists of the cohomology classes $B_i(\alpha, n)$ which are more geometric (Definition 7.3). Moreover, we will prove a stability theorem regarding the cup product in the cohomology of $X^{[n]}$.

We begin with a technical lemma which expresses a Heisenberg monomial class in terms of the cohomology classes $G_i(\alpha, n)$.

LEMMA 7.1. *Fix* $n, s \ge 1$, $n_1, \ldots, n_s \ge 1$, *and* $\alpha_1, \ldots, \alpha_s \in H^*(X)$. *Put* $n_0 = \sum_{i=1}^{s} n_i$. *Then,* $\mathbf{1}_{-(n-n_0)} \left(\prod_{i=1}^{s} \mathfrak{a}_{-n_i}(\alpha_i) \right) |0\rangle$ *is a finite linear combination of*

$$\prod_{j=1}^{t} G_{m_j}(\beta_j, n) \tag{7.1}$$

whose coefficients are independent of $X, \alpha_1, \ldots, \alpha_s$ *and* n. *Here*

$$\sum_{j=1}^{t} (m_j + 1) \le n_0,$$

and $\beta_1, \ldots, \beta_t$ *depend only on* $e_X, K_X, \alpha_1, \ldots, \alpha_s$ *and* τ_{i*} *with* $1 \le i \le n_0$.

PROOF. We use induction on n_0. When $n_0 = 1$, $s = n_1 = 1$. By (4.22),

$$G_0(\alpha_1, n) = \mathbf{1}_{-(n-1)} \mathfrak{a}_{-1}(\alpha_1)|0\rangle.$$

So the lemma holds for $n_0 = 1$. Next, let $n_0 > 1$. Let $k_i = n_i - 1$. Then, $k_i \ge 0$ for every i. By Lemma 4.13, $\prod_{i=1}^{s} G_{k_i}(\alpha_i, n)$ is a finite linear combination of expressions of the form (4.24) such that the coefficients in this linear combination are independent of $X, \alpha_1, \ldots, \alpha_s$ and n. Note that

$$\sum_{i=1}^{\ell(\pi)} \sum_{j=1}^{m_i - r_i} n_{i,j} \le \sum_{i=1}^{\ell(\pi)} \sum_{j \in \pi_i} (k_j + 1) = \sum_{i=1}^{s} (k_i + 1) = n_0.$$

By induction, those expressions (4.24) with

$$\sum_{i=1}^{\ell(\pi)} \sum_{j=1}^{m_i - r_i} n_{i,j} < n_0$$

are linear combinations of the form (7.1) where

$$\sum_{j=1}^{t} (m_j + 1) \le (n_0 - 1),$$

and $\beta_1, \ldots, \beta_t$ depend only on $e_X, K_X, \alpha_1, \ldots, \alpha_s, \tau_{i*}$ with $1 \le i \le (n_0 - 1)$. Moreover, the coefficients in these linear combinations are independent of $X, \alpha_1, \ldots, \alpha_s$ and n. Now our lemma follows from Lemma 4.14. $\square$

REMARK 7.2. By Lemma 7.1, an expression (7.1) in

$$\mathbf{1}_{-(n-n_0)} \left(\prod_{i=1}^{s} \mathfrak{a}_{-n_i}(\alpha_i) \right) |0\rangle$$

satisfies $\sum_{j=1}^{t} (m_j + 1) \le n_0$. In fact, we see from the proof of Lemma 7.1 that an expression (7.1) satisfies the upper bound $\sum_{j=1}^{t} (m_j + 1) = n_0$ if and only if it is equal to $\prod_{i=1}^{s} G_{n_i - 1}(\alpha_i, n)$, whose coefficient is $\prod_{i=1}^{s} ((-1)^{n_i - 1} n_i!)$ by Lemma 4.14 (ii).

DEFINITION 7.3. For $0 \le i < n$ and $\alpha \in H^*(X)$, define

$$B_i(\alpha, n) = \mathbf{1}_{-(n-i-1)}\mathfrak{a}_{-(i+1)}(\alpha)|0\rangle \in H^*(X^{[n]}). \tag{7.2}$$

LEMMA 7.4. *The Chern character $G_k(\alpha, n)$ is a finite linear combination of products*

$$\prod_{j=1}^{t} B_{m_j}(\beta_j, n)$$

whose coefficients are independent of X, α and n. Here $\sum_{j=1}^{t} m_j \le k$, and $\beta_1, \ldots, \beta_t$ depend only on e_X, K_X and α. In addition, $\sum_{j=1}^{t} m_j = k$ if and only if the product $\prod_{j=1}^{t} B_{m_j}(\beta_j, n)$ is equal to $B_k(\alpha, n)$, whose coefficient is equal to $(-1)^k/(k+1)!$.

PROOF. Use induction on k. When $k = 0$, we have $G_0(\alpha, n) = B_0(\alpha, n)$ by (4.22). Next, we assume that the lemma is true for $0, \ldots, k-1$ for some fixed $k \ge 1$. We will prove that the lemma holds for k as well. We apply Lemma 7.1 and Remark 7.2 to $B_k(\alpha, n) = \mathbf{1}_{-(n-k-1)}\mathfrak{a}_{-(k+1)}(\alpha)|0\rangle$. We see that

$$G_k(\alpha, n) - \frac{(-1)^k}{(k+1)!} \cdot B_k(\alpha, n)$$

is a finite linear combination of $\prod_{j=1}^{u} G_{n_j}(\gamma_j, n)$ where

$$\sum_{j=1}^{u} (n_j + 1) < (k+1),$$

and $\gamma_1, \ldots, \gamma_u$ depend only on e_X, K_X and α. Moreover, the coefficients in this linear combination are independent of X, α and n. Note that $n_j < k$ for all $1 \le j \le u$. So by induction hypothesis, the lemma holds for k. □

THEOREM 7.5. *For $n \ge 1$, the cohomology ring $H^*(X^{[n]})$ is generated by the cohomology classes $G_i(\alpha, n)$ (respectively, the cohomology classes $B_i(\alpha, n)$) where $0 \le i < n$ and α runs over a fixed linear basis of $H^*(X)$.*

PROOF. By Lemma 7.1, the cohomology ring $H^*(X^{[n]})$ is generated by the cohomology classes $G_i(\alpha, n)$ where $0 \le i < n$ and α runs over a fixed linear basis of $H^*(X)$. By Lemma 7.4, $H^*(X^{[n]})$ is generated by the cohomology classes $B_i(\alpha, n)$ where $0 \le i < n$ and α runs over a fixed linear basis of $H^*(X)$. □

The following corollary asserts that the cohomology ring of the Hilbert scheme $X^{[n]}$ is uniquely determined by the cohomology ring of the surface X and its canonical class K_X.

COROLLARY 7.6. *Let X and Y be smooth projective complex surfaces. Assume that there exists a ring isomorphism $\Phi : H^*(X) \to H^*(Y)$ with $\Phi(K_X) = K_Y$. Then for every $n \ge 1$, the cohomology rings $H^*(X^{[n]})$ and $H^*(Y^{[n]})$ are isomorphic.*

PROOF. Define a linear map $\Psi : H^*(X^{[n]}) \to H^*(Y^{[n]})$ by

$$\Psi\big(\mathfrak{a}_{-n_1}(\alpha_1)\cdots\mathfrak{a}_{-n_k}(\alpha_k)|0\rangle\big) = \mathfrak{a}_{-n_1}\big(\Phi(\alpha_1)\big)\cdots\mathfrak{a}_{-n_k}\big(\Phi(\alpha_k)\big)|0\rangle.$$

Then Ψ is an isomorphism of vector spaces. Note that $\Phi(e_X) = e_Y$. By Lemma 4.13,

$$\Psi\big(G_{k_1}(\alpha_1, n)\cdots G_{k_s}(\alpha_s, n)\big) = G_{k_1}\big(\Phi(\alpha_1), n\big)\cdots G_{k_s}\big(\Phi(\alpha_s), n\big).$$

Since the classes $G_k(\alpha, n)$ generate $H^*(X^{[n]})$, Ψ is a ring isomorphism. □

Assume that the canonical class K_X of the surface X is numerically trivial. Then by Corollary 7.6, the cohomology ring $H^*(X^{[n]})$ is uniquely determined by the cohomology ring $H^*(X)$. This observation is the foundation of the approach of Lehn and Sorger [**LS2**] to be discussed in Section 7.3 where the cohomology ring $H^*(X)$ will be explicitly constructed from the cohomology ring $H^*(X)$.

Next, we study the cup product of Heisenberg monomial classes

$$\mathbf{1}_{-(n-\sum_{j=1}^{k_i} n_{i,j})}\left(\prod_{j=1}^{k_i}\mathfrak{a}_{-n_{i,j}}(\alpha_{i,j})\right)|0\rangle$$

where $1 \le i \le s$. The following result is known as the stability theorem.

THEOREM 7.7. *Let $s \ge 1$ and $k_i \ge 1$ for $1 \le i \le s$. Fix $n_{i,j} \ge 1$ and $\alpha_{i,j} \in H^*(X)$ for $1 \le j \le k_i$, and fix n with $n \ge \sum_{j=1}^{k_i} n_{i,j}$ for all $1 \le i \le s$. Then the cup product*

$$\prod_{i=1}^{s}\left(\mathbf{1}_{-(n-\sum_{j=1}^{k_i} n_{i,j})}\left(\prod_{j=1}^{k_i}\mathfrak{a}_{-n_{i,j}}(\alpha_{i,j})\right)|0\rangle\right) \tag{7.3}$$

in $H^*(X^{[n]})$ is equal to a finite linear combination of monomials of the form

$$\mathbf{1}_{-(n-\sum_{p=1}^{N} m_p)}\left(\prod_{p=1}^{N}\mathfrak{a}_{-m_p}(\gamma_p)\right)|0\rangle \tag{7.4}$$

whose coefficients are independent of $X, \alpha_{i,j}$ and n. Here

$$\sum_{p=1}^{N} m_p \le \sum_{i=1}^{s}\sum_{j=1}^{k_i} n_{i,j},$$

and $\gamma_1, \ldots, \gamma_N$ depend only on $e_X, K_X, \alpha_{i,j}$, $1 \le i \le s, 1 \le j \le k_i$. In addition, the expression (7.4) satisfies the upper bound

$$\sum_{p=1}^{N} m_p = \sum_{i=1}^{s}\sum_{j=1}^{k_i} n_{i,j}$$

if and only if it is equal to

$$\mathbf{1}_{-(n-\sum_{i=1}^{s}\sum_{j=1}^{k_i} n_{i,j})}\left(\prod_{i=1}^{s}\prod_{j=1}^{k_i}\mathfrak{a}_{-n_{i,j}}(\alpha_{i,j})\right)|0\rangle,$$

whose coefficient is 1.

PROOF. Put $N_i = \sum_{j=1}^{k_i} n_{i,j}$ for $1 \le i \le s$. For each i, we see from Lemma 7.1 that

$$\mathbf{1}_{-(n-\sum_{j=1}^{k_i} n_{i,j})} \left(\prod_{j=1}^{k_i} \mathfrak{a}_{-n_{i,j}}(\alpha_{i,j}) \right) |0\rangle$$

is a finite linear combination of products $\prod_{j=1}^{t_i} G_{m_{i,j}}(\beta_{i,j}, n)$ where $\sum_{j=1}^{t_i} (m_{i,j}+1) \le N_i$, and $\beta_{i,1}, \ldots, \beta_{i,t_i}$ depend only on e_X, K_X, $\alpha_{i,1}, \ldots, \alpha_{i,k_i}$ and τ_{j*} with $1 \le j \le N_i$. Moreover, the coefficients in the linear combinations are independent of $X, \alpha_{i,1}, \ldots, \alpha_{i,k_i}$ and n. By Remark 7.2, the product satisfies the upper bound $\sum_{j=1}^{t_i} (m_{i,j}+1) = N_i$ if and only if it is equal to $\prod_{j=1}^{k_i} G_{n_{i,j}-1}(\alpha_{i,j}, n)$. Furthermore, the coefficient of $\prod_{j=1}^{k_i} G_{n_{i,j}-1}(\alpha_{i,j}, n)$ in

$$\mathbf{1}_{-(n-\sum_{j=1}^{k_i} n_{i,j})} \left(\prod_{j=1}^{k_i} \mathfrak{a}_{-n_{i,j}}(\alpha_{i,j}) \right) |0\rangle$$

is equal to $\prod_{j=1}^{k_i} ((-1)^{n_{i,j}-1} n_{i,j}!)$.

So (7.3) is a universal finite linear combination of products

$$\prod_{i=1}^{s} \prod_{j=1}^{t_i} G_{m_{i,j}}(\beta_{i,j}, n)$$

where

$$\sum_{i=1}^{s} \sum_{j=1}^{t_i} (m_{i,j}+1) \le \sum_{i=1}^{s} N_i = \sum_{i=1}^{s} \sum_{j=1}^{k_i} n_{i,j}.$$

Also,

$$\sum_{i=1}^{s} \sum_{j=1}^{t_i} (m_{i,j}+1) = \sum_{i=1}^{s} \sum_{j=1}^{k_i} n_{i,j}$$

if and only if the product

$$\prod_{i=1}^{s} \prod_{j=1}^{t_i} G_{m_{i,j}}(\beta_{i,j}, n)$$

is equal to

$$\prod_{i=1}^{s} \prod_{j=1}^{k_i} G_{n_{i,j}-1}(\alpha_{i,j}, n)$$

whose coefficient in (7.3) is equal to

$$\prod_{i=1}^{s} \prod_{j=1}^{k_i} ((-1)^{n_{i,j}-1} n_{i,j}!).$$

It follows from Lemma 4.13 that (7.3) is a universal linear combination of expressions (7.4). The statement for the expression (7.4) reaching the upper bound

$$\sum_{p=1}^{N} m_p = \sum_{i=1}^{s} \sum_{j=1}^{k_i} n_{i,j}$$

follows from Lemma 4.14. □

Theorem 7.7 indicates that the cup product on the Hilbert scheme $X^{[n]}$ is independent of n in an appropriate sense. Furthermore, it provides an explicit form of the leading term in the cup product. This result enables us to construct the Hilbert ring in the next section.

7.2. The Hilbert ring

In this section, using the stability theorem (Theorem 7.7), we will introduce and determine *the Hilbert ring* $\mathfrak{H}_X$ associated to a smooth projective complex surface X. On one hand, the Hilbert ring $\mathfrak{H}_X$ encodes all the information about the cohomology ring of the Hilbert scheme $X^{[n]}$ for every $n \geq 1$, and thus offers an approach to describing the cohomology ring of $X^{[n]}$. On the other hand, the ring structure of $\mathfrak{H}_X$ itself is implicit since the structure constants of the product are implicit. This makes $\mathfrak{H}_X$ less appealing.

Given a finite set S which is a disjoint union of subsets S_0 and S_1, we denote by $\mathcal{P}(S)$ the set of partition-valued functions $\rho = (\rho(c))_{c \in S}$ on S such that for every $c \in S_1$, the partition $\rho(c)$ is required to be *strict* in the sense that

$$\rho(c) = (1^{m_1(c)} 2^{m_2(c)} \ldots)$$

with $m_r(c) = 0$ or 1 for all $r \geq 1$.

Now let us take a linear basis

$$S = S_0 \cup S_1$$

of $H^*(X)$ such that $1_X \in S_0$, $S_0 \subset H^{\text{even}}(X)$, and $S_1 \subset H^{\text{odd}}(X)$. If we write $\rho = (\rho(c))_{c \in S}$ and

$$\rho(c) = (r^{m_r(c)})_{r \geq 1} = (1^{m_1(c)} 2^{m_2(c)} \ldots),$$

then we put

$$\begin{aligned} \ell(\rho) &= \sum_{c \in S} \ell(\rho(c)) = \sum_{c \in S, r \geq 1} m_r(c), \\ \|\rho\| &= \sum_{c \in S} |\rho(c)| = \sum_{c \in S, r \geq 1} r \cdot m_r(c), \\ \mathcal{P}_n(S) &= \{\rho \in \mathcal{P}(S) \mid \|\rho\| = n\}. \end{aligned}$$

Given

$$\rho = (\rho(c))_{c \in S} = (r^{m_r(c)})_{c \in S, r \geq 1} \in \mathcal{P}(S)$$

and $n \geq 0$, we define

$$\begin{aligned} \mathfrak{a}_{-\rho(c)}(c) &= \prod_{r\geq 1} \mathfrak{a}_{-r}(c)^{m_r(c)} = \mathfrak{a}_{-1}(c)^{m_1(c)}\mathfrak{a}_{-2}(c)^{m_2(c)}\cdots \\ \mathfrak{a}_\rho(n) &= \mathbf{1}_{-(n-\|\rho\|)} \prod_{c\in S} \mathfrak{a}_{-\rho(c)}(c)\cdot|0\rangle \in H^*(X^{[n]}) \end{aligned}$$

where we fix the order of the elements $c \in S_1$ appearing in $\prod_{c\in S}$ once and for all. Note from Definition 3.18 that $\mathfrak{a}_\rho(n) = 0$ for $0 \leq n < \|\rho\|$.

As ρ runs over all partition-valued functions on S with $\|\rho\| \leq n$, the corresponding $\mathfrak{a}_\rho(n)$ linearly span $H^*(X^{[n]})$ as a corollary to Theorem 3.8. By Theorem 7.7 (for $s = 2$), we have the cup product

$$\mathfrak{a}_\rho(n)\cdot\mathfrak{a}_\sigma(n) = \sum_\nu a^\nu_{\rho\sigma}\mathfrak{a}_\nu(n) \tag{7.5}$$

in $H^*(X^{[n]})$, where $\|\nu\| \leq \|\rho\| + \|\sigma\|$ and the structure coefficients $a^\nu_{\rho\sigma}$ are independent of n. Even though the cohomology classes $\mathfrak{a}_\nu(n)$ with $\|\nu\| \leq n$ in $H^*(X^{[n]})$ are not linearly independent in general, we have the following.

LEMMA 7.8. *The structure constants $a^\nu_{\rho\sigma}$ in the formula (7.5) are uniquely determined by the requirement that they are independent of n.*

PROOF. Assume that there exist a finite subset $I \subset \mathcal{P}(S)$ and some constants $c^\nu \in \mathbb{Q}$ independent of n such that for all $n \geq 0$, we have

$$\sum_{\nu\in I} c^\nu\mathfrak{a}_\nu(n) = 0. \tag{7.6}$$

By Theorem 3.8, the Heisenberg monomial classes

$$\prod_{c\in S}\mathfrak{a}_{-\rho(c)}(c)\cdot|0\rangle,$$

where $\rho = (\rho(c))_c \in \mathcal{P}_n(S)$, are linearly independent in the cohomology $H^*(X^{[n]})$. Therefore, by the definition of $\mathfrak{a}_\nu(n)$, we may assume in (7.6) that any two distinct ν and $\tilde{\nu}$ in I satisfy

$$\begin{aligned} \nu(1_X) &= (1^{m_1}2^{m_2}\cdots), \\ \tilde{\nu}(1_X) &= (1^{m_1+\|\tilde{\nu}|-\|\nu\|}2^{m_2}\cdots), \end{aligned}$$

and $\tilde{\nu}(c) = \nu(c)$ for $c \neq 1_X$ (here we assume for definiteness that $n \geq \|\tilde{\nu}\| > \|\nu\|$). In this case, we have

$$\mathfrak{a}_\nu(n) = \frac{(n-\|\tilde{\nu}\|)!}{(n-\|\nu\|)!}\cdot\mathfrak{a}_{\tilde{\nu}}(n).$$

Letting $n \to +\infty$, we see from (7.6) that $c^{\tilde{\nu}} = 0$ for the $\tilde{\nu} \in I$ with the largest size $\|\tilde{\nu}\|$. So all the constants c^ν are zero. ☐

Now we are ready to introduce the Hilbert ring.

DEFINITION 7.9. The *Hilbert ring* associated to a smooth projective surface X, denoted by $\mathfrak{H}_X$, is defined to be the ring with a linear basis formed by the symbols $\mathfrak{a}_\rho$, $\rho \in \mathcal{P}(S)$ and with the multiplication defined by

$$\mathfrak{a}_\rho\cdot\mathfrak{a}_\sigma = \sum_\nu a^\nu_{\rho\sigma}\mathfrak{a}_\nu \tag{7.7}$$

where the structure constants $a^\nu_{\rho\sigma}$ are from the relations (7.5).

Note that the Hilbert ring does not depend on the choice of a linear basis S of $H^*(X)$ containing 1_X since the operator $\mathfrak{a}_n(\alpha)$ depends on the cohomology class $\alpha \in H^*(X)$ linearly. It follows from the super-commutativity and associativity of the cohomology ring $H^*(X^{[n]})$ that the Hilbert ring $\mathfrak{H}_X$ itself is also super-commutative and associative.

THEOREM 7.10. (Stability) *Let X be a smooth projective complex surface.*

(i) *The cohomology rings $H^*(X^{[n]})$, $n \geq 1$, give rise to a Hilbert ring $\mathfrak{H}_X$ which is super-commutative and associative.*
(ii) *Conversely, the Hilbert ring $\mathfrak{H}_X$ completely determines the cohomology ring structure of $H^*(X^{[n]})$ for each $n \geq 1$.*

PROOF. (i) Follows from Definition 7.9 and the observations in the preceding paragraph.

(ii) Recall that as ρ runs over all partition-valued functions on S with $\|\rho\| \leq n$, the corresponding classes $\mathfrak{a}_\rho(n)$ linearly span $H^*(X^{[n]})$. Given the Hilbert ring $\mathfrak{H}_X$, the cup products (7.5) of these cohomology classes $\mathfrak{a}_\rho(n) \in H^*(X^{[n]})$ can be read from the multiplications (7.7). □

We also have the following result on the structure of the Hilbert ring $\mathfrak{H}_X$. For convenience, in the case when $\ell(\rho) = 1$, that is, when the partition $\rho(c)$ is a one-part partition (r) for some element $c \in S$ and is empty for all the other elements in S, we will simply write $\mathfrak{a}_\rho = \mathfrak{a}_{r,c}$ and $\mathfrak{a}_\rho(n) = \mathfrak{a}_{r,c}(n)$.

THEOREM 7.11. *The Hilbert ring $\mathfrak{H}_X$ is isomorphic to the tensor product $P \otimes E$, where P is the polynomial algebra generated by*

$$\mathfrak{a}_{r,c}, \ c \in S_0, r \geq 1,$$

and E is the exterior algebra generated by

$$\mathfrak{a}_{r,c}, \ c \in S_1, r \geq 1.$$

PROOF. Note that

$$\mathfrak{a}_{r,c}(n) = \mathbf{1}_{-(n-r)}\mathfrak{a}_{-r}(c)|0\rangle = B_{r-1}(c,n).$$

By Lemma 7.1 and Lemma 7.4, the ring $\mathfrak{H}_X$ is generated by the elements $\mathfrak{a}_{r,c}$, where $c \in S = S_0 \cup S_1$ and $r \geq 1$. By the super-commutativity of $\mathfrak{H}_X$, we have $\mathfrak{a}_{r,c}^2 = 0$ for $c \in S_1$ and $r \geq 1$. It remains to show that as $\rho = (r^{m_r(c)})_{s \in S, r \geq 1}$ runs over $\mathcal{P}(S)$, the monomials

$$\prod_{c \in S, r \geq 1} \mathfrak{a}_{r,c}^{m_r(c)}$$

are linearly independent in $\mathfrak{H}_X$. Assume

$$\sum_{i \in I} a_i \prod_{c \in S, r \geq 1} \mathfrak{a}_{r,c}^{m_r^i(c)} = 0$$

where $a_i \in \mathbb{C}$ and $\rho_i = (r^{m_r^i(c)})_{c\in S, r\geq 1}$ runs over a finite set I of distinct elements in $\mathcal{P}(S)$. Since $\mathbf{1}_{-(n-r)}\mathfrak{a}_{-r}(c)|0\rangle = \mathfrak{a}_{r,c}(n)$, we obtain

$$(7.8)\qquad \begin{aligned} &\sum_{i\in I} a_i \cdot \prod_{c\in S, r\geq 1} \big(\mathbf{1}_{-(n-r)}\mathfrak{a}_{-r}(c)|0\rangle\big)^{m_r^i(c)} \\ = &\sum_{i\in I} a_i \cdot \prod_{c\in S, r\geq 1} \mathfrak{a}_{r,c}(n)^{m_r^i(c)} \\ = &\ 0 \end{aligned}$$

by the definition of the structure constants in $\mathfrak{H}_X$.

Take an integer n large enough such that

$$n \geq n_i \stackrel{\text{def}}{=} \sum_{r,c} r m_r^i(c)$$

for all $i \in I$. From Theorem 7.7, we see that (7.8) can be rewritten as

$$(7.9)\qquad \sum_{i\in I} a_i \left(\mathbf{1}_{-(n-n_i)} \prod_{c\in S, r\geq 1} (\mathfrak{a}_{-r}(c))^{m_r^i(c)} \cdot |0\rangle + w_i\right) = 0$$

where each w_i is a finite linear combination of

$$\mathbf{1}_{-(n-\sum_{p=1}^N m_p)} \prod_{p=1}^N \mathfrak{a}_{-m_p}(\gamma_p) \cdot |0\rangle$$

with $\sum_{p=1}^N m_p < n_i$ and $\gamma_p \in S$ for every p. Recall that $\mathbf{1}_{-k} = 1/k! \cdot \mathfrak{a}_{-1}(1_X)^k$, $k \geq 0$. If we multiply (7.9) by $n!$, and locate in the resulting summation those terms whose coefficients contain the largest power of n, then we see that

$$(7.10)\qquad \sum_i a_i \cdot \mathbf{1}_{-(n-n_i)} \prod_{c\in S, r\geq 1} (\mathfrak{a}_{-r}(c))^{m_r^i(c)} \cdot |0\rangle = 0$$

where i satisfies $n_i = \max\{n_j | j \in I\}$. Since all the integers n_i in (7.10) are equal, the Heisenberg monomials in (7.10) are linearly independent as a corollary to Theorem 3.8. Thus all the coefficients a_i in (7.10) are zero. By repeating the above argument, we conclude that $a_i = 0$ for all $i \in I$. □

In Section 7.1 and the present section, the cohomology ring structure of the Hilbert scheme $X^{[n]}$ has been studied and described. However, these descriptions are very implicit in terms of the multiplication operators or explicit generators with implicitly given relations. In [**LS2**, **CoG**], two new approaches to study the cohomology ring of the Hilbert scheme $X^{[n]}$ appeared. In the next two sections, we will survey these two papers respectively.

7.3. Approach of Lehn-Sorger via graded Frobenius algebras

This section is devoted to [**LS2**]. The main idea in the approach of Lehn and Sorger is to construct a certain algebraic model from a graded Frobenius algebra. In the end, this algebraic model will be applied to the cohomology ring $H^*(X)$ of a smooth projective complex surface X, viewed naturally as a graded Frobenius algebra. We remark that the description given in [**LS2**] works only when the canonical divisor K_X of the surface X is numerically trivial. The numerical vanishing of K_X simplifies the description of the cohomology ring of $X^{[n]}$.

We start with the definition of graded Frobenius algebras.

DEFINITION 7.12. A *graded Frobenius algebra* of degree d is a finite dimensional graded (complex) vector space

$$A = \bigoplus_{i=-d}^{d} A^i$$

with a graded commutative and associative multiplication $A \otimes A \to A$ of degree d and unit element 1 (necessarily of degree $-d$) together with a linear form

$$T : A \to \mathbb{C}$$

of degree $-d$ such that the induced symmetric bilinear form

$$\langle a, b \rangle = T(ab)$$

is nondegenerate (and of degree 0).

It follows from $1 \cdot 1 = 1$ that d must be an even number. The degree conventions are chosen in such a way that A is centered around degree 0. The degree of an element $a \in A$ will be denoted by $|a|$.

The tensor product $A^{\otimes n}$ is again a graded Frobenius algebra of degree nd with product

$$(a_1 \otimes \cdots \otimes a_n) \cdot (b_1 \otimes \cdots \otimes b_n) = \epsilon(\underline{a}, \underline{b}) \cdot (a_1 b_1) \otimes \cdots \otimes (a_n b_n)$$

where $\epsilon(\underline{a}, \underline{b})$ is the sign resulting from reordering the a's and b's. The linear form on $A^{\otimes n}$ is given by

$$T(a_1 \otimes \cdots \otimes a_n) = T(a_1) \cdots T(a_n).$$

The n-th symmetric group S_n acts on $A^{\otimes n}$ by

$$\pi(a_1 \otimes \cdots \otimes a_n) = \epsilon(\pi, \underline{a}) \cdot a_{\pi^{-1}(1)} \otimes \cdots \otimes a_{\pi^{-1}(n)}$$

where $\epsilon(\pi, \underline{a})$ is the sign resulting from reordering the a's.

It will be useful to extend the definition of the n-fold tensor product to allow arbitrary unordered finite indexing sets. Let I be a finite set with n elements and let $\{A_i\}_{i \in I}$ be a family of copies of A. Denote the set $\{1, \ldots, n\}$ by $[n]$. Define

$$A^{\otimes I} = \left(\bigoplus_{f:[n] \xrightarrow{\cong} I} A_{f(1)} \otimes \cdots \otimes A_{f(n)} \right) / S_n$$

where the direct sum inherits a Frobenius algebra structure by component-wise defined operations, and S_n acts on this direct sum via the induced operation on the set of bijections $[n] \xrightarrow{\cong} I$. So for each bijection $f : [n] \xrightarrow{\cong} I$, there is a canonical isomorphism

$$A^{\otimes n} \xrightarrow{\cong} A^{\otimes I}$$

which will again be denoted by f.

Let $n_1, \ldots, n_k$ be positive integers, and let $n = n_1 + \ldots + n_k$. Define the ring homomorphism

$$\varphi_{n_\bullet, k} : A^{\otimes n} \to A^{\otimes k}$$

by mapping $a_1 \otimes \cdots \otimes a_n$ to

$$(a_1 \cdots a_{n_1}) \otimes \cdots \otimes (a_{n_1 + \cdots + n_{k-1} + 1} \cdots a_n).$$

To generalize this to get a graded ring homomorphism $\varphi^* : A^{\otimes I} \to A^{\otimes J}$ for any surjective map $\varphi : I \to J$ of finite sets of cardinality $n = |I|$ and $k = |J|$, choose a bijection $g : [k] \to J$ and let $n_i = |\varphi^{-1}(g(i))|$. Then there is a bijection $f : [n] \to I$ such that for each $i \in [k]$, we have

$$\varphi^{-1}(g(i)) = \big\{f(n_1 + \cdots + n_{i-1} + 1), \ldots, f(n_1 + \cdots + n_i)\big\}.$$

The composition

$$\varphi^* : \quad A^{\otimes I} \xrightarrow{f^{-1}} A^{\otimes n} \xrightarrow{\varphi_{n_\bullet,k}} A^{\otimes k} \xrightarrow{g} A^{\otimes J}$$

is well-defined and independent of the choices of f and g. Moreover, if $\varphi : I \to J$ and $\psi : J \to K$ are surjections, then

$$\psi^* \circ \varphi^* = (\psi \circ \varphi)^*.$$

In a different direction, for a surjective map $\varphi : I \to J$, define

$$\varphi_* : A^{\otimes J} \to A^{\otimes I}$$

to be the linear map adjoint to φ^* with respect to the bilinear forms on $A^{\otimes I}$ and $A^{\otimes J}$. Note that φ^* is a ring homomorphism, but φ_* is not. However, the projection formula holds:

$$\varphi_*(a \cdot \varphi^*(b)) = \varphi_*(a) \cdot b$$

for $b \in A^{\otimes I}$ and $a \in A^{\otimes J}$.

Next, consider the composition map

$$A \xrightarrow{\Delta_*} A \otimes A \longrightarrow A$$

where the first map is the co-product and the second map is the product. The image of 1 under the composition map is called the *Euler class* of A and is denoted by $e = e(A)$. If $\dim(A^{-d}) = 1$, then by duality we have $\dim(A^d) = 1$ as well. Hence there is a unique element $\mathrm{vol} \in A^d$ such that $T(\mathrm{vol}) = 1$. In this case

$$e(A) = \chi(A) \cdot \mathrm{vol}$$

where $\chi(A) = \sum_i (-1)^i \dim(A^i)$ is the *Euler-Poincaré characteristic* of A.

For a function $\nu : I \to \{0, 1, 2, \ldots\}$, define

$$e^\nu = \otimes_{i \in I} e^{\nu(i)} \in A^{\otimes I}.$$

Let $\Box$ denote the co-Cartesian diagram of finite sets and surjective maps:

$$\begin{array}{ccc} I & \xrightarrow{\alpha} & K \\ \big\downarrow \beta & & \big\downarrow \gamma \\ J & \xrightarrow{\delta} & L. \end{array}$$

The associated diagram

$$\begin{array}{ccc} A^{\otimes I} & \xrightarrow{\alpha^*} & A^{\otimes K} \\ \big\uparrow \beta_* & & \big\uparrow \gamma_* \\ A^{\otimes J} & \xrightarrow{\delta^*} & A^{\otimes L} \end{array}$$

of linear maps is not commutative. In fact, by [**LS2**, Lemma 2.4], we have

$$\alpha^* \beta_*(a) = \gamma_*\big(e^{\nu_\Box} \cdot \delta^*(a)\big) \tag{7.11}$$

for $a \in A^{\otimes J}$, where the function $\nu_\square : L \to \{0, 1, 2, \ldots\}$ is defined by

$$\nu_\square(\ell) = 1 - |\delta^{-1}(\ell)| - |\gamma^{-1}(\ell)| + |(\delta \circ \beta)^{-1}(\ell)|.$$

The *degree of a permutation* $\pi \in S_n$ of cycle type given by a partition $\lambda = (\lambda_1 \geq \ldots \geq \lambda_\ell > 0)$ is defined to be

$$d(\pi) = d(\lambda) = n - \ell. \tag{7.12}$$

Since $\pi \mapsto \operatorname{sgn}(\pi) = (-1)^{d(\pi)}$ is a homomorphism, the *degree defect*

$$a(\pi_1, \ldots, \pi_t) = \frac{1}{2}\big(d(\pi_1) + \ldots + d(\pi_t) - d(\pi_1 \cdots \pi_t)\big)$$

is a non-negative integer. For a subgroup $H \subset S_n$ and an H-stable subset $B \subset [n]$, let $H \backslash B$ denote the orbit space for the induced action. For $\pi, \rho \in S_n$, the *graph defect*

$$g(\pi, \rho) : \langle \pi, \rho \rangle \backslash [n] \to \mathbb{Q}$$

is defined by

$$g(\pi, \rho)(B) = \frac{1}{2}\big(|B| + 2 - |\langle \pi \rangle \backslash B| - |\langle \rho \rangle \backslash B| - |\langle \pi\rho \rangle \backslash B|\big). \tag{7.13}$$

By [**LS2**, Lemma 2.7], the graph defect g takes value in the non-negative integers. As in the case of the degree defect, the graph defect allows a natural multivariant extension: for $\pi_1, \ldots, \pi_t \in S_n$ and $B \in \langle \pi_1, \ldots, \pi_t \rangle \backslash [n]$, define

$$g(\pi_1, \ldots, \pi_t)(B) = \frac{1}{2}\left(|B| + 2 - \sum_j |\langle \pi_j \rangle \backslash B| - |\langle \pi_1 \cdots \pi_t \rangle \backslash B|\right).$$

Define the graded vector space $A\{S_n\}$ by putting

$$A\{S_n\} = \bigoplus_{\pi \in S_n} A^{\otimes(\langle \pi \rangle \backslash [n])} \cdot \pi \tag{7.14}$$

where the grading of an element $a \cdot \pi$ is $|a \cdot \pi| = |a|$. The symmetric group S_n acts on $A\{S_n\}$ as follows: the action of $\sigma \in S_n$ on $[n]$ induces a bijection

$$\sigma : \langle \pi \rangle \backslash [n] \to \langle \sigma\pi\sigma^{-1} \rangle \backslash [n], \quad x \mapsto \sigma x$$

for each π and hence an isomorphism

$$\tilde{\sigma} : A\{S_n\} \to A\{S_n\}, \quad a \cdot \pi \mapsto \sigma^*(a) \cdot \sigma\pi\sigma^{-1}.$$

Define

$$A^{[n]} = (A\{S_n\})^{S_n}$$

to be the subspace of the fixed elements.

Next, we will define a ring structure on $A\{S_n\}$ such that $A^{[n]}$ becomes a commutative subring. Note that any inclusion $H \subset K$ of subgroups of S_n leads to a surjection $H\backslash[n] \twoheadrightarrow K\backslash[n]$ of orbit spaces and hence to maps

$$f^{H,K} : \quad A^{\otimes H\backslash[n]} \to A^{\otimes K\backslash[n]}$$

and

$$f_{K,H} : \quad A^{\otimes K\backslash[n]} \to A^{\otimes H\backslash[n]}.$$

For $\pi, \rho \in S_n$, define

$$m_{\pi,\rho} : A^{\otimes\langle \pi \rangle\backslash[n]} \otimes A^{\otimes\langle \rho \rangle\backslash[n]} \to A^{\otimes\langle \pi\rho \rangle\backslash[n]}$$

by

$$m_{\pi,\rho}(a \otimes b) = f_{\langle \pi,\rho \rangle,\langle \pi\rho \rangle}\big(f^{\langle \pi \rangle,\langle \pi,\rho \rangle}(a) \cdot f^{\langle \rho \rangle,\langle \pi,\rho \rangle}(b) \cdot e^{g(\pi,\rho)}\big). \tag{7.15}$$

where $g(\pi, \rho)$ is the graph defect defined in (7.13). The next result is [**LS2**, Propositions 2.13 and 2.14], and follows from (7.15) and repeated applications of (7.11).

PROPOSITION 7.13. (i) *The product* $A\{S_n\} \times A\{S_n\} \to A\{S_n\}$ *defined by*

$$a\pi \cdot b\rho = m_{\pi,\rho}(a \otimes b)\pi\rho$$

is associative, S_n*-equivariant, and homogeneous of degree* nd.

(ii) *For any two homogeneous elements* $a\pi, b\rho \in A\{S_n\}$*, we have the following (non)commutativity relation:*

$$a\pi \cdot b\rho = (-1)^{|a|\cdot|b|}\pi^*(b)\pi\rho\pi^{-1} \cdot a\pi = (-1)^{|a|\cdot|b|}\tilde{\pi}^*(b\rho) \cdot a\pi.$$

By Proposition 7.13, $A^{[n]}$ is a subring of the center of $A\{S_n\}$. Let

$$\widetilde{T} : A\{S_n\} \to \mathbb{C}$$

be defined by

$$\widetilde{T}(a\pi) = \begin{cases} T(a), & \text{if } \pi = 1, \\ 0, & \text{else.} \end{cases}$$

Then $\widetilde{T}$ induces a non-degenerate and S_n-invariant bilinear pairing on $A\{S_n\}$. Therefore, the restriction of $\widetilde{T}$ to $A^{[n]}$ defines the structure of a graded Frobenius algebra of degree nd on $A^{[n]}$.

Let

$$\mathcal{V}(A) = \operatorname{Sym}(A \otimes t^{-1}\mathbb{C}[t^{-1}])$$

be the bosonic Fock space modeled on the graded vector space A. Then $\mathcal{V}(A)$ is bi-graded by degree and weight, where an element $a \otimes t^{-m} \in A \otimes t^{-m}$ is given degree $|a|$ and weight m. The component of $\mathcal{V}(A)$ of weight n is the graded vector space

$$\mathcal{V}(A)_n \cong \bigoplus_{\alpha \vdash n} \bigotimes_i \operatorname{Sym}^{\alpha_i} A$$

where $\alpha = (1^{\alpha_1}2^{\alpha_2}\cdots)$ runs through all partitions of n. Let

$$f : \{1, \ldots, N\} \to \langle \pi \rangle \backslash [n]$$

be an enumeration of the orbits of $\pi \in S_n$, and let $\ell_i = |f(i)|$. Define $\Phi' : A^{\otimes N} \to \mathcal{V}(A)$ by

$$a_1 \otimes \cdots \otimes a_N \mapsto \frac{1}{n!}(a_1 \otimes t^{-\ell_1}) \cdots (a_N \otimes t^{-\ell_N}),$$

and let

$$\Phi : \quad \bigoplus_{n=0}^{+\infty} A\{S_n\} \to \mathcal{V}(A)$$

be given on the summand $A^{\otimes \langle \pi \rangle \backslash [n]} \cdot \pi$ by the composition

$$A^{\otimes \langle \pi \rangle \backslash [n]} \xrightarrow{f^{-1}} A^{\otimes N} \xrightarrow{\Phi'} \mathcal{V}(A).$$

Note that Φ is surjective and invariant under the S_n-action, so that its restriction to $A^{[n]}$ is also surjective. Moreover, the vector spaces $A^{[n]}$ and $\mathcal{V}(A)_n$ have the same dimension. These observations lead to the following [**LS2**, Proposition 2.11].

PROPOSITION 7.14. *Φ induces an isomorphism of graded vector spaces*

$$A^{[n]} \to \mathcal{V}(A)_n.$$

Finally, the above constructions are applied to $A = H^*(X)[2]$ for a smooth projective complex surface X, where the linear form $T : A = H^*(X)[2] \to \mathbb{C}$ is given by

$$T(a) = -\int_X a. \tag{7.16}$$

The following is from [**LS2**, Theorem 3.2].

THEOREM 7.15. *Let X be a smooth projective complex surface with a numerically trivial canonical divisor. Then there is a canonical isomorphism of graded rings*

$$\big(H^*(X)[2]\big)^{[n]} \xrightarrow{\cong} H^*(X^{[n]})[2n]. \tag{7.17}$$

The linear form $T : H^*(X^{[n]})[2n] \to \mathbb{C}$ is given by

$$T(a) = (-1)^n \cdot \int_{X^{[n]}} a$$

which is consistent with (7.16) when $n = 1$. We now sketch the main ideas in the proof of Theorem 7.15, and refer to Sections 3 and 4 in [**LS2**] for further details. Put $A = H^*(X)[2]$. By Theorem 3.8, there is an isomorphism of graded vector spaces

$$\Psi : \quad \mathcal{V}(A) \to \bigoplus_{n=0}^{+\infty} H^*(X^{[n]})[2n]$$

sending $(a_1 t^{-n_1}) \cdots (a_s t^{-n_s})$ to $\mathfrak{a}_{-n_1}(a_1) \cdots \mathfrak{a}_{-n_s}(a_s)|0\rangle$. Note that Ψ maps the n-th component $\mathcal{V}(A)_n$ to $H^*(X^{[n]})[2n]$. It was observed in [**Leh1, LQW1**] that the ring structure of $H^*(X^{[n]})[2n], n \ge 1$ is uniquely determined by the linear isomorphism Ψ, Theorem 3.24 (ii) (also (4.19)), and Theorem 4.2. Another observation is the relation [**FW**] between Goulden's operator [**Gou**, Section 3] on the symmetric groups and Lehn's boundary operator (4.19) in the Hilbert schemes. Exploiting these observations shows that the restriction of $\Psi \circ \Phi$ to $A^{[n]}$ is the desired isomorphism (7.17).

7.4. Approach of Costello-Grojnowski via Calogero-Sutherland operators

The approach of Lehn and Sorger [**LS2**] described in Section 7.3 fails when the canonical divisor of the smooth projective complex surface X is not numerically trivial. In [**CoG**], Costello and Grojnowski came up with a different approach to deal with an arbitrary smooth projective complex surface X. In this section, we will outline the main ideas in [**CoG**]. To avoid the formalism and technicalities, our presentation follows that of [**CoG**, Introduction].

The starting point of [**CoG**] is the same as what is outlined for the proof of Theorem 7.15: the ring structure of $H^*(X^{[n]}), n \ge 1$ is uniquely determined by the theorem of Nakajima and Grojnowski (Theorem 3.8), Lehn's boundary operator $\mathfrak{d}$ (Theorem 3.24 (ii) and (4.19)), and Theorem 4.2.

So let X be a smooth projective complex surface. Put

$$H = H^*(X).$$

Regard H as a graded Frobenius algebra over $\mathbb{C}$. By Theorem 3.8, there is a canonical identification

$$\mathcal{F} = \mathrm{Sym}(xH[x]) \cong \mathbb{H}_X = \bigoplus_{n=0}^{+\infty} H^*(X^{[n]}) \tag{7.18}$$

as modules for the Heisenberg algebra modeled on H. Write

$$\mathcal{F} = \bigoplus_{n=0}^{+\infty} \mathcal{F}^n$$

where $\mathcal{F}^n$ is the n-th eigenspace of the energy operator

$$\partial = x\frac{\partial}{\partial x}.$$

Via (7.18), $\mathcal{F}^n$ is linearly identified with $H^*(X^{[n]})$. Cup product with an element

$$a \in \mathcal{F} = \bigoplus_{n=0}^{+\infty} \mathcal{F}^n$$

defines a linear operator on $\mathcal{F}$. For reasonable a, such a linear operator is a differential operator. This differential operator commutes with Lehn's boundary operator $\mathfrak{d}$ since cohomology is a (super)-commutative ring. Hence there are at least as many differential operators on $\mathcal{F}$ which commute with $\mathfrak{d}$ as the dimension of the space $\mathcal{F}$.

Thinking of $\mathfrak{d}$ as a Hamiltonian, this is precisely saying that $\mathfrak{d}$ defines an integrable system. It rephrases the problem of determining the ring structure on $\mathcal{F}$ as the problem of finding integrals of motion, i.e., determining the centralizer of $\mathfrak{d}$ in an appropriate ring of differential operators on $\mathcal{F}$, and showing that the centralizer is commutative and in bijection with $\mathcal{F}$. As noted in [**FW**, Introduction] and [**Mat**, Section 5], $\mathfrak{d}$ is a version of the bosonized Calogero-Sutherland Hamiltonian. So the problem of describing the cohomology ring structure of the Hilbert scheme $X^{[n]}$ becomes an algebraic problem, that of solving a deformed version of the Calogero-Sutherland system.

To carry out the observations in the preceding paragraph, a ring, denoted by

$$\mathcal{D}\mathrm{iff}_{H,K'}\mathcal{F},$$

of continuous differential operators on the Fock space $\mathcal{F}$ is defined. It depends on the ring structure on $H = H^*(X)$ and an additional parameter

$$K' \in H^0(X) \oplus H^2(X) \oplus H^4(X).$$

It is a limit of finitely generated algebras and much smaller than the ring of all differential operators. Moreover, $\mathcal{D}\mathrm{iff}_{H,K'}\mathcal{F}$ contains the Calogero-Sutherland operator

$$\mathcal{L}_{K'}$$

which is a second order differential operator and coincides with $\mathfrak{d}$ if K' is equal to the canonical divisor K_X of X. The Calogero-Sutherland operator $\mathcal{L}_{K'}$ can be written as the inverse limit of r-particle generalized Calogero-Sutherland Hamiltonians. Each r-particle system is integrable via the methods of Dunkl, Cherednik and Opdam.

For generic K', define

$$\mathcal{IM}_{K'}$$

to be the centralizer in $\mathcal{D}\mathrm{iff}_{H,K'}\mathcal{F}$ of $\mathcal{L}_{K'}$ and the energy operator ∂. Since $\mathcal{IM}_{K'}$ commutes with ∂, it projects to give a subalgebra of $\mathrm{End}(\mathcal{F}^n)$ for each $n \geq 1$. An explicit construction of $\mathcal{IM}_{K'}$ as a polynomial algebra is given. The algebra $\mathcal{IM}_{K_X}$ is obtained by degenerating

$$K' = \lambda + K_X$$

to K_X as $\lambda \to 0$. A description of $\mathcal{IM}_{K_X}$ in terms of degenerating families of Dunkl-Cherednik operators is then obtained.

The following is the main result [**CoG**, Theorem 1.0.1].

THEOREM 7.16. *$\mathcal{IM}_{K_X}$ is a commutative algebra, and for every n the image of $\mathcal{IM}_{K_X}$ in $\mathrm{End}(\mathcal{F}^n)$ is precisely the algebra of left multiplication operators (i.e., the image of the natural map $H^*(X^{[n]}) \to \mathrm{End}(\mathcal{F}^n), a \mapsto a\cup$).*

Theorem 7.16 together with the explicit structure of $\mathcal{IM}_{K_X}$ provides a new description of the cohomology ring of $X^{[n]}$. We refer to [**CoG**] for further details.

CHAPTER 8

Ideals of the cohomology rings of Hilbert schemes

In Chapter 7, we studied the cohomology ring of the Hilbert scheme $X^{[n]}$ of points on a smooth projective complex surface X. However, the situation changes dramatically when X is quasi-projective. The understanding of the cohomology ring $H^*(X^{[n]})$ for quasi-projective X has been rather limited with the exception of the affine plane [**ES2**, **Leh1**, **LS1**, **Vas**] and the minimal resolution of a simple singularity $\mathbb{C}^2/\Gamma$ where Γ is a finite group of $\mathrm{SL}_2(\mathbb{C})$ [**QW3**, **Wan3**]. Besides these minimal resolutions, typical important examples of quasi-projective surfaces include the cotangent bundle of a smooth projective curve and the surface obtained from a smooth projective surface by deleting a point. All these surfaces are among a class of quasi-projective surfaces which satisfy the S-property (Definition 8.20). In this chapter, we will establish some general results regarding the cohomology rings of the Hilbert schemes of points for such a large class of quasi-projective surfaces. As a preparation, for a smooth projective complex surface X, we will define the ideal $\mathcal{I}^{[n]}$ of the cohomology ring $H^*(X^{[n]})$ associated to an ideal $\mathcal{I}$ of the cohomology ring $H^*(X)$. We will prove that when

$$\mathcal{I} = \bigoplus_{\ell=1}^{4} H^\ell(X),$$

the quotient ring $H^*(X^{[n]})/\mathcal{I}^{[n]}$ is isomorphic to the cohomology ring $H^*((\mathbb{C}^2)^{[n]})$ of the Hilbert schemes of points on the affine plane $\mathbb{C}^2$. For this purpose, we will review the cohomology ring $H^*((\mathbb{C}^2)^{[n]})$, following [**ES2**, **Leh1**, **LS1**, **Vas**]. Moreover, when $\mathcal{I} = H^4(X)$, we will show that modulo the ideal $\mathcal{I}^{[n]}$, the structure constants in certain cup products of the Hilbert schemes $X^{[n]}$ are independent of n. The main results and their proofs in this chapter are from [**LQW5**].

8.1. The cohomology ring of the Hilbert scheme $(\mathbb{C}^2)^{[n]}$

In this section, we will address the issue of Heisenberg algebra actions on the cohomology of the Hilbert scheme $X^{[n]}$ when X is a smooth quasi-projective complex surface. Then following [**Leh1**, **LS1**, **Vas**], we will see that the cohomology ring of the Hilbert scheme $(\mathbb{C}^2)^{[n]}$ can be identified with the class algebra (with the cup product) of the symmetric group S_n. This part of the discussion may be viewed as a simplified version of Section 7.3 with $A = \mathbb{C}$.

We begin with some discussions about the Heisenberg algebra actions when the complex surface is quasi-projective. The constructions of the Heisenberg algebra actions in Section 3.2 are carried out for smooth projective surfaces. According to [**Nak3**, Section 3], with minor modifications, these constructions hold for a smooth quasi-projective surface X. Let $H_*^{BM}(X)$ be the Borel-Moore homology of X, which is defined by

$$H_*^{BM}(X) = H_*(\hat{X}, \infty)$$

where $\hat{X} = X \cup \{\infty\}$ is the one-point compactification of X, and $H_*(\hat{X}, \infty)$ is the ordinary relative homology of the pair $(\hat{X}, \infty)$. Then the creation operators are modeled on the Borel-Moore homology $H_*^{BM}(X)$, while the annihilation operators are modeled on the ordinary homology $H_*(X)$. The Fock space of the Heisenberg algebra is taken to be the direct sum

$$\bigoplus_{n=0}^{+\infty} H_*^{BM}(X^{[n]})$$

of the Borel-Moore homology groups $H_*^{BM}(X^{[n]})$. Let $H_c^*(X)$ be the cohomology with compact support. Using the the Poincaré dualities

$$\begin{aligned} \text{PD}: &\quad H^{4-i}(X) \to H_i^{BM}(X) \\ \text{PD}: &\quad H_c^{4-i}(X) \to H_i(X), \end{aligned}$$

we can regard the creation operators $\mathfrak{a}_{-n}(\alpha)$ with $n > 0$ as being modeled on $H^*(X)$ (i.e., $\alpha \in H^*(X)$), while we can regard the annihilation operators $\mathfrak{a}_n(\beta)$ with $n > 0$ as being modeled on $H_c^*(X)$ (i.e., $\beta \in H_c^*(X)$). Accordingly, with the help of the Poincaré duality

$$\text{PD}: H^{4n-i}(X^{[n]}) \to H_i^{BM}(X^{[n]}),$$

from now on we can take the Fock space to be the direct sum

$$\mathbb{H}_X = \bigoplus_{n=0}^{+\infty} H^*(X^{[n]}) \tag{8.1}$$

of the ordinary cohomology groups $H^*(X^{[n]})$.

Let $\overline{X}$ be a smooth projective surface compactifying the smooth quasi-projective surface X, and let $\iota: X \to \overline{X}$ be the inclusion map. Then we have induced embeddings $\iota_n: X^{[n]} \to \overline{X}^{[n]}$ for $n \geq 0$, and induced ring homomorphisms

$$\iota_n^*: H^*(\overline{X}^{[n]}) \to H^*(X^{[n]}).$$

The maps ι^* and ι_n^* are related by the following.

Lemma 8.1. *Let notations be as in the preceding paragraph. Then,*

$$\iota_n^*\big(\mathfrak{a}_{-n_1}(\overline{\alpha}_1)\cdots\mathfrak{a}_{-n_k}(\overline{\alpha}_k)|0\rangle\big) = \mathfrak{a}_{-n_1}(\iota^*\overline{\alpha}_1)\cdots\mathfrak{a}_{-n_k}(\iota^*\overline{\alpha}_k)|0\rangle \tag{8.2}$$

where $k \geq 0$, $n_1, \ldots, n_k > 0$, *and* $n_1 + \ldots + n_k = n$.

Proof. For $n \geq 0$, let

$$(\iota_n^*)^{BM}: H_i^{BM}(\overline{X}^{[n]}) \to H_i^{BM}(X^{[n]})$$

be the natural map induced by the embedding $\iota_n : X^{[n]} \to \overline{X}^{[n]}$. Then, it is well-known that

$$(\iota_n^*)^{BM} \circ \text{PD} = \text{PD} \circ (\iota_n^*)^{BM}.$$

Combining with the construction of the creation operators, we obtain

$$\iota_{m_1+m_2}^* \mathfrak{a}_{-m_2}(\overline{\alpha})(A) = \mathfrak{a}_{-m_2}(\iota^*\overline{\alpha})(\iota_{m_1}^* A)$$

for $m_1, m_2 > 0$ and $A \in H^*(\overline{X}^{[m_1]})$. Applying this repeatedly, we get (8.2). □

In the rest of this section, we consider the affine plane $X = \mathbb{C}^2$. Since $H^*(\mathbb{C}^2) = \mathbb{C} \cdot 1_{\mathbb{C}^2}$, we see that the Fock space

$$\mathbb{H}_{\mathbb{C}^2} = \bigoplus_{n=0}^{+\infty} H^*\big((\mathbb{C}^2)^{[n]}\big)$$

has a linear basis consisting of the Heisenberg monomial classes

$$\mathfrak{a}_{-n_1}(1_{\mathbb{C}^2})^{r_1} \cdots \mathfrak{a}_{-n_k}(1_{\mathbb{C}^2})^{r_k}|0\rangle \tag{8.3}$$

where $k \geq 0$ and $n_1, \ldots, n_k, r_1, \ldots, r_k > 0$.

Let $\mathbb{C}[q_1, q_2, \ldots]$ be the polynomial ring in countably infinitely many variables. Then there is a natural isomorphism of Fock spaces:

$$\Psi_1 : \quad \mathbb{C}[q_1, q_2, \ldots] \to \mathbb{H}_{\mathbb{C}^2} \tag{8.4}$$

defined by

$$\Psi_1\big(q_{n_1}^{r_1} \cdots q_{n_k}^{r_k}\big) = \mathfrak{a}_{-n_1}(1_{\mathbb{C}^2})^{r_1} \cdots \mathfrak{a}_{-n_k}(1_{\mathbb{C}^2})^{r_k}|0\rangle. \tag{8.5}$$

Under this linear isomorphism, the creation operators $\mathfrak{a}_{-n}(1_{\mathbb{C}^2}), n > 0$ on $\mathbb{H}_{\mathbb{C}^2}$ correspond to the multiplication by q_n on $\mathbb{C}[q_1, q_2, \ldots]$, while the annihilation operators on $\mathbb{H}_{\mathbb{C}^2}$ correspond to the operators

$$-m\frac{\partial}{\partial q_m} \tag{8.6}$$

on $\mathbb{C}[q_1, q_2, \ldots]$.

As for the ring structure on $H^*((\mathbb{C}^2)^{[n]})$, note that the Chern character operators $\mathfrak{G}_k(1_{\mathbb{C}^2})$ from Definition 4.1 (iii) are well-defined on the Fock space $\mathbb{H}_{\mathbb{C}^2}$.

THEOREM 8.2. *Let* $\partial_m = m\frac{\partial}{\partial q_m}$.

(i) *Via the linear isomorphism* Ψ_1 *in* (8.4),

$$\mathfrak{G}_k(1_{\mathbb{C}^2}) = \frac{(-1)^k}{(k+1)!} \sum_{n_0, \ldots, n_k > 0} q_{n_0+\ldots+n_k} \partial_{n_0} \cdots \partial_{n_k}.$$

(ii) *For each* n, *the cohomology ring* $H^*((\mathbb{C}^2)^{[n]})$ *is generated over* $\mathbb{C}$ *by the Chern characters* $\mathrm{ch}_k\big((\mathcal{O}_{\mathbb{C}^2})^{[n]}\big)$.

Theorem 8.2 (i) is from [**Leh1**, Theorem 4.10], while Theorem 8.2 (ii) is a classical result that follows from [**ES2**, Theorem 1.1]. Intuitively, Theorem 8.2 (i) also follows from Theorem 4.7. Indeed, since $K_{\mathbb{C}^2} = e_{\mathbb{C}^2} = 0$, Theorem 4.7 says that $\mathfrak{G}_k(1_{\mathbb{C}^2})$ is equal to

$$-\sum_{\ell(\lambda)=k+2, |\lambda|=0} \frac{1}{\lambda^!} \mathfrak{a}_\lambda(\tau_* 1_{\mathbb{C}^2}). \tag{8.7}$$

Note that $\mathfrak{a}_\lambda(\tau_* 1_{\mathbb{C}^2})$ is not zero only when the generalized partition λ has exactly one negative part. Moreover, when λ has exactly one negative part, via the linear isomorphism Ψ_1, the operator $\mathfrak{a}_\lambda(\tau_* 1_{\mathbb{C}^2})$ is of the form

$$q_{n_0+\ldots+n_k}(-\partial_{n_0}) \cdots (-\partial_{n_k})$$

where $n_0, \ldots, n_k > 0$ (see (8.6)). Recalling the definition of $\lambda^!$ from Definition 1.6 and keeping track of the coefficient of $q_{n_0+\ldots+n_k}(-\partial_{n_0}) \cdots (-\partial_{n_k})$, we conclude that Theorem 8.2 (i) is a consequence of (8.7).

In [**LS1, Vas**], the cohomology $H^*((\mathbb{C}^2)^{[n]})$ as a $\mathbb{C}$-algebra is further identified with the class algebra (or the algebra of class functions)

$$\mathcal{C}(S_n)$$

of the symmetric group S_n. Let $\mathbb{C}[S_n]$ be the group algebra of S_n over $\mathbb{C}$. An element $\sum_{\pi \in S_n} a_\pi \pi$ may be regarded as a function $f : S_n \to \mathbb{C}$ defined by $f(\pi) = a_\pi$. The convolution product $\star$ on $\mathbb{C}[S_n]$ is defined by

$$(f * g)(\pi) = \sum_{\sigma \in S_n} f(\pi\sigma^{-1})g(\sigma). \tag{8.8}$$

For each partition $\lambda = (\lambda_1 \geq \ldots \geq \lambda_\ell)$ of n, define

$$C_\lambda = \sum_{\pi \text{ of type } \lambda} \pi \tag{8.9}$$

where π runs through all permutations with cycle type λ, i.e., those having a disjoint cycle decomposition with cycle lengths $\lambda_1, \ldots, \lambda_\ell$.

The class algebra $\mathcal{C}(S_n)$ is the center of $\mathbb{C}[S_n]$, and has a linear basis

$$\{C_\lambda | \lambda \vdash n\}.$$

The product on $\mathcal{C}(S_n)$ is the cup product $\cup$ defined by

$$\sigma \cup \pi = \begin{cases} \sigma \star \pi, & \text{if } d(\sigma) + d(\pi) = d(\sigma\pi), \\ 0, & \text{otherwise} \end{cases}$$

where $d(\pi)$ is defined in (7.12). Put

$$\mathcal{C} = \bigoplus_{n=0}^{+\infty} \mathcal{C}(S_n). \tag{8.10}$$

In [**FJW**], an isomorphism of vertex algebras

$$\Psi_2 : \quad \mathcal{C} \to \mathbb{C}[q_1, q_2, \ldots] \tag{8.11}$$

was defined by putting

$$\Psi_2(C_\lambda) = \prod_{i \geq 1} \left(\frac{1}{m_i!} \left(\frac{q_i}{i}\right)^{m_i} \right) \tag{8.12}$$

for $\lambda = (1^{m_1} 2^{m_2} \cdots)$. The following is from [**LS1**, Theorem 1.1].

THEOREM 8.3. *The composite isomorphism of Fock spaces*

$$\mathcal{C} \xrightarrow{\Psi_2} \mathbb{C}[q_1, q_2, \ldots] \xrightarrow{\Psi_1} \mathbb{H}_{\mathbb{C}^2}$$

induces an isomorphism of $\mathbb{C}$-algebras

$$\mathcal{C}(S_n) \longrightarrow H^*((\mathbb{C}^2)^{[n]})$$

for each $n \geq 0$.

The main idea in the proof of Theorem 8.3 is to use the alternating character $\varepsilon_n \in \mathcal{C}(S_n)$ defined by

$$\varepsilon_n = \sum_{\pi \in S_n} (-1)^{d(\pi)} \pi = \sum_{\pi \in S_n} \operatorname{sgn}(\pi)\, \pi.$$

It is known that

$$\sum_{n\geq 0} \Psi_2(\varepsilon_n)z^n = \exp\left(\sum_{m>0}(-1)^{m-1}\frac{z^m}{m}q_m\right). \tag{8.13}$$

On the other hand, it follows from Theorem 8.2 (i) that

$$\sum_{n\geq 0} \Psi_1^{-1}(\gamma_n)z^n = \exp\left(\sum_{m>0}(-1)^{m-1}\frac{z^m}{m}q_m\right).$$

where γ_n denotes the total Chern class of the tautological bundle $(\mathcal{O}_{\mathbb{C}^2})^{[n]}$. Therefore, $\Psi_1\Psi_2$ maps the alternating character $\varepsilon_n \in \mathcal{C}(S_n)$ to the total Chern class of $(\mathcal{O}_{\mathbb{C}^2})^{[n]}$ in $H^*\big((\mathbb{C}^2)^{[n]}\big)$. In addition, via some technical computation, one proves that

$$\Psi_1\Psi_2(\varepsilon_n \cup w) = \gamma_n \cup \Psi_1\Psi_2(w)$$

for every $w \in \mathcal{C}(S_n)$. Now Theorem 8.3 follows since by Theorem 8.2 (ii), the cohomology ring $H^*\big((\mathbb{C}^2)^{[n]}\big)$ is generated by the Chern classes of $(\mathcal{O}_{\mathbb{C}^2})^{[n]}$.

8.2. Ideals in $H^*(X^{[n]})$ for a projective surface X

In this section, X denotes a smooth projective complex surface. We will define the ideal $\mathcal{I}^{[n]}$ of the cohomology ring $H^*(X^{[n]})$ associated to an ideal $\mathcal{I}$ of the cohomology ring $H^*(X)$. Moreover, we will study the properties of $\mathcal{I}^{[n]}$, including its generators.

The first lemma is elementary.

LEMMA 8.4. *Let $\mathcal{I}$ be an ideal in the cohomology ring $H^*(X)$. Let $\alpha \in \mathcal{I}$ and $k \geq 2$. Then, the pushforward $\tau_{k*}\alpha$ can be written as*

$$\sum_j \alpha_{j,1} \otimes \cdots \otimes \alpha_{j,k} \in H^*(X^k)$$

such that for each fixed j, there exists some ℓ with $\alpha_{j,\ell} \in \mathcal{I}$.

PROOF. First of all, note from the third formula in (3.11) that if

$$\tau_{k*}\alpha = \sum_j \alpha_{j,1} \otimes \alpha_{j,2} \otimes \cdots \otimes \alpha_{j,k},$$

then

$$\tau_{(k+1)*}\alpha = \sum_j (\tau_{2*}\alpha_{j,1}) \otimes \alpha_{j,2} \otimes \cdots \otimes \alpha_{j,k}.$$

Therefore, by induction, it suffices to prove the lemma for $k = 2$. Now the case $k = 2$ follows from the first formula in (3.11) that if we write

$$\tau_{2*}(1_X) = \sum_i \alpha_i \otimes \beta_i,$$

then

$$\tau_{2*}(\alpha) = \sum_i (\alpha\alpha_i) \otimes \beta_i$$

with $(\alpha\alpha_i) \in \mathcal{I}$. □

In view of the preceding lemma, we introduce the following important definition.

DEFINITION 8.5. Let X be a smooth projective complex surface, and $\mathcal{I}$ be an ideal in the cohomology ring $H^*(X)$. For $n \geq 1$, define $\mathcal{I}^{[n]}$ to be the subset of $H^*(X^{[n]})$ consisting of the linear spans of Heisenberg monomials of the form

$$\mathfrak{a}_{-n_1}(\alpha_1) \cdots \mathfrak{a}_{-n_b}(\alpha_b)|0\rangle$$

where $\alpha_i \in \mathcal{I}$ for some i, and $n_1, \ldots, n_b$ are positive integers with $\sum_\ell n_\ell = n$.

LEMMA 8.6. *Let $\mathcal{I}$ be an ideal in the cohomology ring $H^*(X)$. Then,*

(i) *the linear subspace $\mathcal{I}^{[n]}$ in $H^*(X^{[n]})$ is an ideal.*

(ii) *$G_k(\alpha, n) \in \mathcal{I}^{[n]}$ if $\alpha \in \mathcal{I}$.*

PROOF. (i) By Theorem 7.5, the cohomology ring $H^*(X^{[n]})$ is generated by the cohomology classes $G_k(\alpha, n)$. So it suffices to prove

$$\mathfrak{G}_k(\alpha)\mathfrak{a}_{-n_1}(\alpha_1) \cdots \mathfrak{a}_{-n_b}(\alpha_b)|0\rangle \in \mathcal{I}^{[n]} \tag{8.14}$$

whenever $\alpha_1 \in \mathcal{I}$, and $n_1, \ldots, n_b$ are positive with $\sum_\ell n_\ell = n$. For simplicity, put $\mathfrak{g} = \mathfrak{G}_k(\alpha)$. Then, the operator $\mathfrak{g}$ is of bi-degree $(0, 2k + |\alpha|)$, and satisfies

$$[[\cdots[\mathfrak{g}, \mathfrak{a}_{m_1}(\beta_1)], \cdots], \mathfrak{a}_{m_{k+2}}(\beta_{k+2})] = 0$$

whenever $m_i < 0$ for all i by Theorem 4.7. Now we see from Lemma 4.9 that $\mathfrak{G}_k(\alpha)\mathfrak{a}_{-n_1}(\alpha_1) \cdots \mathfrak{a}_{-n_b}(\alpha_b)|0\rangle$ is a linear combination of expressions of the form

$$\left(\prod_{\ell \in \sigma_i^0} \mathfrak{a}_{-n_\ell}(\alpha_\ell)\right) [[\cdots[\mathfrak{g}, \mathfrak{a}_{-n_{\sigma_i(1)}}(\alpha_{\sigma_i(1)})], \cdots], \mathfrak{a}_{-n_{\sigma_i(i)}}(\alpha_{\sigma_i(i)})]|0\rangle \tag{8.15}$$

where $0 \leq i \leq k+1$, σ_i maps the set $\{1, \ldots, i\}$ to the set $\{1, \ldots, b\}$ with $\sigma_i(1) < \cdots < \sigma_i(i)$, and

$$\sigma_i^0 = \{\ell \,|\, 1 \leq \ell \leq b, \ell \neq \sigma_i(1), \ldots, \sigma_i(i)\}.$$

If $1 \in \sigma_i^0$, then (8.15) is contained in $\mathcal{I}^{[n]}$. In the following, we assume $1 \notin \sigma_i^0$. So $1 = \sigma_i(1)$ since $\sigma_i(1) < \cdots < \sigma_i(i)$. By Theorem 4.7 and Lemma 3.12 (i), (8.15) is a linear combination of expressions of the form

$$\left(\prod_{\ell \in \sigma_i^0} \mathfrak{a}_{-n_\ell}(\alpha_\ell)\right) \mathfrak{a}_{-\lambda}(\tau_*(\epsilon\alpha\alpha_{\sigma_i(1)} \cdots \alpha_{\sigma_i(i)}))|0\rangle \tag{8.16}$$

where $\epsilon \in \{1_X, e_X, K_X, K_X^2\}$, $\lambda \vdash \sum_{j=1}^i n_{\sigma_i(j)}$, and $\ell(\lambda) = k + 2 - |\epsilon|/2 - i$. By Lemma 8.4, the expression (8.16) is contained in $\mathcal{I}^{[n]}$ since $\alpha_{\sigma_i(1)} = \alpha_1 \in \mathcal{I}$. It follows that (8.15) is contained in $\mathcal{I}^{[n]}$. This proves (8.14).

(ii) This follows from Theorem 4.10 and Lemma 8.4. □

The following technical definition is similar to Definition 4.4.

DEFINITION 8.7. Let X be a smooth projective complex surface, $s \geq 1$, and $t_1, \ldots, t_s \geq 1$. Fix $m_{i,j} \geq 0$ and $\beta_{i,j} \in H^*(X)$ for $1 \leq i \leq s$ and $1 \leq j \leq t_i$. Then, a *universal linear combination* of $\prod_{j=1}^{t_i} G_{m_{i,j}}(\beta_{i,j}, n)$, $1 \leq i \leq s$ is a linear combination of the form

$$\sum_{i=1}^s f_i \prod_{j=1}^{t_i} G_{m_{i,j}}(\beta_{i,j}, n)$$

where the coefficients f_i are independent of X and n.

The next result deals with the generators of the ideal $\mathcal{I}^{[n]}$.

THEOREM 8.8. *Let X be a smooth projective complex surface, and $\mathcal{I}$ be an ideal in the cohomology ring $H^*(X)$. If $\mathcal{I}$ is homogeneous (i.e., $\mathcal{I} = \oplus_{i=0}^4(\mathcal{I} \cap H^i(X))$), then the ideal $\mathcal{I}^{[n]}$ is generated by the classes $G_k(\alpha, n)$ with $\alpha \in \mathcal{I}$.*

PROOF. Note that every Heisenberg monomial class in $\mathcal{I}^{[n]}$ can be written as

$$A = \mathbf{1}_{-(n-n_0)} \left(\prod_{i=1}^{s} \mathfrak{a}_{-n_i}(\alpha_i) \right) |0\rangle$$

where $s \geq 1$, $n_1, \ldots, n_s \geq 1$, $n_0 = \sum_{i=1}^{s} n_i$, and α_ℓ is contained in $\mathcal{I}$ and homogeneous for some ℓ. By Lemma 8.6 (ii), it suffices to show that $A \in \mathcal{I}^{[n]}$ is a universal finite linear combination of expressions of the form

$$\prod_{j=1}^{t} G_{m_j}(\beta_j, n) \tag{8.17}$$

where $\sum_{j=1}^{t} (m_j + 1) \leq n_0$, and $\beta_\ell \in \mathcal{I}$ for some ℓ.

Use induction on n_0. When $n_0 = 1$, $s = n_1 = 1$. So

$$A = \mathbf{1}_{-(n-1)}\mathfrak{a}_{-1}(\alpha_1)|0\rangle = G_0(\alpha_1, n)$$

by (4.22). Hence the statement in the previous paragraph holds for $n_0 = 1$.

Next assume $n_0 > 1$. Let $k_i = n_i - 1$ for $1 \leq i \leq s$. Then, $k_i \geq 0$ for every i. By Lemma 4.13 and Lemma 4.14, the cup product $\prod_{i=1}^{s} G_{k_i}(\alpha_i, n)$ is equal to

$$\left(\prod_{i=1}^{s} \frac{(-1)^{k_i}}{(k_i + 1)!} \right) A$$

(defined to be *the leading term*) plus a universal finite linear combination of expressions

$$\mathbf{1}_{-(n-\tilde{n}_0)} \left(\prod_{i=1}^{\tilde{s}} \mathfrak{a}_{-\tilde{n}_i}(\tilde{\alpha}_i) \right) |0\rangle$$

where $\tilde{\alpha}_\ell \in \mathcal{I}$ for some ℓ, $\tilde{s} \geq 1$, $\tilde{n}_1, \ldots, \tilde{n}_{\tilde{s}} \geq 1$, and

$$\sum_{i=1}^{\tilde{s}} \tilde{n}_i = \tilde{n}_0 < \sum_{i=1}^{s} (k_i + 1) = n_0.$$

By induction, A is a universal finite linear combination of expressions of the form (8.17). □

REMARK 8.9. The assumption in Theorem 8.8 that the ideal $\mathcal{I} \subset H^*(X)$ is homogeneous can be dropped when the surface X is simply connected.

REMARK 8.10.

(i) Let X be a smooth projective complex surface, and $\mathcal{I}$ be a homogeneous ideal in the cohomology ring $H^*(X)$. Assume further that for every $k \geq 1$ and $\alpha \in \mathcal{I}$, the pushforward $\tau_{k*}\alpha$ can be written as

$$\sum_j \alpha_{j,1} \otimes \cdots \otimes \alpha_{j,k} \in H^*(X^k)$$

such that $\alpha_{j,s} \in \mathcal{I}$ for all j and s. Let

$$A = \mathbf{1}_{-(n-n_0)} \left(\prod_{i=1}^{s} \mathfrak{a}_{-n_i}(\alpha_i) \right) |0\rangle$$

where $s \geq 1$, $n_1, \ldots, n_s \geq 1$, $n_0 = \sum_{i=1}^{s} n_i$, and $\alpha_1, \ldots, \alpha_s \in \mathcal{I}$. Then the same proof of Theorem 8.8 shows that A can be written as a polynomial of the classes $G_k(\alpha, n)$, $k \geq 0$ and $\alpha \in \mathcal{I}$. Moreover, the coefficients in the polynomial are independent of n.

(ii) Under certain conditions, the conclusion in (i) can be sharpened. For instance, let $\lambda \vdash n_0$, $\alpha \in H^*(X)$ with $|\alpha| = 2$, and $m \geq 1$.

(ii-a) The class $\mathbf{1}_{-(n-n_0)}\mathfrak{a}_{-\lambda}(x)|0\rangle$ can be written as a polynomial of the classes $G_k(x, n)$, $k \geq 0$. Moreover, the coefficients are independent of n and X.

(ii-b) If the odd Betti numbers of the surface X are equal to zero, then

$$\begin{aligned} &\mathbf{1}_{-(n-n_0-m)}\mathfrak{a}_{-\lambda}(x)\mathfrak{a}_{-m}(\alpha)|0\rangle \\ = \ &\langle \alpha, K_X \rangle \cdot F_1(n) + \sum_i G_{k_i}(\alpha, n) \cdot F_{2,i}(n) \end{aligned}$$

where $F_1(n)$ and $F_{2,i}(n)$ are polynomials of the classes $G_k(x, n)$, $k \geq 0$. Moreover, the coefficients of $F_1(n)$ and $F_{2,i}(n)$ are independent of n, α and X.

The statements (ii-a) and (ii-b) follow from the same proof of Theorem 8.8 by setting $\mathcal{I} = \mathbb{C} \cdot x \subset H^*(X)$ and $\mathcal{I} = \mathbb{C} \cdot x + \mathbb{C} \cdot \alpha \subset H^*(X)$ respectively.

A natural problem to consider is to interpret the quotient ring

$$H^*(X^{[n]})/\mathcal{I}^{[n]}$$

for an ideal $\mathcal{I}$ of the cohomology ring $H^*(X)$. In the next two sections, we will do this for $\mathcal{I} = \oplus_{\ell=1}^4 H^\ell(X)$ and $\mathcal{I} = H^4(X)$ respectively.

8.3. Relation with the cohomology ring of the Hilbert scheme $(\mathbb{C}^2)^{[n]}$

Let X denote a smooth projective complex surface. In this section, we will study the quotient ring $H^*(X^{[n]})/\mathcal{I}^{[n]}$ when

$$\mathcal{I} = \bigoplus_{\ell=1}^{4} H^\ell(X),$$

and prove that it is isomorphic to the cohomology ring $H^*((\mathbb{C}^2)^{[n]})$ which has been dealt with in Section 8.1.

First of all, note that $H^*(X^{[n]})/\mathcal{I}^{[n]}$ has a linear basis consisting of Heisenberg monomials of the form

$$\mathfrak{a}_{-n_1}(1_X)^{r_1} \cdots \mathfrak{a}_{-n_k}(1_X)^{r_k}|0\rangle$$

where $r_1, \ldots, r_k \geq 1$, and $0 < n_1 < \ldots < n_k$ with $\sum_{\ell=1}^k r_\ell n_\ell = n$. Therefore, we have an isomorphism of vector spaces

$$\Phi : \bigoplus_{n \geq 0} H^*(X^{[n]})/\mathcal{I}^{[n]} \to \mathbb{C}[q_1, q_2, \ldots] \tag{8.18}$$

defined by

$$\Phi\big(\mathfrak{a}_{-n_1}(1_X)^{r_1} \cdots \mathfrak{a}_{-n_k}(1_X)^{r_k}|0\rangle\big) = q_{n_1}^{r_1} \cdots q_{n_k}^{r_k}.$$

Setting the degree of the variable q_i to be i, we see that Φ maps $H^*(X^{[n]})/\mathcal{I}^{[n]}$ to the homogeneous component of $\mathbb{C}[q_1, q_2, \ldots]$ of degree n.

LEMMA 8.11. *Let*

$$\mathcal{I} = \bigoplus_{\ell=1}^{4} H^\ell(X).$$

Then, the quotient ring $H^*(X^{[n]})/\mathcal{I}^{[n]}$ is generated by the Chern character classes $G_k(1_X, n)$, $k = 0, 1, \ldots, n-1$. Moreover,

$$G_k(1_X, n) \equiv \frac{(-1)^k}{(k+1)!} \cdot \mathbf{1}_{-(n-k-1)}\mathfrak{a}_{-(k+1)}(1_X)|0\rangle \pmod{\mathcal{I}^{[n]}}. \tag{8.19}$$

PROOF. Since the ring $H^*(X^{[n]})$ is generated by the classes $G_k(\alpha, n)$ with $0 \leq k < n$ and $\alpha \in H^*(X)$, the first statement follows from Lemma 8.6 (ii). To prove (8.19), we note from Theorem 4.10 that the leading term in $G_k(1_X, n)$ is

$$\frac{(-1)^k}{(k+1)!} \cdot \mathbf{1}_{-(n-k-1)}\mathfrak{a}_{-(k+1)}(1_X)|0\rangle$$

corresponding to $j = k$, $\lambda \vdash (j+1) = k+1$, and $\ell(\lambda) = k - j + 1 = 1$. The other terms in $G_k(1_X, n)$ contain $\tau_{i*}(\epsilon)$ with either $i \geq 2$ or $\epsilon = e_X, K_X, K_X^2 \in \mathcal{I}$, and hence are contained in $\mathcal{I}^{[n]}$. This proves (8.19). ☐

THEOREM 8.12. *Let X be a smooth projective complex surface, and*

$$\mathcal{I} = \bigoplus_{\ell=1}^{4} H^\ell(X).$$

Then, the quotient $H^*(X^{[n]})/\mathcal{I}^{[n]}$ is isomorphic to the cohomology ring $H^*((\mathbb{C}^2)^{[n]})$.

PROOF. By Lemma 8.11, the quotient ring $H^*(X^{[n]})/\mathcal{I}^{[n]}$ is generated by the classes $G_k(1_X, n)$, $k = 0, 1, \ldots, n-1$. So by Theorem 8.2, it suffices to show that via the isomorphism Φ in (8.18), the linear operator $\mathfrak{g}_k$ on $\mathbb{C}[q_1, q_2, \ldots]$ induced by the operator $\mathfrak{G}_k(1_X)$ on $\bigoplus_{n \geq 0} H^*(X^{[n]})$ is given by

$$\mathfrak{g}_k = \frac{(-1)^k}{(k+1)!} \sum_{n_1, \ldots, n_{k+1} > 0} q_{n_1 + \ldots + n_{k+1}} \partial_{n_1} \cdots \partial_{n_{k+1}}. \tag{8.20}$$

Indeed, let

$$A = \mathfrak{a}_{-n_1}(1_X) \cdots \mathfrak{a}_{-n_b}(1_X)|0\rangle \in H^*(X^{[n]})$$

where $n_1, \ldots, n_b > 0$ with $\sum_\ell n_\ell = n$. By Lemma 4.9, $\mathfrak{G}_k(1_X)(A)$ is equal to
(8.21)

$$\sum_{i=0}^{k+1} \sum_{\sigma_i} \prod_{\ell \in \sigma_i^0} \mathfrak{a}_{-n_\ell}(1_X) \cdot [[\cdots [\mathfrak{G}_k(1_X), \mathfrak{a}_{-n_{\sigma_i(1)}}(1_X)], \cdots], \mathfrak{a}_{-n_{\sigma_i(i)}}(1_X)]|0\rangle$$

where for each fixed i, σ_i runs over all the maps $\{1, \ldots, i\} \to \{1, \ldots, b\}$ satisfying $\sigma_i(1) < \cdots < \sigma_i(i)$, and

$$\sigma_i^0 = \{\ell \,|\, 1 \le \ell \le b, \ell \ne \sigma_i(1), \ldots, \sigma_i(i)\}.$$

Note from Lemma 3.12 that

$$[[\cdots [\mathfrak{a}_{t_1} \cdots \mathfrak{a}_{t_r}(\tau_{r*}\alpha), \mathfrak{a}_{-n_{\sigma_i(1)}}(1_X)], \cdots], \mathfrak{a}_{-n_{\sigma_i(i)}}(1_X)]|0\rangle \in \mathcal{I}^{[n]}$$

if $i \le r-2$ or $\alpha \in \mathcal{I}$. Hence by (8.21) and Theorem 4.7, $\mathfrak{G}_k(1_X)(A)$ equals

$$\sum_{\sigma_{k+1}} \left(\prod_{\ell \in \sigma_{k+1}^0} \mathfrak{a}_{-n_\ell}(1_X) \right) \cdot$$

$$\left[\left[\cdots\left[- \sum_{\ell(\lambda)=k+2, |\lambda|=0} \frac{1}{\lambda^!} \mathfrak{a}_\lambda(\tau_* 1_X), \mathfrak{a}_{-n_{\sigma_{k+1}(1)}}(1_X)\right], \cdots\right], \mathfrak{a}_{-n_{\sigma_{k+1}(k+1)}}(1_X)\right] |0\rangle$$

modulo $\mathcal{I}^{[n]}$. By Lemma 3.12 (i) again, the preceding expression is equal to

$$\sum_{\sigma_{k+1}} \left(- \prod_{\ell=1}^{k+1} (-n_{\sigma_{k+1}(\ell)})\right) \left(\prod_{\ell \in \sigma_{k+1}^0} \mathfrak{a}_{-n_\ell}(1_X) \right) \mathfrak{a}_{-n_{\sigma_{k+1}(1)} - \ldots - n_{\sigma_{k+1}(k+1)}}(1_X)|0\rangle.$$

Therefore, we conclude that for the induced operator $\mathfrak{g}_k$, we have

$$\begin{aligned}
&\mathfrak{g}_k(q_{n_1} \cdots q_{n_b}) \\
= \;& (-1)^k \sum_{\sigma_{k+1}} \left(\prod_{\ell=1}^{k+1} n_{\sigma_{k+1}(\ell)} \right) \left(\prod_{\ell \in \sigma_{k+1}^0} q_{n_\ell} \right) q_{n_{\sigma_{k+1}(1)} + \ldots + n_{\sigma_{k+1}(k+1)}} \\
= \;& \frac{(-1)^k}{(k+1)!} \sum_{\tilde{\sigma}_{k+1}} \left(\prod_{\ell=1}^{k+1} n_{\tilde{\sigma}_{k+1}(\ell)} \right) \left(\prod_{\ell \in \tilde{\sigma}_{k+1}^0} q_{n_\ell} \right) q_{n_{\tilde{\sigma}_{k+1}(1)} + \ldots + n_{\tilde{\sigma}_{k+1}(k+1)}}
\end{aligned}$$

where $\tilde{\sigma}_{k+1}$ runs over all injective maps $\{1, \ldots, k+1\} \to \{1, \ldots, b\}$ and

$$\tilde{\sigma}_{k+1}^0 = \{\ell \,|\, 1 \le \ell \le b, \ell \ne \tilde{\sigma}_{k+1}(1), \ldots, \tilde{\sigma}_{k+1}(k+1)\}$$

(so the only difference between $\tilde{\sigma}_{k+1}$ and σ_{k+1} is that we have dropped the condition $\sigma_{k+1}(1) < \cdots < \sigma_{k+1}(k+1)$). Finally, the above formula for $\mathfrak{g}_k(q_{n_1} \cdots q_{n_b})$ is equivalent to (8.20). □

8.4. Partial n-independence of structure constants for X projective

Let X be a smooth projective complex surface. In this section, we will define a linear basis of $H^*(X^{[n]})$ consisting of the classes $\mathfrak{b}_\rho(n)$ where ρ denotes certain type of partitions. Let $\mathcal{I} = H^4(X)$. The main theorem states that modulo the ideal $\mathcal{I}^{[n]}$, all the structure constants in the cup product $\mathfrak{b}_\rho(n) \cdot \mathfrak{b}_\sigma(n)$ are independent of n.

First of all, take a linear basis

$$S = S_0 \cup S_1$$

of $H^*(X)$ such that $1_X, x \in S_0$, $S_0 \subset H^{\text{even}}(X)$ and $S_1 \subset H^{\text{odd}}(X)$. Let notations such as $\mathcal{P}(S)$ be the same as in Section 7.2. If we write $\rho = (\rho(c))_{c\in S}$ and

$$\rho(c) = (r^{m_r(\rho(c))})_{r\geq 1} = (1^{m_1(\rho(c))}2^{m_2(\rho(c))}\ldots),$$

then we put

$$\ell(\rho) = \sum_{c\in S} \ell(\rho(c)) = \sum_{c\in S, r\geq 1} m_r(\rho(c))$$

and

$$\|\rho\| = \sum_{c\in S} |\rho(c)| = \sum_{c\in S, r\geq 1} r\cdot m_r(\rho(c)).$$

Given $\rho \in \mathcal{P}(S)$ and $n \geq \|\rho\| + \ell(\rho(1_X))$, we define $\tilde{\rho} \in \mathcal{P}(S)$ by putting

$$m_r(\tilde{\rho}(c)) = \begin{cases} m_r(\rho(c)), & \text{if } c \in S - \{1_X\}, \\ m_{r-1}(\rho(1_X)), & \text{if } c = 1_X \text{ and } r \geq 2, \\ n - \|\rho\| - \ell(\rho(1_X)), & \text{if } c = 1_X \text{ and } r = 1. \end{cases}$$

Note that $\|\tilde{\rho}\| = n$. We define $\mathfrak{b}_\rho(n) \in H^*(X^{[n]})$ by

$$\mathfrak{b}_\rho(n) = \frac{1}{\prod_{r\geq 1}(r^{m_r(\tilde{\rho}(1_X))}m_r(\tilde{\rho}(1_X))!)} \prod_{c\in S}\prod_{r\geq 1} \mathfrak{a}_{-r}(c)^{m_r(\tilde{\rho}(c))} \cdot |0\rangle \tag{8.22}$$

$$= \frac{\mathbf{1}_{-(n-\|\rho\|-\ell(\rho(1_X)))}}{\prod_{r\geq 2}(r^{m_r(\tilde{\rho}(1_X))}m_r(\tilde{\rho}(1_X))!)} \prod_{\substack{c\in S, r\geq 1 \\ c\neq 1_X \text{ or } r>1}} \mathfrak{a}_{-r}(c)^{m_r(\tilde{\rho}(c))} \cdot |0\rangle. \tag{8.23}$$

where we fix the order of the elements $c \in S_1$ appearing in the product $\prod_{c\in S}$ once and for all. For $0 \leq n < \|\rho\| + \ell(\rho(1_X))$, we set $\mathfrak{b}_\rho(n) = 0$. This is consistent with (8.23) and Definition 3.18. We remark that the only part in $\mathfrak{b}_\rho(n)$ involving n is the factor $\mathbf{1}_{-(n-\|\rho\|-\ell(\rho(1_X)))}$ in (8.23) when $n \geq \|\rho\| + \ell(\rho(1_X))$.

REMARK 8.13. The Heisenberg monomial class $\mathfrak{b}_\rho(n)$ is different from the Heisenberg monomial class $\mathfrak{a}_\rho(n)$ defined in Section 7.2.

By Theorem 3.8, $H^*(X^{[n]})$ has a linear basis consisting of the classes:

$$\mathfrak{b}_\rho(n), \qquad \rho \in \mathcal{P}(S) \text{ and } \|\rho\| + \ell(\rho(1_X)) \leq n. \tag{8.24}$$

Fix a positive integer n and choose $\rho, \sigma \in \mathcal{P}(S)$ satisfying $\|\rho\| + \ell(\rho(1_X)) \leq n$ and $\|\sigma\| + \ell(\sigma(1_X)) \leq n$. Then we can write the cup product $\mathfrak{b}_\rho(n) \cdot \mathfrak{b}_\sigma(n)$ as

$$\mathfrak{b}_\rho(n) \cdot \mathfrak{b}_\sigma(n) = \sum_{\nu\in\mathcal{P}(S)} b_{\rho\sigma}^\nu(n)\ \mathfrak{b}_\nu(n) \tag{8.25}$$

where we have used $b_{\rho\sigma}^\nu(n) \in \mathbb{C}$ to denote the structure constants.

PROPOSITION 8.14. *Let X be a smooth projective complex surface. The structure constants $b_{\rho\sigma}^\nu(n)$ of the cohomology $H^*(X^{[n]})$ are polynomials in n of degree at most*

$$(\|\rho\| + \ell(\rho(1_X))) + (\|\sigma\| + \ell(\sigma(1_X))) - (\|\nu\| + \ell(\nu(1_X))). \tag{8.26}$$

PROOF. For simplicity, let

$$f(\rho) = \prod_{r \geq 2} (r^{m_r(\tilde{\rho}(1_X))} m_r(\tilde{\rho}(1_X))!).$$

Then, $f(\rho)$ is independent of n. By Theorem 7.7, the cup product

$$f(\rho)\mathfrak{b}_\rho(n) \cdot f(\sigma)\mathfrak{b}_\sigma(n)$$

is a linear combination of expressions of the form:

$$\frac{(n - \|\nu\| - \ell(\nu(1_X)))!}{(n - \|\nu\| - \ell(\nu(1_X)) - i)!} \cdot f(\nu)\mathfrak{b}_\nu(n),$$

such that $i \geq 0$,

$$(\|\nu\| + \ell(\nu(1_X)) + i) \leq (\|\rho\| + \ell(\rho(1_X))) + (\|\sigma\| + \ell(\sigma(1_X))),$$

and all the coefficients in this linear combination are independent of n. It follows that all the structure constants $b_{\rho\sigma}^{\nu}(n)$ are polynomials in n of degree at most (8.26). □

In the following, we will study various cup products in the cohomology of $X^{[n]}$ modulo the ideal $\mathcal{I}^{[n]}$ with $\mathcal{I} = H^4(X)$. We begin with an elementary topological lemma.

LEMMA 8.15. *If $k \geq 2$ and α is homogeneous, then*

$$\tau_{k*}\alpha = \sum_j \alpha_{j,1} \otimes \cdots \otimes \alpha_{j,k}$$

where for each j, either $|\alpha_{j,\ell}| = 4$ for some ℓ, or $0 < |\alpha_{j,\ell}| < 4$ for every ℓ.

PROOF. Assume $|\alpha_{j,\ell}| < 4$ for every ℓ. If $|\alpha_{j,\ell}| = 0$ for some ℓ, then

$$4(k-1) + |\alpha| = |\tau_{k*}\alpha| = \sum_{i=1}^{k} |\alpha_{j,i}| \leq 3(k-1).$$

So $(k-1) + |\alpha| \leq 0$, contradicting $k \geq 2$. □

LEMMA 8.16. *Let $\mathcal{I} = H^4(X)$, $s \geq 0$, $n_1, \ldots, n_s > 0$, $\tilde{n} = \sum_{\ell=1}^{s} n_\ell$, and $n \geq \tilde{n}$. Let $\alpha, \alpha_1, \ldots, \alpha_s \in H^*(X)$ be homogeneous. Assume that $k + |\alpha| \geq 1$ and $n_\ell + |\alpha_\ell| \geq 2$ for every ℓ. Then modulo $\mathcal{I}^{[n]}$, the cup product*

$$G_k(\alpha, n) \cdot \big(\mathbf{1}_{-(n-\tilde{n})}\mathfrak{a}_{-n_1}(\alpha_1) \cdots \mathfrak{a}_{-n_s}(\alpha_s)|0\rangle\big)$$

is a universal linear combination of the basis (8.24).

PROOF. By Lemma 8.6 (i) and (ii), the statement is trivial if one of the classes $\alpha, \alpha_1, \ldots, \alpha_s \in H^*(X)$ is contained in $\mathcal{I}$. So in the rest of the proof, we assume that none of the classes $\alpha, \alpha_1, \ldots, \alpha_s$ is contained in $\mathcal{I}$.

Our argument is similar to the proof of Lemma 8.6 (i). Put $\mathfrak{g} = \mathfrak{G}_k(\alpha)$ and

$$B = G_k(\alpha, n) \cdot \mathbf{1}_{-(n-\tilde{n})}\mathfrak{a}_{-n_1}(\alpha_1) \cdots \mathfrak{a}_{-n_s}(\alpha_s)|0\rangle.$$

Then, B is equal to

$$\frac{1}{(n-\tilde{n})!}\mathfrak{g}\mathfrak{a}_{-1}(1_X)^{n-\tilde{n}}\mathfrak{a}_{-n_1}(\alpha_1) \cdots \mathfrak{a}_{-n_s}(\alpha_s)|0\rangle.$$

By Lemma 4.9, B is a universal finite linear combination of expressions of the form:

$$\frac{1}{(n-\tilde{n})!}\binom{n-\tilde{n}}{j}\mathfrak{a}_{-1}(1_X)^{n-\tilde{n}-j}\prod_{\ell\in\sigma_i^0}\mathfrak{a}_{-n_\ell}(\alpha_\ell) \tag{8.27}$$
$$\cdot[[\cdots[[[[\mathfrak{g},\underbrace{\mathfrak{a}_{-1}(1_X)],\ldots],\mathfrak{a}_{-1}(1_X)]}_{j \text{ times}},\mathfrak{a}_{-n_{\sigma_i(1)}}(\alpha_{\sigma_i(1)})],\cdots],\mathfrak{a}_{-n_{\sigma_i(i)}}(\alpha_{\sigma_i(i)})]]|0\rangle$$

where $0\le j\le n-\tilde{n}$, $0\le i\le s$, $0\le j+i\le k+1$, σ_i maps the set $\{1,\ldots,i\}$ to the set $\{1,\ldots,s\}$ with $\sigma_i(1)<\cdots<\sigma_i(i)$, and

$$\sigma_i^0=\{\ell\,|\,1\le\ell\le s,\ell\ne\sigma_i(1),\ldots,\sigma_i(i)\}.$$

By Theorem 4.7 and Lemma 3.12 (i), we conclude that

$$[[\cdots[[[[\mathfrak{g},\underbrace{\mathfrak{a}_{-1}(1_X)],\ldots],\mathfrak{a}_{-1}(1_X)]}_{j \text{ times}},\mathfrak{a}_{-n_{\sigma_i(1)}}(\alpha_{\sigma_i(1)})],\cdots],\mathfrak{a}_{-n_{\sigma_i(i)}}(\alpha_{\sigma_i(i)})]]|0\rangle$$

is a universal finite linear combination of expressions

$$\mathfrak{a}_{-\lambda}(\tau_*(\epsilon\alpha\alpha_{\sigma_i(1)}\cdots\alpha_{\sigma_i(i)}))|0\rangle,$$

where $\epsilon\in\{1_X,e_X,K_X,K_X^2\}$, $\lambda\vdash j+n_{\sigma_i(1)}+\cdots+n_{\sigma_i(i)}$ and

$$\ell(\lambda)=k+2-|\epsilon|/2-(j+i).$$

So B is a universal finite linear combination of expressions

$$\mathbf{1}_{-(n-\tilde{n}-j)}\left(\prod_{\ell\in\sigma_i^0}\mathfrak{a}_{-n_\ell}(\alpha_\ell)\right)\mathfrak{a}_{-\lambda}(\tau_*(\epsilon\alpha\alpha_{\sigma_i(1)}\cdots\alpha_{\sigma_i(i)}))|0\rangle \tag{8.28}$$

where $\lambda\vdash j+n_{\sigma_i(1)}+\cdots+n_{\sigma_i(i)}$ and $\ell(\lambda)=k+2-|\epsilon|/2-(j+i)$.

To prove our lemma, we see from (8.23) that it suffices to show that modulo $\mathcal{I}^{[\tilde{n}+j]}$, the part

$$\left(\prod_{\ell\in\sigma_i^0}\mathfrak{a}_{-n_\ell}(\alpha_\ell)\right)\mathfrak{a}_{-\lambda}(\tau_*(\epsilon\alpha\alpha_{\sigma_i(1)}\cdots\alpha_{\sigma_i(i)}))|0\rangle$$

in (8.28) does not contain $\mathfrak{a}_{-1}(1_X)$. Since $n_\ell+|\alpha_\ell|\ge 2$ for every ℓ, this is equivalent to show that modulo $\mathcal{I}^{[\ell(\lambda)]}$, the part $\mathfrak{a}_{-\lambda}(\tau_*(\epsilon\alpha\alpha_{\sigma_i(1)}\cdots\alpha_{\sigma_i(i)}))|0\rangle$ in (8.28) does not contain $\mathfrak{a}_{-1}(1_X)$. By Lemma 8.15, this is true if $\ell(\lambda)\ge 2$. So let $\ell(\lambda)=1$. Then, we have

$$\mathfrak{a}_{-\lambda}(\tau_*(\epsilon\alpha\alpha_{\sigma_i(1)}\cdots\alpha_{\sigma_i(i)}))|0\rangle=\mathfrak{a}_{-t}(\epsilon\alpha\alpha_{\sigma_i(1)}\cdots\alpha_{\sigma_i(i)})|0\rangle$$

where $t=|\lambda|=j+n_{\sigma_i(1)}+\cdots+n_{\sigma_i(i)}$ and $k+2-|\epsilon|/2-(j+i)=1$.

If $\mathfrak{a}_{-\lambda}(\tau_*(\epsilon\alpha\alpha_{\sigma_i(1)}\cdots\alpha_{\sigma_i(i)}))|0\rangle$ contains $\mathfrak{a}_{-1}(1_X)$, then we must have $t=1$ and $|\epsilon\alpha\alpha_{\sigma_i(1)}\cdots\alpha_{\sigma_i(i)}|=0$. So $j+n_{\sigma_i(1)}+\cdots+n_{\sigma_i(i)}=1$, $\epsilon=1_X$, and

$$|\alpha|=|\alpha_{\sigma_i(1)}|=\ldots=|\alpha_{\sigma_i(i)}|=0.$$

Thus, either $j=0$, $i=1$, $n_{\sigma_1(1)}=1$ and $|\alpha_{\sigma_1(1)}|=0$, or $j=1$ and $i=0$. The first case contradicts $n_{\sigma_1(1)}+|\alpha_{\sigma_1(1)}|\ge 2$. In the second case, we see from $k+2-|\epsilon|/2-(j+i)=1$ that $k=0$, contradicting $k+|\alpha|\ge 1$. So $\mathfrak{a}_{-\lambda}(\tau_*(\epsilon\alpha\alpha_{\sigma_i(1)}\cdots\alpha_{\sigma_i(i)}))|0\rangle$ can not contain $\mathfrak{a}_{-1}(1_X)$. □

LEMMA 8.17. *Let $\mathcal{I} = H^4(X)$, $s \ge 1$, $k_1, \ldots, k_s \ge 0$, $k_i + |\alpha_i| \ge 1$ for every i. Put*

$$n_0 = \sum_{i=1}^{s} (k_i + 1).$$

(i) *Modulo $\mathcal{I}^{[n]}$, $\prod_{i=1}^{s} G_{k_i}(\alpha_i, n)$ is a universal linear combination of (8.24).*

(ii) *When $n \ge n_0$, the leading term in $\prod_{i=1}^{s} G_{k_i}(\alpha_i, n)$ is equal to*

$$\left(\prod_{i=1}^{s} \frac{(-1)^{k_i}}{(k_i+1)!}\right) \mathbf{1}_{-(n-n_0)} \left(\prod_{i=1}^{s} \mathfrak{a}_{-(k_i+1)}(\alpha_i)\right) |0\rangle$$

which is equal to a universal multiple of $\mathfrak{b}_\nu(n)$ for some $\nu \in \mathcal{P}(S)$.

PROOF. (i) Use induction on s. When $s = 1$, our statement follows from Lemma 8.16 (take the integer s there to be 0). Now, assume that the statement is true for $s - 1$ with $s \ge 2$, i.e., $\prod_{i=1}^{s-1} G_{k_i}(\alpha_i, n)$ is a universal finite linear combination of the basis classes $\mathfrak{b}_\nu(n)$, $\nu \in \mathcal{P}(S)$ and $\|\nu\| + \ell(\rho(1_X)) \le n$. Note from (8.23) that up to some universal factor, every basis class $\mathfrak{b}_\nu(n)$ is of the form

$$\mathbf{1}_{-(n-\tilde{n})} \mathfrak{a}_{-n_1}(\beta_1) \cdots \mathfrak{a}_{-n_s}(\beta_s)|0\rangle,$$

where $n_1, \ldots, n_s > 0$, $\tilde{n} = n_1 + \ldots + n_s$, $n \ge \tilde{n}$, and $n_\ell + |\beta_\ell| \ge 2$ for every ℓ. So by Lemma 8.16, our statement for s follows.

(ii) The statement about the leading term comes from the proof of Theorem 8.8, while the other statement about the universal multiple follows from (8.23). $\square$

The next lemma studies the basis element $\mathfrak{b}_\rho(n)$ modulo $\mathcal{I}^{[n]}$ when $\mathcal{I} = H^4(X)$.

LEMMA 8.18. *Let $\mathcal{I} = H^4(X)$. Then modulo $\mathcal{I}^{[n]}$, the basis element $\mathfrak{b}_\rho(n)$ is a universal finite linear combination of products of the form $\prod_{j=1}^{t} G_{m_j}(\beta_j, n)$ where $m_j + |\beta_j| \ge 1$ for each j, and*

$$\sum_{j=1}^{t} (m_j + 1) \le \|\rho\| + \ell(\rho(1_X)).$$

PROOF. By (8.23), up to some universal factor, $\mathfrak{b}_\rho(n)$ is of the form

$$\mathbf{1}_{-(n-n_0)} \left(\prod_{i=1}^{s} \mathfrak{a}_{-(k_i+1)}(\alpha_i)\right) |0\rangle,$$

where $n \ge n_0 \stackrel{\text{def}}{=} \|\rho\| + \ell(\rho(1_X)) = \sum_{i=1}^{s} (k_i + 1)$, and $k_i \ge 0$ and $k_i + |\alpha_i| \ge 1$ for every i. Now our result follows from an induction on n_0 the same way as in the proof of Theorem 8.8, and Lemma 8.17 (i) and (ii). $\square$

To state our main result in this section, let $\mathcal{I} = H^4(X)$ and

$$S_{\mathcal{I}} = \{[x]\} \subset S.$$

Regard $\mathcal{P}(S - S_{\mathcal{I}}) \subset \mathcal{P}(S)$. Then, (8.25) implies that

$$\mathfrak{b}_\rho(n) \cdot \mathfrak{b}_\sigma(n) \equiv \sum_{\nu \in \mathcal{P}(S-S_{\mathcal{I}})} b^\nu_{\rho\sigma}(n)\ \mathfrak{b}_\nu(n) \pmod{\mathcal{I}^{[n]}}. \tag{8.29}$$

THEOREM 8.19. *Let X be a smooth projective complex surface, and $\mathcal{I} = H^4(X)$. Then, all the structure constants $b^\nu_{\rho\sigma}(n)$ in (8.29) are independent of n.*

PROOF. By Lemma 8.18, $\mathfrak{b}_\rho(n)$ is a universal finite linear combination of expressions of the form

$$\prod_{j=1}^{t_1} G_{m_{1,j}}(\beta_{1,j}, n)$$

where $m_{1,j} + |\beta_{1,j}| \geq 1$ for every j. Similarly, $\mathfrak{b}_\sigma(n)$ is a universal finite linear combination of expressions of the form

$$\prod_{j=1}^{t_2} G_{m_{2,j}}(\beta_{2,j}, n)$$

where $m_{1,j} + |\beta_{1,j}| \geq 1$ for every j. Therefore, $\mathfrak{b}_\rho(n) \cdot \mathfrak{b}_\sigma(n)$ is a universal finite linear combination of expressions of the form

$$\prod_{j=1}^{t} G_{m_j}(\beta_j, n)$$

where $m_j + |\beta_j| \geq 1$ for every j. Now our result follows from Lemma 8.17 (i). □

8.5. Applications to quasi-projective surfaces with the S-property

In this section, we will apply our results in previous sections to smooth quasi-projective surfaces satisfying the S-property. Our terminology "quasi-projective" here means "quasi-projective but not projective".

DEFINITION 8.20. A smooth quasi-projective complex surface X satisfies the *S-property* if it can be embedded in a smooth projective complex surface $\overline{X}$ such that the induced ring homomorphism $H^*(\overline{X}) \to H^*(X)$ is surjective.

EXAMPLE 8.21. Let $\overline{X}$ be a projective surface and let X be the quasi-projective surface obtained from X with a point removed. It is easy to see that this smooth quasi-projective surface X satisfies the S-property.

EXAMPLE 8.22. Let Γ be a finite subgroup of $\mathrm{SL}_2(\mathbb{C})$. Let X be the minimal resolution of the simple singularity $\mathbb{C}^2/\Gamma$. It is known that this smooth quasi-projective surface X satisfies the S-property. Moreover, $K_X = 0$.

EXAMPLE 8.23. (The cotangent bundle of a smooth projective curve) Consider the ruled surface

$$\overline{X} = \mathbb{P}(\mathcal{O}_C(-K_C) \oplus \mathcal{O}_C)$$

where C is a smooth projective curve. Let σ be the section (to the projection $\overline{X} \to C$) corresponding to the natural surjection

$$\mathcal{O}_C(-K_C) \oplus \mathcal{O}_C \to \mathcal{O}_C(-K_C) \to 0,$$

and put $X = \overline{X} - \sigma$. Then, X is the total space of the cotangent bundle of C, and $K_X = 0$. We claim that X satisfies the S-property. In fact, the following general statement is true. Let $\overline{X} = \mathbb{P}(\mathcal{L}_1 \oplus \mathcal{L}_2)$ where $\mathcal{L}_1$ and $\mathcal{L}_2$ are two invertible sheaves over C. Let σ (respectively, σ') be the section of $\overline{X} \to C$ corresponding to the natural surjection $\mathcal{L}_1 \oplus \mathcal{L}_2 \to \mathcal{L}_1 \to 0$ (respectively, $\mathcal{L}_1 \oplus \mathcal{L}_2 \to \mathcal{L}_2 \to 0$). Put $X = \overline{X} - \sigma$. Then X satisfies the S-property. To see this, let $X' = \overline{X} - \sigma'$, and notice that X and X' are affine bundles over C. Hence X is homotopic to C, and $H^i(X) \cong H^i(C)$ for every i. Therefore, to verify the S-property of X, it remains to verify the surjectivities of the induced homomorphisms $r_i : H^i(\overline{X}) \to H^i(X)$ for $i = 1, 2$. Consider the relative cohomology group $H^2(\overline{X}, X)$. We have

$$H^2(\overline{X}, X) \cong H^2(\overline{X} - \sigma', X - \sigma') = H^2(X', X' - \sigma)$$

by the excision theorem. By the Thom isomorphism, since X' is an affine bundle over C with σ being the zero section, we have

$$H^2(X', X' - \sigma) \cong H^0(C) \cong \mathbb{C}.$$

Hence $H^2(\overline{X}, X) \cong \mathbb{C}$. Now consider the exact sequence

$$H^1(\overline{X}) \xrightarrow{r_1} H^1(X) \xrightarrow{\delta} H^2(\overline{X}, X) \longrightarrow H^2(\overline{X}) \xrightarrow{r_2} H^2(X).$$

Since $H^2(X) \cong H^2(C) \cong \mathbb{C}$ and $H^2(\overline{X}) \cong \mathbb{C} \oplus \mathbb{C}$, we conclude that the map δ must be zero and the map r_2 must be surjective. Therefore, r_1 is also surjective.

In the following, let X be a smooth quasi-projective surface satisfying the S-property. So X can be embedded in a smooth projective surface $\overline{X}$ such that the induced ring homomorphism $\iota^* : H^*(\overline{X}) \to H^*(X)$ is surjective where $\iota : X \to \overline{X}$ is the inclusion map. Let $\mathcal{I} = \ker(\iota^*)$. Fix a linear basis S of $H^*(\overline{X})$ as in Section 8.4 such that S contains a linear basis $S_{\mathcal{I}}$ of $\mathcal{I}$ and $\iota^*(S - S_{\mathcal{I}})$ is a linear basis of $H^*(X)$. Put

$$S_X = \iota^*(S - S_{\mathcal{I}})$$

By Lemma 8.1, $\ker(\iota_n^*) = \mathcal{I}^{[n]}$ which is defined in Definition 8.5. Also, a linear basis of $H^*(X^{[n]})$ is given by

$$\mathfrak{b}_{\rho_X}(n), \qquad \rho_X \in \mathcal{P}(S_X) \text{ and } \|\rho_X\| + \ell(\rho_X(1_X)) \le n \tag{8.30}$$

where $\mathfrak{b}_{\rho_X}(n)$ is defined in a similar way as in (8.22) and (8.23). So for $\rho_X, \sigma_X \in \mathcal{P}(S_X)$, we can write the cup product $\mathfrak{b}_{\rho_X}(n) \cdot \mathfrak{b}_{\sigma_X}(n)$ as

$$\mathfrak{b}_{\rho_X}(n) \cdot \mathfrak{b}_{\sigma_X}(n) = \sum_{\nu_X \in \mathcal{P}(S_X)} b_{\rho_X \sigma_X}^{\nu_X}(n)\, \mathfrak{b}_{\nu_X}(n) \tag{8.31}$$

where $b_{\rho_X \sigma_X}^{\nu_X}(n) \in \mathbb{C}$ stands for the structure constants.

THEOREM 8.24. *Let X be a smooth quasi-projective surface satisfying the S-property. Then the structure constants $b_{\rho_X \sigma_X}^{\nu_X}(n)$ in (8.31) are independent of n.*

PROOF. Let notations be as above. Note that $\iota^*(S - S_{\mathcal{I}}) = S_X$. Define $\rho \in \mathcal{P}(S)$ by

$$m_r(\rho(\overline{c})) = \begin{cases} m_r(\rho_X(\iota^*\overline{c})), & \text{if } \overline{c} \in (S - S_{\mathcal{I}}), \\ 0, & \text{if } \overline{c} \in S_{\mathcal{I}}. \end{cases}$$

Similarly, define $\sigma \in \mathcal{P}(S)$ by

$$m_r(\sigma(\overline{c})) = \begin{cases} m_r(\sigma_X(\iota^*\overline{c})), & \text{if } \overline{c} \in (S - S_{\mathcal{I}}), \\ 0, & \text{if } \overline{c} \in S_{\mathcal{I}}. \end{cases}$$

By Theorem 8.19,

$$\mathfrak{b}_\rho(n) \cdot \mathfrak{b}_\sigma(n) \equiv \sum_{\nu \in \mathcal{P}(S - \{[x]\})} b^\nu_{\rho\sigma}(n)\ \mathfrak{b}_\nu(n) \pmod{\overline{\mathcal{I}}^{[n]}}$$

where $\overline{\mathcal{I}} = H^4(\overline{X})$ and all the $b^\nu_{\rho\sigma}(n)$ are independent of n. Since $\overline{\mathcal{I}} \subset \mathcal{I}$,

$$\mathfrak{b}_\rho(n) \cdot \mathfrak{b}_\sigma(n) \equiv \sum_{\nu \in \mathcal{P}(S - S_{\mathcal{I}})} b^\nu_{\rho\sigma}(n)\ \mathfrak{b}_\nu(n) \pmod{\mathcal{I}^{[n]}} \tag{8.32}$$

where all the structure constants $b^\nu_{\rho\sigma}(n)$ are independent of n. By Lemma 8.1, $\iota_n^*(\mathfrak{b}_\rho(n)) = \mathfrak{b}_{\rho_X}(n)$ and $\iota_n^*(\mathfrak{b}_\sigma(n)) = \mathfrak{b}_{\sigma_X}(n)$. Therefore, applying ι_n^* to (8.32), we see that all the structure constants $b^{\nu_X}_{\rho_X\, \sigma_X}(n)$ in (8.31) are independent of n. □

DEFINITION 8.25. Let X be a smooth quasi-projective surface satisfying the S-property. We define the *FH ring $\mathcal{G}_X$ associated to X* to be the ring with a linear basis given by the symbols $\mathfrak{b}_{\rho_X}$, $\rho_X \in \mathcal{P}(S_X)$, with the multiplication given by

$$\mathfrak{b}_{\rho_X} \cdot \mathfrak{b}_{\sigma_X} = \sum_{\nu_X \in \mathcal{P}(S_X)} b^{\nu_X}_{\rho_X\, \sigma_X}\ \mathfrak{b}_{\nu_X}$$

where the structure constants $b^{\nu_X}_{\rho_X\, \sigma_X}$ come from (8.31).

THEOREM 8.26. *Let X be a smooth quasi-projective surface which satisfies the S-property. Then the FH ring $\mathcal{G}_X$ is isomorphic to the tensor product $P \otimes E$, where P is the polynomial algebra generated by*

$$\mathfrak{b}_{r,c},\ c \in S_X \cap H^{\mathrm{even}}(X), r \geq 1,$$

and E is the exterior algebra generated by

$$\mathfrak{b}_{r,c},\ c \in S_X \cap H^{\mathrm{odd}}(X), r \geq 1.$$

The proof of Theorem 8.26 is similar to the proof of Theorem 7.11. We refer to [**LQW5**, Section 4] for the details.

CHAPTER 9

Integral cohomology of Hilbert schemes

The cohomology $H^*(X^{[n]})$ of the Hilbert scheme $X^{[n]}$ with complex coefficients has been investigated in previous chapters. By Theorem 3.8, it is linearly spanned over $\mathbb{C}$ by the Heisenberg monomial classes

$$\mathfrak{a}_{-n_1}(\alpha_1) \cdots \mathfrak{a}_{-n_k}(\alpha_k)|0\rangle$$

where $n_1, \ldots, n_k > 0$, $n_1 + \ldots + n_k = n$, and the cohomology classes $\alpha_1, \ldots, \alpha_k$ run over a fixed linear basis of $H^*(X)$. Its ring structure was discussed in Chapters 7 and 8. A natural problem is to study the integral cohomology of the Hilbert scheme $X^{[n]}$, its integral linear basis, and its integral generators. This chapter is devoted to these issues. We will define integral operators, show that some familiar operators are integral, and construct some integral operators by using integral classes in $H^2(X)$. Moreover, we will prove that the intersection matrix of certain integral classes in the middle cohomology $H^{2n}(X^{[n]})$ has determinant ± 1. Then we will write down a linear basis of the integral cohomology of $X^{[n]}$, modulo the torsion part. As an application, we will compare two integral bases of the integral cohomology of the Hilbert scheme $(\mathbb{P}^2)^{[n]}$ of points on the complex projective plane $\mathbb{P}^2$, and prove that when the basis elements are arranged properly, the transition matrix is a triangular matrix. Further applications can also be found in [**Kap, KaM**]. Theorem 9.12, Theorem 9.22 and their proofs are from [**QW2**], while Theorem 9.14, Theorem 9.23 and their proofs are from [**LQ3**]. The proof of Theorem 9.35 follows the presentation of [**Mur**], while we refer to [**Mur**, Subsection 3.3] for the proof of Proposition 9.32.

9.1. Integral operators

In this section, we will introduce the concept of integral operators, and discuss the methods to construct these operators. Moreover, we will prove that many of the familiar operators, such as $1/\mathfrak{z}_\lambda \cdot \mathfrak{a}_{-\lambda}(1_X)$, are integral operators.

Let X be a smooth projective complex surface. Recall from Definition 3.2 (i) that

$$\mathbb{H}_X = \bigoplus_{n=0}^{+\infty} H^*(X^{[n]}).$$

DEFINITION 9.1. (i) A class $A \in H^*(X^{[n]})$ is *integral* if it is contained in

$$H^*(X^{[n]}; \mathbb{Z})/\mathrm{Tor} \subset H^*(X^{[n]})$$

where by abusing notation, Tor denotes the torsion part of $H^*(X^{[n]}; \mathbb{Z})$.

(ii) A linear basis of $H^*(X^{[n]})$ is *integral* if its members are integral classes and form a $\mathbb{Z}$-basis of the lattice $H^*(X^{[n]}; \mathbb{Z})/\mathrm{Tor}$;

(iii) A linear operator $\mathfrak{f} \in \mathrm{End}(\mathbb{H}_X)$ is *integral* if $\mathfrak{f}(A) \in \mathbb{H}_X$ is an integral class whenever $A \in \mathbb{H}_X$ is an integral cohomology class.

A linear basis of $H^*(X^{[n]})$ is integral if and only if its members are integral classes and the matrix formed by the pairings of its members is unimodular. In the next two lemmas, we prove that the Heisenberg operators $\mathfrak{a}_n(\alpha)$ and the operators $\mathbf{1}_{-n}$ are integral.

LEMMA 9.2. (i) *If $\mathfrak{f} \in \mathrm{End}(\mathbb{H}_X)$ is integral, then so is its adjoint $\mathfrak{f}^\dagger$;*
(ii) *The Heisenberg operators $\mathfrak{a}_n(\alpha), n \in \mathbb{Z}$ are integral if $\alpha \in H^*(X)$ is integral.*

PROOF. (i) Note that a class $A \in \mathbb{H}_X$ is integral if and only if $\langle A, B\rangle$ is an integer whenever $B \in \mathbb{H}_X$ is an integral class. It follows that $\mathfrak{f} \in \mathrm{End}(\mathbb{H}_X)$ is integral if and only if its adjoint operator $\mathfrak{f}^\dagger \in \mathrm{End}(\mathbb{H}_X)$ is integral.

(ii) Recall that $\mathfrak{a}_0(\alpha) = 0$. Next, fix $n > 0$. By Definition 3.3, $\mathfrak{a}_{-n}(\alpha)$ is integral. By Lemma 3.7 and (i), $\mathfrak{a}_n(\alpha)$ is integral as well. □

LEMMA 9.3. *For $n \geq 0$, the operator $\mathbf{1}_{-n}$ is integral.*

PROOF. Recall from Definition 3.18 that $\mathbf{1}_{-n} = 1/n! \cdot \mathfrak{a}_{-1}(1_X)^n$. Fix any integer $m \geq 0$ and an integral class $A \in H^j(X^{[m]})$. For $i = m, \ldots, m+n-1$, recall the subscheme $Q^{[i+1,i]}$ of $X^{[i+1]} \times X \times X^{[i]}$ defined by (3.6), and let Q_i be the image of $Q^{[i+1,i]}$ under the natural projection

$$X^{[i+1]} \times X \times X^{[i]} \to X^{[i+1]} \times X^{[i]}.$$

Set-theoretically,

$$Q_i = \{(\xi_{i+1}, \xi_i) \in X^{[i+1]} \times X^{[i]} \mid \xi_{i+1} \supset \xi_i\}.$$

Let $\phi_{i,1}$ and $\phi_{i,2}$ be the two projections of $X^{[i+1]} \times X^{[i]}$. Let

$$Y = X^{[m+n]} \times \cdots \times X^{[m+1]} \times X^{[m]}.$$

By Definition 3.3,

$$\mathfrak{a}_{-1}(1_X)(A) = (\phi_{m,1})_*([Q_m] \cdot \phi_{m,2}^* A).$$

Repeating this process and using the projection formula, we conclude that

$$\mathfrak{a}_{-1}(1_X)^n(A) = (\tilde{\phi}_{m+n})_*([Q] \cdot \tilde{\phi}_m^* A)$$

where $\tilde{\phi}_i$ denotes the projection of Y to $X^{[i]}$ and

$$Q = \{(\xi_{m+n}, \ldots, \xi_{m+1}, \xi_m) \in Y \mid \xi_{m+n} \supset \cdots \supset \xi_{m+1} \supset \xi_m\}$$

(the scheme structure on Q can be described similarly as that on Q_i). To show that $\mathbf{1}_{-n}(A) \in H^j(X^{[m+n]})$ is integral, it suffices to prove that the intersection number

$$(\tilde{\phi}_{m+n})_*\big([Q] \cdot \tilde{\phi}_m^* A\big) \cdot B$$

is divisible by $n!$ for any integral class $B \in H^{4(m+n)-j}(X^{[m+n]})$.

Represent the integral class A by a piecewise smooth cycle $W_A \subset X^{[m]}$. Let

$$Q_A = \{(\xi_{m+n}, \ldots, \xi_{m+1}, \xi_m) \in Y \mid \xi_{m+n} \supset \cdots \supset \xi_{m+1} \supset \xi_m \text{ and } \xi_m \in W_A\}.$$

Then, the integral class $[Q] \cdot \tilde{\phi}_m^* A$ is represented by Q_A. Note that

$$\tilde{\phi}_{m+n}|_{Q_A} : Q_A \to \tilde{\phi}_{m+n}(Q_A)$$

is generically finite, and a generic element in $\tilde{\phi}_{m+n}(Q_A)$ is of the form $\xi_m + x_1 + \ldots + x_n$ where $\xi_m \in W_A$ is generic, the points $x_1, \ldots, x_n$ are distinct, and

$$\mathrm{Supp}(\xi_m) \cap \{x_1, \ldots, x_n\} = \emptyset.$$

Represent B by a piecewise smooth cycle W_B such that W_B and $\tilde{\phi}_{m+n}(Q_A)$ intersect transversely at generic points $P_1, \ldots, P_s$. Write P_i as $\xi_m + x_1 + \ldots + x_n$. Let

$$\xi_m = \eta_1 + \ldots + \eta_t + x_1' + \ldots + x_u' + x_1'' + \ldots + x_v''$$

where each η_i is supported at one point with $\ell(\eta_i) \geq 2$, the following subsets of X:

$$\text{Supp}(\eta_1), \ldots, \text{Supp}(\eta_t), \{x_1'\}, \ldots, \{x_u'\}, \{x_1''\}, \ldots, \{x_v''\}$$

are mutually disjoint, and only the points $x_1', \ldots, x_u'$ can move in Zariski open subsets of the surface X (In other words, if we put

$$U = X - \text{Supp}(\eta_1) \cup \ldots \cup \text{Supp}(\eta_t) \cup \{x_1', \ldots, x_u', x_1'', \ldots, x_v''\},$$

then for all distinct points $\tilde{x}_1', \ldots, \tilde{x}_u'$ in U, we have

$$\eta_1 + \ldots + \eta_t + \tilde{x}_1' + \ldots + \tilde{x}_u' + x_1'' + \ldots + x_v'' \in W_A).$$

So $\left(\tilde{\phi}_{m+n}|_{Q_A}\right)^{-1}(P_i)$ consists of $(u+n)(u+n-1)\cdots(u+1)$ distinct points. It follows that the intersection number $(\tilde{\phi}_{m+n})_*([Q] \cdot \tilde{\phi}_m^* A) \cdot B$ is divisible by $n!$. □

Let $A \in H^*(X^{[n]})$. By (3.10), A is a linear combination of classes of the form

$$\mathfrak{a}_{-1}(1_X)^m \mathfrak{a}_{-m_1}(\alpha_1) \cdots \mathfrak{a}_{-m_\ell}(\alpha_\ell)|0\rangle$$

where $\alpha_1, \ldots, \alpha_\ell \in \bigoplus_{i \geq 1} H^i(X)$, $m \geq 0$, $\ell \geq 0$, and $m_1, \ldots, m_\ell > 0$. By (3.9),

$$\mathfrak{a}_1(x)\big(\mathfrak{a}_{-m_1}(\alpha_1) \cdots \mathfrak{a}_{-m_\ell}(\alpha_\ell)|0\rangle\big) = 0.$$

It follows that the class $A \in H^*(X^{[n]})$ can be written as

$$A = \mathbf{1}_{-n_1}(A_1) + \ldots + \mathbf{1}_{-n_k}(A_k) \tag{9.1}$$

where $0 \leq n_1 < \ldots < n_k$, and $A_i \in H^*(X^{[n-n_i]})$ with $\mathfrak{a}_1(x)(A_i) = 0$ for every i.

LEMMA 9.4. *Let $A \in H^*(X^{[n]})$ be expressed as in (9.1). Then,*

(i) *the classes A_i are uniquely determined by A;*
(ii) *A is integral if and only if all the classes A_i are integral.*

PROOF. First of all, by Lemma 9.3, A is integral if all the classes A_i are integral. So it remains to prove (i) and the "only if" part of (ii).

Next, applying the operator $\mathfrak{a}_1(x)^{n_k}$ to both sides of (9.1), we obtain

$$\mathfrak{a}_1(x)^{n_k}(A) = \mathfrak{a}_1(x)^{n_k}\big(\mathbf{1}_{-n_1}(A_1) + \ldots + \mathbf{1}_{-n_k}(A_k)\big) = (-1)^{n_k} \cdot A_k.$$

Thus the class A_k is uniquely determined by A. Moreover, by Lemma 9.2 (ii), if A is an integral class, then so is the class A_k.

Finally, repeating the above process to the class $A' = (A - \mathbf{1}_{-n_k}(A_k))$, we conclude that (i) and the "only if" part of (ii) hold for all the classes A_i. □

The next results enables us to construct integral operators.

PROPOSITION 9.5. *Let $A \in \mathbb{H}_X$ be an integral class. Write A as $\mathfrak{a}_A|0\rangle$ where $\mathfrak{a}_A$ is a unique polynomial of creation operators. Then, $\mathfrak{a}_A$ is an integral operator.*

PROOF. Let $A \in H^*(X^{[n]})$. Fix an integral class $B \in H^*(X^{[m]})$. We want to show that the cohomology class $\mathfrak{a}_A(B) \in H^*(X^{[m+n]})$ is still integral.

First of all, decomposing A as in (9.1), we see from Lemma 9.4 (ii) and Lemma 9.3 that we may assume $\mathfrak{a}_1(x)(A) = 0$. Similarly, write

$$B = \mathbf{1}_{-m_1}(B_1) + \ldots + \mathbf{1}_{-m_k}(B_k)$$

as in (9.1), where $0 \le m_1 < \ldots < m_k \le m$, and $B_i \in H^*(X^{[m-m_i]})$ with $\mathfrak{a}_1(x)(B_i) = 0$ for every i. By Lemma 9.4 (ii), each class B_i is integral. Now

$$\mathfrak{a}_A(B) = \sum_{i=1}^{k} \mathfrak{a}_A\big(\mathbf{1}_{-m_i}(B_i)\big) = \sum_{i=1}^{k} \mathbf{1}_{-m_i}\big(\mathfrak{a}_A(B_i)\big).$$

It follows from Lemma 9.3 that we may assume $\mathfrak{a}_1(x)(B) = 0$ as well.

By Lemma 9.3 again, we have two integral classes $\mathbf{1}_{-m}A, \mathbf{1}_{-n}B \in H^*(X^{[m+n]})$. By Theorem 7.7, the cup product $(\mathbf{1}_{-m}A) \cdot (\mathbf{1}_{-n}B)$ equals

$$\mathfrak{a}_A(B) + \sum_{i \ge 1} \mathbf{1}_{-i}(A_i)$$

where $\mathfrak{a}_1(x)(A_i) = 0$ for every $i \ge 1$. By our assumption on A and B, we have $\mathfrak{a}_1(x)\big(\mathfrak{a}_A(B)\big) = 0$. Hence $\mathfrak{a}_A(B) \in H^*(X^{[m+n]})$ is integral by Lemma 9.4. $\square$

In view of Proposition 9.5, to obtain integral operators, we need to construct integral classes. We begin with the Chern classes of some tautological vector bundles over the Hilbert scheme $X^{[n]}$. Recall the notation $L^{[n]}$ from (1.25). For a partition $\lambda = (1^{m_1}2^{m_2}\cdots)$ and a cohomology class $\alpha \in H^*(X)$, define

$$\mathfrak{a}_{-\lambda}(\alpha) = \prod_{r \ge 1} \mathfrak{a}_{-r}(\alpha)^{m_r}. \tag{9.2}$$

LEMMA 9.6. *Let $0 \le i \le n$. Let L be a line bundle on X with $\alpha = c_1(L)$. Then,*

$$c_i(L^{[n]}) = \sum_{\substack{|\lambda|+|\mu|=n \\ \ell(\lambda)=n-i}} \frac{(-1)^{i-\ell(\mu)}\mathfrak{a}_{-\lambda}(1_X)\mathfrak{a}_{-\mu}(\alpha)|0\rangle}{\mathfrak{z}_\lambda \mathfrak{z}_\mu}. \tag{9.3}$$

PROOF. Let

$$c_\hbar(L^{[n]}) = \sum_{i=0}^{n} c_i(L^{[n]})\hbar^i$$

be the generating function for the total Chern classes of $L^{[n]}$, where $\hbar$ is a formal variable. By Proposition 4.19,

$$\sum_{n=0}^{+\infty} c_\hbar(L^{[n]})w^n = \exp\left(\sum_{r \ge 1} \frac{(-\hbar)^{r-1}}{r}\mathfrak{a}_{-r}(c_\hbar(L))w^r\right) \cdot |0\rangle \tag{9.4}$$

where w is another formal variable. Expanding the right-hand-side yields

$$(9.5)\quad \exp\left(\sum_{r\geq 1}\frac{(-\hbar)^{r-1}}{r}\mathfrak{a}_{-r}(c_\hbar(L))w^r\right)\cdot|0\rangle$$
$$= \exp\left(\sum_{r\geq 1}\frac{(-1)^{r-1}}{r}\mathfrak{a}_{-r}(1_X)\hbar^{r-1}w^r\right)\cdot\exp\left(\sum_{r\geq 1}\frac{(-1)^{r-1}}{r}\mathfrak{a}_{-r}(\alpha)\hbar^r w^r\right)\cdot|0\rangle$$
$$= \sum_\lambda \frac{(-1)^{|\lambda|-\ell(\lambda)}\mathfrak{a}_{-\lambda}(1_X)}{\mathfrak{z}_\lambda}\hbar^{|\lambda|-\ell(\lambda)}w^{|\lambda|}\cdot\sum_\mu\frac{(-1)^{|\mu|-\ell(\mu)}\mathfrak{a}_{-\mu}(\alpha)}{\mathfrak{z}_\mu}\hbar^{|\mu|}w^{|\mu|}|0\rangle$$
$$= \sum_{(\lambda,\mu)}\frac{(-1)^{|\lambda|+|\mu|-\ell(\lambda)-\ell(\mu)}\mathfrak{a}_{-\lambda}(1_X)\mathfrak{a}_{-\mu}(\alpha)|0\rangle}{\mathfrak{z}_\lambda\mathfrak{z}_\mu}\hbar^{|\lambda|+|\mu|-\ell(\lambda)}w^{|\lambda|+|\mu|}.$$

Now the Chern class $c_i(L^{[n]})$ is the coefficient of $\hbar^i w^n$ in (9.5). □

LEMMA 9.7. *For every partition λ, we have*
(i) $1/\mathfrak{z}_\lambda\cdot\mathfrak{a}_{-\lambda}(1_X)|0\rangle$ *is an integral class;*
(ii) $1/\mathfrak{z}_\lambda\cdot\mathfrak{a}_{-\lambda}(1_X)$ *is an integral operator.*

PROOF. (ii) follows from (i) and Proposition 9.5. To prove (i), we let $n=|\lambda|$ and use induction on n. Our result is trivially true when $n=0,1$. In the following, assume that $1/\mathfrak{z}_\mu\cdot\mathfrak{a}_{-\mu}(1_X)|0\rangle$ is integral whenever $|\mu|<n$.

Let $\lambda=(1^{m_1}2^{m_2}3^{m_3}\cdots)$ so that $n=\sum_r rm_r$. First of all, assume $m_r>0$ for at least two different r's. Then, $rm_r<n$ for every r. Putting $\mu^r=(r^{m_r})$ for $r\geq 1$ and applying induction to the partitions μ^r, we obtain integral classes $1/\mathfrak{z}_{\mu^r}\cdot\mathfrak{a}_{-\mu^r}(1_X)|0\rangle$. Put

$$A_r=\frac{1}{\mathfrak{z}_{\mu^r}}\cdot\mathfrak{a}_{-\mu^r}(1_X)|0\rangle.$$

By Proposition 9.5, the operators $\mathfrak{a}_{A_r}=1/\mathfrak{z}_{\mu^r}\cdot\mathfrak{a}_{-\mu^r}(1_X)$ are integral. Thus,

$$1/\mathfrak{z}_\lambda\cdot\mathfrak{a}_{-\lambda}(1_X)|0\rangle=\prod_{r\geq 1}\left(1/\mathfrak{z}_{\mu^r}\cdot\mathfrak{a}_{-\mu^r}(1_X)\right)\cdot|0\rangle$$

is integral.

We are left with the case when $m_r>0$ for a unique r. In this case, $|\lambda|=rm_r=n$ and $\ell(\lambda)=m_r$. Applying (9.3) to $L=\mathcal{O}_X$ and $i=n-m_r$, we have

$$(9.6)\quad c_i(\mathcal{O}_X^{[n]}) = (-1)^i\cdot\sum_{|\nu|=rm_r,\ \ell(\nu)=m_r}\frac{\mathfrak{a}_{-\nu}(1_X)|0\rangle}{\mathfrak{z}_\nu}$$
$$= (-1)^i\cdot\frac{\mathfrak{a}_{-\lambda}(1_X)|0\rangle}{\mathfrak{z}_\lambda}+(-1)^i\cdot\sum_{\substack{\nu\neq\lambda,|\nu|=rm_r\\ \ell(\nu)=m_r}}\frac{\mathfrak{a}_{-\nu}(1_X)|0\rangle}{\mathfrak{z}_\nu}.$$

Note that for any partition $\nu=(1^{t_1}2^{t_2}\cdots)$ satisfying $\nu\neq\lambda$, $|\nu|=rm_r$ and $\ell(\nu)=m_r$, there are at least two i's with $t_i>0$. By the previous paragraph, $1/\mathfrak{z}_\nu\cdot\mathfrak{a}_{-\nu}(1_X)|0\rangle$ is integral. Hence, $1/\mathfrak{z}_\lambda\cdot\mathfrak{a}_{-\lambda}(1_X)|0\rangle$ is integral by (9.6). □

9.2. Integral operators involving only divisors in $H^2(X)$

In this section, we will define the linear operator $\mathfrak{m}_{\lambda,\alpha} \in \mathrm{End}(\mathbb{H}_X)$ for a partition λ and an arbitrary class $\alpha \in H^2(X)$. We will show that when α is a divisor, $\mathfrak{m}_{\lambda,\alpha}$ is an integral operator.

To begin with, let C be a smooth irreducible curve in the smooth projective complex surface X. By abusing notations, we also use C to denote the corresponding divisor and cohomology class. For every partition $\lambda = (\lambda_1 \geq \lambda_2 \geq \ldots)$ of n, we define the subvariety

$$L^\lambda C = \overline{(\rho_n)^{-1}(S^n_\lambda C)} \tag{9.7}$$

of $X^{[n]}$, where

$$S^n_\lambda C = \left\{ \sum_i \lambda_i x_i \in C^{(n)} |\ x_i \in C, x_i \neq x_j \text{ for } i \neq j \right\}, \tag{9.8}$$

and ρ_n is the Hilbert-Chow morphism. By Remark 1.19, $L^\lambda C$ is irreducible and of dimension n. For $n \geq 0$, let

$$\mathbb{H}_{n,C}$$

be the $\mathbb{C}$-linear span of all the classes $\mathfrak{a}_{-\lambda}(C)|0\rangle, \lambda \vdash n$. By the Theorem 9.14 in [**Nak5**], the integral class $[L^\lambda C] \in H^{2n}(X^{[n]};\mathbb{Z})$ is contained in $\mathbb{H}_{n,C} \subset H^*(X^{[n]})$; moreover,

$$\mathfrak{a}_{-i}(C)[L^\lambda C] = \sum_\mu a_{\lambda,\mu}[L^\mu C] \tag{9.9}$$

where μ and $a_{\lambda,\mu}$ are the same as defined in (1.5).

Define

$$\mathbb{H}_C = \bigoplus_{n\geq 0} \mathbb{H}_{n,C}. \tag{9.10}$$

Next, we recall some notations from Section 1.2. Let Λ be the ring of symmetric functions in infinitely many variables, and

$$\Lambda_\mathbb{C} = \Lambda \otimes_\mathbb{Z} \mathbb{C}.$$

Let Λ^n and $\Lambda^n_\mathbb{C}$ be the degree-n parts in Λ and $\Lambda_\mathbb{C}$ respectively. For a partition λ, let p_λ, m_λ and s_λ be the power-sum symmetric function, the monomial symmetric function and the Schur function respectively.

In [**Nak5**, (9.13)], Nakajima defined a linear isomorphism

$$\Phi_C: \quad \Lambda_\mathbb{C} = \bigoplus_{n\geq 0} \Lambda^n_\mathbb{C} \quad \to \quad \mathbb{H}_C \tag{9.11}$$

which satisfies the following two properties:

$$\Phi_C(p_\lambda) = \mathfrak{a}_{-\lambda}(C)|0\rangle, \qquad \Phi_C(m_\lambda) = [L^\lambda C]. \tag{9.12}$$

By Proposition 9.5, the operator $\mathfrak{a}_{[L^\lambda C]} \in \mathrm{End}(\mathbb{H}_X)$ is integral. Define

$$\mathfrak{m}_{\lambda,C} = \mathfrak{a}_{[L^\lambda C]} \in \mathrm{End}(\mathbb{H}_X).$$

By (9.12), $\mathfrak{m}_{\lambda,C}$ is a polynomial of the creation Heisenberg operators $\mathfrak{a}_{-i}(C)$, $i > 0$ with rational coefficients.

DEFINITION 9.8. For a partition λ and an arbitrary class $\alpha \in H^2(X)$, we define

$$\mathfrak{m}_{\lambda,\alpha} \in \mathrm{End}(\mathbb{H}_X)$$

by replacing the creation operators $\mathfrak{a}_{-i}(C)$ in $\mathfrak{m}_{\lambda,C}$ by the corresponding creation operators $\mathfrak{a}_{-i}(\alpha)$. Define the cohomology class $[L^\lambda \alpha] \in H^*(X^{[|\lambda|]})$ by putting

$$[L^\lambda \alpha] = \mathfrak{m}_{\lambda,\alpha}|0\rangle. \tag{9.13}$$

Similarly, we can define the subspaces $\mathbb{H}_{n,\alpha}$ and $\mathbb{H}_\alpha$ of $\mathbb{H}_X$ as in (9.10). Moreover, in view of (9.11) and (9.12), we have a linear isomorphism

$$\Phi_\alpha: \quad \Lambda_\mathbb{C} \quad \to \quad \mathbb{H}_\alpha \tag{9.14}$$

which satisfies the following two properties:

$$\Phi_\alpha(p_\lambda) = \mathfrak{a}_{-\lambda}(\alpha)|0\rangle, \qquad \Phi_\alpha(m_\lambda) = [L^\lambda \alpha] = \mathfrak{m}_{\lambda,\alpha}|0\rangle. \tag{9.15}$$

Applying Φ_α to both sides of (1.4) and using (9.15), we obtain

$$\mathfrak{m}_{(1^n),\alpha}|0\rangle = [L^{(1^n)}\alpha] = \sum_{|\mu|=n} \frac{(-1)^{n-\ell(\mu)}\mathfrak{a}_{-\mu}(\alpha)|0\rangle}{\mathfrak{z}_\mu}. \tag{9.16}$$

Our goal in this section is to prove that $\mathfrak{m}_{\lambda,\alpha} \in \mathrm{End}(\mathbb{H}_X)$ is an integral operator if α is a divisor. In view of (9.13) and Proposition 9.5, it suffices to show that $[L^\lambda \alpha]$ is an integral class. In the next three lemmas, we will study the properties of the class $[L^\lambda \alpha]$.

LEMMA 9.9. *Let $\alpha \in H^2(X)$ and $i > 0$. Then, we have*

$$\mathfrak{a}_{-i}(\alpha)[L^\lambda \alpha] = \sum_\mu a_{\lambda,\mu}[L^\mu \alpha] \tag{9.17}$$

where μ and $a_{\lambda,\mu}$ are the same as defined in (1.5).

PROOF. Fix a smooth irreducible curve C in X. Define a linear map

$$\Psi_{C,\alpha}: \mathbb{H}_C \to \mathbb{H}_\alpha$$

by sending the basis elements $\mathfrak{a}_{-\lambda}(C)|0\rangle$ of $\mathbb{H}_C$ to the elements $\mathfrak{a}_{-\lambda}(\alpha)|0\rangle$ in $\mathbb{H}_\alpha$. Note that the creation operators $\mathfrak{a}_{-j}(C)$ and $\mathfrak{a}_{-j}(\alpha)$ preserve $\mathbb{H}_C$ and $\mathbb{H}_\alpha$ respectively. Moreover,

$$\Psi_{C,\alpha} \circ \mathfrak{a}_{-j}(C) = \mathfrak{a}_{-j}(\alpha) \circ \Psi_{C,\alpha}.$$

It follows from the definition of $\mathfrak{m}_{\lambda,\alpha}$ that $\Psi_{C,\alpha} \circ \mathfrak{m}_{\lambda,C} = \mathfrak{m}_{\lambda,\alpha} \circ \Psi_{C,\alpha}$. Thus

$$\Psi_{C,\alpha}([L^\lambda C]) = \Psi_{C,\alpha} \circ \mathfrak{m}_{\lambda,C}|0\rangle = \mathfrak{m}_{\lambda,\alpha} \circ \Psi_{C,\alpha}|0\rangle = \mathfrak{m}_{\lambda,\alpha}|0\rangle = [L^\lambda \alpha]. \tag{9.18}$$

Now applying $\Psi_{C,\alpha}$ to both sides of (9.9), we obtain (9.17). □

LEMMA 9.10. *If $[L^\lambda \alpha]$ is integral for every λ, then so is $[L^\lambda(-\alpha)]$.*

PROOF. Fix a smooth irreducible curve C in X, and let notations be the same as in the proof of Lemma 9.9. Recall from (9.12) that the linear isomorphism Φ_C sends m_λ and p_λ to $[L^\lambda C]$ and $\mathfrak{a}_{-\lambda}(C)|0\rangle$ respectively. Put

$$m_\lambda = \sum_{|\mu|=|\lambda|} d_\mu p_\mu \tag{9.19}$$

where $d_\mu \in \mathbb{C}$. Applying $\Psi_{C,\alpha} \circ \Phi_C$ to both sides and using (9.18), we obtain

$$[L^\lambda \alpha] = \sum_{|\mu|=|\lambda|} d_\mu \mathfrak{a}_{-\mu}(\alpha)|0\rangle.$$

It follows from the definition of $[L^\lambda(-\alpha)]$ that

$$[L^\lambda(-\alpha)] = \sum_{|\mu|=|\lambda|} d_\mu (-1)^{\ell(\mu)} \mathfrak{a}_{-\mu}(\alpha)|0\rangle. \tag{9.20}$$

Next, recall from (1.8) the involution ω on Λ and from (1.9) the forgotten symmetric function $f_\lambda = \omega(m_\lambda)$ associated to a partition λ. The forgotten symmetric function f_λ is an integral linear combination of the monomial symmetric functions m_μ, $\mu \vdash |\lambda|$. Thus $\Psi_{C,\alpha} \circ \Phi_C(f_\lambda)$ is an integral linear combination of the classes $\Psi_{C,\alpha} \circ \Phi_C(m_\lambda) = [L^\mu\alpha]$, $\mu \vdash |\lambda|$, and hence is an integral class. By (9.19),

$$\begin{aligned} \Psi_{C,\alpha} \circ \Phi_C(f_\lambda) &= \Psi_{C,\alpha} \circ \Phi_C \circ \omega(m_\lambda) \\ &= \Psi_{C,\alpha} \circ \Phi_C \circ \omega\left(\sum_{|\mu|=|\lambda|} d_\mu p_\mu\right) \\ &= \Psi_{C,\alpha} \circ \Phi_C \left(\sum_{|\mu|=|\lambda|} d_\mu (-1)^{|\mu|-\ell(\mu)} p_\mu\right) \\ &= (-1)^{|\lambda|} \cdot \sum_{|\mu|=|\lambda|} d_\mu (-1)^{\ell(\mu)} \mathfrak{a}_{-\mu}(\alpha)|0\rangle. \end{aligned} \tag{9.21}$$

Combining (9.20) and (9.21), we conclude that $[L^\lambda(-\alpha)]$ is integral as well. □

LEMMA 9.11. *Let $\alpha_1, \alpha_2 \in H^2(X)$, and $\alpha = \alpha_1 + \alpha_2$. Then,*

$$[L^\lambda \alpha] = \sum_{(\lambda^1,\lambda^2):\lambda^1 \cup \lambda^2 = \lambda} \mathfrak{m}_{\lambda^1,\alpha_1} \mathfrak{m}_{\lambda^2,\alpha_2}|0\rangle \tag{9.22}$$

where (λ^1, λ^2) stands for ordered pairs of partitions.

PROOF. We start with some notations. For the partitions μ obtained from λ as in (9.17), we will denote $\mu = \lambda \uparrow^i$. If we specify further that such a μ is obtained from adding i to a part of λ equal to j (here j is allowed to be 0), then we denote $\mu = \lambda \uparrow^i_j$. Given a partition λ, we denote by $m_k(\lambda)$ the multiplicity of the parts of λ equal to k. In these notations, the coefficient $a_{\lambda,\mu}$ in Lemma 9.9 for $\mu = \lambda \uparrow^i_j$ is simply equal to $m_{i+j}(\mu)$. Denote the right-hand-side of (9.22) by $R^\lambda(\alpha)$.

Claim. $\mathfrak{a}_{-i}(\alpha) R^\lambda(\alpha) = \sum_{\mu=\lambda\uparrow^i} a_{\lambda,\mu} R^\mu(\alpha)$.

PROOF. The right-hand-side in the Claim is equal to

$$\sum_{\mu=\lambda\uparrow^i} \sum_{(\mu^1,\mu^2):\mu^1\cup\mu^2=\mu} a_{\lambda,\mu} \mathfrak{m}_{\mu^1,\alpha_1} \mathfrak{m}_{\mu^2,\alpha_2}|0\rangle. \tag{9.23}$$

On the other hand, by Lemma 9.9, the left-hand-side in the claim is

$$\begin{aligned}
&\mathfrak{a}_{-i}(\alpha)R^\lambda(\alpha) \\
= &\sum_{(\lambda^1,\lambda^2):\lambda^1\cup\lambda^2=\lambda} (\mathfrak{m}_{\lambda^2,\alpha_2}\mathfrak{a}_{-i}(\alpha_1)\mathfrak{m}_{\lambda^1,\alpha_1}|0\rangle + \mathfrak{m}_{\lambda^1,\alpha_1}\mathfrak{a}_{-i}(\alpha_2)\mathfrak{m}_{\lambda^2,\alpha_2}|0\rangle) \\
= &\sum_{(\lambda^1,\lambda^2):\lambda^1\cup\lambda^2=\lambda}\ \sum_{\rho^1=\lambda^1\uparrow^i} a_{\lambda^1,\rho^1}\ \mathfrak{m}_{\rho^1,\alpha_1}\mathfrak{m}_{\lambda^2,\alpha_2}|0\rangle \qquad (9.24)\\
&+ \sum_{(\lambda^1,\lambda^2):\lambda^1\cup\lambda^2=\lambda}\ \sum_{\rho^2=\lambda^2\uparrow^i} a_{\lambda^2,\rho^2}\ \mathfrak{m}_{\lambda^1,\alpha_1}\mathfrak{m}_{\rho^2,\alpha_2}|0\rangle. \qquad (9.25)
\end{aligned}$$

Note that the partitions $\rho^1\cup\lambda^2$ and $\lambda^1\cup\rho^2$ associated to $\lambda^1, \lambda^2, \rho^1$ and ρ^2 appearing above are of the form $\lambda\uparrow^i$. Thus, the same types of terms appear on both sides of the Claim. It remains to identify the coefficients of a given term.

Fix $\mu = \lambda\uparrow^i_j$ for some part j of λ, and fix μ^1, μ^2 such that $\mu^1\cup\mu^2 = \mu$. From the above computation, the contributions to the term $\mathfrak{m}_{\mu^1,\alpha_1}\mathfrak{m}_{\mu^2,\alpha_2}|0\rangle$ in the left-hand-side of the Claim come from two places: the term in (9.24) for $\rho^1 = \mu^1, \lambda^2 = \mu^2$ whose coefficient is $m_{i+j}(\mu^1)$, and the term in (9.25) for $\lambda^1 = \mu^1, \rho^2 = \mu^2$ whose coefficient is $m_{i+j}(\mu^2)$. Therefore in view of (9.23), the coefficients of the term $\mathfrak{m}_{\mu^1,\alpha_1}\mathfrak{m}_{\mu^2,\alpha_2}|0\rangle$ in both sides of the Claim coincide thanks to

$$m_{i+j}(\mu^1) + m_{i+j}(\mu^2) = m_{i+j}(\mu^1\cup\mu^2) = m_{i+j}(\mu) = a_{\lambda,\mu}. \qquad (9.26)$$

This completes the proof of the above Claim. □

Next, we continue the proof of (9.22) by using induction on n and the reverse dominance ordering of partitions λ of n (see Definition 1.2 for the dominance ordering). For $n = 1$, formula (9.22) is clear. Assume that formula (9.22) holds for all partitions of size less than n.

For $\lambda = (n)$, (9.22) holds since $\mathfrak{m}_{\lambda,\alpha} = \mathfrak{a}_{-n}(\alpha)$ and $[L^\lambda\alpha] = \mathfrak{a}_{-n}(\alpha)|0\rangle$.

For a general partition λ of n with a part equal to, say, i, we denote by $\widetilde{\lambda}$ the partition obtained from λ with a part equal to i removed. Now replacing λ in (9.17) and the above Claim by $\widetilde{\lambda}$ respectively, we obtain

$$\mathfrak{a}_{-i}(\alpha)[L^{\widetilde{\lambda}}\alpha] = \sum_{\mu=\widetilde{\lambda}\uparrow^i} a_{\widetilde{\lambda},\mu}[L^\mu\alpha], \qquad (9.27)$$

$$\mathfrak{a}_{-i}(\alpha)R^{\widetilde{\lambda}}(\alpha) = \sum_{\mu=\widetilde{\lambda}\uparrow^i} a_{\widetilde{\lambda},\mu}R^\mu(\alpha). \qquad (9.28)$$

Note that λ appears among the above μ's as the maximum in the reverse dominance ordering. By induction hypothesis, the left-hand-sides of (9.27) and (9.28) coincide, and all the terms on the right-hand-sides of (9.27) and (9.28) involving μ not equal to λ coincide. Thus (9.22) follows since $a_{\widetilde{\lambda},\lambda} = m_i(\lambda) \neq 0$. □

THEOREM 9.12. *For every divisor α on X and every partition λ, $[L^\lambda\alpha]$ is an integral class and $\mathfrak{m}_{\lambda,\alpha} \in \mathrm{End}(\mathbb{H}_X)$ is an integral operator.*

PROOF. Every divisor α can be written as $C_1 - C_2$ for some very ample divisors C_1 and C_2. Represent C_1 and C_2 by smooth irreducible curves. Then the operators $\mathfrak{m}_{\mu,C_i}$ are integral. By Lemma 9.10 and Lemma 9.11, $[L^\lambda\alpha]$ is an integral class. By Proposition 9.5, $\mathfrak{m}_{\lambda,\alpha} \in \mathrm{End}(\mathbb{H}_X)$ is an integral operator. □

Theorem 9.12 is not sufficient in applications. In the next section, we will extend it to cover the case of an arbitrary integral class $\alpha \in H^2(X)$.

9.3. Integrality of $\mathfrak{m}_{\lambda,\alpha}$ for integral α

In this section, we will prove that if $\alpha \in H^2(X)$ is integral, then the class $[L^\lambda \alpha]$ and the operator $\mathfrak{m}_{\lambda,\alpha}$ are integral.

We begin with the case when $\lambda = (1^n)$.

LEMMA 9.13. *If $\alpha \in H^2(X)$ is integral, then so is the operator $\mathfrak{m}_{(1^n),\alpha}$.*

PROOF. Let p_1 and p_2 be the projections of $X^{[n]} \times X$ to $X^{[n]}$ and X respectively. By the Theorem on p.20 in [**AH**], p_1 induces a homomorphism

$$(p_1)_! : K^0_{\text{top}}(X^{[n]} \times X) \to K^0_{\text{top}}(X^{[n]})$$

where $K^0_{\text{top}}(\cdot)$ denotes the ring of *complex* vector bundles; moreover, for an element $V \in K^0_{\text{top}}(X^{[n]} \times X)$, the Riemann-Roch formula holds:

$$\text{ch}\,\big((p_1)_!(V)\big) = (p_1)_*\big(\text{ch}(V) \cdot p_2^*\text{td}(X)\big). \tag{9.29}$$

Next, by the Theorem E.5 in [**FU**], there exists a *complex* (not necessarily algebraic) line bundle L_α on X with $c_1(L_\alpha) = \alpha$. Put

$$V_\alpha = \mathcal{O}_{\mathcal{Z}_n} \otimes p_2^* L_\alpha \in K^0_{\text{top}}(X^{[n]} \times X)$$

where $\mathcal{Z}_n$ is the universal codimension-2 subscheme. Applying (9.29) to the element $V = V_\alpha$, we obtain

$$\begin{aligned} \text{ch}\,\big((p_1)_!(V_\alpha)\big) &= p_{1*}\big(\text{ch}(\mathcal{O}_{\mathcal{Z}_n}) \cdot p_2^*\text{td}(X) \cdot p_2^*\,\text{ch}(L_\alpha)\big) \\ &= G(1_X, n) + G(\alpha, n) + \langle \alpha, \alpha \rangle/2 \cdot G(x, n) \end{aligned} \tag{9.30}$$

where we have used the definition of $G(\alpha, n)$ and the fact that

$$\text{ch}(L_\alpha) = 1_X + \alpha + \langle \alpha, \alpha \rangle/2 \cdot x.$$

In particular, we conclude that the integral class $c_i\big((p_1)_!(V_\alpha)\big)$ is a polynomial P_i of $G_{i_1}(1_X, n)$, $G_{i_2}(\alpha, n)$, $\langle \alpha, \alpha \rangle$ and $G_{i_3}(x, n)$ with $0 \le i_1, i_2, i_3 \le 2n$, i.e.,

$$\begin{aligned} &c_i\big((p_1)_!(V_\alpha)\big) \\ &= P_i\Big(G_{i_1}(1_X, n), G_{i_2}(\alpha, n), \langle \alpha, \alpha \rangle, G_{i_3}(x, n) \big| \, 0 \le i_1, i_2, i_3 \le 2n\Big). \end{aligned} \tag{9.31}$$

Note that the coefficients in P_i are independent of α and X.

When α is a divisor, L_α is an algebraic line bundle on X, and the element

$$(p_1)_!(V_\alpha) \in K^0_{\text{top}}(X^{[n]})$$

is represented precisely by the tautological rank-n vector bundle $(L_\alpha)^{[n]}$ defined in (1.25). Therefore, we obtain from Lemma 9.6 that

$$c_i\big((p_1)_!(V_\alpha)\big) = \sum_{\substack{|\lambda|+|\mu|=n \\ \ell(\lambda)=n-i}} \frac{(-1)^{i-\ell(\mu)} \mathfrak{a}_{-\lambda}(1_X)\mathfrak{a}_{-\mu}(\alpha)|0\rangle}{\mathfrak{z}_\lambda \mathfrak{z}_\mu}. \tag{9.32}$$

Finally, we conclude from (9.31) and Lemma 4.13 that (9.32) holds for a general integral class $\alpha \in H^2(X)$. Setting $i = n$ in (9.32) and using (9.16), we obtain the integral class

$$c_n\big((p_1)_!(V_\alpha)\big) = \sum_{|\mu|=n} \frac{(-1)^{n-\ell(\mu)}\mathfrak{a}_{-\mu}(\alpha)|0\rangle}{\mathfrak{z}_\mu} = [L^{(1^n)}\alpha].$$

By Proposition 9.5 and (9.13), the operator $\mathfrak{m}_{(1^n),\alpha}$ is integral. □

THEOREM 9.14. *Let the class $\alpha \in H^2(X)$ be integral. Then for every partition λ, the class $[L^\lambda\alpha]$ and the operator $\mathfrak{m}_{\lambda,\alpha}$ are integral.*

PROOF. Recall from Section 1.2 that the monomial symmetric functions

$$m_{(1^i)}, \quad i \geq 1$$

generate the integral ring Λ. So the monomial symmetric function m_λ can be written as an integral polynomial P of the functions $m_{(1^i)}$, $i \geq 1$, i.e.,

$$m_\lambda = P\big(m_{(1^i)}\big|\, i \geq 1\big).$$

Combining this with the two formulas in (9.15), we see that

$$(9.33) \qquad [L^\lambda\alpha] = \Phi_\alpha(m_\lambda) = \Phi_\alpha\left(P\big(m_{(1^i)}\big|\, i \geq 1\big)\right) = P\big(\mathfrak{m}_{(1^i),\alpha}\big|\, i \geq 1\big)\,|0\rangle.$$

Hence, the class $[L^\lambda\alpha]$ is integral by Lemma 9.13. Finally, we conclude from Proposition 9.5 and (9.13) that the operator $\mathfrak{m}_{\lambda,\alpha}$ is integral. □

9.4. Unimodularity

Now Lemma 9.4, Lemma 9.7 and Theorem 9.14 have provided us with three types of integral operators:

$$\frac{1}{\mathfrak{z}_\lambda}\,\mathfrak{a}_{-\lambda}(1_X), \quad \mathfrak{a}_{-\mu}(x), \quad \mathfrak{m}_{\nu,\alpha}$$

where $\alpha \in H^2(X)$ is integral. These operators enable us to write down integral classes for the Hilbert scheme $X^{[n]}$. To ensure that a certain set of these integral classes is an integral linear basis, we must show that their intersection matrix is unimodular. In this section, we will prove that if the intersection matrix of the cohomology classes $\alpha_1, \cdots, \alpha_k \in H^2(X)$ has determinant ± 1, then so does the intersection matrix of the following cohomology classes in $H^{2n}(X^{[n]})$:

$$\mathfrak{m}_{\lambda^1,\alpha_1}\cdots\mathfrak{m}_{\lambda^k,\alpha_k}|0\rangle, \quad |\lambda^1| + \ldots + |\lambda^k| = n.$$

First of all, we begin with some linear algebra preparations. Let V be a k-dimensional complex vector space with a symmetric bilinear form

$$\langle \cdot, \cdot \rangle : V \times V \to \mathbb{C}.$$

Fix a linear basis $\mathbf{v} = \{v_1, \ldots, v_k\}$ of V, and let

$$M_{\mathbf{v}} = \big(\langle v_i, v_j\rangle\big)_{1\leq i,j\leq k}$$

be the matrix formed by the pairings $\langle v_i, v_j\rangle$. The following is elementary.

LEMMA 9.15. *Let $\tilde{\mathbf{v}} = \{\tilde{v}_1, \ldots, \tilde{v}_k\}$ be another basis of V. Let T be the transition matrix from $\tilde{\mathbf{v}}^t$ to $\mathbf{v}^t$ (i.e., $\mathbf{v}^t = T\tilde{\mathbf{v}}^t$). Then, $M_{\mathbf{v}} = TM_{\tilde{\mathbf{v}}}T^t$.* □

For a given $n \geq 0$, the symmetric power $S^n(V)$ admits a monomial basis

$$\{v_{i_1} \cdots v_{i_n} \mid 1 \leq i_1 \leq \ldots \leq i_n \leq k\}. \tag{9.34}$$

Define a bilinear form, still denoted by $\langle \cdot, \cdot \rangle$, on $S^n(V)$ by letting

$$\langle v_{i_1} \cdots v_{i_n}, v_{j_1} \cdots v_{j_n} \rangle = \sum_{\sigma \in S_n} \prod_{a=1}^{n} \langle v_{i_a}, v_{j_{\sigma(a)}} \rangle$$

where S_n denotes the n-th symmetric group. Denote by

$$M_{n,\mathbf{v}}$$

the matrix of the pairings of the monomial basis elements of $S^n(V)$ (so we have $M_{1,\mathbf{v}} = M_{\mathbf{v}}$). The following lemma expresses $\det M_{n,\mathbf{v}}$ in terms of $\det M_{\mathbf{v}}$.

LEMMA 9.16. *For some constant $c(n,k)$, we have*

$$\det M_{n,\mathbf{v}} = c(n,k) \cdot (\det M_{\mathbf{v}})^{\binom{n+k-1}{k}}. \tag{9.35}$$

PROOF. First we prove the formula under the assumption that

$$\det M_{\mathbf{v}_i} \neq 0 \quad \text{for } 1 \leq i \leq k-1 \tag{9.36}$$

where $\mathbf{v}_i = \{v_1, \ldots, v_i\}$ for $1 \leq i \leq k$ (so $\mathbf{v}_k = \mathbf{v}$).

Use induction on k. When $k = 1$, our formula is clearly true.

Fix $k > 1$. Since $\det M_{\mathbf{v}_{k-1}} \neq 0$, there is a vector

$$\tilde{v}_k = e_1 v_1 + \ldots + e_{k-1} v_{k-1} + v_k$$

such that $e_1, \ldots, e_{k-1} \in \mathbb{C}$ and $\langle v_1, \tilde{v}_k \rangle = \ldots = \langle v_{k-1}, \tilde{v}_k \rangle = 0$. Let

$$\tilde{\mathbf{v}} = \{v_1, \ldots, v_{k-1}, \tilde{v}_k\}.$$

Then the transition matrix from the basis $\tilde{\mathbf{v}}^t$ to the basis $\mathbf{v}^t$ is lower triangular with all diagonal entries being 1. By Lemma 9.15, we have

$$\det M_{\mathbf{v}} = \det M_{\tilde{\mathbf{v}}} = \det M_{\mathbf{v}_{k-1}} \cdot \langle \tilde{v}_k, \tilde{v}_k \rangle. \tag{9.37}$$

Next, we apply Lemma 9.15 to $S^n(V)$ for the two monomial bases:

$$\{v_1^{i_1} \cdots v_{k-1}^{i_{k-1}} v_k^{i_k} \mid i_1 + \ldots + i_{k-1} + i_k = n\},$$
$$\{v_1^{i_1} \cdots v_{k-1}^{i_{k-1}} \tilde{v}_k^{i_k} \mid i_1 + \ldots + i_{k-1} + i_k = n\}.$$

Since the transition matrix from the second basis to the first basis is lower triangular with all diagonal entries being 1, we conclude that

$$\det M_{n,\mathbf{v}} = \det M_{n,\tilde{\mathbf{v}}}. \tag{9.38}$$

Now note that $\langle v_1^{i_1} \cdots v_{k-1}^{i_{k-1}} \tilde{v}_k^{i_k},\ v_1^{j_1} \cdots v_{k-1}^{j_{k-1}} \tilde{v}_k^{j_k} \rangle$ is equal to

$$\langle v_1^{i_1} \cdots v_{k-1}^{i_{k-1}},\ v_1^{j_1} \cdots v_{k-1}^{j_{k-1}} \rangle \cdot \delta_{i_k, j_k} \cdot i_k! \cdot \langle \tilde{v}_k, \tilde{v}_k \rangle^{i_k}.$$

It follows that

$$M_{n,\tilde{\mathbf{v}}} = \operatorname{diag}\big(\cdots,\ M_{n-m,\mathbf{v}_{k-1}} \cdot m! \cdot \langle \tilde{v}_k, \tilde{v}_k \rangle^m,\ \cdots\big)$$

where m runs from 0 to n. The matrix $M_{n-m,\mathbf{v}_{k-1}}$ has $\binom{n-m+k-2}{k-2}$ rows. So

$$\det M_{n,\tilde{\mathbf{v}}} = c_1(n,k) \cdot \prod_{m=0}^{n} \left(\det M_{n-m,\mathbf{v}_{k-1}} \cdot \langle \tilde{v}_k, \tilde{v}_k \rangle^{m \cdot \binom{n-m+k-2}{k-2}} \right)$$

for some constant $c_1(n,k)$. Applying induction to $\det M_{n-m,\mathbf{v}_{k-1}}$ yields

$$\begin{aligned}(9.39)\qquad & \det M_{n,\tilde{\mathbf{v}}} \\ &= c(n,k)\cdot\prod_{m=0}^{n}\Big(\big(\det M_{\mathbf{v}_{k-1}}\big)^{\binom{n-m+k-2}{k-1}}\cdot\langle\tilde{v}_k,\tilde{v}_k\rangle^{m\cdot\binom{n-m+k-2}{k-2}}\Big) \\ &= c(n,k)\cdot\big(\det M_{\mathbf{v}_{k-1}}\big)^{\sum_{m=0}^{n}\binom{n-m+k-2}{k-1}}\cdot\langle\tilde{v}_k,\tilde{v}_k\rangle^{\sum_{m=0}^{n}m\cdot\binom{n-m+k-2}{k-2}} \\ &= c(n,k)\cdot\big(\det M_{\mathbf{v}_{k-1}}\cdot\langle\tilde{v}_k,\tilde{v}_k\rangle\big)^{\binom{n+k-1}{k}}\end{aligned}$$

for some constant $c(n,k)$, where we have used the combinatorial identities:

$$(9.40)\qquad \sum_{m=0}^{n}\binom{n-m+k-2}{k-1}=\sum_{m=0}^{n}m\cdot\binom{n-m+k-2}{k-2}=\binom{n+k-1}{k}.$$

Indeed, all the three terms in (9.40) compute the dimension of the space of degree-$(n-1)$ homogeneous polynomials in $(k+1)$-variables.

Combining (9.38), (9.39) and (9.37), we conclude that

$$\begin{aligned}\det M_{n,\mathbf{v}} &= c(n,k)\cdot\big(\det M_{\mathbf{v}_{k-1}}\cdot\langle\tilde{v}_k,\tilde{v}_k\rangle\big)^{\binom{n+k-1}{k}} \\ &= c(n,k)\cdot\big(\det M_{\mathbf{v}}\big)^{\binom{n+k-1}{k}}.\end{aligned}$$

Finally, we come to the general case. Assume that the bilinear form $\langle\cdot,\cdot\rangle$ on V is not identically zero (otherwise the lemma trivially holds). Set $z_{ij}=\langle v_i,v_j\rangle$ for $1\le i\le j\le k$. Then both sides of (9.35) are easily seen to be polynomials in the variables z_{ij}, $1\le i\le j\le k$. Under the assumption (9.36), the proof above implies that the (polynomial) identity (9.35) holds for a Zariski open subset $(z_{ij})_{1\le i\le j\le k}$ of $\mathbb{C}^{k(k+1)/2}$. Thus, (9.35) holds for an arbitrary $(z_{ij})_{1\le i\le j\le k}$. □

For $r\ge 1$, we denote by $V[r]$ a copy of V with bilinear form

$$(9.41)\qquad \langle\cdot,\cdot\rangle_r=(-1)^{r-1}r\cdot\langle\cdot,\cdot\rangle.$$

We will denote by $\langle\cdot,\cdot\rangle_r$ as well the induced bilinear form on the symmetric power $S^*(V[r])$. In particular, $V[1]=V$ with $\langle\cdot,\cdot\rangle_1=\langle\cdot,\cdot\rangle$.

For a partition $\mu=(r^{m_r})_{r\ge1}$, we form the vector space

$$S^\mu V=\bigotimes_{r\ge1}S^{m_r}(V[r])$$

with a bilinear form $\langle\cdot,\cdot\rangle_\mu$ given by

$$\langle\otimes_r u_{I_r},\otimes_r v_{J_r}\rangle_\mu=\prod_r\langle u_{I_r},v_{J_r}\rangle_r$$

for $u_{I_r},v_{J_r}\in S^{m_r}(V[r])$. Denote by $M_{\mu,\mathbf{v}}$ the matrix of the pairings of the induced monomial basis for $S^\mu V$ (see (9.34) for the monomial basis of $S^n(V)$).

LEMMA 9.17. *For some constant $c(\mu,k)$ and some integer $d(\mu,k)\ge1$, we have*

$$\det M_{\mu,\mathbf{v}}=c(\mu,k)\cdot\big(\det M_{\mathbf{v}}\big)^{d(\mu,k)}.$$

PROOF. Follows immediately from our definitions and Lemma 9.16. □

Now let $\alpha_1, \cdots, \alpha_k \in H^2(X)$ be linearly independent classes, and let $M_{\underline{\alpha}}$ be the intersection matrix of $\alpha_1, \cdots, \alpha_k$. Fix a positive integer n. Let

$$M_{n,\underline{\alpha}}$$

be the intersection matrix of the classes in $H^{2n}(X^{[n]})$:

$$\mathfrak{m}_{\lambda^1,\alpha_1} \cdots \mathfrak{m}_{\lambda^k,\alpha_k}|0\rangle, \quad |\lambda^1| + \ldots + |\lambda^k| = n. \tag{9.42}$$

Let

$$V = \mathbb{C}\alpha_1 \oplus \cdots \oplus \mathbb{C}\alpha_k$$

with the pairing induced from the one on $H^2(X)$. Let $\mathbb{H}_{n,V} \subset H^*(X^{[n]})$ be the $\mathbb{C}$-linear span of the classes (9.42). Then, $\mathbb{H}_{n,V}$ has two linear bases:

$$\mathfrak{m}_{n,\underline{\alpha}} = \{\mathfrak{m}_{\lambda^1,\alpha_1} \cdots \mathfrak{m}_{\lambda^k,\alpha_k}|0\rangle\}, \tag{9.43}$$

$$\mathfrak{a}_{n,\underline{\alpha}} = \{\mathfrak{a}_{-\lambda^1}(\alpha_1) \cdots \mathfrak{a}_{-\lambda^k}(\alpha_k)|0\rangle\} \tag{9.44}$$

where $\lambda^1, \ldots, \lambda^k$ are partitions satisfying $|\lambda^1| + \cdots + |\lambda^k| = n$.

Our goal is to compute $\det M_{n,\underline{\alpha}}$. To accomplish this, we will first take an orthogonal basis $\{\beta_1, \cdots, \beta_k\}$ of V, and calculate $\det M_{n,\underline{\beta}}$ in Lemma 9.19 below. Via the elementary Lemma 9.18, the matrix $M_{n,\underline{\alpha}}$ is expressed in terms of $M_{n,\underline{\beta}}$ and certain transition matrices A and B. Then $(\det A)^2$ is determined in Lemma 9.21. Summarizing all these yields $\det M_{n,\underline{\alpha}}$ in Theorem 9.22 which is the main result in this section.

So take an orthogonal basis $\{\beta_1, \cdots, \beta_k\}$ of V with $\langle \beta_i, \beta_i \rangle = \pm 1$ for every i. We similarly define bases $\mathfrak{m}_{n,\underline{\beta}}$ and $\mathfrak{a}_{n,\underline{\beta}}$ for $\mathbb{H}_{n,V}$. Observe that the transition matrix B from the basis $\mathfrak{m}_{n,\underline{\alpha}}$ to $\mathfrak{a}_{n,\underline{\alpha}}$ is independent of the basis $\{\alpha_1, \cdots, \alpha_k\}$ of $H^2(X)$, that is, it is the same as the transition matrix from the basis $\mathfrak{m}_{n,\underline{\beta}}$ to $\mathfrak{a}_{n,\underline{\beta}}$.

Denote by A the transition matrix from the basis $\mathfrak{a}_{n,\underline{\beta}}$ to $\mathfrak{a}_{n,\underline{\alpha}}$. Denote by $M_{n,\underline{\beta}}$ the intersection matrix of the basis $\mathfrak{m}_{n,\underline{\beta}}$. Since the transition matrix from $\mathfrak{m}_{n,\underline{\beta}}$ to $\mathfrak{m}_{n,\underline{\alpha}}$ is $B^{-1}AB$, we obtain from Lemma 9.15 the following.

LEMMA 9.18. $M_{n,\underline{\alpha}} = (B^{-1}AB)\, M_{n,\underline{\beta}}\, (B^{-1}AB)^t$. □

Next, we prove a special case of Theorem 9.22 for the basis $\{\beta_1, \cdots, \beta_k\}$.

LEMMA 9.19. $\det M_{n,\underline{\beta}} = \pm 1$.

PROOF. As complex vector spaces, we have a linear isomorphism:

$$\mathbb{H}_{n,V} \cong \bigoplus_{\underline{n}} \bigotimes_{i=1}^{k} \mathbb{H}_{n_i,\mathbb{C}\beta_i} \tag{9.45}$$

where the sum is over $\underline{n} = (n_1, \cdots, n_k) \in (\mathbb{Z}_{\geq 0})^k$ such that $n_1 + \ldots + n_k = n$. Denote by M_{n_i,β_i} the matrix of pairings among the basis $\{\mathfrak{m}_{\lambda^i,\beta_i}|0\rangle \mid \lambda^i \vdash n_i\}$ for $\mathbb{H}_{n_i,\mathbb{C}\beta_i}$. For given $1 \leq i \leq k$ and λ^i, the operator $\mathfrak{m}_{\lambda^i,\beta_i}$ is a linear combination of the operators $\mathfrak{a}_{-\mu^i}(\beta_i)$, $\mu^i \vdash |\lambda^i|$. Since the classes $\beta_1, \cdots, \beta_k$ are orthogonal, we see from (3.9) that

$$\begin{aligned} &\big\langle \mathfrak{m}_{\lambda^1,\beta_1} \cdots \mathfrak{m}_{\lambda^k,\beta_k}|0\rangle, \mathfrak{m}_{\mu^1,\beta_1} \cdots \mathfrak{m}_{\mu^k,\beta_k}|0\rangle\big\rangle \\ = &\prod_{i=1}^{k} \big(\delta_{|\lambda^i|,|\mu^i|} \cdot \big\langle \mathfrak{m}_{\lambda^i,\beta_i}|0\rangle, \mathfrak{m}_{\mu^i,\beta_i}|0\rangle\big\rangle\big). \end{aligned}$$

This orthogonality together with (9.45) implies by standard linear algebra that

$$\det M_{n,\underline{\beta}} = \prod_{\underline{n}} \prod_{i=1}^{k} (\det M_{n_i,\beta_i})^{\prod_{1\le j\le k, j\ne i} \dim \mathbb{H}_{n_j,\mathbb{C}\beta_j}}.$$

Thus it suffices to prove our lemma for $k = 1$.

Put $\beta = \beta_1$ and $V = \mathbb{C}\beta$. By (9.43) and (9.44), $\mathbb{H}_{n,V}$ has two linear bases:

$$\begin{aligned} \mathfrak{m}_{n,\underline{\beta}} &= \{\mathfrak{m}_{\lambda,\beta}|0\rangle \mid \lambda \vdash n\}, \\ \mathfrak{a}_{n,\underline{\beta}} &= \{\mathfrak{a}_{-\lambda}(\beta)|0\rangle \mid \lambda \vdash n\}. \end{aligned}$$

Since $\langle \beta, \beta \rangle = \pm 1$, we obtain by Lemma 3.7 and (3.9) that

$$(9.46) \qquad \big\langle \mathfrak{a}_{-\lambda}(\beta)|0\rangle, \mathfrak{a}_{-\mu}(\beta)|0\rangle \big\rangle = \pm\delta_{\lambda,\mu} \cdot \mathfrak{z}_\lambda \cdot \langle \beta, \beta \rangle^{\ell(\lambda)} = \pm\delta_{\lambda,\mu} \cdot \mathfrak{z}_\lambda.$$

Recall that the transition matrix from the basis $\mathfrak{a}_{n,\underline{\beta}}$ to $\mathfrak{m}_{n,\underline{\beta}}$ is B^{-1}. So we conclude from Lemma 9.15 and (9.46) that

$$M_{n,\underline{\beta}} = B^{-1} \cdot \mathrm{diag}(\cdots, \pm\mathfrak{z}_\lambda, \cdots) \cdot (B^{-1})^t$$

where λ runs over all partitions of n. Hence we have

$$(9.47) \qquad \det M_{n,\underline{\beta}} = \pm(\det B)^{-2} \cdot \prod_{\lambda \vdash n} \mathfrak{z}_\lambda.$$

By (9.12), Lemma 9.9, and (1.5), we see that the transition matrix from the basis $\{p_\lambda \mid \lambda \vdash n\}$ of $\Lambda^n_{\mathbb{C}} \subset \Lambda_{\mathbb{C}}$ to the basis $\{m_\lambda \mid \lambda \vdash n\}$ is B^{-1} as well. Introduce the standard bilinear form $\langle \cdot, \cdot \rangle$ on $\Lambda^n_{\mathbb{C}}$ by letting

$$(9.48) \qquad \langle p_\lambda, p_\mu \rangle = \delta_{\lambda,\mu} \cdot \mathfrak{z}_\lambda.$$

The matrix M_n formed by the pairings of the monomial symmetric functions m_λ, $\lambda \vdash n$ is unimodular, and thus $\det M_n = \pm 1$. By Lemma 9.15 and (9.48),

$$M_n = B^{-1} \cdot \mathrm{diag}(\cdots, \mathfrak{z}_\lambda, \cdots) \cdot (B^{-1})^t$$

where λ runs over all partitions of n. It follows immediately that

$$(9.49) \qquad (\det B)^{-2} \cdot \prod_{\lambda \vdash n} \mathfrak{z}_\lambda = \det M_n = \pm 1.$$

Combining this with (9.47), we finally obtain $\det M_{n,\underline{\beta}} = \pm 1$. □

For $\mu \vdash n$, let $\mathbb{H}_{\mu,\underline{\alpha}} \subset \mathbb{H}_{n,V}$ be the span of $\mathfrak{a}_{-\mu^1}(\alpha_1) \cdots \mathfrak{a}_{-\mu^k}(\alpha_k)|0\rangle$ where $\mu^1, \ldots, \mu^k$ are partitions such that $\mu^1 \cup \cdots \cup \mu^k = \mu$. Note that

$$\mathbb{H}_{\mu,\underline{\alpha}} \subset \mathbb{H}_{n,V} \subset H^{2n}(X^{[n]}).$$

So $\mathbb{H}_{\mu,\underline{\alpha}}$ carries a pairing induced from the one on $H^{2n}(X^{[n]})$.

LEMMA 9.20. *For every partition μ, we have an isometry $\mathbb{H}_{\mu,\underline{\alpha}} \cong S^\mu(V)$.*

PROOF. Follows from the definitions, (9.41), and (3.9). □

LEMMA 9.21. *If $\det M_{\underline{\alpha}} = \pm 1$, then $(\det A)^2 = \pm 1$.*

PROOF. By (3.9), $\mathfrak{a}_{-\mu^1}(*)\cdots\mathfrak{a}_{-\mu^k}(*)$ and $\mathfrak{a}_{-\nu^1}(*)\cdots\mathfrak{a}_{-\nu^\ell}(*)|0\rangle$ are orthogonal unless $\mu^1\cup\cdots\cup\mu^k=\nu^1\cup\cdots\cup\nu^\ell$. Here, $*$'s denote unspecified classes in $H^2(X)$. Thus, we have an orthogonal direct sum

$$\mathbb{H}_{n,V}=\bigoplus_{\mu\vdash n}\mathbb{H}_{\mu,\underline{\alpha}}.$$

By Lemma 9.20, the intersection matrix $\widetilde{M}_{n,\underline{\alpha}}$ of the basis $\mathfrak{a}_{n,\underline{\alpha}}$ for $\mathbb{H}_{n,V}$ is given by the diagonal block matrix whose diagonal consists of $M_{\mu,\underline{\alpha}}$, $\mu\vdash n$. So

$$\det\widetilde{M}_{n,\underline{\alpha}}=\prod_{\mu\vdash n}\det M_{\mu,\underline{\alpha}}. \tag{9.50}$$

Similarly, by repeating the above with the α's replaced by the β's, we obtain

$$\det\widetilde{M}_{n,\underline{\beta}}=\prod_{\mu\vdash n}\det M_{\mu,\underline{\beta}}. \tag{9.51}$$

By assumption, $\det M_{\underline{\alpha}}=\pm1$ and $\det M_{\underline{\beta}}=\pm1$. Thus by Lemma 9.17, $\det M_{\mu,\underline{\alpha}}=\pm\det M_{\mu,\underline{\beta}}$ for every μ. Therefore, we see from (9.50) and (9.51) that

$$\det\widetilde{M}_{n,\underline{\alpha}}=\pm\det\widetilde{M}_{n,\underline{\beta}}. \tag{9.52}$$

By Lemma 9.15 and the definition of the matrix A, we have $\widetilde{M}_{n,\underline{\alpha}}=A\,\widetilde{M}_{n,\underline{\beta}}A^t$. Combining this with (9.52) yields $(\det A)^2=\pm1$. □

THEOREM 9.22. *Let $\alpha_1,\cdots,\alpha_k\in H^2(X)$ be linearly independent classes, and let $M_{\underline{\alpha}}$ be the intersection matrix of $\alpha_1,\cdots,\alpha_k$. Fix a positive integer n. Let $M_{n,\underline{\alpha}}$ be the intersection matrix of the classes in $H^{2n}(X^{[n]})$:*

$$\mathfrak{m}_{\lambda^1,\alpha_1}\cdots\mathfrak{m}_{\lambda^k,\alpha_k}|0\rangle,\quad |\lambda^1|+\ldots+|\lambda^k|=n.$$

If $\det M_{\underline{\alpha}}=\pm1$, then we have $\det M_{n,\underline{\alpha}}=\pm1$ as well.

PROOF. Follows immediately from Lemmas 9.18, 9.19, and 9.21. □

9.5. Integral bases for the cohomology of Hilbert schemes

The following theorem is the main result in this chapter.

THEOREM 9.23. *Let X be a smooth projective surface with vanishing odd Betti numbers. Let $\alpha_1,\cdots,\alpha_k$ be an integral basis for $H^2(X;\mathbb{Z})/\mathrm{Tor}$. Then the classes*

$$\frac{1}{\mathfrak{z}_\lambda}\mathfrak{a}_{-\lambda}(1_X)\mathfrak{a}_{-\mu}(x)\mathfrak{m}_{\nu^1,\alpha_1}\cdots\mathfrak{m}_{\nu^k,\alpha_k}|0\rangle,\quad |\lambda|+|\mu|+\sum_{i=1}^k|\nu^i|=n \tag{9.53}$$

are integral, and furthermore, they form an integral basis for $H^*(X^{[n]};\mathbb{Z})/\mathrm{Tor}$.

PROOF. By Lemma 9.4, Lemma 9.7 and Theorem 9.14, we have three types of integral operators:

$$\frac{1}{\mathfrak{z}_\lambda}\mathfrak{a}_{-\lambda}(1_X),\quad \mathfrak{a}_{-\mu}(x),\quad \mathfrak{m}_{\nu_i,\alpha_i}$$

where λ,μ,ν_i stand for partitions. Therefore, we obtain the integral classes (9.53). Since $H^1(X)=H^3(X)=0$, we see from (3.10) that the number of classes in (9.53) is equal to the rank of $H^*(X^{[n]};\mathbb{Z})/\mathrm{Tor}$. Now consider the intersection number of two classes from (9.53):

$$\frac{1}{\mathfrak{z}_\lambda}\mathfrak{a}_{-\lambda}(1_X)\mathfrak{a}_{-\mu}(x)\mathfrak{m}_{\nu^1,\alpha_1}\cdots\mathfrak{m}_{\nu^k,\alpha_k}|0\rangle,$$

$$\frac{1}{\mathfrak{z}_{\tilde\lambda}}\mathfrak{a}_{-\tilde\lambda}(1_X)\mathfrak{a}_{-\tilde\mu}(x)\mathfrak{m}_{\tilde\nu^1,\alpha_1}\cdots\mathfrak{m}_{\tilde\nu^k,\alpha_k}|0\rangle.$$

By the Heisenberg algebra commutation relation (3.9), the intersection number is

$$\pm\delta_{\lambda,\tilde\mu}\cdot\delta_{\tilde\lambda,\mu}\cdot\big\langle\mathfrak{m}_{\nu^1,\alpha_1}\cdots\mathfrak{m}_{\nu^k,\alpha_k}|0\rangle,\ \mathfrak{m}_{\tilde\nu^1,\alpha_1}\cdots\mathfrak{m}_{\tilde\nu^k,\alpha_k}|0\rangle\big\rangle.$$

By Theorem 9.22, the intersection matrix formed by the pairings of the integral classes in (9.53) is unimodular. Therefore, the classes in (9.53) form an integral basis of $H^*(X^{[n]};\mathbb{Z})/\mathrm{Tor}$. □

REMARK 9.24. Under the same assumptions as in Theorem 9.23, one can further prove that the integral ring $H^*(X^{[n]};\mathbb{Z})/\mathrm{Tor}$ is generated over $\mathbb{Z}$ by all the integral classes:

$$c_i(\mathcal{O}_X^{[n]}),\quad \mathbf{1}_{-(n-j)}\mathfrak{m}_{(1^j),\alpha_s}|0\rangle,\quad \mathbf{1}_{-(n-j)}\mathfrak{a}_{-j}(x)|0\rangle$$

where $1\le i\le n-1$, $1\le j\le n$, and $1\le s\le k$. We refer to [**LQ3**, Section 4] for the details.

REMARK 9.25. When X is a $K3$ surface, the ring generators for the integral cohomology ring of the Hilbert scheme $X^{[n]}$ are also obtained by Markman [**Mar2**].

9.6. Comparison of two integral bases of $H^*((\mathbb{P}^2)^{[n]};\mathbb{Z})$

In this section, as an application of Theorem 9.23, we will compare two integral bases of the integral cohomology of the Hilbert scheme $(\mathbb{P}^2)^{[n]}$ of points on the complex projective plane $\mathbb{P}^2$. One integral basis comes from the torus action on $(\mathbb{P}^2)^{[n]}$, and the other one comes from Theorem 9.23. We will prove that when the basis elements are arranged properly, the transition matrix is a triangular matrix.

First of all, we recall the integral basis of $H^*((\mathbb{P}^2)^{[n]};\mathbb{Z})$ from [**ES1**, Sections 1 and 2]. Let

$$T_0, T_1, T_2$$

be a system of homogeneous coordinates for $\mathbb{P}^2$. Let

$$H=\{T_2=0\}\subset\mathbb{P}^2$$

be a coordinator line. Let $G\subset \mathrm{SL}_3(\mathbb{C})$ be the maximal torus consisting of all diagonal matrices $g=\mathrm{diag}(g_0,g_1,g_2)$ with $g_0,g_1,g_2\in\mathbb{C}^*$ and $g_0g_1g_2=1$. The group G acts on $\mathbb{P}^2$ via

$$g(T_i)=g_iT_i,\qquad i=0,1,2.$$

On points $(a_0,a_1,a_2)\in\mathbb{P}^2$, this action is given by

$$g(a_0,a_1,a_2)=(g_0^{-1}a_0,\ g_1^{-1}a_1,\ g_2^{-1}a_2). \tag{9.54}$$

The fixed points are

$$P_0=(1,0,0),\qquad P_1=(0,1,0),\qquad P_2=(0,0,1).$$

Let

$$F_0=\{P_0\},\qquad F_1=H-\{P_0\},\qquad F_2=\mathbb{P}^2-H.$$

Then, F_0, F_1, F_2 define a cellular decomposition of $\mathbb{P}^2$. The one-parameter subgroups

$$\phi : \mathbb{C}^* \to G \tag{9.55}$$

inducing this cellular decomposition are those of the type

$$\phi(t) = \mathrm{diag}(t^{w_0}, t^{w_1}, t^{w_2}) \tag{9.56}$$

with $w_0 < w_1 < w_2$ and $w_0 + w_1 + w_2 = 0$.

The fixed points of the G-action on $(\mathbb{P}^2)^{[n]}$ are in one-to-one correspondence with ordered triples (λ, μ, ν) of partitions satisfying $|\lambda| + |\mu| + |\nu| = n$. More precisely, the fixed points are of the form

$$\xi_\lambda + \xi_\mu + \xi_\nu$$

where $\xi_\lambda \in (\mathbb{P}^2)^{[|\lambda|]}$, $\xi_\mu \in (\mathbb{P}^2)^{[|\mu|]}$, $\xi_\nu \in (\mathbb{P}^2)^{[|\nu|]}$ are fixed points supported at P_0, P_1, P_2 respectively. Define

$$\xi_{\lambda,\mu,\nu} = \xi_\lambda + \xi_\mu + \xi_\nu. \tag{9.57}$$

If u and v are the affine coordinates at P_0 and if $\lambda = (\lambda_1 \ge \lambda_2 \ge \cdots \ge \lambda_k)$, then the 0-dimensional closed subscheme ξ_λ is defined by the following ideal in $\mathbb{C}[u, v]$:

$$I_{\xi_\lambda} = \langle v^{\lambda_1}, uv^{\lambda_2}, \ldots, u^{k-1}v^{\lambda_k}, u^k \rangle. \tag{9.58}$$

Take a generic one-parameter subgroup ϕ in (9.55). Then, the fixed points of ϕ are the same as those of G, and $(\mathbb{P}^2)^{[n]}$ has a cellular decomposition with cells

$$C_{\lambda,\mu,\nu} = \left\{ \xi \in (\mathbb{P}^2)^{[n]} \mid \lim_{t\to 0} \phi(t)\xi = \xi_{\lambda,\mu,\nu} \right\} \tag{9.59}$$

where $|\lambda| + |\mu| + |\nu| = n$. It follows that the integral cohomology $H^*((\mathbb{P}^2)^{[n]}; \mathbb{Z})$ is a $\mathbb{Z}$-module freely generated by the fundamental classes of the closures:

$$[\overline{C}_{\lambda,\mu,\nu}], \qquad |\lambda| + |\mu| + |\nu| = n. \tag{9.60}$$

Moreover, it is known that

$$\dim(C_{0,\mu,0}) = |\mu|; \tag{9.61}$$

if $(w_2 - w_0) \gg (w_2 - w_1)$, then

$$\dim(C_{0,0,\nu}) = |\nu| + \ell(\nu); \tag{9.62}$$

if $(w_2 - w_0) \gg (w_1 - w_0)$, then

$$\dim(C_{\lambda,0,0}) = |\lambda| - \ell(\lambda). \tag{9.63}$$

Since the integral cohomology $H^*((\mathbb{P}^2)^{[n]}; \mathbb{Z})$ is torsion free, we conclude from Theorem 9.23 that another integral basis of $H^*((\mathbb{P}^2)^{[n]}; \mathbb{Z})$ consists of the classes

$$\frac{1}{\mathfrak{z}_\lambda} \mathfrak{a}_{-\lambda}(1)\mathfrak{m}_{\mu,H}\mathfrak{a}_{-\nu}(x)|0\rangle, \quad |\lambda| + |\mu| + |\nu| = n \tag{9.64}$$

where 1 denotes $1_{\mathbb{P}^2}$ and $x \in H^4(\mathbb{P}^2; \mathbb{Z})$ is the class corresponding to a point in $\mathbb{P}^2$.

We will study the transition matrix between the two integral bases (9.60) and (9.64) of $H^*((\mathbb{P}^2)^{[n]}; \mathbb{Z})$. To begin with, we will express the three special types of classes $[\overline{C}_{0,\mu,0}]$, $[\overline{C}_{0,0,\nu}]$ and $[\overline{C}_{\lambda,0,0}]$ in terms of the basis (9.64) . For the cohomology classes $[\overline{C}_{0,\mu,0}]$, we put

$$u = \frac{T_0}{T_1}, \qquad v = \frac{T_2}{T_1}.$$

Then u, v form a system of affine coordinates at P_1.

PROPOSITION 9.26. $[\overline{C}_{0,\mu,0}] = \mathfrak{m}_{\mu,H}|0\rangle$ *for every partition* μ.

PROOF. Let $n = |\mu|$. By the definition of $\mathfrak{m}_{\mu,H}$, we have $\mathfrak{m}_{\mu,H}|0\rangle = [L^\mu H]$. By (9.61), we have

$$\dim(C_{0,\mu,0}) = |\mu| = n.$$

Now both $L^\mu H$ and $\overline{C}_{0,\mu,0}$ are irreducible and have the same dimension. It suffices to show that generic elements $\xi \in L^\mu H$ are contained in the cell $C_{0,\mu,0}$, i.e.,

$$\lim_{t\to 0} \phi(t)\xi = \xi_{0,\mu,0}. \tag{9.65}$$

Assume that $\mu = (\mu_1 \geq \mu_2 \geq \ldots \geq \mu_k)$. The fixed point $\xi_{0,\mu,0}$, regarded as a point in $(\mathbb{C}^2)^{[n]}$, is defined by the following ideal in $\mathbb{C}[u,v]$:

$$I_{\xi_\mu} = \langle v^{\mu_1}, uv^{\mu_2}, \ldots, u^{k-1}v^{\mu_k}, u^k\rangle. \tag{9.66}$$

In view of (9.7), a generic element $\xi \in L^\mu H$ is defined by an ideal of the form

$$I = \prod_{i=1}^{k} \Big\langle (u - a_0^{(i)}) + b_1^{(i)} v + \ldots + b_{\mu_i - 1}^{(i)} v^{\mu_i - 1},\ v^{\mu_i} \Big\rangle$$

where $a_0^{(1)}, \ldots, a_0^{(k)}$ are distinct points in $\mathbb{C}$, and $b_1^{(i)}, \ldots, b_{\mu_i-1}^{(i)} \in \mathbb{C}$ for every i. For each j with $0 \leq j < k$, the ideal I contains the polynomials:

$$\left[\prod_{i=1}^{j}(u - a_0^{(i)}) \cdot \prod_{i=j+1}^{s-1} (u - a_0^{(i)}) \cdot \prod_{i=s+1}^{k} (u - a_0^{(i)}) + f_s(u,v)\, v\right] v^{\mu_s}$$

where $s = j+1, \ldots, k$, and $f_s(u,v)$ denotes polynomials of u and v. Since $\mu_{j+1} \geq \ldots \geq \mu_k$, we see that I contains the polynomials:

$$\left[\prod_{i=1}^{j}(u - a_0^{(i)}) \cdot \prod_{i=j+1}^{s-1} (u - a_0^{(i)}) \cdot \prod_{i=s+1}^{k} (u - a_0^{(i)}) + f_s(u,v)\, v\right] v^{\mu_{j+1}} \tag{9.67}$$

where $s = j+1, \ldots, k$. The $(k-j)$ degree-$(k-j-1)$ polynomials

$$\prod_{i=j+1}^{s-1} (u - a_0^{(i)}) \cdot \prod_{i=s+1}^{k} (u - a_0^{(i)}), \qquad s = j+1, \ldots, k$$

are linearly independent. Hence a certain linear combination of them is equal to 1. It follows from (9.67) that there exists a polynomial $f(u,v)$ of u and v such that

$$\left[\prod_{i=1}^{j}(u - a_0^{(i)}) + f(u,v)\, v\right] v^{\mu_{j+1}} \in I. \tag{9.68}$$

Note from (9.56) that $\phi(t)u = t^{w_0 - w_1}u$ with $w_1 - w_0 > 0$, and $\phi(t)v = t^{w_2 - w_1}v$ with $w_2 - w_1 > 0$. Choosing $(w_2 - w_1) \gg (w_1 - w_0)$, we see from (9.68) that

$$u^j v^{\mu_{j+1}} \in \lim_{t\to 0} \phi(t)I$$

for with $0 \leq j < k$. Similar, we have $u^k \in \lim_{t\to 0}\phi(t)I$. Since $\mu_1 + \ldots + \mu_k = |\mu| = n$ and the flat limit $\lim_{t\to 0}\phi(t)I$ is of colength-n in $\mathbb{C}[u,v]$, we must have

$$\lim_{t\to 0}\phi(t)I = \langle v^{\mu_1}, uv^{\mu_2}, \ldots, u^{k-1}v^{\mu_k}, u^k\rangle.$$

In view of (9.66), we obtain $\lim_{t\to 0}\phi(t)I = I_{\xi_\mu}$. This proves (9.65). □

Next, to handle the classes $[\overline{C}_{0,0,\nu}]$, we let

$$u = \frac{T_1}{T_2}, \qquad v = \frac{T_0}{T_2}$$

(note that the proof of the Proposition 4.1 in [**ES1**] took $u = T_0/T_2$ and $v = T_1/T_2$). Then u and v form a system of affine coordinates at P_2. Let $(w_2 - w_0) \gg (w_2 - w_1)$. The following lemma determines generic elements in the cell $C_{0,0,\nu}$.

LEMMA 9.27. *Let $\nu = (\nu_1 \ge \ldots \ge \nu_k) \vdash n$. Then generic elements in $C_{0,0,\nu}$ are:*

$$\sum_{i=1}^{\nu_1} (a_1, b_{1,i}) + \ldots + \sum_{i=1}^{\nu_k} (a_k, b_{k,i}) \in (\mathbb{C}^2)^{[n]} \subset (\mathbb{P}^2)^{[n]} \tag{9.69}$$

where $a_1, b_{1,1} \ldots, b_{1,\nu_1}, \ldots, a_k, b_{k,1} \ldots, b_{k,\nu_k} \in \mathbb{C}$ are distinct.

PROOF. By (9.62),

$$\dim(C_{0,0,\nu}) = |\nu| + \ell(\nu) = n + k.$$

So it suffices to show that every element (9.69) is contained in the cell $C_{0,0,\nu}$, i.e.,

$$\lim_{t \to 0} \phi(t) I = \langle v^{\nu_1}, uv^{\nu_2}, \ldots, u^{k-1}v^{\nu_k}, u^k \rangle \tag{9.70}$$

where I denotes the ideal in $\mathbb{C}[u, v]$ defining the element (9.69). Note that

$$\begin{aligned} & I \\ = & \prod_{i=1}^{\nu_1} \langle u - a_1, v - b_{1,i} \rangle \cdots \prod_{i=1}^{\nu_k} \langle u - a_k, v - b_{k,i} \rangle \\ = & \big\langle u - a_1, (v - b_{1,1}) \cdots (v - b_{1,\nu_1}) \big\rangle \cdots \big\langle u - a_k, (v - b_{k,1}) \cdots (v - b_{k,\nu_k}) \big\rangle. \end{aligned}$$

Since $(u - a_1) \cdots (u - a_k) \in I$, we get $u^k \in \lim_{t \to 0} \phi(t) I$. Next, fix j with $0 \le j \le (k-1)$. Then the ideal I contains the $(k-j)$ polynomials:

$$\prod_{i=1}^{j} (u - a_i) \cdot \prod_{i=1}^{\nu_s} (v - b_{s,i}) \cdot \prod_{j+1 \le i \le k,\, i \ne s} (u - a_i)$$

where $j + 1 \le s \le k$. Since $\nu_{j+1} \ge \cdots \ge \nu_k$, we see that there exist polynomials $f_s(v)$, $j + 1 \le s \le k$, in v of degree less than ν_{j+1} such that

$$\prod_{i=1}^{j} (u - a_i) \cdot (v^{\nu_{j+1}} + f_s(v)) \cdot \prod_{j+1 \le i \le k,\, i \ne s} (u - a_i) \in I \tag{9.71}$$

where $j + 1 \le s \le k$. Since $a_{j+1}, \ldots, a_k \in \mathbb{C}$ are distinct, a certain linear combination of the $(k-j)$ degree-$(k-j-1)$ polynomials

$$\prod_{j+1 \le i \le k,\, i \ne s} (u - a_i), \qquad j + 1 \le s \le k$$

is equal to 1. It follows from (9.71) that there exists a polynomial $f(u, v)$ with

$$\prod_{i=1}^{j} (u - a_i) \cdot (v^{\nu_{j+1}} + f(u, v)) \in I \tag{9.72}$$

and with the degree of $f(u,v)$ in v being less than ν_{j+1}. Since $(w_2-w_0) \gg (w_2-w_1)$, we conclude that $u^j v^{\nu_{j+1}} \in \lim_{t\to 0}\phi(t)I$ for $0 \le j \le (k-1)$. In summary,

$$\langle v^{\nu_1}, uv^{\nu_2}, \ldots, u^{k-1}v^{\nu_k}, u^k\rangle \subset \lim_{t\to 0}\phi(t)I.$$

Since both the ideals $\langle v^{\nu_1}, uv^{\nu_2}, \ldots, u^{k-1}v^{\nu_k}, u^k\rangle$ and $\lim_{t\to 0}\phi(t)I$ have co-length n in the polynomial ring $\mathbb{C}[u,v]$, we see that (9.70) holds. □

The next lemma studies the intersection between the integral classes $\mathfrak{a}_{-\lambda}(x)|0\rangle$ and $[\overline{C}_{0,0,\nu}]$ where $\lambda, \nu \vdash n$.

LEMMA 9.28. (i) *Let $\lambda, \nu \vdash n$. If the intersection pairing between the cohomology classes $\mathfrak{a}_{-\lambda}(x)|0\rangle$ and $[\overline{C}_{0,0,\nu}]$ is nonzero, then $\lambda = \nu$;*

(ii) *The intersection pairing between $\mathfrak{a}_{-\nu}(x)|0\rangle$ and $[\overline{C}_{0,0,\nu}]$ is positive.*

PROOF. (i) Let $\lambda = (\lambda_1 \ge \ldots \ge \lambda_\ell)$ and $\nu = (\nu_1 \ge \ldots \ge \nu_k)$. Let $a_1, b_1, \ldots, a_\ell, b_\ell \in \mathbb{C}$ be different numbers. Put

$$x_1 = (a_1, b_1), \ldots, x_\ell = (a_\ell, b_\ell) \in \mathbb{C}^2.$$

Then, the points $x_1, \ldots, x_\ell$ are distinct. Now $\mathfrak{a}_{-\lambda}(x)|0\rangle$ is represented by the subvariety

$$M_{\lambda_1}(x_1) \times \cdots \times M_{\lambda_\ell}(x_\ell)$$

whose dimension is equal to $(n-\ell)$. Since $\dim(C_{0,0,\nu}) = n+k$, we must have $\ell = k$. Since the pairing between $\mathfrak{a}_{-\lambda}(x)|0\rangle$ and $[\overline{C}_{0,0,\nu}]$ is nonzero, the intersection between $M_{\lambda_1}(x_1) \times \cdots \times M_{\lambda_\ell}(x_\ell)$ and $\overline{C}_{0,0,\nu}$ is nonempty. Let ξ be a point in the intersection. Then in view of (9.69), $\xi = \xi_1 + \ldots + \xi_k$ where for each i with $1 \le i \le k$,

$$\xi_i \in M_i(x_i) \quad \text{and} \quad \lambda_i = \ell(\xi_i) = \nu_{j_i}$$

for some j_i such that $\{j_1, \ldots, j_k\} = \{1, \ldots, k\}$. Therefore, $\lambda = \nu$.

(ii) Let notations be the same as in the previous paragraph. Then the only point in the intersection between $M_{\nu_1}(x_1) \times \cdots \times M_{\nu_k}(x_k)$ and $\overline{C}_{0,0,\nu}$ is

$$\xi = \xi_1 + \ldots + \xi_k$$

where each element $\xi_i \in M_{\nu_i}(x_i)$ is defined by the ideal $\langle u-a_i, (v-b_i)^{\nu_i}\rangle$ in the polynomial ring $\mathbb{C}[u,v]$. In particular, $\langle \mathfrak{a}_{-\nu}(x)|0\rangle, [\overline{C}_{0,0,\nu}]\rangle > 0$. □

DEFINITION 9.29. For $n \ge 1$, define $\mathcal{I}^{[n]}$ to be the subset of $H^*((\mathbb{P}^2)^{[n]};\mathbb{Z})$ consisting of the integral linear spans of the basis classes:

$$\frac{1}{\mathfrak{z}_\lambda}\mathfrak{a}_{-\lambda}(1)\mathfrak{m}_{\mu,H}\mathfrak{a}_{-\nu}(x)|0\rangle$$

where $|\lambda| + |\mu| + |\nu| = n$, and either $|\mu| > 0$ or $|\nu| > 0$.

As in Lemma 8.6 (i), $\mathcal{I}^{[n]}$ is an ideal of the cohomology ring $H^*((\mathbb{P}^2)^{[n]};\mathbb{Z})$. In fact,

$$\mathcal{I}^{[n]} = (\mathcal{I}_0)^{[n]} \cap H^*((\mathbb{P}^2)^{[n]};\mathbb{Z})$$

where $\mathcal{I}_0 = H^2(\mathbb{P}^2) \oplus H^4(\mathbb{P}^2)$ and $(\mathcal{I}_0)^{[n]}$ is from Definition 8.5. Lemma 9.28 enables us to determine $[\overline{C}_{0,0,\nu}]$ modulo the ideal $\mathcal{I}^{[n]}$.

PROPOSITION 9.30. *For every partition ν, we have*

$$[\overline{C}_{0,0,\nu}] \equiv (-1)^{|\nu|-\ell(\nu)} \frac{\tilde{f}_\nu}{\mathfrak{z}_\nu} \mathfrak{a}_{-\nu}(1)|0\rangle \pmod{\mathcal{I}^{[|\nu|]}} \tag{9.73}$$

where $\tilde{f}_\nu$ is a positive integer.

PROOF. Let $n = |\nu|$. By (9.64), $[\overline{C}_{0,0,\nu}]$ is an integral linear combination of the classes:

$$\frac{1}{\mathfrak{z}_\lambda} \mathfrak{a}_{-\lambda}(1)\mathfrak{m}_{\mu,H}\mathfrak{a}_{-\tilde{\nu}}(x)|0\rangle, \quad |\lambda| + |\mu| + |\tilde{\nu}| = n.$$

By Lemma 9.28 (i), the coefficient of $1/\mathfrak{z}_\lambda\, \mathfrak{a}_{-\lambda}(1)|0\rangle$ is 0 if $\lambda \neq \nu$. In view of Lemma 9.28 (ii), the coefficient of $1/\mathfrak{z}_\nu\, \mathfrak{a}_{-\nu}(1)|0\rangle$ is $(-1)^{|\nu|-\ell(\nu)}\tilde{f}_\nu$ for some positive integer $\tilde{f}_\nu$. □

Next, we determine the class $[\overline{C}_{0,0,\nu}]$ when $\nu = (1^n)$ or $(1^{n-2}2)$.

LEMMA 9.31. (i) *If $\nu = (1^n)$, then*

$$[\overline{C}_{0,0,\nu}] = \frac{1}{\mathfrak{z}_\nu} \cdot \mathfrak{a}_{-\nu}(1)|0\rangle = 1_{(\mathbb{P}^2)^{[n]}};$$

(ii) *If $\nu = (1^{n-2}2)$ where $n \geq 2$, then we have*

$$[\overline{C}_{0,0,\nu}] = (n-1) \cdot \frac{1}{\mathfrak{z}_{(1^{n-1})}} \mathfrak{a}_{-(1^{n-1})}(1)\, \mathfrak{m}_{(1),H}|0\rangle - \frac{1}{\mathfrak{z}_\nu} \mathfrak{a}_{-\nu}(1)|0\rangle. \tag{9.74}$$

PROOF. (i) Recall that $1_{(\mathbb{P}^2)^{[n]}} = 1/n! \cdot \mathfrak{a}_{-1}(1)^n|0\rangle$. By (9.62), the dimension of the cell $C_{0,0,\nu}$ is $2n$. So $\overline{C}_{0,0,\nu} = (\mathbb{P}^2)^{[n]}$ and $[\overline{C}_{0,0,\nu}] = 1_{(\mathbb{P}^2)^{[n]}}$.

(ii) A basis of $H^2((\mathbb{P}^2)^{[n]};\mathbb{Z})$ consists of the two classes:

$$\frac{1}{\mathfrak{z}_{(1^{n-1})}} \mathfrak{a}_{-(1^{n-1})}(1)\, \mathfrak{m}_{(1),H}|0\rangle, \quad \frac{1}{\mathfrak{z}_\nu} \mathfrak{a}_{-\nu}(1)|0\rangle.$$

So the divisor $[\overline{C}_{0,0,\nu}]$ is an integral linear combination of these two classes. Using the proof of Proposition 9.30 (ii), we see that $\tilde{f}_\nu = 1$. So it remains to show that

$$\big\langle [\overline{C}_{0,0,\nu}], \mathfrak{a}_{-1}(H)\mathfrak{a}_{-1}(x)^{n-1}|0\rangle \big\rangle = (n-1). \tag{9.75}$$

Fix a line H' in $\mathbb{P}^2$ such that H', viewed as a line in $\mathbb{C}^2$, is not parallel to the v-axis. Choose distinct numbers $a_1, b_1, \ldots, a_{n-1}, b_{n-1} \in \mathbb{C}$ such that none of the $(n-1)$ points $x_i = (a_i, b_i)$, $1 \leq i \leq (n-1)$ in $\mathbb{C}^2$ is contained in the line H'. Then the cohomology class $\mathfrak{a}_{-1}(H)\mathfrak{a}_{-1}(x)^{n-1}|0\rangle$ is represented by the curve:

$$H' + x_1 + \ldots + x_{n-1} \subset (\mathbb{P}^2)^{[n]}.$$

For $1 \leq i \leq (n-1)$, let H_i be the line in $\mathbb{P}^2$ defined by $u = a_i$ when H_i is viewed as a line in $\mathbb{C}^2$. Using the description (9.69) of generic elements in $\overline{C}_{0,0,\nu}$, we see that the intersection $\overline{C}_{0,0,\nu} \cap (H' + x_1 + \ldots + x_{n-1})$ consists of $(n-1)$ elements:

$$(H' \cap H_i) + x_1 + \ldots + x_{n-1}, \qquad 1 \leq i \leq (n-1).$$

Moreover, the intersection is transversal at these points. Therefore, the pairing between the classes $[\overline{C}_{0,0,\nu}]$ and $\mathfrak{a}_{-1}(H)\mathfrak{a}_{-1}(x)^{n-1}|0\rangle$ is equal to $(n-1)$. □

To study the classes $[\overline{C}_{\lambda,0,0}]$, a system of affine coordinates at P_0 is chosen to be

$$u = \frac{T_1}{T_0}, \qquad v = \frac{T_2}{T_0}.$$

Let $(w_2 - w_0) \gg (w_1 - w_0)$. Since the closure $\overline{C}_{\lambda,0,0}$ is supported at one point, we conclude that $[\overline{C}_{\lambda,0,0}]$ is a linear combination of the integral classes

$$\mathfrak{a}_{-\nu}(x)|0\rangle$$

where ν denotes partitions with $\nu \vdash n$ and $\ell(\nu) = \ell(\lambda)$. Moreover, in view of (9.64), all the coefficients in the linear combination are integers. Hence

$$[\overline{C}_{\lambda,0,0}] = \sum_{|\lambda|=|\nu|,\, \ell(\lambda)=\ell(\nu)} e_\nu^\lambda \, \mathfrak{a}_{-\nu}(x)|0\rangle \tag{9.76}$$

where $e_\nu^\lambda \in \mathbb{Z}$. Recall from Definition 1.1 the total ordering $\prec$ on the set of all partitions.

PROPOSITION 9.32. *For every partition λ, there exist integers $e_\nu^\lambda \in \mathbb{Z}$ for partitions ν with $|\lambda| = |\nu|, \ell(\lambda) = \ell(\nu), \lambda \prec \nu$ such that*

$$[\overline{C}_{\lambda,0,0}] = e_\lambda^\lambda \, \mathfrak{a}_{-\lambda}(x)|0\rangle + \sum_{\substack{|\lambda|=|\nu|,\, \ell(\lambda)=\ell(\nu) \\ \lambda \prec \nu}} e_\nu^\lambda \, \mathfrak{a}_{-\nu}(x)|0\rangle. \tag{9.77}$$

Moreover, e_λ^λ is a positive integer.

The proof of Proposition 9.32 is straightforward but extremely lengthy, and follows from similar arguments as in the proofs of Lemma 9.27 and Proposition 9.30. We refer to [**Mur**, Subsection 3.3] for details.

Now we compare the two integral bases (9.60) and (9.64) of $H^*((\mathbb{P}^2)^{[n]};\mathbb{Z})$. Let $n = |\lambda| + |\mu| + |\nu|$. Then there exist $e^{\lambda,\mu,\nu}_{\lambda',\mu',\nu'} \in \mathbb{Z}$ such that

$$[\overline{C}_{\lambda,\mu,\nu}] = \sum_{\lambda',\mu',\nu'} e^{\lambda,\mu,\nu}_{\lambda',\mu',\nu'} \frac{1}{\mathfrak{z}_{\nu'}} \mathfrak{a}_{-\nu'}(1)\mathfrak{m}_{\mu',H}\mathfrak{a}_{-\lambda'}(x)|0\rangle \tag{9.78}$$

where $|\lambda'| + |\mu'| + |\nu'| = n$. Assume that

$$e^{\lambda,\mu,\nu}_{\lambda',\mu',\nu'} \neq 0 \tag{9.79}$$

for some λ', μ', ν'.

The cell $C_{\lambda,\mu,\nu} \subset (\mathbb{P}^2)^{[n]}$ consists of elements of the form

$$\xi_1 + \xi_2 + \xi_3 \tag{9.80}$$

with $\xi_1 \in C_{\lambda,0,0}$, $\xi_2 \in C_{0,\mu,0}$ and $\xi_3 \in C_{0,0,\nu}$. In particular, $\mathrm{Supp}(\xi_1) = F_0 = \{P_0\}$, $\mathrm{Supp}(\xi_2) \subset F_1 = H - \{P_0\}$, and $\mathrm{Supp}(\xi_2) \subset F_2 = \mathbb{P}^2 - H$.

Choose a partition $\tilde{\mu}$ with $|\tilde{\mu}| = |\mu'|$. Let $\tilde{H} \subset \mathbb{P}^2$ be a line different from H. Fix distinct points $x_1, \ldots, x_{\ell(\nu')} \in \mathbb{P}^2$ not lying on the lines H and $\tilde{H}$. Let

$$Z_{\lambda',\tilde{\mu},\nu'} \subset (\mathbb{P}^2)^{[n]}$$

be the subset consisting of the elements of the form

$$\sum_{i=1}^{\ell(\lambda')} \eta_{1,i} + \eta_2 + \sum_{i=1}^{\ell(\nu')} \eta_{3,i} \tag{9.81}$$

where $\eta_{3,i} \in M_{\nu'_i}(x_i)$, $\eta_2 \in L^{\tilde{\mu}}\tilde{H}$, and $\eta_{1,i} \in M_{\lambda'_i}(y_i)$ such that $y_1, \ldots, y_{\ell(\lambda')} \in \mathbb{P}^2$ are distinct and not in the subset $\{x_1, \ldots, x_{\ell(\nu')}\} \cup \tilde{H}$. Let $\overline{Z}_{\lambda',\tilde{\mu},\nu'}$ be the closure of $Z_{\lambda',\tilde{\mu},\nu'}$. Let $\lambda' = (1^{m_1(\lambda')}2^{m_2(\lambda')}\cdots)$. As in Proposition 3.16, we have

$$[\overline{Z}_{\lambda',\tilde{\mu},\nu'}] = \prod_{i=1}^{\ell(\lambda')} i^{m_i(\lambda')} \cdot \frac{1}{\mathfrak{z}_{\lambda'}} \mathfrak{a}_{-\lambda'}(1)\mathfrak{m}_{\tilde{\mu},\tilde{H}}\mathfrak{a}_{-\nu'}(x)|0\rangle. \tag{9.82}$$

By (9.78) and (9.82), the pairing $\langle [\overline{C}_{\lambda,\mu,\nu}], [\overline{Z}_{\lambda',\tilde{\mu},\nu'}]\rangle$ is equal to

$$\prod_{i=1}^{\ell(\lambda')} i^{m_i(\lambda')} \cdot (-1)^{(|\lambda'|-\ell(\lambda'))+(|\nu'|-\ell(\nu'))} \cdot \left\langle \sum_{\mu'} e^{\lambda,\mu,\nu}_{\lambda',\mu',\nu'} \mathfrak{m}_{\mu',H}|0\rangle, \mathfrak{m}_{\tilde{\mu},\tilde{H}}|0\rangle \right\rangle. \tag{9.83}$$

The next lemma determines the leading coefficient in the expression (9.78). The main idea in its proof is to exploit the non-emptiness of the intersecton $\overline{C}_{\lambda,\mu,\nu} \cap \overline{Z}_{\lambda',\tilde{\mu},\nu'}$ for a suitable $\tilde{\mu}$, whenever $e^{\lambda,\mu,\nu}_{\lambda',\mu',\nu'} \neq 0$.

LEMMA 9.33. *Let λ, μ, ν be three partitions. Then, $[\overline{C}_{\lambda,\mu,\nu}]$ is equal to*

$$\frac{(-1)^{|\nu|-\ell(\nu)}\, e^{\lambda}_{\lambda}\tilde{f}_{\nu}}{\mathfrak{z}_{\nu}} \mathfrak{a}_{-\nu}(1)\mathfrak{m}_{\mu,H}\mathfrak{a}_{-\lambda}(x)|0\rangle + \sum_{\lambda',\mu',\nu'} \frac{e^{\lambda,\mu,\nu}_{\lambda',\mu',\nu'}}{\mathfrak{z}_{\nu'}} \mathfrak{a}_{-\nu'}(1)\mathfrak{m}_{\mu',H}\mathfrak{a}_{-\lambda'}(x)|0\rangle$$

where $e^{\lambda,\mu,\nu}_{\lambda',\mu',\nu'} \in \mathbb{Z}$, $|\lambda'|+|\mu'|+|\nu'| = |\lambda|+|\mu|+|\nu|$, and either $|\nu'|<|\nu|$, or $\nu'=\nu$ and $|\mu'|<|\mu|$, or $\nu'=\nu$, $\mu'=\mu$, $\ell(\lambda')=\ell(\lambda)$ and $\lambda \prec \lambda'$.

PROOF. Recall the expression (9.78) and other notations introduced in the previous paragraphs. As in (9.79), assume $e^{\lambda,\mu,\nu}_{\lambda',\mu',\nu'} \neq 0$ for some λ', μ', ν'. By Theorem 9.22, there exists a partition $\tilde{\mu}$ such that $|\tilde{\mu}| = |\mu'|$ and

$$\left\langle \sum_{\mu'} e^{\lambda,\mu,\nu}_{\lambda',\mu',\nu'} \mathfrak{m}_{\mu',H}|0\rangle, \mathfrak{m}_{\tilde{\mu},\tilde{H}}|0\rangle \right\rangle \neq 0. \tag{9.84}$$

By (9.83), $\langle [\overline{C}_{\lambda,\mu,\nu}], [\overline{Z}_{\lambda',\tilde{\mu},\nu'}]\rangle \neq 0$. So $\overline{C}_{\lambda,\mu,\nu} \cap \overline{Z}_{\lambda',\tilde{\mu},\nu'} \neq \emptyset$. Let

$$\xi \in \overline{C}_{\lambda,\mu,\nu} \cap \overline{Z}_{\lambda',\tilde{\mu},\nu'}.$$

Since $\{x_1, \ldots, x_{\ell(\nu')}\} \cap H = \emptyset$, we see from (9.80) and (9.81) that

$$\xi = \xi' + \xi'_3$$

where $\mathrm{Supp}(\xi') \subset H$, $\mathrm{Supp}(\xi'_3) \cap H = \emptyset$, $\ell(\xi') \geq |\lambda|+|\mu|$, and $|\nu| \geq \ell(\xi'_3) \geq |\nu'|$. If $|\nu| > |\nu'|$, we are done. In the following, we assume $|\nu| = |\nu'|$.

Since $|\nu| = |\nu'|$, we obtain $\ell(\xi'_3) = |\nu| = |\nu'|$, and $\ell(\xi') = |\lambda|+|\mu|$. Moreover,

$$\xi'_3 \in \overline{C}_{0,0,\nu} \cap \prod_{i=1}^{\ell(\nu')} M_{\nu'_i}(x_i).$$

From the proof of Lemma 9.28 (i), we see that $\nu = \nu'$. Also, by the proof of Lemma 9.28 (ii), $\overline{C}_{0,0,\nu}$ and $\prod_{i=1}^{\ell(\nu)} M_{\nu_i}(x_i)$ intersect at a unique point which will be denoted by ξ_3' for simplicity. Now, $\xi \in \overline{C}_{\lambda,\mu,\nu} \cap \overline{Z}_{\lambda',\tilde{\mu},\nu}$ if and only if

$$\xi = \xi' + \xi_3'$$

with $\xi' \in \overline{C}_{\lambda,\mu,0} \cap \overline{Z}_{\lambda',\tilde{\mu},0}$. This decomposition allows us to split $(\mathbb{P}^2)^{[n]}$ locally into the product $(\mathbb{P}^2)^{[|\lambda|+|\mu|]} \times (\mathbb{P}^2)^{[|\nu|]}$, i.e., we obtain

$$\big\langle [\overline{C}_{\lambda,\mu,\nu}], [\overline{Z}_{\lambda',\tilde{\mu},\nu}] \big\rangle = \big\langle [\overline{C}_{\lambda,\mu,0}], [\overline{Z}_{\lambda',\tilde{\mu},0}] \big\rangle \cdot \big\langle [\overline{C}_{0,0,\nu}], \mathfrak{a}_{-\nu}(x)|0\rangle \big\rangle. \tag{9.85}$$

By Proposition 9.30, $\big\langle [\overline{C}_{0,0,\nu}], \mathfrak{a}_{-\nu}(x)|0\rangle \big\rangle = \tilde{f}_\nu$. Therefore,

$$\big\langle [\overline{C}_{\lambda,\mu,\nu}], [\overline{Z}_{\lambda',\tilde{\mu},\nu}] \big\rangle = \tilde{f}_\nu \, \big\langle [\overline{C}_{\lambda,\mu,0}], [\overline{Z}_{\lambda',\tilde{\mu},0}] \big\rangle. \tag{9.86}$$

Note that (9.86) is independent of the nonvanishing condition (9.84), i.e., it holds as long as $|\tilde{\mu}| = |\mu'|$. Combining (9.86) and (9.83), we see that for every partition $\tilde{\mu}$ with $|\tilde{\mu}| = |\mu'|$, the number $\tilde{f}_\nu \, \big\langle [\overline{C}_{\lambda,\mu,0}], [\overline{Z}_{\lambda',\tilde{\mu},0}] \big\rangle$ is equal to

$$\prod_{i=1}^{\ell(\lambda')} i^{m_i(\lambda')} \cdot (-1)^{(|\lambda'|-\ell(\lambda'))+(|\nu|-\ell(\nu))} \cdot \Big\langle \sum_{\mu'} e^{\lambda,\mu,\nu}_{\lambda',\mu',\nu'} \mathfrak{m}_{\mu',H}|0\rangle, \mathfrak{m}_{\tilde{\mu},\tilde{H}}|0\rangle \Big\rangle. \tag{9.87}$$

In the rest of the proof, we work with $\big\langle [\overline{C}_{\lambda,\mu,0}], [\overline{Z}_{\lambda',\tilde{\mu},0}] \big\rangle$, i.e., both ν and ν' are empty partitions. Now (9.80) and (9.81) are simplified to:

$$\xi_1 + \xi_2, \qquad \sum_{i=1}^{\ell(\lambda')} \eta_{1,i} + \eta_2$$

where $\xi_1 \in C_{\lambda,0,0}$, $\xi_2 \in C_{0,\mu,0}$, $\eta_2 \in L^{\tilde{\mu}}\tilde{H}$, and $\eta_{1,i} \in M_{\lambda_i'}(y_i)$ such that the points $y_1, \ldots, y_{\ell(\lambda')} \in \mathbb{P}^2$ are distinct and not on the line $\tilde{H}$. Assuming (9.84), we see from (9.87) that

$$\big\langle [\overline{C}_{\lambda,\mu,0}], [\overline{Z}_{\lambda',\tilde{\mu},0}] \big\rangle \neq 0.$$

So $\overline{C}_{\lambda,\mu,0} \cap \overline{Z}_{\lambda',\tilde{\mu},0} \neq \emptyset$. Since $P_0 \notin \tilde{H}$, $\xi \in \overline{C}_{\lambda,\mu,0} \cap \overline{Z}_{\lambda',\tilde{\mu},0}$ is of the form:

$$\xi = \xi_1' + \xi_2'$$

where $\mathrm{Supp}(\xi_1') = \{P_0\}$, $P_0 \notin \mathrm{Supp}(\xi_2')$, $\ell(\xi_1') \geq |\lambda|$, and $|\mu| \geq \ell(\xi_2') \geq |\tilde{\mu}|$. Since $|\tilde{\mu}| = |\mu'|$, we get $|\mu| \geq |\mu'|$. If $|\mu| > |\mu'|$, we are done.

Assume that $|\mu| = |\mu'|$. Then, we have $|\mu| = |\tilde{\mu}| = \ell(\xi_2')$ and $|\lambda| = |\lambda'| = \ell(\xi_1')$. Moreover, we see that $\xi \in \overline{C}_{\lambda,\mu,0} \cap \overline{Z}_{\lambda',\tilde{\mu},0}$ if and only if

$$\xi = \xi_1' + \xi_2'$$

where $\xi_1' \in \overline{C}_{\lambda,0,0} \cap \overline{Z}_{\lambda',0,0}$ and $\xi_2' \in \overline{C}_{0,\mu,0} \cap L^{\tilde{\mu}}\tilde{H}$. As in (9.85), we obtain

$$\big\langle [\overline{C}_{\lambda,\mu,0}], [\overline{Z}_{\lambda',\tilde{\mu},0}] \big\rangle = \big\langle [\overline{C}_{\lambda,0,0}], [\overline{Z}_{\lambda',0,0}] \big\rangle \cdot \big\langle [\overline{C}_{0,\mu,0}], [L^{\tilde{\mu}}\tilde{H}] \big\rangle \tag{9.88}$$

for every $\tilde{\mu}$ with $|\tilde{\mu}| = |\mu|$. Note that $[\overline{C}_{0,\mu,0}] = \mathfrak{m}_{\mu,H}|0\rangle$ by Proposition 9.26, and $[L^{\tilde{\mu}}\tilde{H}] = \mathfrak{m}_{\tilde{\mu},\tilde{H}}|0\rangle = \mathfrak{m}_{\tilde{\mu},H}|0\rangle$. Combining with (9.87) and (9.88) yields

$$\begin{aligned} &\prod_{i=1}^{\ell(\lambda')} i^{m_i(\lambda')} \cdot (-1)^{(|\lambda'|-\ell(\lambda'))+(|\nu|-\ell(\nu))} \cdot \left\langle \sum_{\mu'} e^{\lambda,\mu,\nu}_{\lambda',\mu',\nu'} \mathfrak{m}_{\mu',H}|0\rangle, \mathfrak{m}_{\tilde{\mu},\tilde{H}}|0\rangle \right\rangle \\ = \;& \tilde{f}_\nu \left\langle [\overline{C}_{\lambda,0,0}], [\overline{Z}_{\lambda',0,0}] \right\rangle \cdot \left\langle \mathfrak{m}_{\mu,H}|0\rangle, \mathfrak{m}_{\tilde{\mu},H}|0\rangle \right\rangle. \end{aligned}$$

In view of (9.82), we conclude for all the partitions $\tilde{\mu}$ with $|\tilde{\mu}| = |\mu|$ that

$$\begin{aligned} (9.89) \qquad &(-1)^{(|\lambda'|-\ell(\lambda'))+(|\nu|-\ell(\nu))} \cdot \left\langle \sum_{\mu'} e^{\lambda,\mu,\nu}_{\lambda',\mu',\nu'} \mathfrak{m}_{\mu',H}|0\rangle, \mathfrak{m}_{\tilde{\mu},\tilde{H}}|0\rangle \right\rangle \\ = \;& \tilde{f}_\nu \left\langle [\overline{C}_{\lambda,0,0}], \frac{1}{\mathfrak{z}_{\lambda'}} \mathfrak{a}_{-\lambda'}(1)|0\rangle \right\rangle \cdot \left\langle \mathfrak{m}_{\mu,H}|0\rangle, \mathfrak{m}_{\tilde{\mu},H}|0\rangle \right\rangle. \end{aligned}$$

Let $\mathbb{L} \subset H^{2n}((\mathbb{P}^2)^{[n]};\mathbb{Z})$ be the $\mathbb{Z}$-span of all the classes $\mathfrak{m}_{\tilde{\mu},H}|0\rangle, \tilde{\mu} \vdash |\mu|$. Since $\langle H, H\rangle = 1$, we see from Theorem 9.22 that $\mathbb{L}$, together with the bilinear form $\langle \cdot,\cdot\rangle$, is an integral lattice, and the classes $\mathfrak{m}_{\tilde{\mu},H}|0\rangle, \tilde{\mu} \vdash |\mu|$ form a basis of $\mathbb{L}$. It follows from (9.89) that

$$\begin{aligned} &(-1)^{(|\lambda'|-\ell(\lambda'))+(|\nu|-\ell(\nu))} \cdot \sum_{\mu'} e^{\lambda,\mu,\nu}_{\lambda',\mu',\nu'} \mathfrak{m}_{\mu',H}|0\rangle \\ = \;& \tilde{f}_\nu \left\langle [\overline{C}_{\lambda,0,0}], \frac{1}{\mathfrak{z}_{\lambda'}} \mathfrak{a}_{-\lambda'}(1)|0\rangle \right\rangle \cdot \mathfrak{m}_{\mu,H}|0\rangle. \end{aligned}$$

Therefore, we must have $\mu = \mu'$ and

$$(-1)^{(|\lambda'|-\ell(\lambda'))+(|\nu|-\ell(\nu))} \cdot e^{\lambda,\mu,\nu}_{\lambda',\mu,\nu} = \tilde{f}_\nu \left\langle [\overline{C}_{\lambda,0,0}], \frac{1}{\mathfrak{z}_{\lambda'}} \mathfrak{a}_{-\lambda'}(1)|0\rangle \right\rangle.$$

In particular, the right-hand-side is nonzero. By Proposition 9.32, $\ell(\lambda) = \ell(\lambda')$ and $\lambda \preceq \lambda'$. If $\lambda \prec \lambda'$, then we are done. If $\lambda = \lambda'$, then

$$(-1)^{(|\lambda|-\ell(\lambda))+(|\nu|-\ell(\nu))} \cdot e^{\lambda,\mu,\nu}_{\lambda,\mu,\nu} = \tilde{f}_\nu \left\langle [\overline{C}_{\lambda,0,0}], \frac{1}{\mathfrak{z}_{\lambda}} \mathfrak{a}_{-\lambda}(1)|0\rangle \right\rangle.$$

By Proposition 9.32 again, the right-hand-side is equal to $(-1)^{|\lambda|-\ell(\lambda)} e^\lambda_\lambda \tilde{f}_\nu$. Hence we obtain $e^{\lambda,\mu,\nu}_{\lambda,\mu,\nu} = (-1)^{|\nu|-\ell(\nu)} e^\lambda_\lambda \tilde{f}_\nu$. □

Inspired by Lemma 9.33, we introduce the following definition.

DEFINITION 9.34. Let (λ,μ,ν) and (λ',μ',ν') be 3-tuples of partitions. We define $(\lambda',\mu',\nu') \prec (\lambda,\mu,\nu)$ if one of the three following conditions holds:

(i) $\nu' \prec \nu$,
(ii) $\nu' = \nu$ and $\mu' \prec \mu$,
(iii) $\nu' = \nu$, $\mu' = \mu$ and $\lambda \prec \lambda'$.

This defines a total ordering for all 3-tuples (λ,μ,ν) of partitions. Combining with Lemma 9.33 yields the main result in this section.

THEOREM 9.35. *Let λ,μ,ν be three partitions. Then, $[\overline{C}_{\lambda,\mu,\nu}]$ is equal to*

$$\frac{(-1)^{|\nu|-\ell(\nu)}}{\mathfrak{z}_\nu}\mathfrak{a}_{-\nu}(1_{\mathbb{P}^2})\mathfrak{m}_{\mu,H}\mathfrak{a}_{-\lambda}(x)|0\rangle$$
$$+\sum_{\lambda',\mu',\nu'}\frac{e^{\lambda,\mu,\nu}_{\lambda',\mu',\nu'}}{\mathfrak{z}_{\nu'}}\mathfrak{a}_{-\nu'}(1)\mathfrak{m}_{\mu',H}\mathfrak{a}_{-\lambda'}(x)|0\rangle$$

where $e^{\lambda,\mu,\nu}_{\lambda',\mu',\nu'}\in\mathbb{Z}$, $|\lambda'|+|\mu'|+|\nu'|=|\lambda|+|\mu|+|\nu|$, and $(\lambda',\mu',\nu')\prec(\lambda,\mu,\nu)$.

PROOF. By Lemma 9.33, it remains to show that $e^\lambda_\lambda=\tilde{f}_\nu=1$. Let $n=|\lambda|+|\mu|+|\nu|$. Consider the two integral bases of of $H^*((\mathbb{P}^2)^{[n]};\mathbb{Z})$:

$$\begin{aligned}\mathcal{A}_n &= \left\{[\overline{C}_{\lambda,\mu,\nu}]\,\big|\,|\lambda|+|\mu|+|\nu|=n\right\},\\ \mathcal{B}_n &= \left\{\frac{1}{\mathfrak{z}_\nu}\mathfrak{a}_{-\nu}(1)\mathfrak{m}_{\mu,H}\mathfrak{a}_{-\lambda}(x)|0\rangle\,\big|\,|\lambda|+|\mu|+|\nu|=n\right\}.\end{aligned}$$

Order the basis elements in both $\mathcal{A}_n$ and $\mathcal{B}_n$ by using the total ordering for all 3-tuples of partitions defined in Definition 9.34. Let T be the transition matrix between the ordered bases $\mathcal{A}_n$ and $\mathcal{B}_n$. By Lemma 9.33, T is an integral triangular matrix whose diagonal entries are $(-1)^{|\nu|-\ell(\nu)}e^\lambda_\lambda\tilde{f}_\nu$ with $|\lambda|+|\nu|\le n$. Since $\det T=\pm1$ and both e^λ_λ and $\tilde{f}_\nu$ are positive integers, we must have $e^\lambda_\lambda=\tilde{f}_\nu=1$. □

We refer to [**Mur**, Section 5] for the transition matrices between the two integral bases (9.60) and (9.64) of $H^*((\mathbb{P}^2)^{[n]};\mathbb{Z})$ in the special cases of $n=2,3$.

CONJECTURE 9.36. $[\overline{C}_{\lambda,0,0}]=\mathfrak{a}_{-\lambda}(x)|0\rangle$ *for every partition λ.*

By Theorem 9.35, Conjecture 9.36 holds when $\lambda=(1^n),(1^{n-2}2)$, or (n).

COROLLARY 9.37. *Let $\mathbb{C}^*$ act on the affine coordinates u and v of $\mathbb{C}^2$ by*

$$t(u,v)=(t^{w_1-w_2}u,t^{w_0-w_2}v),\quad t\in\mathbb{C}^*\text{ and }w_1\gg w_0.$$

For $\nu\vdash n$, let C_ν be the cell in $(\mathbb{C}^2)^{[n]}$ corresponding to the fixed point ξ_ν. Then,

$$\frac{1}{\mathfrak{z}_\nu}\mathfrak{a}_{-\nu}(1_{\mathbb{C}^2})|0\rangle=(-1)^{|\nu|-\ell(\nu)}\,[\overline{C}_\nu]\tag{9.90}$$

in the integral cohomology $H^*((\mathbb{C}^2)^{[n]};\mathbb{Z})$ of the Hilbert scheme $(\mathbb{C}^2)^{[n]}$.

PROOF. We see from (8.2) and (9.73) that in $H^*((\mathbb{C}^2)^{[n]};\mathbb{Z})$, we have

$$\frac{1}{\mathfrak{z}_\nu}\mathfrak{a}_{-\nu}(1_{\mathbb{C}^2})|0\rangle=(-1)^{|\nu|-\ell(\nu)}\,\iota_n^*\big([\overline{C}_{0,0,\nu}]\big).$$

Now (9.90) follows from $\iota_n^*\big([\overline{C}_{0,0,\nu}]\big)=[\iota_n^{-1}(\overline{C}_{0,0,\nu})]=[\overline{C}_\nu]$. □

DEFINITION 9.38. Fix a point x on an arbitrary smooth projective surface X. Identify $M_n(x)$ with

$$M_n(P_0)\subset(\mathbb{C}^2)^{[n]}\subset(\mathbb{P}^2)^{[n]},$$

where P_0 denotes the origin of $\mathbb{C}^2$. For $\lambda\vdash n$, we define $\overline{M}_\lambda(x)$ to be the closed subvariety of $M_n(x)\subset X^{[n]}$ corresponding to $\overline{C}_{\lambda,0,0}\subset M_n(P_0)$ under this identification.

COROLLARY 9.39. *Let x be a point on a smooth projective surface X. Then,*

$$[\overline{M}_\lambda(x)] = \mathfrak{a}_{-\lambda}(x)|0\rangle + \sum_{\substack{|\lambda|=|\nu|,\, \ell(\lambda)=\ell(\nu) \\ \lambda \prec \nu}} e_\nu^\lambda\, \mathfrak{a}_{-\nu}(x)|0\rangle. \tag{9.91}$$

Moreover, the coefficients e_ν^λ are integers and independent of X and x.

PROOF. Since the elements in $\overline{M}_\lambda(x)$ are supported at x, we see as in (9.76) that

$$[\overline{M}_\lambda(x)] = \sum_{|\lambda|=|\nu|,\, \ell(\lambda)=\ell(\nu)} e_\nu^\lambda(x)\, \mathfrak{a}_{-\nu}(x)|0\rangle, \qquad e_\nu^\lambda(x) \in \mathbb{Z}. \tag{9.92}$$

To analyze the coefficients $e_\nu^\lambda(x)$, put $n = |\lambda|$. For $\nu \vdash n$ with $\ell(\nu) = \ell(\lambda)$, let $\overline{Z}_\nu(X) \subset X^{[n]}$ be the closure of the subset consisting of the elements of the form

$$\sum_{i=1}^{\ell(\nu)} \xi_i$$

where $\xi_i \in M_{\nu_i}(x_i)$, and $x_1, \ldots, x_{\ell(\nu)} \in X$ are distinct. Then,

$$[\overline{Z}_\nu(X)] = \prod_{i=1}^{\ell(\nu)} i^{m_i} \cdot \frac{1}{\mathfrak{z}_\nu} \mathfrak{a}_{-\nu}(1)|0\rangle \tag{9.93}$$

where $\nu = (1^{m_1} 2^{m_2} \cdots)$. Note that the pairing $\langle [\overline{M}_\lambda(x)], [\overline{Z}_\nu(X)] \rangle$ depends only on an analytic neighborhood of x, which can be chosen to be independent of X and x. Setting $X = \mathbb{P}^2$ and $x = P_0$, we obtain from Theorem 9.35 that

$$\langle [\overline{M}_\lambda(x)], [\overline{Z}_\nu(X)] \rangle = (-1)^{n-\ell(\nu)} \prod_{i=1}^{\ell(\nu)} i^{m_i} \cdot e_\nu^\lambda \tag{9.94}$$

where $e_\lambda^\lambda = 1$ and $e_\nu^\lambda = 0$ for $\nu \prec \lambda$. On the other hand, by (9.92) and (9.93),

$$\langle [\overline{M}_\lambda(x)], [\overline{Z}_\nu(X)] \rangle = (-1)^{n-\ell(\nu)} \prod_{i=1}^{\ell(\nu)} i^{m_i} \cdot e_\nu^\lambda(x).$$

Now it follows immediately from (9.94) that $e_\nu^\lambda(x) = e_\nu^\lambda$. □

CHAPTER 10

The ring structure of $H^*_{\mathrm{orb}}(X^{(n)})$

By Remark 1.25, the Hilbert scheme $X^{[n]}$ of points on a surface X is a crepant resolution of singularities of the symmetric product $X^{(n)} = X^n/S_n$ where S_n is the n-th symmetric group. Inspired by orbifold string theory [**DHVW**, **VW**], the geometry of the Hilbert scheme $X^{[n]}$ should be "equivalent" to the S_n-equivariant geometry of X^n. It is stated in the footnote 3 of [**Groj**] that the direct sum

$$\bigoplus_{n=0}^{+\infty} K_{S_n}(X^n) \otimes \mathbb{C}$$

of equivariant K-groups can be identified with a Fock space of a Heisenberg algebra [**Seg**, **Wan1**]. In particular, the size of the cohomology group $H^*(X^{[n]})$ coincides with that of the equivariant K-group $K_{S_n}(X^{[n]})$ or the orbifold cohomology group of $X^{(n)}$. (Equivariant K-groups are related to orbifold cohomology groups by a decomposition theorem due to [**BC**] and independently [**Kuhn**]). However the "equivalence" on the level of ring structures is more subtle.

An *orbifold* is a topological space locally modeled on the quotient of Euclidean space by a finite group. In [**CR2**, Section 4], Chen and Ruan introduced the orbifold cohomology ring $H^*_{\mathrm{orb}}(\cdot)$. For the symmetric product $X^{(n)}$ which is an orbifold, the ring structure of $H^*_{\mathrm{orb}}(X^{(n)})$ has been obtained in [**FG1**, **Uri**]. Combined with [**LS2**], which was surveyed in Section 7.3, the results in [**FG1**, **Uri**] yield the proof of Ruan's Cohomological Crepant Resolution Conjecture [**Rua2**, Section 2] identifying the cohomology ring of the Hilbert scheme $X^{[n]}$ of points on a surface X with a numerically trivial canonical class and the orbifold ring $H^*_{\mathrm{orb}}(X^{(n)})$. The isomorphisms in [**FG1**, **LS2**, **Uri**] are quite indirect. To deal with Ruan's conjecture in Chapter 16 for an arbitrary smooth projective complex surface X, an alternative approach is needed to obtain a direction connection between $H^*_{\mathrm{orb}}(X^{(n)})$ and vertex algebras. This is achieved in [**QW1**] via Heisenberg algebra actions. In this chapter, we will review $H^*_{\mathrm{orb}}(M/G)$ for a global orbifold M/G and recall the results from [**QW1**]. Since the results here are parallel to those in previous chapters for the Hilbert schemes $X^{[n]}$, many proofs will be omitted.

10.1. Generalities

In this section, we will review the orbifold cohomology $H^*_{\mathrm{orb}}(M/G)$ of the global orbifold M/G where M is a complex manifold of complex dimension d with a finite group G action.

First of all, we introduce the space

$$M \diamond G = \{(g, x) \in G \times M \mid gx = x\} = \coprod_{g \in G} M^g,$$

and G acts on $M \diamond G$ naturally by $h.(g,x) = (hgh^{-1}, hx)$. As a vector space, we define $H^*(M,G)$ to be the cohomology group of $M \diamond G$ with complex coefficients, or equivalently,

$$H^*(M,G) = \bigoplus_{g\in G} H^*(M^g).$$

The space $H^*(M,G)$ has a naturally induced G action, which is denoted by

$$\mathrm{ad}\, h : H^*(M^g) \to H^*(M^{hgh^{-1}}).$$

As a vector space, the orbifold cohomology group

$$H^*_{\mathrm{orb}}(M/G)$$

is the G-invariant part of $H^*(M,G)$, which is isomorphic to

$$\bigoplus_{[g]\in G_*} H^*(M^g/Z(g))$$

where G_* denotes the set of conjugacy classes of G and $Z(g) = Z_G(g)$ denotes the centralizer of g in G.

For $g \in G$ and $x \in M^g$, write the eigenvalues of the action of g on the complex tangent space $T_{x,M}$ to be $\mu_k = e^{2\pi i r_k}$, where $0 \le r_k < 1$ and $k = 1, \dots, d$. The *degree shift number* (or *age* in the terminology of Ito and Reid) is the rational number

$$F^g_x = \sum_{k=1}^d r_k.$$

It depends only on the connected component Z which contains x, so we can denote it by F^g_Z. Then associated to a cohomology class in $H^r(Z)$, we assign the corresponding element in $H^*(M,G)$ (and thus in $H^*_{\mathrm{orb}}(M/G)$) the degree $r + 2F^g_Z$.

A ring structure on $H^*_{\mathrm{orb}}(M/G)$ was introduced in [**CR2**, Section 4]. This was subsequently clarified in [**FG1**] by introducing a ring structure on $H^*(M,G)$ first and then passing to $H^*_{\mathrm{orb}}(M/G)$ by restriction. We will use $\bullet$ to denote this product. The ring structure on $H^*(M,G)$ is degree-preserving, and has the property that $\alpha \bullet \beta$ lies in $H^*(M^{gh})$ for $\alpha \in H^*(M^g)$ and $\beta \in H^*(M^h)$.

For $1 \in G$,

$$H^*(M^1/Z(1)) \cong H^*(M/G),$$

and thus we can regard $\alpha \in H^*(M/G)$ to be $\alpha \in H^*_{\mathrm{orb}}(M/G)$ by this isomorphism. Also, given

$$a = \sum_{g\in G} a_g g$$

in the group algebra $\mathbb{C}[G]$ (respectively, $\mathbb{C}[G]^G$), we may regard a as an element in $H^*(M,G)$ (respectively, $H^*_{\mathrm{orb}}(M/G)$) whose component in each $H^*(M^g)$ is

$$a_g \cdot 1_{M^g} \in H^0(M^g).$$

If K is a subgroup of G, then we can define the restriction map from $H^*(M,G)$ to $H^*(M,K)$ by projection to the component $\oplus_{g\in K} H^*(M^g)$ which, when restricted to the G-invariant part, induces naturally a *degree-preserving* linear map

$$\mathrm{Res}^G_K : H^*_{\mathrm{orb}}(M/G) \to H^*_{\mathrm{orb}}(M/K).$$

We define the induction map

$$\mathrm{Ind}^G_K : H^*(M,K) \to H^*_{\mathrm{orb}}(M/G)$$

by sending $\alpha \in H^*(M^h)$, where $h \in K$, to

$$\mathrm{Ind}_K^G(\alpha) = \frac{1}{|K|}\sum_{g\in G} \mathrm{ad}g(\alpha).$$

Note that $\mathrm{Ind}_K^G(\alpha)$ is clearly G-invariant. When restricted to the invariant part, we obtain a *degree-preserving* linear map

$$\mathrm{Ind}_K^G : H^*_{\mathrm{orb}}(M/K) \to H^*_{\mathrm{orb}}(M/G).$$

We often write the restriction and induction maps as $\mathrm{Res}_K, \mathrm{Res}$ and $\mathrm{Ind}^G, \mathrm{Ind}$, when the groups involved are clear from the context. In particular, when M is a point, $H^*_{\mathrm{orb}}(pt/G)$ reduces to the Grothendieck ring $R_{\mathbb{C}}(G)$ of G, and we recover the induction and restriction functors in the theory of finite groups.

10.2. The Heisenberg algebra

In this section, we will discuss the Heisenberg algebra action on the orbifold cohomology $H^*_{\mathrm{orb}}(X^{(n)})$ of the symmetric products $X^{(n)}$. We will construct the linear operators $\mathfrak{p}_n(\alpha)$ $(n \in \mathbb{Z}, \alpha \in H^*(X))$, and write down their Heisenberg commutation relations.

Let X be a closed complex manifold of complex dimension d. Our main objects are the ring $H^*_{\mathrm{orb}}(X^{(n)})$ and the non-commutative ring $H^*(X^n, S_n)$. We denote

$$\mathcal{F}_X = \bigoplus_{n=0}^{+\infty} H^*_{\mathrm{orb}}(X^{(n)}).$$

We introduce a linear map

$$\omega_n : H^*(X) \to H^*_{\mathrm{orb}}(X^{(n)})$$

as follows. Let σ_n be any permutation in the conjugacy class $[n] \in (S_n)_*$ which consists of the n-cycles, and let $\tau_n : H^*(X) \to H^*((X^n)^{\sigma_n})$ be the obvious isomorphism. Given $\alpha \in H^*(X)$, we define

$$\omega_n(\alpha) = \tau_n(n\alpha).$$

Note that if $\alpha \in H^r(X)$, then $\omega_n(\alpha) \in H^{r+d(n-1)}_{\mathrm{orb}}(X^{(n)})$. We also define

$$\mathrm{ch}_n : H^*_{\mathrm{orb}}(X^{(n)}) \to H^*(X)$$

as the composition of the isomorphism $H^*((X^n)^{\sigma_n}) \cong H^*(X)$ with the projection from $H^*_{\mathrm{orb}}(X^{(n)})$ to $H^*((X^n)^{\sigma_n})$.

Let $\alpha \in H^*(X)$. Set $\mathfrak{p}_0(\alpha) = 0$. For $n > 0$, we define the creation operator $\mathfrak{p}_{-n}(\alpha) \in \mathrm{End}(\mathcal{F}_X)$ given by the composition $(k \geq 0)$:

$$\begin{aligned}
H^*_{\mathrm{orb}}(X^k/S_k) &\overset{\omega_n(\alpha)\otimes\cdot}{\longrightarrow} H^*_{\mathrm{orb}}(X^n/S_n) \bigotimes H^*_{\mathrm{orb}}(X^k/S_k) \\
&\overset{\cong}{\longrightarrow} H^*_{\mathrm{orb}}(X^{n+k}/(S_n \times S_k)) \\
&\overset{\mathrm{Ind}}{\longrightarrow} H^*_{\mathrm{orb}}(X^{n+k}/S_{n+k}),
\end{aligned}$$

and the annihilation operator $\mathfrak{p}_n(\alpha) \in \mathrm{End}(\mathcal{F}_X)$ given by the composition ($k \geq 0$):

$$\begin{aligned} H^*_{\mathrm{orb}}(X^{n+k}/S_{n+k}) &\xrightarrow{\mathrm{Res}} H^*_{\mathrm{orb}}(X^{n+k}/(S_n \times S_k)) \\ &\xrightarrow{\cong} H^*_{\mathrm{orb}}(X^n/S_n) \bigotimes H^*_{\mathrm{orb}}(X^k/S_k) \\ &\xrightarrow{\mathrm{ch}_n} H^*(X) \bigotimes H^*_{\mathrm{orb}}(X^k/S_k) \\ &\xrightarrow{(\alpha,\cdot)} H^*_{\mathrm{orb}}(X^k/S_k). \end{aligned}$$

THEOREM 10.1. *The operators $\mathfrak{p}_n(\alpha) \in \mathrm{End}(\mathcal{F}_X)$ ($n \in \mathbb{Z}, \alpha \in H^*(X)$) generate a Heisenberg (super)algebra with commutation relations given by*

$$[\mathfrak{p}_m(\alpha), \mathfrak{p}_n(\beta)] = m\delta_{m,-n}(\alpha, \beta) \cdot \mathrm{Id}_{\mathcal{F}_X}$$

where $n, m \in \mathbb{Z}$, $\alpha, \beta \in H^*(X)$. Furthermore, $\mathcal{F}_X$ is an irreducible representation of the Heisenberg algebra with the vacuum vector $|0\rangle = 1 \in H^*(pt) \cong \mathbb{C}$.

This theorem can be proved in the same way as an analogous theorem formulated by using the equivariant K-group $K_{S_n}(X^n) \otimes \mathbb{C}$. This analogous theorem was established in [**Seg**] (we refer to [**Wan1**, Theorem 4] and its proof for the details). Note that there is a fundamental sign difference in the two commutators of Theorems 3.8 and Theorems 10.1.

In particular, for a given $y \in H^{|y|}_{\mathrm{orb}}(X^{(n-1)})$, by the definition of $\mathfrak{p}_{-1}(\alpha)$ (where $\alpha \in H^{|\alpha|}(X)$) and the induction map, we can write that

$$\begin{aligned} \mathfrak{p}_{-1}(\alpha)(y) &= \frac{1}{(n-1)!} \sum_{g \in S_n} \mathrm{ad}g\,(\alpha \otimes y) \\ &= \frac{(-1)^{|\alpha|\cdot|y|}}{(n-1)!} \sum_{g \in S_n} \mathrm{ad}g\,(y \otimes \alpha). \end{aligned}$$

For $0 \leq i < n$, we introduce the following cohomology class in $H^*_{\mathrm{orb}}(X^{(n)})$:

$$P_i(\alpha, n) = \frac{1}{(n-i-1)!} \cdot \mathfrak{p}_{-i-1}(\alpha)\mathfrak{p}_{-1}(1_X)^{n-i-1}|0\rangle. \tag{10.1}$$

This is the analogue of the class $B_i(\alpha, n)$ defined in Definition 7.3. The analogue of the class $G_i(\alpha, n)$ (in the setting of Hilbert schemes) will be studied in the next section.

10.3. The cohomology classes $\eta_n(\gamma)$ and $O_k(\alpha, n)$

In this section, we will define the classes $\eta_n(\gamma)$ and $O_k(\alpha, n)$ in the orbifold cohomology $H^*_{\mathrm{orb}}(X^{(n)})$. The classes $\eta_n(\gamma)$ play the role of the total Chern classes $c(L^{[n]*})$ of the dual tautological bundles $L^{[n]*}$ over the Hilbert scheme $X^{[n]}$, where L denotes a line bundle over X. The classes $O_k(\alpha, n)$ play the role of the classes $G_k(\alpha, n)$ for the Hilbert scheme $X^{[n]}$ which are defined in Definition 4.1.

Let

$$\mathrm{ch} : \bigoplus_{n=0}^{+\infty} \mathcal{C}(S_n) \longrightarrow \Lambda_{\mathbb{C}} = \Lambda \otimes_{\mathbb{Z}} \mathbb{C}$$

be the Frobenius characteristic map from the direct sum of class algebras of the symmetric group S_n to the ring $\Lambda_{\mathbb{C}}$ of symmetric functions in infinitely many variables (with complex coefficients). Denote by η_n and ε_n the trivial and alternating characters of S_n. Then ch sends η_n and ε_n to the n-th complete and elementary symmetric functions in $\Lambda_{\mathbb{C}}$ respectively. We denote by p_r the r-th power sum symmetric function.

Following [**Juc**] and [**Mu**, p. 287], we define the Jucys-Murphy elements $\xi_{j;n}$ of the symmetric group S_n to be the sums of transpositions:

$$\xi_{j;n} = \sum_{i<j} (i,j), \quad j = 1, \ldots, n.$$

When it is clear from the text, we may simply write $\xi_{j;n}$ as ξ_j. Put

$$\Xi_n = \{\xi_1, \ldots, \xi_n\}.$$

Then the k-th elementary symmetric function $e_k(\Xi_n)$ of Ξ_n is equal to the sum of all permutations in S_n having exactly $(n-k)$ cycles. Therefore, we obtain

$$\varepsilon_n = \sum_{\sigma \in S_n} (-1)^{d(\sigma)} \sigma = \sum_{k=0}^{n} (-1)^k e_k(\Xi_n) = \prod_{i=1}^{n} (1 - \xi_i) \tag{10.2}$$

where $d(\sigma)$ is the degree of σ defined by (7.12). Define

$$\varepsilon_n(\hbar) = \prod_{i=1}^{n} (1 - \hbar \xi_i),$$

where $\hbar$ is a formal parameter here and below. We have

$$\sum_{n=0}^{+\infty} \mathrm{ch}(\varepsilon_n(\hbar)) z^n = \exp\left(\sum_{r \geq 1} (-\hbar)^{r-1} \frac{p_r}{r} z^r \right). \tag{10.3}$$

Noting that $\varepsilon_n(1)$ (respectively, $\varepsilon_n(-1)$) coincides with the alternating character ε_n (respectively, the trivial character η_n), we obtain two classical identities involving η_n, ε_n, and p_r by setting $\hbar = \pm 1$ in (10.3). See also (8.13).

In the rest of this chapter, we will assume that X is a closed complex manifold of *even* complex dimension d. Given $\gamma \in H^*(X)$, we denote

$$\gamma^{(i)} = 1^{\otimes i-1} \otimes \gamma \otimes 1^{\otimes n-i} \in H^*(X^n),$$

and regard it to be a cohomology class in $H^*(X^n, S_n)$ associated to the identity conjugacy class. We define

$$\xi_i(\gamma) = \xi_i + \gamma^{(i)} \in H^*(X^n, S_n).$$

We sometimes write $\xi_i(\gamma)$ as $\xi_{i;n}(\gamma)$ to specify its dependence on n when necessary.

DEFINITION 10.2. Given $\gamma \in H^*(X)$, we define $\eta_n(\gamma)$ (respectively, $\varepsilon_n(\gamma)$) to be the cohomology class in $H^*_{\mathrm{orb}}(X^{(n)})$ whose component associated to an element σ in the conjugacy class of partition λ of n is given by

$$\gamma^{\otimes \ell(\lambda)} \in H^*((X^n)^\sigma) \cong H^{\otimes \ell(\lambda)}$$

(respectively, by $(-1)^{d(\lambda)} \gamma^{\otimes \ell(\lambda)}$). We further define an operator $\eta(\gamma)$ (respectively, $\varepsilon(\gamma)$) in $\mathrm{End}(\mathcal{F}_X)$ by letting it act on $H^*_{\mathrm{orb}}(X^{(n)})$ by the orbifold product with $\eta_n(\gamma)$ (respectively, $\varepsilon_n(\gamma)$) for every n.

This definition is motivated by its counterpart in terms of equivariant K-groups [**Seg, Wan1**]. Parallel to the Proposition 4 in [**Wan1**], we have

$$\sum_{n=0}^{+\infty} \eta_n(\gamma) z^n = \exp\left(\sum_{r \geq 1} \frac{1}{r} \mathfrak{p}_{-r}(\gamma) z^r \right) \cdot |0\rangle.$$

PROPOSITION 10.3. *Given $\gamma \in H^*(X)$, the orbifold cup product of the n elements $\xi_i(\gamma)$ $(i = 1, \ldots, n)$ in $H^*(X^n, S_n)$ lies in $H^*_{orb}(X^{(n)})$, and furthermore the following identity holds:*

$$\eta_n(\gamma) = \prod_{i=1}^{n} \xi_i(\gamma) = \xi_1(\gamma) \bullet \xi_2(\gamma) \bullet \ldots \bullet \xi_n(\gamma). \tag{10.4}$$

PROOF. It suffices to prove (10.4), since the first claim follows from (10.4) and the fact that $\eta_n(\gamma)$ is S_n-invariant.

A typical monomial on the right-hand side of (10.4) is of the form

$$(\xi_{i_1} \cdots \xi_{i_k}) \bullet (\gamma^{(j_1)} \ldots \gamma^{(j_{n-k})})$$

where $i_1 < \ldots < i_k$, $j_1 < \ldots < j_{n-k}$, and

$$\{i_1, \ldots, i_k, j_1, \ldots, j_{n-k}\} = \{1, \ldots, n\}.$$

Here we have used the observation that $\xi_{i_1} \bullet \cdots \bullet \xi_{i_k}$ is just the usual multiplication of permutations $\xi_{i_1} \cdots \xi_{i_k}$ and $\gamma^{(j_1)} \bullet \cdots \bullet \gamma^{(j_{n-k})}$ is just the ordinary cup product $\gamma^{(j_1)} \ldots \gamma^{(j_{n-k})}$ in $H^*(X^n) \cong H^*(X)^{\otimes n}$. Note that every cycle of each permutation σ appearing in $\xi_{i_1} \cdots \xi_{i_k}$ has exactly one number which does not belong to $i_1, \ldots, i_k$, and in addition, $\ell(\sigma) = n - k$ and $d(\sigma) = k$. Using the definition of the orbifold cup product, we see that the product $\sigma \bullet \gamma^{(j_1)} \ldots \gamma^{(j_{n-k})}$ does not involve the obstruction bundles (or the group defects are trivial in the terminology of Lehn-Sorger) and is equal to

$$\gamma^{\otimes \ell(\sigma)} \in (H^*(X))^{\otimes \ell(\sigma)} \cong H^*((X^n)^\sigma).$$

This proves (10.4). □

Put

$$\varepsilon_n(\gamma, \hbar) = \prod_{i=1}^{n} (\gamma^{(i)} - \hbar \xi_i).$$

Then we have

$$\sum_{n=0}^{+\infty} \varepsilon_n(\gamma, \hbar) z^n = \exp\left(\sum_{r \geq 1} (-\hbar)^{r-1} \frac{\mathfrak{p}_{-r}(\gamma)}{r} z^r \right). \tag{10.5}$$

Regarding $\xi_i = \xi_i(0) \in H^*(X^n, S_n)$, we denote

$$\xi_i^{\bullet k} = \underbrace{\xi_i \bullet \ldots \bullet \xi_i}_{k \text{ times}} \in H^*(X^n, S_n),$$

and define

$$e^{-\xi_i} = \sum_{k \geq 0} \frac{1}{k!} (-\xi_i)^{\bullet k} \in H^*(X^n, S_n).$$

DEFINITION 10.4. For a homogeneous class $\alpha \in H^{|\alpha|}(X)$, we define the class $O_k(\alpha, n) \in H^*_{\text{orb}}(X^{(n)})$ to be

$$O_k(\alpha, n) = \frac{1}{k!} \sum_{i=1}^{n} (-\xi_i)^{\bullet k} \bullet \alpha^{(i)} \in H^{dk+|\alpha|}_{\text{orb}}(X^{(n)}). \tag{10.6}$$

We extends the notion $O_k(\alpha, n)$ linearly to all $\alpha \in H^*(X)$, and put

$$\begin{aligned} O(\alpha, n) &= \sum_{k \geq 0} O_k(\alpha, n) = \sum_{i=1}^{n} e^{-\xi_i} \bullet \alpha^{(i)}, \\ O_\hbar(\alpha, n) &= \sum_{k \geq 0} \hbar^k O_k(\alpha, n). \end{aligned}$$

We further define the operator

$$\mathfrak{O}_k(\alpha) \in \mathrm{End}(\mathcal{F}_X)$$

(respectively, $\mathfrak{O}(\alpha)$, or $\mathfrak{O}_\hbar(\alpha)$) to be the orbifold cup product with $O_k(\alpha, n)$ (respectively, $O(\alpha, n)$, or $O_\hbar(\alpha, n)$) in $H^*_{\mathrm{orb}}(X^{(n)})$ for every $n \geq 0$.

REMARK 10.5. We can see that $O_k(\alpha, n) \in H^*(X^n, S_n)$ is S_n-invariant (and thus lies in $H^*_{\mathrm{orb}}(X^{(n)})$) as follows. For $\gamma \in H^*(X)$, note that $e_j(\xi_1(\gamma), \ldots, \xi_n(\gamma))$ lies in $H^*_{\mathrm{orb}}(X^{(n)})$, where $e_j(\xi_1(\gamma), \ldots, \xi_n(\gamma))$ $(1 \leq j \leq n)$ is the j-th elementary symmetric function in $\xi_i(\gamma)$'s. So $H^*_{\mathrm{orb}}(X^{(n)})$ contains all symmetric functions in $\xi_i(\gamma)$'s. In particular,

$$O(e^{-\gamma}, n) = \sum_i (e^{-\xi_i} \bullet (e^{-\gamma})^{(i)}) = \sum_i e^{-\xi_i(\gamma)} \in H^*_{\mathrm{orb}}(X^{(n)}).$$

Letting γ vary, we see that $O(\alpha, n)$ and, similarly, $O_k(\alpha, n)$ lie in $H^*_{\mathrm{orb}}(X^{(n)})$.

In particular, $\mathfrak{O}_1(1_X) \in \mathrm{End}(\mathcal{F}_X)$ is the generalized Goulden's operator [**Gou**, Section 3]. The reason for the convention of putting the sign in front of ξ_i in (10.6) is to make the comparison with the Hilbert scheme side easier. Also, the power sums of Jucys-Murphy elements have been studied in [**LT**, Section 4] which corresponds to the present case when X is a point.

Note that the generalized Goulden's operator $\mathfrak{O}_1(1_X)$ is defined to be the orbifold cup product with $O_1(1_X, n)$ in $H^*_{\mathrm{orb}}(X^{(n)})$. Given an operator $\mathfrak{f} \in \mathrm{End}(\mathcal{F}_X)$, we define

$$\mathfrak{f}^{\{1\}} = \mathfrak{f}' = [\mathfrak{O}_1(1_X), f],$$

and $\mathfrak{f}^{\{k+1\}} = (\mathfrak{f}^{\{k\}})'$. We have the following.

THEOREM 10.6. $\mathfrak{O}_1(1_X) = -\dfrac{1}{6} : \mathfrak{p}^3 :_0 (\tau_* 1_X)$.

REMARK 10.7. This theorem is a counterpart of (4.18). The proof is essentially the same as the proof in the case when X is a point, which was treated in [**Gou**, Section 3] and [**FW**, Section 3] (also [**LS2**]). For example, if we look at the proof of [**LS2**, Proposition 4.4], the Δ_* and e there should be replaced by our τ_* (which equals $-\Delta_*$) and $-e$ respectively, since we are using the orbifold cup product of [**CR2**].

10.4. Interactions between Heisenberg algebra and $\mathfrak{O}_k(\gamma)$

In this section, we will prove that the commutator $[\mathfrak{O}_k(\gamma), \mathfrak{p}_{-1}(\alpha)]$ is equal to $1/k! \cdot \mathfrak{p}_{-1}^{\{k\}}(\gamma\alpha)$. Moreover, we will study the relations between the operators $\eta(\gamma), \varepsilon(\gamma)$ and the operator $\mathfrak{p}_{-1}(\alpha)$.

The following is parallel to formula (4.7).

THEOREM 10.8. *Let $\gamma, \alpha \in H^*(X)$. Then we have*

$$\begin{aligned}
[\mathfrak{O}_\hbar(\gamma), \mathfrak{p}_{-1}(\alpha)] &= \exp(\hbar \cdot \mathrm{ad}\, \mathfrak{O}_1(1_X))(\mathfrak{p}_{-1}(\gamma\alpha)) \\
[\mathfrak{O}_k(\gamma), \mathfrak{p}_{-1}(\alpha)] &= \frac{1}{k!}\, \mathfrak{p}_{-1}^{\{k\}}(\gamma\alpha), \quad k \geq 0.
\end{aligned}$$

PROOF. For simplicity of signs, we assume that the cohomology classes γ, α are of even degree. It suffices to prove the second identity.

Recall that for $y \in H^*_{\mathrm{orb}}(X^{(n-1)})$, we have

$$\mathfrak{p}_{-1}(\alpha)(y) = \frac{1}{(n-1)!} \sum_{g \in S_n} \mathrm{ad} g\, (y \otimes \alpha).$$

Regarding S_{n-1} as the subgroup $S_{n-1} \times 1$ of S_n, we introduce an injective ring homomorphism

$$\iota : H^*(X^{n-1}, S_{n-1}) \to H^*(X^n, S_n)$$

by sending α_σ to $\alpha_\sigma \otimes 1_X$, where $\sigma \in S_{n-1}$. Thus

$$\begin{aligned}
& (n-1)!\; [\mathfrak{O}_k(\gamma), \mathfrak{p}_{-1}(\alpha)](y) \\
= \;& (n-1)!\; [\mathfrak{O}_k(\gamma) \cdot \mathfrak{p}_{-1}(\alpha)(y) - \mathfrak{p}_{-1}(\alpha) \cdot \mathfrak{O}_k(\gamma)(y)] \\
= \;& O_k(\gamma, n) \bullet \sum_{g \in S_n} \mathrm{ad} g\, (y \otimes \alpha) - \sum_{g \in S_n} \mathrm{ad} g\, ((O_k(\gamma, n-1) \bullet y) \otimes \alpha) \\
= \;& \sum_g \mathrm{ad} g\, [(O_k(\gamma, n) - \iota(O_k(\gamma, n-1))) \bullet (y \otimes \alpha)]
\end{aligned}$$

where we used the fact that $O_k(\gamma, n)$ is S_n-invariant. By definition, we have

$$O_k(\gamma, n) - \iota(O_k(\gamma, n-1)) = \frac{1}{k!} \cdot (-\xi_{n;n})^{\bullet k} \bullet \gamma^{(n)}.$$

Thus, we obtain

$$\begin{aligned}
(n-1)!\; [\mathfrak{O}_k(\gamma), \mathfrak{p}_{-1}(\alpha)](y) &= \frac{1}{k!} \cdot \sum_g \mathrm{ad} g\, [(-\xi_{n;n})^{\bullet k} \bullet \gamma^{(n)} \bullet (y \otimes \alpha)] \\
&= \frac{1}{k!} \cdot \sum_g \mathrm{ad} g\, [(-\xi_{n;n})^{\bullet k} \bullet (y \otimes \gamma\alpha)].
\end{aligned}$$

It remains to prove that

$$\sum_{g \in S_n} \mathrm{ad} g\, [(-\xi_{n;n})^{\bullet k} \bullet (y \otimes \gamma\alpha)] = (n-1)!\; \mathfrak{p}_{-1}^{\{k\}}(\gamma\alpha)(y). \tag{10.7}$$

We will prove this by induction. It is clearly true for $k = 0$. Note that

$$O_1(1_X, n) - \iota(O_1(1_X, n-1)) = -\xi_{n;n}.$$

Under the assumption that the formula (10.7) is true for k, we have

$$\begin{aligned}
& \sum_g \mathrm{ad}g\,[(-\xi_{n;n})^{\bullet(k+1)} \bullet (y \otimes \gamma\alpha)] \\
= \; & \sum_g \mathrm{ad}g\,[(O_1(1_X,n) - \iota(O_1(1_X,n-1))) \bullet (-\xi_{n;n})^{\bullet k} \bullet (y \otimes \gamma\alpha)] \\
= \; & O_1(1_X,n) \bullet \sum_g \mathrm{ad}g\,[(-\xi_{n;n})^{\bullet k} \bullet (y \otimes \gamma\alpha)] \\
& - \sum_g \mathrm{ad}g\,[\iota(O_1(1_X,n-1)) \bullet (-\xi_{n;n})^{\bullet k} \bullet (y \otimes \gamma\alpha)],
\end{aligned}$$

since $O_1(\gamma,n)$ is S_n-invariant. By using the induction assumption twice, we get

$$\begin{aligned}
& \sum_g \mathrm{ad}g\,[(-\xi_{n;n})^{\bullet(k+1)} \bullet (y \otimes \gamma\alpha)] \\
= \; & (n-1)!\, O_1(1_X,n) \bullet \mathfrak{p}_{-1}^{\{k\}}(\gamma\alpha)(y) \\
& - \sum_g \mathrm{ad}g\,[(-\xi_{n;n})^{\bullet k} \bullet ((O_1(1_X,n-1) \bullet y) \otimes \gamma\alpha)] \\
= \; & (n-1)!\, [\mathfrak{D}_1(1_X) \cdot \mathfrak{p}_{-1}^{\{k\}}(\gamma\alpha)(y) - \mathfrak{p}_{-1}^{\{k\}}(\gamma\alpha)(O_1(1_X,n-1) \bullet y)] \\
= \; & (n-1)!\, \mathfrak{p}_{-1}^{\{k+1\}}(\gamma\alpha)(y).
\end{aligned}$$

So by induction, we have established (10.7) and thus the theorem. □

We also have a theorem concerning the operator $\eta(\gamma)$. It generalizes [**LS2**, Proposition 4.6] which corresponds to the special case when $\gamma = 1_X$ (the assumption there that X is a surface is unnecessary).

THEOREM 10.9. *Let $\gamma, \alpha \in H^*(X)$. We further assume that γ can be written as a sum of classes of even degree. Then we have the following operator identities:*

$$\begin{aligned}
\eta(\gamma) \cdot \mathfrak{p}_{-1}(\alpha) &= \mathfrak{p}_{-1}(\gamma\alpha) \cdot \eta(\gamma) - \mathfrak{p}'_{-1}(\alpha) \cdot \eta(\gamma), \\
\varepsilon(\gamma) \cdot \mathfrak{p}_{-1}(\alpha) &= \mathfrak{p}_{-1}(\gamma\alpha) \cdot \varepsilon(\gamma) + \mathfrak{p}'_{-1}(\alpha) \cdot \varepsilon(\gamma).
\end{aligned}$$

PROOF. The proof of the second formula is similar to that of the first, so we will prove the first one only. For simplicity of signs in the proof, we assume that the cohomology class α is of even degree.

By definition,

$$\eta_n(\gamma) = \prod_{i=1}^{n} (\xi_{i;n} + \gamma^{(i)}).$$

It follows that

$$\eta_n(\gamma) - \eta_{n-1}(\gamma) \otimes \gamma = \xi_{n;n} \bullet \iota(\eta_{n-1}(\gamma)).$$

Given $y \in H^*_{\rm orb}(X^{(n-1)})$, we have

$$\begin{aligned}
& (n-1)!\, [\eta(\gamma)\cdot\mathfrak{p}_{-1}(\alpha)(y) - \mathfrak{p}_{-1}(\gamma\alpha)\cdot \eta(\gamma)(y)] \\
= \;& \eta_n(\gamma) \bullet \sum_{g\in S_n} \mathrm{ad}g\,(y\otimes\alpha) - \sum_{g\in S_n} \mathrm{ad}g\,[(\eta_{n-1}(\gamma)\bullet y)\otimes\gamma\alpha] \\
= \;& \sum_g \mathrm{ad}g\,[(\eta_n(\gamma) - \eta_{n-1}(\gamma)\otimes\gamma)\bullet(y\otimes\alpha)] \\
= \;& \sum_g \mathrm{ad}g\,[\xi_{n;n}\bullet\iota(\eta_{n-1}(\gamma))\bullet(y\otimes\alpha)] \\
= \;& \sum_g \mathrm{ad}g\,[\xi_{n;n}\bullet((\eta_{n-1}(\gamma)\bullet y)\otimes\alpha)] \\
= \;& -O_1(1_X,n)\bullet\sum_g \mathrm{ad}g\,[(\eta_{n-1}(\gamma)\bullet y)\otimes\alpha] \\
& +\sum_g \mathrm{ad}g\,[(O_1(1_X,n-1)\bullet\eta_{n-1}(\gamma)\bullet y)\otimes\alpha] \\
= \;& (n-1)!\,[-\mathfrak{O}_1(1_X)\cdot\mathfrak{p}_{-1}(\alpha)\cdot\eta(\gamma)(y) + \mathfrak{p}_{-1}(\alpha)\cdot\mathfrak{O}_1(1_X)\cdot\eta(\gamma)(y)] \\
= \;& -(n-1)!\,\mathfrak{p}'_{-1}(\alpha)\cdot\eta(\gamma)(y).
\end{aligned}$$

This finishes the proof. □

10.5. The ring structure of $H^*_{\rm orb}(X^{(n)})$

Using Theorem 10.1, Theorem 10.6 and Theorem 10.8, we can follow the approaches in Chapter 3, Chapter 4 and Chapter 7 to study the ring structure of $H^*_{\rm orb}(X^{(n)})$. Note that the terms involving the canonical class K_X of X in various formulas for the Hilbert schemes $X^{[n]}$ will disappear because there is no K_X-term in Theorem 10.6. Also, the fact $e_X^2 = 0$ was used in previous chapters (where X is a surface and e_X is the Euler class of X). Thus, when dealing with the ring $H^*_{\rm orb}(X^{(n)})$ in this section, we sometimes need to treat separately and carefully the case when X is a point (i.e. when $e_X^2 \neq 0$).

Recall the classes $P_i(\alpha,n)$ and $O_i(\alpha,n)$ from (10.1) and (10.6) respectively. The next result is parallel to Theorem 7.5.

THEOREM 10.10. *Let X be a closed complex manifold of even dimension. Then, the ring $H^*_{orb}(X^{(n)})$ is generated by the cohomology classes $O_i(\alpha,n)$ (respectively, the cohomology classes $P_i(\alpha,n)$) where $0 \le i < n$ and α runs over a fixed linear basis of $H^*(X)$.*

THEOREM 10.11. *Let X be a closed complex manifold of even dimension. Then the ring $H^*_{orb}(X^{(n)})$ is determined uniquely by the ring $H^*(X)$.*

We refer to Theorem 10.11, and Proposition 10.12 and Proposition 10.13 below as the universality of the ring $H^*_{\rm orb}(X^{(n)})$. Theorem 10.11 follows from the more quantitative descriptions of the orbifold cup product of ring generators of $H^*_{\rm orb}(X^{(n)})$ in Proposition 10.12 and Proposition 10.13. It also follows from the results in [**FG1**]. Let $\mathbf{1}_{-k} = 0$ if $k < 0$, and

$$\mathbf{1}_{-k} = \frac{1}{k!}\,\mathfrak{p}_{-1}(1_X)^k$$

if $k \ge 0$. The following proposition is parallel to Lemma 4.13.

PROPOSITION 10.12. *(Universality) Let X be a closed complex manifold of even dimension $d > 0$. Let $n, s \geq 1$, $k_1, \ldots, k_s \geq 0$, and let $\alpha_1, \ldots, \alpha_s \in H^*(X)$ be homogeneous. Then, the orbifold product $O_{k_1}(\alpha_1, n) \bullet \cdots \bullet O_{k_s}(\alpha_s, n)$ in $H^*_{orb}(X^{(n)})$ is a finite linear combination of expressions of the form*

$$\mathrm{sign}(\alpha, \pi) \cdot \mathbf{1}_{-\left(n - \sum_{i=1}^{\ell(\pi)} \sum_{j=1}^{m_i - r_i} n_{i,j}\right)} \prod_{i=1}^{\ell(\pi)} \left(\prod_{j=1}^{m_i - r_i} \mathfrak{p}_{-n_{i,j}} \right) (\tau_{(m_i - r_i)*}(\epsilon_i \alpha_{\pi_i})) \cdot |0\rangle$$

where π runs over all partitions of $\{1, \ldots, s\}$, $\epsilon_i \in \{1_X, e_X\}$,

$$r_i = |\epsilon_i|/d \leq m_i \leq 2 + \sum_{j \in \pi_i} k_j,$$

$0 < n_{i,1} \leq \ldots \leq n_{i, m_i - r_i}$, $\sum_{j=1}^{m_i - r_i} n_{i,j} \leq \sum_{j \in \pi_i} (k_j + 1)$ *for every i, and*

$$\sum_{i=1}^{\ell(\pi)} \left(m_i - 2 + \sum_{j=1}^{m_i - r_i} n_{i,j} \right) = \sum_{i=1}^{s} k_i.$$

Moreover, all the coefficients in this linear combination are independent of the manifold X, the cohomology classes $\alpha_1, \ldots, \alpha_s$, and the integer n.

For the case $d = 0$ (i.e., X is a point), we adopt the simplified notations $\mathfrak{p}_m$ and $O_k(n)$ for $\mathfrak{p}_m(1_X)$ and $O_k(1_X, n)$ respectively. We have the following analog of Proposition 10.12.

PROPOSITION 10.13. *Let $n, s \geq 1$, $k_1, \ldots, k_s \geq 0$. Then, $O_{k_1}(n) \bullet \cdots \bullet O_{k_s}(n)$ is a finite linear combination of expressions of the form*

$$\mathbf{1}_{-\left(n - \sum_{i=1}^{\ell(\pi)} \sum_{j=1}^{m_i - 2r_i} n_{i,j}\right)} \prod_{i=1}^{\ell(\pi)} \left(\prod_{j=1}^{m_i - 2r_i} \mathfrak{p}_{-n_{i,j}} \right) \cdot |0\rangle$$

where π runs over all partitions of $\{1, \ldots, s\}$, $m_i, r_i \in \mathbb{Z}_+$ such that

$$2r_i \leq m_i \leq 2 + \sum_{j \in \pi_i} k_j,$$

$0 < n_{i,1} \leq \ldots \leq n_{i, m_i - 2r_i}$, $\sum_{j=1}^{m_i - 2r_i} n_{i,j} \leq \sum_{j \in \pi_i} (k_j + 1)$ *for every i, and*

$$\sum_{i=1}^{\ell(\pi)} \left(m_i - 2 + \sum_{j=1}^{m_i - 2r_i} n_{i,j} \right) = \sum_{i=1}^{s} k_i.$$

Moreover, all the coefficients in this linear combination are independent of n.

The stability of the ring $H^*_{\rm orb}(X^{(n)})$ is given by the following which is parallel to Theorem 7.7.

THEOREM 10.14. *Let X be a closed complex manifold of even dimension d. Let $s \geq 1$ and $k_i \geq 1$ for $1 \leq i \leq s$. Fix $n_{i,j} \geq 1$ and $\alpha_{i,j} \in H^*(X)$ for $1 \leq j \leq k_i$, and* *fix n with $n \ge \sum_{j=1}^{k_i} n_{i,j}$ for all $1 \le i \le s$. Then the orbifold cup product*

$$\prod_{i=1}^{s} \left(\mathbf{1}_{-(n-\sum_{j=1}^{k_i} n_{i,j})} \prod_{j=1}^{k_i} \mathfrak{p}_{-n_{i,j}}(\alpha_{i,j}) \cdot |0\rangle \right)$$

in $H^*_{orb}(X^{(n)})$ is equal to a finite linear combination of monomials of the form

$$\mathbf{1}_{-(n-\sum_{a=1}^{N} m_a)} \prod_{a=1}^{N} \mathfrak{p}_{-m_a}(\gamma_a) \cdot |0\rangle$$

where

$$\sum_{a=1}^{N} m_a \le \sum_{i=1}^{s} \sum_{j=1}^{k_i} n_{i,j},$$

and $\gamma_1, \ldots, \gamma_N$ depend only on $e_X, \alpha_{i,j}$, $1 \le i \le s, 1 \le j \le k_i$. Moreover, the coefficients in this linear combination are independent of $\alpha_{i,j}$ and n. These coefficients are also independent of X provided $d > 0$.

Parallel to Section 7.2, we can use Theorem 10.14 to define a ring $\mathfrak{R}_X$, called *the stable ring*, which completely encodes the ring structure of $H^*_{\rm orb}(X^{(n)})$ for each n. The stable ring $\mathfrak{R}_X$ depends only on the cohomology ring $H^*(X)$. We refer to [**QW1**, Subsection 4.4] for details.

10.6. The $\mathcal{W}$ algebras

In this section, we will assume that X is a closed complex manifold of even dimension $d > 0$. Results here are counterparts of those in Section 4.4. However, some signs have been modified due to the sign difference between the two Heisenberg algebra commutators in the setups of Hilbert schemes and symmetric products (see Theorem 3.8 and Theorem 10.1).

For $p \ge 0$, $m \in \mathbb{Z}$ and $\alpha \in H^*(X)$, define $\widetilde{\mathfrak{J}}^p_m(\alpha) \in \mathrm{End}(\mathcal{F}_X)$ to be

$$p! \cdot \left(\sum_{\ell(\lambda)=p+1, |\lambda|=m} \frac{1}{\lambda^!} \mathfrak{p}_\lambda(\tau_*\alpha) + \sum_{\ell(\lambda)=p-1, |\lambda|=m} \frac{s(\lambda)+m^2-2}{24\lambda^!} \mathfrak{p}_\lambda(\tau_*(e_X\alpha)) \right)$$

where the λ's are generalized partitions. Note that $\widetilde{\mathfrak{J}}^0_m(\alpha) = \mathfrak{p}_m(\alpha)$. We define $\widetilde{\mathcal{W}}_X$ to be the linear span of the identity operator $\mathrm{Id}_{\mathcal{F}_X}$ and the operators $\widetilde{\mathfrak{J}}^p_m(\alpha)$ in $\mathrm{End}(\mathcal{F}_X)$, where $p \ge 0, m \in \mathbb{Z}$ and $\alpha \in H^*(X)$.

The following theorem describes the operator $\mathfrak{D}_k(\alpha)$ in terms of the Heisenberg generators explicitly. It is a counterpart of Theorem 4.7.

THEOREM 10.15. *Let $k \ge 0$, and $\alpha \in H^*(X)$. Then, $\mathfrak{D}_k(\alpha)$ is equal to*

$$(-1)^k \cdot \left(\sum_{\ell(\lambda)=k+2, |\lambda|=0} \frac{1}{\lambda^!} \mathfrak{p}_\lambda(\tau_*\alpha) + \sum_{\ell(\lambda)=k, |\lambda|=0} \frac{s(\lambda)-2}{24\lambda^!} \mathfrak{p}_\lambda(\tau_*(e_X\alpha)) \right).$$

In view of Theorem 10.15 and the definition of $\widetilde{\mathfrak{J}}^p_m(\alpha) \in \mathrm{End}(\mathcal{F}_X)$, we have

$$\mathfrak{D}_k(\alpha) = \frac{(-1)^k}{(k+1)!} \cdot \widetilde{\mathfrak{J}}^{k+1}_0(\alpha).$$

Since $O_k(\alpha, n) = \mathfrak{O}_k(\alpha)\mathfrak{p}_{-1}(1_X)^n|0\rangle/n!$, we conclude from Theorem 10.15 that $O_k(\alpha, n)$ is equal to

$$(10.8) \qquad (-1)^k \cdot \left(\sum_{\substack{0\le j\le k, \lambda\vdash(j+1)\\ \ell(\lambda)=k-j+1}} \frac{1}{\lambda^! \cdot |\lambda|!} \cdot \mathbf{1}_{-(n-j-1)}\mathfrak{p}_{-\lambda}(\tau_*\alpha)|0\rangle \right.$$

$$\left. + \sum_{0\le j\le k} \sum_{\substack{\lambda\vdash(j+1)\\ \ell(\lambda)=k-j-1}} \frac{1}{\lambda^! \cdot |\lambda|!} \cdot \frac{|\lambda| + s(\lambda) - 2}{24} \cdot \mathbf{1}_{-(n-j-1)}\mathfrak{p}_{-\lambda}(\tau_*(e_X\alpha))|0\rangle \right).$$

In terms of vertex operators, the operator $\widetilde{\mathfrak{J}}^p_m(\alpha)$ can be rewritten as:

$$(10.9) \qquad \frac{1}{(p+1)} :\mathfrak{p}^{p+1}:_m (\tau_*\alpha) + \frac{1}{24}p(m^2 - 3m - 2p) :\mathfrak{p}^{p-1}:_m (\tau_*(e_X\alpha))$$
$$+\frac{p(p-1)}{24} :(\partial^2\mathfrak{p})\,\mathfrak{p}^{p-2}:_m (\tau_*(e_X\alpha)).$$

If we want the coefficients above to be independent of m, we can further rewrite

$$\begin{aligned}
\widetilde{\mathfrak{J}}^p_m(\alpha) &= \frac{1}{(p+1)} :\mathfrak{p}^{p+1}:_m (\tau_*\alpha) + \frac{p}{24}(\partial^2 :\mathfrak{p}^{p-1}:)_m(\tau_*(e_X\alpha)) \\
&\quad +\frac{(p+1)p}{12}(\partial :\mathfrak{p}^{p-1}:)_m(\tau_*(e_X\alpha)) \\
&\quad +\frac{p(p^2-p-2)}{24} :\mathfrak{p}^{p-1}:_m (\tau_*(e_X\alpha)) \\
&\quad +\frac{p(p-1)}{24} :(\partial^2\mathfrak{p})\mathfrak{p}^{p-2}:_m (\tau_*(e_X\alpha)).
\end{aligned}$$

The operators $\mathfrak{O}_p(\alpha), \mathfrak{p}_n(\alpha)$, and $\widetilde{\mathfrak{J}}^p_n(\alpha)$ are related in the following way.

PROPOSITION 10.16. *Given $p \ge 0$, $\alpha, \beta \in H^*(X)$, we have*

$$[\mathfrak{O}_p(\alpha), \mathfrak{p}_n(\beta)] = -n \cdot \frac{(-1)^p}{p!} \cdot \widetilde{\mathfrak{J}}^p_n(\alpha\beta).$$

Next, recall the number $\Omega^{p,q}_{m,n}$ defined by (4.48). The following is parallel to Theorem 4.25.

THEOREM 10.17. *Let X be a closed complex manifold of even dimension $d > 0$. The vector space $\widetilde{\mathcal{W}}_X$ is closed under the Lie bracket. More explicitly, for $m, n \in \mathbb{Z}$, and $\alpha, \beta \in H^*(X)$, we have*

$$[\widetilde{\mathfrak{J}}^p_m(\alpha), \widetilde{\mathfrak{J}}^q_n(\beta)] = (qm - pn) \cdot \widetilde{\mathfrak{J}}^{p+q-1}_{m+n}(\alpha\beta) + \frac{\Omega^{p,q}_{m,n}}{12} \cdot \widetilde{\mathfrak{J}}^{p+q-3}_{m+n}(e_X\alpha\beta)$$

where $(p,q) \in \mathbb{Z}_+^2$ *except for the* unordered *pairs* $(0,0),(1,0),(2,0)$ *and* $(1,1)$. *In addition, for these four exceptional cases, we have*

$$\begin{aligned}
[\widetilde{\mathfrak{J}}^0_m(\alpha), \widetilde{\mathfrak{J}}^0_n(\beta)] &= m\delta_{m,-n}\int_X (\alpha\beta)\cdot \mathrm{Id}_{\mathcal{F}_X},\\
[\widetilde{\mathfrak{J}}^1_m(\alpha), \widetilde{\mathfrak{J}}^0_n(\beta)] &= -n\cdot \widetilde{\mathfrak{J}}^0_{m+n}(\alpha\beta),\\
[\widetilde{\mathfrak{J}}^2_m(\alpha), \widetilde{\mathfrak{J}}^0_n(\beta)] &= -2n\cdot \widetilde{\mathfrak{J}}^1_{m+n}(\alpha\beta) + \frac{m^3-m}{6}\delta_{m,-n}\int_X (e_X\alpha\beta)\cdot \mathrm{Id}_{\mathcal{F}_X},\\
[\widetilde{\mathfrak{J}}^1_m(\alpha), \widetilde{\mathfrak{J}}^1_n(\beta)] &= (m-n)\cdot \widetilde{\mathfrak{J}}^1_{m+n}(\alpha\beta) + \frac{m^3-m}{12}\delta_{m,-n}\int_X (e_X\alpha\beta)\cdot \mathrm{Id}_{\mathcal{F}_X}.
\end{aligned}$$

REMARK 10.18. The $\mathcal{W}$ algebra $\widetilde{\mathcal{W}}_X$ should be viewed as a generalization of the $\mathcal{W}_\infty$ algebra. The assumption $d>0$ above ensures that $e_X^2=0$. The case when $d=0$ (i.e. X is a point) has been treated in [**LT**].

Part 4

Equivariant cohomology of the Hilbert schemes of points

CHAPTER 11

Equivariant cohomology of Hilbert schemes

There exist fundamental and deep connections between the geometry of the Hilbert schemes $X^{[n]}$ of points on a (quasi-)projective surface X and combinatorics of symmetric functions. The monomial symmetric functions can be realized as certain ordinary cohomology classes of the Hilbert schemes associated to an embedded curve in a surface ((9.12) and [**Nak5**, (9.13)]). Nakajima [**Nak7**] further showed that the Jack polynomials whose Jack parameter is a positive integer γ can be realized as certain $\mathbb{T}$-equivariant cohomology classes of the Hilbert schemes of points on the surface $X(\gamma)$ which is the total space of the line bundle $\mathcal{O}_{\mathbb{P}^1}(-\gamma)$ over the complex projective line $\mathbb{P}^1$. Here $\mathbb{T}$ stands for the one-dimensional complex torus, and the Jack parameter is interpreted as minus the self-intersection number of the zero-section in $X(\gamma)$. With very different motivations, Haiman [**Hai1**, **Hai2**, **Hai3**] developed connections between the Macdonald polynomials and the geometry of Hilbert schemes, and in particular realized the Macdonald polynomials as certain $\mathbb{T}$-equivariant K-homology classes of the Hilbert schemes of points on the affine plane $\mathbb{C}^2$. Moreover, the equivariant cohomology of Hilbert schemes are related to Toda hierarchies, Hurwitz theory, equivariant (local) Gromov-Witten theory, and equivariant (local) Donaldson-Thomas theory. These connections will be covered in Chapters 12 and 14.

In this chapter, we will deal with the equivariant cohomology of the Hilbert schemes $X^{[n]}$ where X is either the complex affine plane $\mathbb{C}^2$ or the total space of a line bundle over $\mathbb{P}^1$, and is equipped with a torus action. As in Section 3.2, Heisenberg algebra actions on the equivariant cohomology of the Hilbert schemes $X^{[n]}$ will be constructed. These actions enable the interpretation of the equivariant cohomology of $X^{[n]}$ in terms of the ring of symmetric functions. In addition, the equivariant cohomology classes corresponding to the Jack symmetric functions will be identified. This chapter follows the presentation of [**LQW7**] which generalizes the results in [**Nak6**, **Vas**]. We refer to [**Nak7**] for an excellent expository on further topics such as geometric proofs of the norm formula and Pieri formula of Jack symmetric functions, and a representation of the Virasoro algebra on the equivariant cohomology of the Hilbert schemes $(\mathbb{C}^2)^{[n]}$ (which is a special case of Lehn's result, Theorem 3.24).

11.1. Equivariant cohomology rings of Hilbert schemes

In this section, we will review the basics of equivariant cohomology and the localization theorem. Let X be either the complex affine plane $\mathbb{C}^2$ or the total space of a line bundle over $\mathbb{P}^1$. We will set up the torus actions on X, the fixed points of the induced actions on the Hilbert scheme $X^{[n]}$, and the distinguished equivariant cohomology classes associated to the fixed points. Moreover, we will introduce a nondegenerate bilinear form on the space

$$\mathbb{H}_X = \bigoplus_{n=0}^{+\infty} \mathbb{H}_n$$

where $\mathbb{H}_n$ denotes the middle equivariant cohomology $H^{2n}_{\mathbb{T}}(X^{[n]})$.

We begin with a brief review of equivariant cohomology and the localization theorem. Let G be a connected Lie group, and let $EG \to BG$ be the universal principal G-bundle where the total space EG is contractible. Let M be a topological space with a G-action. Then the *equivariant cohomology* of X is defined to be

$$H^*_G(M) = H^*(M \times_G EG).$$

Note that $H^*_G(\mathrm{pt}) = H^*(BG)$ where pt denotes a point. Via the trivial map $M \to \mathrm{pt}$, we see that $H^*_G(M)$ is an $H^*(BG)$-module.

Let $\mathbb{T} = (\mathbb{C}^*)^n$. For an algebraic variety M with a $\mathbb{T}$-action, the equivariant cohomology $H^*_{\mathbb{T}}(M)$ is a $\mathbb{C}[t_1, \ldots, t_n]$-module if we identify $H^*_{\mathbb{T}}(\mathrm{pt})$ with $\mathbb{C}[t_1, \ldots, t_n]$ (each t_i is an element of degree 2). For a $\mathbb{T}$-equivariant and proper morphism $f: N \to M$ of algebraic varieties, there is a Gysin homomorphism

$$f_!: H^*_{\mathbb{T}}(N) \to H^*_{\mathbb{T}}(M)$$

of $\mathbb{T}$-equivariant cohomology groups. If N is a $\mathbb{T}$-equivariant codimension-k closed subvariety of M and $\iota : N \to M$ is the inclusion map, define

$$[N] = \iota_!(1_N) \in H^{2k}_{\mathbb{T}}(M)$$

where $1_N \in H^0_{\mathbb{T}}(N)$ is the unit of the algebra $H^*_{\mathbb{T}}(N)$.

Let M be a smooth algebraic variety with a $\mathbb{T}$-action. It is known that the fixed point set $M^{\mathbb{T}}$ is a union of smooth connected components Z_j. Let

$$\iota_j : Z_j \to M$$

be the inclusion map, and N_j be the normal bundle of Z_j in M. Since N_j is an equivariant vector bundle, it has an equivariant Euler class:

$$e_{\mathbb{T}}(N_j) \in H^*_{\mathbb{T}}(Z_j).$$

Then for every $\alpha \in H^*_{\mathbb{T}}(Z_j)$, we have

$$\iota_j^*(\iota_{j!}(\alpha)) = \alpha \cup e_{\mathbb{T}}(N_j).$$

THEOREM 11.1. (Localization Theorem) *There is an isomorphism*

$$H^*_{\mathbb{T}}(X) \otimes \mathbb{C}(t_1, \ldots, t_n) \xrightarrow{\cong} \bigoplus_j H^*_{\mathbb{T}}(Z_j) \otimes \mathbb{C}(t_1, \ldots, t_n)$$

induced by the map $\alpha \mapsto \left(\iota_j^*\alpha / e_{\mathbb{T}}(N_j)\right)_j$. *Moreover, the inverse is induced by* $(\alpha_j)_j \mapsto \sum_j \iota_{j!}(\alpha_j)$. *In particular, for every* $\alpha \in H^*_{\mathbb{T}}(X) \otimes \mathbb{C}(t_1, \ldots, t_n)$, *we have*

$$\alpha = \sum_j \iota_{j!} \left(\frac{\iota_j^* \alpha}{e_{\mathbb{T}}(N_j)} \right).$$

In the rest of this chapter, we will let

$$\mathbb{T} = \mathbb{C}^*,$$

and θ be the 1-dimensional standard $\mathbb{T}$-module. We will consider two types of surfaces with $\mathbb{T}$-actions and the Hilbert schemes of points on these surfaces. The first one is the complex plane $\mathbb{C}^2$, while the second is the total space of a line bundle over $\mathbb{P}^1$. The $\mathbb{T}$-actions on these surfaces are specified in the following two examples.

EXAMPLE 11.2. Fix two nonzero integers α and β with the same signs. Let u, v be the standard coordinate functions on $\mathbb{C}^2$. We define the action of $\mathbb{T}$ on $\mathbb{C}^2$ by

$$s \cdot (u, v) = (s^{\alpha}u, s^{-\beta}v), \quad s \in \mathbb{T}. \tag{11.1}$$

The origin of $\mathbb{C}^2$ is the only fixed point, which will be denoted by x. Let Σ and Σ' be the u-axis and v-axis respectively in $\mathbb{C}^2$. As $\mathbb{T}$-modules, we have $T_{x,\Sigma} = \theta^{-\alpha}$ and $T_{x,\Sigma'} = \theta^{\beta}$. By the localization theorem, we get

$$[\Sigma] = -\alpha^{-1}t^{-1}[x], \quad [\Sigma'] = \beta^{-1}t^{-1}[x]. \tag{11.2}$$

EXAMPLE 11.3. Fix an integer $\gamma > 1$. Let $X(\gamma)$ be the total space of the line bundle $\mathcal{O}_{\mathbb{P}^1}(-\gamma)$ over $\mathbb{P}^1$. The quasi-projective surface $X(\gamma)$ can be regarded as the quotient space of $\mathbb{C} \times (\mathbb{C}^2 - \{0\})$ by the $\mathbb{C}^*$-action defined by

$$s \cdot (b, b_1, b_2) = (s^{-\gamma}b, sb_1, sb_2), \quad s \in \mathbb{C}^*. \tag{11.3}$$

We use $[(b, b_1, b_2)]$ to denote the equivalence class. Define a $\mathbb{T}$-action on $X(\gamma)$ by

$$s \cdot [(b, b_1, b_2)] = [(sb, s^{-1}b_1, b_2)], \quad s \in \mathbb{T}. \tag{11.4}$$

For $i = 1, 2$, let X_i be the open subset of $X(\gamma)$ given by

$$X_1 = \{[(b, b_1, b_2)] \mid b_2 = 1\}, \quad X_2 = \{[(b, b_1, b_2)] \mid b_1 = 1\}. \tag{11.5}$$

Then X_1 and X_2 form an affine open cover of $X(\gamma)$. Moreover, each X_i is $\mathbb{T}$-invariant. For simplicity, denote the point $[(b, b_1, 1)] \in X_1$ by (b, b_1). Similarly, denote $[(b, 1, b_2)] \in X_2$ by (b, b_2). Then $\mathbb{T}$ acts on the points of X_1 by

$$s \cdot (b, b_1) = (sb, s^{-1}b_1),$$

i.e., $\mathbb{T}$ acts on the coordinate functions u_1 and v_1 of X_1 by

$$s \cdot (u_1, v_1) = (s^{-1}u_1, sv_1), \quad s \in \mathbb{T}. \tag{11.6}$$

Similarly, $\mathbb{T}$ acts on the coordinate functions u_2 and v_2 of X_2 by

$$s \cdot (u_2, v_2) = (s^{\gamma-1}u_2, s^{-1}v_2), \quad s \in \mathbb{T}. \tag{11.7}$$

Let x_i be the origin of X_i. Then $X(\gamma)^{\mathbb{T}} = \{x_1, x_2\}$. Let

$$\rho : X(\gamma) \to \mathbb{P}^1$$

be the projection sending $[(b, b_1, b_2)]$ to $[b_1, b_2]$. Let $\Sigma_0 \cong \mathbb{P}^1$ be the zero section of ρ, and

$$\Sigma_1 = \rho^{-1}([0, 1]), \qquad \Sigma_2 = \rho^{-1}([1, 0]). \tag{11.8}$$

Then as $\mathbb{T}$-modules, we have

$$T_{x_1,\Sigma_1} = \theta, \quad T_{x_1,\Sigma_0} = \theta^{-1}, \quad T_{x_2,\Sigma_0} = \theta, \quad T_{x_2,\Sigma_2} = \theta^{1-\gamma}.$$

By the localization theorem, we get

$$[\Sigma_1] = t^{-1}[x_1], \; [\Sigma_0] = -t^{-1}[x_1] + t^{-1}[x_2], \; [\Sigma_2] = (1-\gamma)^{-1}t^{-1}[x_2]. \tag{11.9}$$

REMARK 11.4. (i) Let $X(1)$ be the total space of $\mathcal{O}_{\mathbb{P}^1}(-1)$. Using $[(b, b_1, b_2)] \in X(1)$ to denote the equivalence class defined by (11.3), we define a $\mathbb{T}$-action on $X(1)$ by

$$s \cdot [(b, b_1, b_2)] = [(sb, s^{-2}b_1, b_2)], \quad s \in \mathbb{T}.$$

Then the methods and results below apply to $X(1)$ as well.

(ii) Other $\mathbb{T}$-actions on the surfaces $X(\gamma)$, $\gamma \geq 1$, with isolated fixed points can be treated similarly.

In the rest of this chapter, let X be a surface in Example 11.2 or Example 11.3. Next we will define some distinguished equivariant cohomology classes for the Hilbert scheme $X^{[n]}$. The $\mathbb{T}$-action on X induces a $\mathbb{T}$-action on $X^{[n]}$. The support of a $\mathbb{T}$-fixed point in $X^{[n]}$ is contained in $X^{\mathbb{T}}$. As in Section 9.6, the $\mathbb{T}$-fixed points of $X^{[n]}$ are isolated and parametrized in terms of (multi-)partitions.

First of all, let $X = \mathbb{C}^2$ as in Example 11.2. The $\mathbb{T}$-fixed points of $X^{[n]}$ are supported in $X^{\mathbb{T}} = \{x\}$ and indexed by partitions λ of n. We use ξ_λ to denote the fixed point in $(X^{[n]})^{\mathbb{T}}$ corresponding to a partition λ of n, and use $T_{\xi_\lambda, X^{[n]}}$ to denote the $\mathbb{T}$-equivariant tangent space of $X^{[n]}$ at the fixed point ξ_λ. As in (9.58), we adopt the convention that ξ_λ is defined by the ideal

$$I_{\xi_\lambda} = \langle v^{\lambda_1}, uv^{\lambda_2}, \ldots, u^{k-1}v^{\lambda_k}, u^k \rangle \tag{11.10}$$

when $\lambda = (\lambda_1 \geq \lambda_2 \geq \cdots \geq \lambda_k)$. Then by [**Nak6**, Lemma 6.2] (also [**ES1**, **Got2**]), for each $\lambda \vdash n$, we have

$$T_{\xi_\lambda, X^{[n]}} = \bigoplus_{\square \in D_\lambda} \left(\theta^{\alpha(\ell(\square)+1)+\beta a(\square)} \oplus \theta^{-\alpha\ell(\square)-\beta(a(\square)+1)} \right) \tag{11.11}$$

where D_λ is the Young diagram associated to the partition λ, $\square$ is a cell in D_λ, $\ell(\square)$ is the leg length of $\square$, and $a(\square)$ is the arm length of $\square$ (we refer to Definition 1.4 for these notations). So

$$e_{\mathbb{T}}(T_{\xi_\lambda, X^{[n]}}) = (-1)^n c_\lambda(\alpha, \beta) c'_\lambda(\alpha, \beta) t^{2n} \tag{11.12}$$

where $e_{\mathbb{T}}(\cdot)$ stands for the equivariant Euler class and

$$c_\lambda(\alpha, \beta) = \prod_{\square \in D_\lambda} \big(\alpha(\ell(\square)+1) + \beta a(\square)\big), \tag{11.13}$$

$$c'_\lambda(\alpha, \beta) = \prod_{\square \in D_\lambda} \big(\alpha\ell(\square) + \beta(a(\square)+1)\big). \tag{11.14}$$

Note that $[\xi_\lambda] \in H^{4n}_{\mathbb{T}}(X^{[n]})$. We define the following distinguished class:

$$[\lambda] = \frac{(-1)^n}{c_\lambda(\alpha, \beta)} t^{-n} [\xi_\lambda]. \tag{11.15}$$

Now let $X = X(\gamma)$ as in Example 11.3. In view of (11.1) and (11.6), there is a $\mathbb{T}$-equivariant identification between X_1 and the complex plane in Example 11.2 with $\alpha = \beta = -1$. Similarly, there is a $\mathbb{T}$-equivariant identification between X_2 and the complex plane in Example 11.2 with $\alpha = \gamma - 1$ and $\beta = 1$. Then the $\mathbb{T}$-fixed points of $X^{[n]}$ are of the form $\xi_{\lambda^1} + \xi_{\lambda^2}$ where λ^1 and λ^2 are partitions with $|\lambda^1| + |\lambda^2| = n$, and ξ_{λ^1} and ξ_{λ^2} are defined in the previous paragraph as we identify X_1 and X_2 with $\mathbb{C}^2$ respectively. For simplicity, put

$$\xi_{\lambda^1, \lambda^2} = \xi_{\lambda^1} + \xi_{\lambda^2}.$$

We have a $\mathbb{T}$-equivariant splitting of tangent spaces:

$$T_{\xi_{\lambda^1,\lambda^2}, X^{[n]}} \cong T_{\xi_{\lambda^1}, X_1^{[|\lambda^1|]}} \oplus T_{\xi_{\lambda^2}, X_2^{[|\lambda^2|]}}.$$

By (11.12),

$$e_{\mathbb{T}}(T_{\xi_{\lambda^1,\lambda^2}, X^{[n]}}) = (-1)^n c_{\lambda^1}(-1,-1) c'_{\lambda^1}(-1,-1) c_{\lambda^2}(\gamma-1,1) c'_{\lambda^2}(\gamma-1,1) t^{2n}.$$

Also, as in (11.15), we introduce the distinguished class

$$(11.16) \qquad [\lambda^1, \lambda^2] = \frac{(-1)^n}{c_{\lambda^1}(-1,-1) c_{\lambda^2}(\gamma-1,1)} t^{-n} [\xi_{\lambda^1,\lambda^2}].$$

In the following, we study bilinear forms on the equivariant cohomology of the Hilbert schemes. Define

$$(11.17) \qquad \mathbb{H}_n = H^{2n}_{\mathbb{T}}(X^{[n]}).$$

Since $X^{[n]}$ admits a cell decomposition, its odd Betti numbers are equal to zero. By calculating the dimensions of the cells, we conclude that $H^k(X^{[n]}) = 0$ for $k > 2n$. Therefore, the spectral sequence associated with the fibration $X^{[n]} \times_{\mathbb{T}} E\mathbb{T} \to B\mathbb{T}$ degenerates at the E_2-term. We now have

$$H^{2k}_{\mathbb{T}}(X^{[n]}) = t^{k-n} \cup H^{2n}_{\mathbb{T}}(X^{[n]})$$

for $k \geq n$, which shows that the classes defined in (11.15) and (11.16) are contained in $\mathbb{H}_n$. Furthermore, following [**Vas**, Section B], we define a product structure $\star$ on $\mathbb{H}_n$ by

$$(11.18) \qquad t^n \cup (A \star B) = A \cup B \in H^{4n}_{\mathbb{T}}(X^{[n]})$$

for A, $B \in H^{2n}_{\mathbb{T}}(X^{[n]})$. We see that $(\mathbb{H}_n, \star)$ is a ring.

Let

$$H^*_{\mathbb{T}}(\cdot)' = H^*_{\mathbb{T}}(\cdot) \otimes_{\mathbb{C}[t]} \mathbb{C}(t)$$

be the localization, and let

$$(11.19) \qquad \mathbb{H}'_X = \bigoplus_{n=0}^{+\infty} H^*_{\mathbb{T}}(X^{[n]})'.$$

Let

$$\iota : (X^{[n]})^{\mathbb{T}} \to X^{[n]}$$

be the inclusion map. By abusing notations, we also use $\iota_!$ to denote the induced Gysin map on the localized equivariant cohomology groups:

$$(11.20) \qquad \iota_! : H^*_{\mathbb{T}}((X^{[n]})^{\mathbb{T}})' \to H^*_{\mathbb{T}}(X^{[n]})',$$

which is an isomorphism by the localization theorem.

Define a bilinear form $\langle \cdot, \cdot \rangle : H^*_{\mathbb{T}}(X^{[n]})' \times H^*_{\mathbb{T}}(X^{[n]})' \to \mathbb{C}(t)$ by putting

$$(11.21) \qquad \langle A, B \rangle = (-1)^n p_! \iota_!^{-1}(A \cup B)$$

where p is the projection $(X^{[n]})^{\mathbb{T}} \to \mathrm{pt}$. This induces a bilinear form $\langle \cdot, \cdot \rangle$ on $\mathbb{H}'_X$.

REMARK 11.5. Passing from $H^*_{\mathbb{T}}(X^{[n]})'$ to $\mathbb{H}_n$ does not lead to loss of information since we can recover the cup product and bilinear form on $H^*_{\mathbb{T}}(X^{[n]})'$ from those on $\mathbb{H}_n$. Thus, understanding the ring $\mathbb{H}_n$ is the same as understanding the equivariant intersection theory on $X^{[n]}$.

Next, we study the restriction of the bilinear form $\langle\cdot,\cdot\rangle$ to $\mathbb{H}_n = H^{2n}_{\mathbb{T}}(X^{[n]})$. When $X = \mathbb{C}^2$, we see from the projection formula and (11.12) that

$$\begin{aligned}[\xi_\lambda]\cup[\xi_\mu] &= i_{\lambda!}(1_{\xi_\lambda})\cup i_{\mu!}(1_{\xi_\mu}) \\ &= i_{\lambda!}\big(1_{\xi_\lambda}\cup i_\lambda^* i_{\mu!}(1_{\xi_\mu})\big) \\ &= \delta_{\lambda,\mu} e_{\mathbb{T}}(T_{\xi_\lambda,X^{[n]}})[\xi_\lambda] \\ &= \delta_{\lambda,\mu}(-1)^n c_\lambda(\alpha,\beta)c'_\lambda(\alpha,\beta)t^{2n}[\xi_\lambda]\end{aligned} \tag{11.22}$$

where $i_\lambda : \xi_\lambda \to X$ is the inclusion map. It follows from (11.15) and (11.21) that for $\lambda,\mu \vdash n$, we have

$$\langle[\lambda],[\mu]\rangle = \delta_{\lambda,\mu}\frac{c'_\lambda(\alpha,\beta)}{c_\lambda(\alpha,\beta)}. \tag{11.23}$$

By the localization theorem, we see that the classes $[\lambda]$, $\lambda \vdash n$ form a linear basis of the $\mathbb{C}$-vector space $\mathbb{H}_n$. Similarly, when $X = X(\gamma)$ is from Example 11.3,

$$\begin{aligned}\langle[\lambda^1,\lambda^2],[\mu^1,\mu^2]\rangle &= \delta_{\lambda^1,\mu^1}\delta_{\lambda^2,\mu^2}\frac{c'_{\lambda^2}(1-\gamma,-1)}{c_{\lambda^2}(1-\gamma,-1)} \\ &= \langle[\lambda^1],[\mu^1]\rangle\cdot\langle[\lambda^2],[\mu^2]\rangle,\end{aligned} \tag{11.24}$$

and the classes $[\lambda^1,\lambda^2]$, where $|\lambda^1|+|\lambda^2| = n$, form a linear basis of $\mathbb{H}_n$.

It follows that the restriction to $\mathbb{H}_n$ of the bilinear form $\langle\cdot,\cdot\rangle$ on $H^*_{\mathbb{T}}(X^{[n]})'$ is a nondegenerate bilinear form $\langle\cdot,\cdot\rangle : \mathbb{H}_n\times\mathbb{H}_n \to \mathbb{C}$. This induces a nondegenerate bilinear form

$$\langle\cdot,\cdot\rangle : \mathbb{H}_X\times\mathbb{H}_X \to \mathbb{C}$$

where the space $\mathbb{H}_X$ is defined by

$$\mathbb{H}_X = \bigoplus_{n=0}^{+\infty}\mathbb{H}_n. \tag{11.25}$$

11.2. Heisenberg algebras in equivariant setting

In this section, following [**Vas**], we will define Heisenberg algebra actions on the equivariant cohomology $\mathbb{H}_X$ of the Hilbert schemes $X^{[n]}$.

Let X be a surface in Example 11.2 or Example 11.3. Fix a positive integer i. For a $\mathbb{T}$-invariant closed subscheme $Y \subset X$, we define

$$Y_{n,i} = \Big\{(\xi,\eta)\in X^{[n+i]}\times X^{[n]} \,|\, \eta\subset\xi,\ \mathrm{Supp}(I_\eta/I_\xi) = \{y\},\ y\in Y\Big\}$$

where I_η and I_ξ are the sheaves of ideals corresponding to η and ξ respectively. Let p_1 and p_2 be the projections of $X^{[n+i]}\times X^{[n]}$ to the two factors. As in [**Vas**, B.4], we define a linear operator

$$\mathfrak{p}_{-i}([Y]) \in \mathrm{End}(\mathbb{H}'_X)$$

by

$$\mathfrak{p}_{-i}([Y])(A) = p_{1!}\Big(p_2^*A\cup[Y_{n,i}]\Big) \tag{11.26}$$

for $A\in H^*_{\mathbb{T}}(X^{[n]})'$. Note that the restriction of p_1 to $Y_{n,i}$ is proper. We define

$$\mathfrak{p}_i([Y]) \in \mathrm{End}(\mathbb{H}'_X)$$

to be the adjoint operator of $\mathfrak{p}_{-i}([Y])$ with respect to the bilinear form $\langle\cdot,\cdot\rangle$ on $\mathbb{H}'_X$. For $A \in H^*_{\mathbb{T}}(X^{[n]})'$, we have

$$\mathfrak{p}_i([Y])(A) = (-1)^i p'_{2!}\Big(\iota \times \mathrm{Id})_!^{-1}(p_1^* A \cup [Y_{n-i,i}]\Big) \tag{11.27}$$

where p'_2 is the projection of $(X^{[n]})^{\mathbb{T}} \times X^{[n-i]}$ to $X^{[n-i]}$. Finally we put

$$\mathfrak{p}_0([Y]) = 0.$$

In the following, assume that $Y \subset X$ denotes a $\mathbb{T}$-invariant closed curve. By the definition of $\mathfrak{p}_{-i}([Y])$ for $i > 0$, its restriction to $\mathbb{H}_X$ gives a linear operator in $\mathrm{End}(\mathbb{H}_X)$, denoted by $\mathfrak{p}_{-i}([Y])$ as well. Next, we recall from the previous section that there is a nondegenerate bilinear form

$$\langle\cdot,\cdot\rangle \colon \mathbb{H}_X \otimes_{\mathbb{C}} \mathbb{H}_X \to \mathbb{C},$$

which is the restriction of the bilinear form $\langle\cdot,\cdot\rangle$ on $\mathbb{H}'_X$. Thus, the restriction of $\mathfrak{p}_i([Y])$ to $\mathbb{H}_X$ is the adjoint operator of $\mathfrak{p}_{-i}([Y])$ with respect to the bilinear form $\langle\cdot,\cdot\rangle$ on $\mathbb{H}_X$, and hence is an operator in $\mathrm{End}(\mathbb{H}_X)$ which will also be denoted by $\mathfrak{p}_i([Y])$.

By (11.2) and (11.9), $H^2_{\mathbb{T}}(X)$ is linearly spanned by the classes $[Y]$ where Y denotes $\mathbb{T}$-invariant closed curves in X. So we can extend the notion $\mathfrak{p}_k([Y])$ linearly to obtain the operator $\mathfrak{p}_k(\omega) \in \mathrm{End}(\mathbb{H}_X)$ for an arbitrary class $\omega \in H^2_{\mathbb{T}}(X)$. Note from Example 11.2 (respectively, Example 11.3) that Σ and Σ' (respectively, Σ_0 and Σ_i where $i = 1$ or 2) intersect transversely at one point. This simple but crucial observation, together with an argument parallel to the proof of [**Vas**, Lemma 1], leads to the following.

PROPOSITION 11.6. *The operators $\mathfrak{p}_k(\omega)$, $k \in \mathbb{Z}$ and $\omega \in \mathbb{H}_1 = H^2_{\mathbb{T}}(X)$, acting on $\mathbb{H}_X$ satisfy the following Heisenberg commutation relation:*

$$[\mathfrak{p}_k(\omega_1), \mathfrak{p}_\ell(\omega_2)] = k\delta_{k,-\ell}\langle\omega_1,\omega_2\rangle\ \mathrm{Id}. \tag{11.28}$$

Furthermore, $\mathbb{H}_X$ becomes the Fock space over the Heisenberg algebra modeled on $H^2_{\mathbb{T}}(X)$ with the unit $|0\rangle \in H^0_{\mathbb{T}}(X^{[0]})$ of $H^*_{\mathbb{T}}(X^{[0]})$ being a highest weight vector. □

As in the non-equivariant setting, the Heisenberg algebra actions on the Fock space $\mathbb{H}_X$ can be used to study its structure. In the next section, we will describe the ring structure of $\mathbb{H}_X$ via the ring of symmetric functions.

11.3. Equivariant cohomology and Jack polynomials

Let X be either the complex affine plane $\mathbb{C}^2$ as in Example 11.2 or the total space of a line bundle over $\mathbb{P}^1$ as in Example 11.3. In this section, we will interpret the Fock space $\mathbb{H}_X$ in terms of the ring of symmetric functions. In particular, we will identify the equivariant cohomology classes corresponding to the monomial symmetric functions and the Jack symmetric functions.

11.3.1. The case of the complex plane $\mathbb{C}^2$. In this subsection, we consider $X = \mathbb{C}^2$ from Example 11.2. Let $\mathfrak{p}_i = \mathfrak{p}_i([\Sigma])$ for $i \in \mathbb{Z}$. Given a partition $\lambda = (\lambda_1, \ldots, \lambda_\ell) = (1^{m_1} 2^{m_2} \ldots)$, define

$$\tilde{\mathfrak{p}}_{-\lambda} = \frac{1}{\mathfrak{z}_\lambda} \prod_{i \geq 1} \mathfrak{p}_{-i}^{m_i} \tag{11.29}$$

where $\mathfrak{z}_\lambda$ is from (1.1). By (11.2), (11.15) and (11.23), we have

$$\langle [\Sigma], [\Sigma] \rangle = \beta/\alpha.$$

So by Proposition 11.6,

$$\Big\langle \tilde{\mathfrak{p}}_{-\lambda}|0\rangle, \tilde{\mathfrak{p}}_{-\mu}|0\rangle \Big\rangle = \delta_{\lambda,\mu} \, \frac{1}{\mathfrak{z}_\lambda} \, (\beta/\alpha)^{\ell(\lambda)}. \tag{11.30}$$

Let $X^{(n)}$ (respectively, $\Sigma^{(n)}$) be the n-th symmetric product of X (respectively, of Σ). Let $\rho_n \colon X^{[n]} \to X^{(n)}$ be the Hilbert-Chow morphism. As in (9.8), for $\lambda \vdash n$, define

$$S^n_\lambda \Sigma = \left\{ \sum_{i=1}^{\ell} \lambda_i y_i \in \Sigma^{(n)} \mid y_i \in \Sigma, \text{ and } y_i \neq y_j \text{ if } i \neq j \right\}.$$

As in (9.7), let $L^\lambda \Sigma$ be the closure of $\rho_n^{-1}(S^n_\lambda \Sigma)$ in $X^{[n]}$. An argument parallel to the proof of the Corollary 6.10 in [**Nak6**] (also [**Vas**, Lemma 2]) shows that

$$[L^\lambda \Sigma] = [\lambda] + \sum_{\mu < \lambda} c_{\lambda,\mu} [\mu] \tag{11.31}$$

where $c_{\lambda,\mu} \in \mathbb{C}$ and "$<$" denotes the dominance partial ordering of partitions.

Recall from Section 1.2 the ring Λ of symmetric functions in infinitely many variables, the space Λ^n of degree-n symmetric functions, the monomial symmetric function m_λ associated to a partition λ, the power sum symmetric function p_λ associated to λ, and the Jack symmetric function $P^{(\alpha)}_\lambda$. Put

$$\begin{aligned} \Lambda_{\mathbb{C}} &= \Lambda \otimes_{\mathbb{Z}} \mathbb{C}, \\ \Lambda^n_{\mathbb{C}} &= \Lambda^n \otimes_{\mathbb{Z}} \mathbb{C}, \\ \tilde{p}_\lambda &= \frac{1}{\mathfrak{z}_\lambda} \, p_\lambda. \end{aligned} \tag{11.32}$$

The symmetric functions $\tilde{p}_\lambda$ form a linear basis of $\Lambda_{\mathbb{C}}$ since $\{p_\lambda\}_\lambda$ is a linear basis of $\Lambda_{\mathbb{C}}$. So we can define a bilinear form $\langle \cdot, \cdot \rangle$ on $\Lambda_{\mathbb{C}}$ by

$$\langle \tilde{p}_\lambda, \tilde{p}_\mu \rangle = \delta_{\lambda,\mu} \, \frac{1}{\mathfrak{z}_\lambda} \, (\beta/\alpha)^{\ell(\lambda)}. \tag{11.33}$$

We introduce a ring structure $\circ$ on $\Lambda^n_{\mathbb{C}}$ defined by

$$\frac{P^{(\beta/\alpha)}_\lambda}{c'_\lambda(\alpha,\beta)} \circ \frac{P^{(\beta/\alpha)}_\mu}{c'_\mu(\alpha,\beta)} = \delta_{\lambda,\mu} \frac{P^{(\beta/\alpha)}_\lambda}{c'_\lambda(\alpha,\beta)}. \tag{11.34}$$

THEOREM 11.7. *There exists a linear isomorphism $\phi : \mathbb{H}_X \to \Lambda_{\mathbb{C}}$ preserving bilinear forms such that*

$$\begin{aligned} \phi(\tilde{\mathfrak{p}}_{-\lambda}|0\rangle) &= \tilde{p}_\lambda, \\ \phi([L^\lambda \Sigma]) &= m_\lambda, \\ \phi([\lambda]) &= P^{(\beta/\alpha)}_\lambda. \end{aligned}$$

Furthermore, the restriction ϕ_n of ϕ to $\mathbb{H}_n$ is an isomorphism of rings, i.e.,

$$\phi_n : (\mathbb{H}_n, \star) \cong (\Lambda^n_{\mathbb{C}}, \circ).$$

PROOF. The linear isomorphism $\phi : \mathbb{H}_X \to \Lambda_{\mathbb{C}}$ is defined by mapping $\tilde{\mathfrak{p}}_{-\lambda}|0\rangle$ to $\tilde{p}_\lambda$. So

$$\phi(\mathfrak{p}_{-i}A) = p_i\, \phi(A), \qquad i > 0,\ A \in \mathbb{H}_X. \tag{11.35}$$

Next, an argument similar to the proof of (9.9) verifies that for all $i > 0$, we have the identity

$$\mathfrak{p}_{-i} \cdot [L^\lambda \Sigma] = \sum_\mu a_{\lambda,\mu}[L^\mu \Sigma] \tag{11.36}$$

where μ and $a_{\lambda,\mu}$ are the same as defined in (1.5). We conclude from an induction, (11.36) and (11.35) that

$$\phi([L^\lambda \Sigma]) = m_\lambda.$$

By (11.30) and (11.33), ϕ preserves the bilinear forms. So we see from (11.23) that

$$\langle \phi([\lambda]), \phi([\mu])\rangle = \delta_{\lambda,\mu}\frac{c'_\lambda(\alpha,\beta)}{c_\lambda(\alpha,\beta)} = \delta_{\lambda,\mu}\frac{c'_\lambda(1,\beta/\alpha)}{c_\lambda(1,\beta/\alpha)}.$$

By (11.31), we have

$$[\lambda] = [L^\lambda \Sigma] + \sum_{\mu<\lambda} d_{\lambda,\mu}[L^\mu \Sigma].$$

It follows that

$$\phi([\lambda]) = m_\lambda + \sum_{\mu<\lambda} d_{\lambda,\mu} m_\mu.$$

By the characterization of the Jack symmetric functions in Section 1.2,

$$\phi([\lambda]) = P_\lambda^{(\beta/\alpha)}.$$

Finally, by (11.22) and the definition of $\star$, we get

$$\frac{[\lambda]}{c'_\lambda(\alpha,\beta)} \star \frac{[\mu]}{c'_\mu(\alpha,\beta)} = \delta_{\lambda,\mu}\frac{[\lambda]}{c'_\lambda(\alpha,\beta)}, \qquad \lambda,\mu \vdash n.$$

In view of (11.34), $\phi_n : (\mathbb{H}_n, \star) \to (\Lambda_{\mathbb{C}}^n, \circ)$ is an isomorphism of rings. ☐

11.3.2. The case of the surface $X(\gamma)$. Let $X = X(\gamma)$ be the surface from Example 11.3. Recall that there is a $\mathbb{T}$-equivariant identification between the affine open subset X_1 of X and the complex plane in Example 11.2 with $\alpha = \beta = -1$; similarly for X_2 and the complex plane in Example 11.2 with $\alpha = \gamma - 1$ and $\beta = 1$. By Proposition 11.6, $\mathbb{H}_X$ is the Fock space of the Heisenberg algebra generated by $\mathfrak{p}_i([\Sigma_1])$ and $\mathfrak{p}_i([\Sigma_2])$ with $i \in \mathbb{Z}$, where Σ_1 and Σ_2 are the two $\mathbb{T}$-equivariant fibers in X defined by (11.8). Note also that $\mathbb{H}_{X_1}$ is the Fock space of the Heisenberg algebra generated by $\mathfrak{p}_i([\Sigma_1])$ for $i \in \mathbb{Z}$, where Σ_1 is considered as a $\mathbb{T}$-equivariant closed curve in the affine open subset $X_1 \subset X$. Similarly, $\mathbb{H}_{X_2}$ is the Fock space of the Heisenberg algebra generated by $\mathfrak{p}_i([\Sigma_2])$ for $i \in \mathbb{Z}$. To avoid confusion, we use $\mathfrak{p}_i^{X_1}$ to denote the operators $\mathfrak{p}_i([\Sigma_1])$ acting on $\mathbb{H}_{X_1}$, $\mathfrak{p}_i^{X_2}$ to denote the operators $\mathfrak{p}_i([\Sigma_2])$ acting on $\mathbb{H}_{X_2}$, and $\mathfrak{p}_i([\Sigma_1])$ and $\mathfrak{p}_i([\Sigma_2])$ for the Heisenberg operators acting on $\mathbb{H}_X$. As in (11.29), for a partition $\lambda = (1^{m_1}2^{m_2}\dots)$ and for $j = 1$ or 2, we define

$$\tilde{\mathfrak{p}}_{-\lambda}([\Sigma_j]) = \frac{1}{\mathfrak{z}_\lambda}\prod_{i\geq 1}\mathfrak{p}_{-i}([\Sigma_j])^{m_i}.$$

Define a linear map $\Psi : \mathbb{H}_{X_1} \otimes_{\mathbb{C}} \mathbb{H}_{X_2} \to \mathbb{H}_X$ as follows:

$$\Psi([\lambda^1] \otimes [\lambda^2]) = [\lambda^1, \lambda^2]. \tag{11.37}$$

The linear map Ψ is an isomorphism since the classes $[\lambda^1], [\lambda^2], [\lambda^1, \lambda^2]$ defined in (11.15) and (11.16) form linear bases of $\mathbb{H}_{X_1}, \mathbb{H}_{X_2}, \mathbb{H}_X$ respectively. In addition, we see from (11.24) that Ψ preserves the bilinear forms.

LEMMA 11.8. *The isomorphism Ψ commutes with Heisenberg operators, i.e.,*

$$\begin{aligned} \Psi \circ \left(\mathfrak{p}_{-i}^{X_1} \otimes \mathrm{Id}\right) &= \mathfrak{p}_{-i}([\Sigma_1]) \circ \Psi, \\ \Psi \circ \left(\mathrm{Id} \otimes \mathfrak{p}_{-i}^{X_2}\right) &= \mathfrak{p}_{-i}([\Sigma_2]) \circ \Psi. \end{aligned}$$

PROOF. By the symmetry between Σ_1 and Σ_2, we need only to prove the first identity. Also, since Ψ preserves the bilinear forms and $\mathfrak{p}_i$ is the adjoint operator of $\mathfrak{p}_{-i}$, it suffices to prove the first identity for $i > 0$.

Since the two fibers Σ_1 and Σ_2 do not intersect, we see that for partitions λ^1 and λ^2 with $|\lambda^1| + |\lambda^2| = n$, the $\mathbb{T}$-equivariant closed subvariety

$$L^{\lambda^1}\Sigma_1 \times L^{\lambda^2}\Sigma_2 \subset X^{[n]}$$

is the closure of $\rho_n^{-1}\left(S_{\lambda^1}^{|\lambda^1|}\Sigma_1 \times S_{\lambda^2}^{|\lambda^2|}\Sigma_2\right)$ in $X^{[n]}$, where $\rho_n \colon X^{[n]} \to X^{(n)}$ denotes the Hilbert-Chow morphism. We conclude that

$$\Psi\Big([L^{\lambda^1}\Sigma_1] \otimes [L^{\lambda^2}\Sigma_2]\Big) = [L^{\lambda^1}\Sigma_1 \times L^{\lambda^2}\Sigma_2]$$

by writing $[L^{\lambda^1}\Sigma_1 \times L^{\lambda^2}\Sigma_2]$ in terms of $[\lambda^1, \lambda^2]$ and $[\mu^1, \mu^2]$ where $\mu^1 < \lambda^1$ or $\mu^2 < \lambda^2$, similar to (11.31). It implies that the classes $[L^{\lambda^1}\Sigma_1 \times L^{\lambda^2}\Sigma_2]$ form a linear basis of $\mathbb{H}_X$. By (11.36) and a similar computation for $\mathfrak{p}_{-i}([\Sigma_1])[L^{\lambda^1}\Sigma_1 \times L^{\lambda^2}\Sigma_2]$, we have

$$\begin{aligned} &\Psi\Big(\big(\mathfrak{p}_{-i}^{X_1} \otimes \mathrm{Id}\big)\big([L^{\lambda^1}\Sigma_1] \otimes [L^{\lambda^2}\Sigma_2]\big)\Big) \\ = \;& \Psi\Big(\big(\mathfrak{p}_{-i}^{X_1}[L^{\lambda^1}\Sigma_1]\big) \otimes [L^{\lambda^2}\Sigma_2]\Big) \\ = \;& \mathfrak{p}_{-i}([\Sigma_1])[L^{\lambda^1}\Sigma_1 \times L^{\lambda^2}\Sigma_2] \\ = \;& \mathfrak{p}_{-i}([\Sigma_1])\ \Psi\big([L^{\lambda^1}\Sigma_1] \otimes [L^{\lambda^2}\Sigma_2]\big). \end{aligned}$$

It follows that $\Psi \circ \left(\mathfrak{p}_{-i}^{X_1} \otimes \mathrm{Id}\right) = \mathfrak{p}_{-i}([\Sigma_1]) \circ \Psi$ for $i > 0$. □

Let $\Lambda_{\mathbb{C},1}$ be the ring $\Lambda_{\mathbb{C}}$ of symmetric functions with the bilinear form

$$\langle \tilde{p}_\lambda, \tilde{p}_\mu \rangle = \delta_{\lambda,\mu} \frac{1}{\mathfrak{z}_\lambda},$$

$\Lambda_{\mathbb{C},2}$ be the same ring $\Lambda_{\mathbb{C}}$ equipped with a different bilinear form

$$\langle \tilde{p}_\lambda, \tilde{p}_\mu \rangle = \delta_{\lambda,\mu}\ \frac{1}{\mathfrak{z}_\lambda}\ \big(1/(\gamma - 1)\big)^{\ell(\lambda)},$$

and $\Lambda_{\mathbb{C},i}^{n_i}$ be the space of degree-n_i symmetric polynomials in $\Lambda_{\mathbb{C},i}$. The tensor product $\Lambda_{\mathbb{C},1} \otimes_{\mathbb{C}} \Lambda_{\mathbb{C},2}$ has an induced bilinear form. Let

$$(\Lambda_{\mathbb{C},1} \otimes_{\mathbb{C}} \Lambda_{\mathbb{C},2})^n = \bigoplus_{n_1+n_2=n} \Lambda_{\mathbb{C},1}^{n_1} \otimes_{\mathbb{C}} \Lambda_{\mathbb{C},2}^{n_2}.$$

We define a ring structure on $(\Lambda_{\mathbb{C},1} \otimes_{\mathbb{C}} \Lambda_{\mathbb{C},2})^n$ by declaring the elements

$$\frac{P^{(1)}_{\lambda^1}}{c'_{\lambda^1}(-1,-1)} \otimes \frac{P^{(1/(\gamma-1))}_{\lambda^2}}{c'_{\lambda^2}(\gamma-1,1)} \in (\Lambda_{\mathbb{C},1} \otimes_{\mathbb{C}} \Lambda_{\mathbb{C},2})^n,$$

where $|\lambda^1| + |\lambda^2| = n$, to be idempotents.

THEOREM 11.9. *There exists a linear isomorphism*

$$\Phi \colon \mathbb{H}_X \to \Lambda_{\mathbb{C},1} \otimes_{\mathbb{C}} \Lambda_{\mathbb{C},2}$$

preserving bilinear forms such that

$$\begin{aligned} \Phi\big(\tilde{\mathfrak{p}}_{-\lambda^1}([\Sigma_1])\tilde{\mathfrak{p}}_{-\lambda^2}([\Sigma_2])|0\rangle\big) &= \tilde{p}_{\lambda^1} \otimes \tilde{p}_{\lambda^2}, \\ \Phi\big([\Sigma_1^{(\lambda^1} \times \Sigma_2^{(\lambda^2)}]\big) &= m_{\lambda^1} \otimes m_{\lambda^2}, \\ \Phi([\lambda^1, \lambda^2]) &= P^{(1)}_{\lambda^1} \otimes P^{(1/(\gamma-1))}_{\lambda^2}. \end{aligned}$$

Furthermore, the restriction of the map Φ *to the component* $\mathbb{H}_n$ *is a ring isomorphism onto* $(\Lambda_{\mathbb{C},1} \otimes_{\mathbb{C}} \Lambda_{\mathbb{C},2})^n$.

PROOF. The statements follow from Lemma 11.8 and arguments similar to the proof of Theorem 11.7. □

CHAPTER 12

Hilbert/Gromov-Witten correspondence

It is evident from Part 2 and Part 3 that the Chern character operators introduced in Definition 4.1 (ii) are very essential to the understanding of the cohomology ring structure and intersection theory of the Hilbert scheme of points on a surface. In this chapter, we will discuss their analogues in the equivariant setting. More precisely, using the setups and results from Chapter 11, we will define equivariant Chern character operators on the equivariant cohomology of the Hilbert schemes $(\mathbb{C}^2)^{[n]}$ of points on the complex affine plane $\mathbb{C}^2$. These operators give rise to a master operator $\mathfrak{H}(z)$ acting on certain Fock space. In the study of the class algebras of symmetric groups, a counterpart of this operator has also played a distinguished role [**LT**, **Wan4**]. Via the standard boson-fermion correspondence, the operator $\mathfrak{H}(z)$ turns out to be related to another operator $\varepsilon_0(z)$, which has played a key role in the study of stationary Gromov-Witten theory of curves [**OP1**, **OP2**], by the following formula:

$$(12.1) \qquad \mathfrak{H}(z) = \frac{1}{e^{z/2} - e^{-z/2}} \left(\varepsilon_0(z) - \frac{1}{e^{z/2} - e^{-z/2}} \mathrm{I} \right).$$

Formula (12.1) defines a Gromov-Witten/Hilbert correspondence. Combining this with the Gromov-Witten/Hurwitz correspondence in [**OP1**, **OP2**] gives us a Hurwitz/Hilbert correspondence. In fact, the same combinatorics of the symmetric groups underlies these distinct geometric studies.

We will begin this chapter with a brief introduction to Gromov-Witten theory. Then using the equivariant Chern character operators, we will define the N-point function, the multi-point trace function and the τ-functions as the generating series of suitable equivariant intersection theory on $(\mathbb{C}^2)^{[n]}$, and reformulate them in terms of the operator formalism. We will compute these functions, and prove that the τ-functions satisfy the 2-Toda hierarchies. In addition, we will study the numerical aspects of the Hilbert/Gromov-Witten correspondence by proving that certain structure constants in the cup products of the cohomology ring of the Hilbert schemes are related to the 1-point disconnected series of the stationary Gromov-Witten theory of the projective line $\mathbb{P}^1$ relative to $0, \infty \in \mathbb{P}^1$. We will also establish a precise identity between some general structure constants under the cup product in $H^*((\mathbb{C}^2)^{[n]})$ and certain Hurwitz numbers for $\mathbb{P}^1$. We refer to [**QW3**] for generalizations to the Hilbert schemes of points on the minimal resolution of the quotient $\mathbb{C}^2/\Gamma$ where Γ is a finite cyclic subgroup of $\mathrm{SL}_2(\mathbb{C})$ and to their connections with the class algebras of wreath products. Theorem 12.5, Theorem 12.7, Theorem 12.10, Theorem 12.12 and their proofs are from [**LQW6**]; while Theorem 12.13, Theorem 12.16 and their proofs are from [**LQW8**].

12.1. A brief introduction to Gromov-Witten theory

In this section, we will give a brief introduction to Gromov-Witten theory. We will establish the basic notations and definitions in Gromov-Witten theory, and summarize some properties regarding the virtual fundamental class.

Let Y be a smooth projective variety. A k-pointed *stable map* to Y consists of a complete nodal curve D with k distinct ordered smooth points $p_1, \ldots, p_k$ and a morphism $\mu : D \to Y$ such that the data $(\mu, D, p_1, \ldots, p_k)$ has only finitely many automorphisms. In this case, the stable map is denoted by

$$[\mu : (D; p_1, \ldots, p_k) \to Y].$$

For a fixed homology class $\beta \in H_2(Y, \mathbb{Z})$, let

$$\overline{\mathfrak{M}}_{g,k}(Y, \beta)$$

be the coarse moduli space parameterizing all the stable maps $[\mu : (D; p_1, \ldots, p_k) \to Y]$ such that $\mu_*[D] = \beta$ and the arithmetic genus of D is g. Then, we have the i-th evaluation map :

$$\mathrm{ev}_i : \overline{\mathfrak{M}}_{g,k}(Y, \beta) \to Y \tag{12.2}$$

defined by

$$\mathrm{ev}_i([\mu : (D; p_1, \ldots, p_k) \to Y]) = \mu(p_i) \in Y.$$

It is known that the coarse moduli space $\overline{\mathfrak{M}}_{g,k}(Y, \beta)$ is projective [**FP**, Theorem 1] and has a virtual fundamental class [**LT1**, **LT2**, **Beh1**, **BF1**]

$$[\overline{\mathfrak{M}}_{g,k}(Y, \beta)]^{\mathrm{vir}} \in A_{\mathfrak{d}}(\overline{\mathfrak{M}}_{g,k}(Y, \beta))$$

where

$$\mathfrak{d} = -(K_Y \cdot \beta) + (\dim(Y) - 3)(1 - g) + k \tag{12.3}$$

is the expected (complex) dimension of $\overline{\mathfrak{M}}_{g,k}(Y, \beta)$, and $A_{\mathfrak{d}}(\overline{\mathfrak{M}}_{g,k}(Y, \beta))$ is the Chow group of $\mathfrak{d}$-dimensional cycles in the moduli space $\overline{\mathfrak{M}}_{g,k}(Y, \beta)$.

The Gromov-Witten invariants are defined by using the virtual fundamental class $[\overline{\mathfrak{M}}_{g,k}(Y, \beta)]^{\mathrm{vir}}$. Recall that an element

$$\alpha \in H^*(Y, \mathbb{C}) = \bigoplus_{j=0}^{2\dim_{\mathbb{C}}(Y)} H^j(Y, \mathbb{C})$$

is *homogeneous* if $\alpha \in H^j(Y, \mathbb{C})$ for some j; in this case, we take $|\alpha| = j$. Let $\alpha_1, \ldots, \alpha_k \in H^*(Y, \mathbb{C})$ such that every α_i is homogeneous and

$$\sum_{i=1}^{k} |\alpha_i| = 2\mathfrak{d}. \tag{12.4}$$

Then, we have the k-point Gromov-Witten invariant defined by:

$$\langle \alpha_1, \ldots, \alpha_k \rangle_{g,\beta} = \int_{[\overline{\mathfrak{M}}_{g,k}(Y,\beta)]^{\mathrm{vir}}} (\mathrm{ev}_1 \times \cdots \times \mathrm{ev}_k)^*(\alpha_1 \otimes \ldots \otimes \alpha_k). \tag{12.5}$$

The *Fundamental Class Axiom* of Gromov-Witten theory asserts that

$$\langle \alpha_1, \ldots, \alpha_{k-1}, 1_Y \rangle_{g,\beta} = 0 \tag{12.6}$$

if either $k + 2g \geq 4$ or $\beta \neq 0$ and $k \geq 1$. The *Divisor Axiom* states that

$$(12.7) \qquad \langle \alpha_1, \ldots, \alpha_{k-1}, \alpha_k \rangle_{g,\beta} = \int_\beta \alpha_k \cdot \langle \alpha_1, \ldots, \alpha_{k-1} \rangle_{g,\beta}$$

if $\alpha_k \in H^2(Y)$, and if either $k + 2g \geq 4$ or $\beta \neq 0$ and $k \geq 1$. A special case of the *Composition Law* [**KM**, (3.3) and (3.6)] states that

$$(12.8) \qquad \begin{aligned} &\langle \alpha_1\alpha_2, \alpha_3, \alpha_4 \rangle_{0,\beta} + \langle \alpha_1, \alpha_2, \alpha_3\alpha_4 \rangle_{0,\beta} \\ &\quad + \sum_{\beta_1+\beta_2=\beta,\, \beta_1,\beta_2\neq 0} \sum_a \langle \alpha_1, \alpha_2, \Delta_a \rangle_{0,\beta_1} \langle \Delta^a, \alpha_3, \alpha_4 \rangle_{0,\beta_2} \\ = \; &\langle \alpha_1\alpha_3, \alpha_2, \alpha_4 \rangle_{0,\beta} + \langle \alpha_1, \alpha_3, \alpha_2\alpha_4 \rangle_{0,\beta} \\ &\quad + \sum_{\beta_1+\beta_2=\beta,\, \beta_1,\beta_2\neq 0} \sum_a \langle \alpha_1, \alpha_3, \Delta_a \rangle_{0,\beta_1} \langle \Delta^a, \alpha_2, \alpha_4 \rangle_{0,\beta_2} \end{aligned}$$

where $\alpha_1, \alpha_2, \alpha_3, \alpha_4 \in H^*(Y)$ are cohomology classes of even degrees, $\{\Delta_a\}_a$ denotes a basis of $H^*(Y)$, and $\{\Delta^a\}_a$ is the basis of $H^*(Y)$ dual to $\{\Delta_a\}_a$ with respect to the intersection pairing of Y.

Next, we summarize certain properties concerning the virtual fundamental class. To begin with, we recall that the *excess dimension* is the difference between the dimension of $\overline{\mathfrak{M}}_{g,k}(Y, \beta)$ and the expected dimension $\mathfrak{d}$ in (12.3). For $0 \leq i < k$, use

$$(12.9) \qquad f_{k,i} : \overline{\mathfrak{M}}_{g,k}(Y, \beta) \to \overline{\mathfrak{M}}_{g,i}(Y, \beta)$$

to stand for the forgetful map obtained by forgetting the last $(k - i)$ marked points and contracting all the unstable components. It is known that $f_{k,i}$ is flat when $\beta \neq 0$ and $0 \leq i < k$. The following can be found in [**LT1**, Theorem 4.3], [**Get**, Proposition 2.5] and [**CK**, Proposition 7.1.8].

PROPOSITION 12.1. *Let $\beta \in H_2(Y, \mathbb{Z})$ and $\beta \neq 0$. Let e be the excess dimension of $\overline{\mathfrak{M}}_{g,k}(Y, \beta)$, and $\mathfrak{M} \subset \overline{\mathfrak{M}}_{g,k}(Y, \beta)$ be a closed subscheme. Then,*

(i) $[\overline{\mathfrak{M}}_{g,k}(Y, \beta)]^{\text{vir}} = (f_{k,0})^*[\overline{\mathfrak{M}}_{g,0}(Y, \beta)]^{\text{vir}}$;

(ii) $[\overline{\mathfrak{M}}_{g,k}(Y, \beta)]^{\text{vir}} = c_e(R^1(f_{k+1,k})_*(\text{ev}_{k+1})^*T_Y)$ *if* $R^1(f_{k+1,k})_*(\text{ev}_{k+1})^*T_Y$ *is a rank-e locally free sheaf over the moduli space* $\overline{\mathfrak{M}}_{g,k}(Y, \beta)$;

(iii) $[\overline{\mathfrak{M}}_{g,k}(Y, \beta)]^{\text{vir}}|_{\mathfrak{M}} = c_e((R^1(f_{k+1,k})_*(\text{ev}_{k+1})^*T_Y)|_{\mathfrak{M}})$ *if there exists an open subset $\mathfrak{O}$ of $\overline{\mathfrak{M}}_{g,k}(Y, \beta)$ such that $\mathfrak{M} \subset \mathfrak{O}$ (i.e, $\mathfrak{O}$ is an open neighborhood of $\mathfrak{M}$) and the restriction* $(R^1(f_{k+1,k})_*(\text{ev}_{k+1})^*T_Y)|_{\mathfrak{O}}$ *is a rank-e locally free sheaf over $\mathfrak{O}$.*

12.2. The Hilbert/Gromov-Witten correspondence

In this section, we will consider the action of $\mathbb{T} = \mathbb{C}^*$ on $X = \mathbb{C}^2$ by

$$(12.10) \qquad s \cdot (u, v) = (su, s^{-1}v), \quad s \in \mathbb{T}.$$

This is a special case of Example 11.2 with $\alpha = \beta = 1$. We will establish the Hilbert/Gromov-Witten correspondence by identifying some operators on the equivarian cohomology of the Hilbert schemes $X^{[n]}$ with some operators of Okounkov and Pandharipande which play a fundamental role in the study of the stationary Gromov-Witten invariants of curves [**OP1, OP2**].

We will adopt the notations from Chapter 11 relevant to $X = \mathbb{C}^2$. In particular, we have the Heisenberg operators $\mathfrak{p}_i, i \in \mathbb{Z}$ from Subsection 11.3.1 acting on the Fock space (11.25):

$$\mathbb{H}_X = \bigoplus_{n=0}^{+\infty} \mathbb{H}_n = \bigoplus_{n=0}^{+\infty} H^{2n}_{\mathbb{T}}(X^{[n]}),$$

and satisfying the commutation relations (11.28):

$$[\mathfrak{p}_k, \mathfrak{p}_\ell] = k\delta_{k,-\ell} \cdot \mathrm{Id}. \tag{12.11}$$

Next, for a partition $\lambda = (\lambda_1 \geq \ldots \geq \lambda_\ell)$ of n, the corresponding fixed point $\xi_\lambda \in X^{[n]}$ is defined by the ideal

$$I_{\xi_\lambda} = \langle u^{\lambda_1}, u^{\lambda_2}v, \ldots, u^{\lambda_\ell}v^{\ell-1}, v^\ell \rangle. \tag{12.12}$$

For simplicity, let $\mathcal{O} = \mathcal{O}_X$. The tautological rank-$n$ vector bundle $\mathcal{O}^{[n]}$ over $X^{[n]}$ is $\mathbb{T}$-equivariant. The fiber of $\mathcal{O}^{[n]}$ over the fixed point $\xi_\lambda \in X^{[n]}$ is canonically

$$\mathcal{O}^{[n]}|_{\xi_\lambda} = H^0(\mathcal{O}_{\xi_\lambda}) = \mathbb{C}[u,v]/\langle u^{\lambda_1}, u^{\lambda_2}v, \ldots, u^{\lambda_\ell}v^{\ell-1}, v^\ell \rangle. \tag{12.13}$$

It follows that as a $\mathbb{T}$-module, we have

$$\mathcal{O}^{[n]}|_{\xi_\lambda} = \bigoplus_{\square \in D_\lambda} \theta^{c(\square)} \tag{12.14}$$

where $c(\square)$ is the content of the cell $\square$ in the Young diagram D_λ, and is defined in Definition 1.4. Denote by

$$\mathrm{ch}^{[n]}_{k,\mathbb{T}}$$

the k-th $\mathbb{T}$-equivariant Chern character of $\mathcal{O}^{[n]}$. In particular the zero-th Chern character $\mathrm{ch}^{[n]}_{0,\mathbb{T}}$ equals the rank of the vector bundle $\mathcal{O}^{[n]}$, which is n. Then

$$\mathrm{ch}^{[n]}_{k,\mathbb{T}}|_{\xi_\lambda} = \frac{1}{k!}\sum_{\square \in D_\lambda} (c(\square)t)^k.$$

By the projection formula, we have in $H^*_{\mathbb{T}}(X^{[n]})'$ that

$$\mathrm{ch}^{[n]}_{k,\mathbb{T}} \cup [\xi_\lambda] = \frac{1}{k!} i_{\lambda !}\left(\sum_{\square \in D_\lambda} c(\square)^k t^k\right) = \frac{1}{k!}\sum_{\square \in D_\lambda} c(\square)^k t^k [\xi_\lambda]. \tag{12.15}$$

Let k be a nonnegative integer. Denote

$$\widetilde{\mathrm{ch}}^{[n]}_k = t^{n-k}\, \mathrm{ch}^{[n]}_{k,\mathbb{T}} \in \mathbb{H}_n \tag{12.16}$$

where $\mathbb{H}_n$ is from (11.17).

REMARK 12.2. The Theorem 5.10 in [**Wan4**] implies that $\widetilde{\mathrm{ch}}^{[n]}_k$, $0 \leq k < n$, form a set of ring generators of $\mathbb{H}_n = H^{2n}_{\mathbb{T}}(X^{[n]})$.

Recall from (11.18) the ring product $\star$ on $\mathbb{H}_n$. We define an operator $\mathfrak{G}$ (respectively, $\mathfrak{G}_k$) in $\mathrm{End}(\mathbb{H}_X)$ by sending $a \in \mathbb{H}_n$ to

$$a \star \sum_{k \geq 0} \widetilde{\mathrm{ch}}^{[n]}_k$$

(respectively, to $a \star \widetilde{\mathrm{ch}}_k^{[n]}$) in $\mathbb{H}_n$ for each n. Similarly, we define an operator $\mathfrak{G}_z$ by sending $a \in \mathbb{H}_n$ to

$$a \star \sum_{k \geq 0} z^k \widetilde{\mathrm{ch}}_k^{[n]}$$

for each n, where z is a variable. By definition, we have

$$\mathfrak{G}_z = \sum_{k \geq 0} \mathfrak{G}_k z^k.$$

Formula (12.15) is equivalent to

$$\mathfrak{G}_z([\lambda]) = \sum_{\square \in D_\lambda} e^{zc(\square)} \cdot [\lambda] \tag{12.17}$$

where $[\lambda]$ is defined by (11.15) with $\alpha = \beta = 1$.

LEMMA 12.3. *Let $\varsigma(z) = e^{z/2} - e^{-z/2}$ and $\lambda = (\lambda_1 \geq \lambda_2 \geq \ldots)$. Then,*

$$\mathfrak{G}_z([\lambda]) = \frac{1}{\varsigma(z)} \left(\sum_{i=1}^{+\infty} e^{z(\lambda_i - i + 1/2)} - \frac{1}{\varsigma(z)} \right) \cdot [\lambda].$$

PROOF. The contents $c(\square)$'s of λ are:

$$-i+1, -i+2, \ldots, -i+\lambda_i, \quad i = 1, \ldots, \ell(\lambda).$$

Noting that

$$e^{z(-i+1)} + \ldots + e^{z(-i+\lambda_i)} = \frac{e^{z(\lambda_i - i + 1)} - e^{z(-i+1)}}{e^z - 1},$$

formula (12.17) can be written as

$$\mathfrak{G}_z([\lambda]) = \frac{1}{\varsigma(z)} \sum_{i=1}^{+\infty} (e^{z(\lambda_i - i + 1/2)} - e^{z(-i+1/2)}) \cdot [\lambda]$$

where we have used $\lambda_i = 0$ for $i > \ell(\lambda)$. In view of the identity

$$\sum_{i=1}^{+\infty} e^{z(-i+1/2)} = \frac{1}{\varsigma(z)},$$

our lemma follows. □

Next, denote by $\mathcal{O}_m$ the $\mathbb{T}$-equivariant line bundle over $X = \mathbb{C}^2$ associated to the $\mathbb{T}$-character t^m, where $m \in \mathbb{Z}$. Let

$$\mathcal{M}(m, n)$$

be the moduli space which parameterizes all rank-1 subsheaves of $\mathcal{O}_m$ such that the quotients are supported at finitely many points of X and have length n. Given $\mathcal{I} \in X^{[n]}$, then $\mathcal{O}_m \otimes \mathcal{I}$ is an element in $\mathcal{M}(m, n)$. Define

$$\mathbb{H}_n^{(m)} = H_{\mathbb{T}}^{2n}(\mathcal{M}(m, n)).$$

As before, the study of the equivariant cohomology ring $H_{\mathbb{T}}^*(\mathcal{M}(m, n))$ leads to the ring $\mathbb{H}_n^{(m)}$ whose product is denoted by $\star$. The natural identification

$$\mathcal{M}(m, n) \cong X^{[n]}$$

leads to the natural identification of the rings $\mathbb{H}_n^{(m)} \cong \mathbb{H}_n$, which induces a bilinear form $\langle \cdot, \cdot \rangle_n^{(m)}$ on $\mathbb{H}_n^{(m)}$ from $\mathbb{H}_n$. We introduce

$$\begin{aligned} \mathcal{F}^{(m)} &= \bigoplus_{n=0}^{+\infty} \mathbb{H}_n^{(m)}, \quad m \in \mathbb{Z} \\ \mathcal{F} &= \bigoplus_{m \in \mathbb{Z}} \mathcal{F}^{(m)}. \end{aligned}$$

In particular we identify $\mathcal{M}(0, n) = X^{[n]}$ and $\mathbb{H}_n^{(0)} = \mathbb{H}_n$. We denote by S the isomorphism

$$S : \mathbb{H}_n^{(m)} \to \mathbb{H}_n^{(m+1)}.$$

This induces isomorphisms (which will be denoted by S again)

$$S : \mathcal{F} \to \mathcal{F}$$

and

$$S : \mathcal{F}^{(m)} \to \mathcal{F}^{(m+1)}$$

for all $m \in \mathbb{Z}$. The bilinear form on $\mathcal{F}$ induced from $\langle \cdot, \cdot \rangle_n^{(m)}$ on $\mathbb{H}_n^{(m)}$ will be again denoted by $\langle \cdot, \cdot \rangle$.

By the standard boson-fermion correspondence in [**MJD**, Theorem 5.2], $\mathcal{F}$ can be identified with the fermionic Fock space (or equivalently, the infinite wedge space). The operator S is exactly the *shift* operator on the fermionic Fock space. Given an operator $\mathfrak{f} \in \mathrm{End}(\mathcal{F})$, define the *vacuum expectation* $\langle \mathfrak{f} \rangle$ of $\mathfrak{f}$ by

$$\langle \mathfrak{f} \rangle = \langle |0\rangle, \mathfrak{f}|0\rangle \rangle.$$

It is well-known that the completed infinite-rank general linear Lie algebra $\widehat{\mathfrak{sl}}_\infty$ (whose standard basis is denoted by $E_{i,j}$, $i, j \in \mathbb{Z} + 1/2$) acts on the fermionic Fock space. As will become clear below, the study of the equivariant intersection theory on Hilbert schemes naturally leads to the following operator in $\mathrm{End}(\mathcal{F})$:

$$\mathfrak{H}(z) = \frac{1}{\varsigma(z)} \sum_{k \in \mathbb{Z} + \frac{1}{2}} e^{kz} E_{k,k}.$$

The operator is further expanded as

$$\mathfrak{H}(z) = \sum_{k=0}^{+\infty} \mathfrak{H}_k z^k.$$

We remark that the operator $\mathfrak{H}(z)$ in a somewhat different form has appeared in the study of the class algebras of the symmetric groups and wreath products [**LT**, **Wan4**], and it affords a compact expression in terms of vertex operators.

On the other hand, the following operators

$$\varepsilon_r(z) = \sum_{k \in \mathbb{Z} + \frac{1}{2}} e^{z(k - r/2)} E_{k-r,k} + \frac{\delta_{r,0}}{\varsigma(z)}, \quad r \in \mathbb{Z} \tag{12.18}$$

have been introduced by Okounkov and Pandharipande. These operators have played a fundamental role in the study of the stationary Gromov-Witten invariants of curves [**OP1**, **OP2**], where *stationary* means that every marked point of the stable maps is mapped to some prescribed point on the curve. To state the results, for two partitions λ and μ with $|\lambda| = |\mu|$, let

$$F^{\bullet}_{\lambda,\mu}(z_1, \ldots, z_N) \tag{12.19}$$

be the N-point disconnected series of stationary Gromov-Witten invariants of $\mathbb{P}^1$ relative to $0, \infty \in \mathbb{P}^1$, where *disconnected* means that the domain curves of the stable maps to $\mathbb{P}^1$ are allowed to be disconnected. Then, given two partitions λ and μ of n, we have

$$F^{\bullet}_{\lambda,\mu}(z_1, \ldots, z_N) = \langle \tilde{\mathfrak{p}}_{-\lambda}, \varepsilon_0(z_1) \ldots \varepsilon_0(z_N) \tilde{\mathfrak{p}}_{-\mu} \rangle_n \tag{12.20}$$

where the notation $\tilde{\mathfrak{p}}_{-\lambda}$ is introduced in (11.29). Moreover, define the q-trace $\operatorname{Tr}_q(\mathfrak{f})$ of an operator $\mathfrak{f} \in \operatorname{End}(\mathbb{H}_X)$ by

$$\operatorname{Tr}_q(\mathfrak{f}) = \sum_{\lambda} \mathfrak{z}_\lambda \langle \tilde{\mathfrak{p}}_{-\lambda}, \mathfrak{f}(\tilde{\mathfrak{p}}_{-\lambda}) \rangle q^{|\lambda|}, \tag{12.21}$$

and let

$$F_E(z_1, \ldots, z_n; q)$$

be the N-point disconnected series (with an additional variable q inserted) of the stationary Gromov-Witten invariants of an elliptic curve E. Since Gromov-Witten invariants are deformation invariant, $F_E(z_1, \ldots, z_n; q)$ is independent of the elliptic curve E. We have

$$F_E(z_1, \ldots, z_n; q) = \operatorname{Tr}_q(\varepsilon_0(z_1) \cdots \varepsilon_0(z_N)). \tag{12.22}$$

The operators $\varepsilon_r(z)$, $r \in \mathbb{Z}$, satisfy the following identities:

$$[\mathfrak{p}_k, \varepsilon_r(z)] = \varsigma(kz) \varepsilon_{k+r}(z) \tag{12.23}$$

$$S^{-1} \varepsilon_0(z) S = e^z \varepsilon_0(z) \tag{12.24}$$

$$\varepsilon_0(z)([\lambda]) = \sum_{i=1}^{+\infty} e^{z(\lambda_i - i + 1/2)} \cdot [\lambda] \tag{12.25}$$

for all partitions $\lambda = (\lambda_1 \geq \lambda_2 \geq \ldots)$ (here we recall $[\lambda] \in \mathbb{H}_X = \mathcal{F}^{(0)}$).

LEMMA 12.4. *The following identities hold for operators on $\mathcal{F}$:*

$$\mathfrak{H}(z) = \frac{1}{\varsigma(z)} \left(\varepsilon_0(z) - \frac{1}{\varsigma(z)} \mathrm{Id} \right) \tag{12.26}$$

$$S^{-m} \mathfrak{H}(z) S^m = e^{mz} \mathfrak{H}(z) + \frac{e^{mz} - 1}{\varsigma(z)^2} \mathrm{Id}, \quad m \in \mathbb{Z}. \tag{12.27}$$

PROOF. The first identity follows from the definitions. The second one follows from the first one and (12.24). □

THEOREM 12.5. *As operators on $\mathcal{F}^{(0)} = \mathbb{H}_X$, the following identification holds*

$$\mathfrak{G}_z = \mathfrak{H}(z).$$

PROOF. Follows from Lemma 12.3, formula (12.25) and Lemma 12.4. □

In light of the interpretation of $\mathfrak{H}(z)$ in Theorem 12.5 and the role of $\varepsilon_0(z)$ in the study of the stationary Gromov-Witten invariants (12.20) and (12.22), the identity (12.26) defines *the Hilbert/Gromov-Witten correspondence.*

12.3. The N-point functions and the multi-point trace functions

Let $X = \mathbb{C}^2$ and consider the $\mathbb{T}$-action (12.10). In view of Remark 11.5 and Remark 12.2, we see that one way to understand the equivariant intersection theory on $X^{[n]}$ is to study the products

$$\widetilde{\mathrm{ch}}_{k_1}^{[n]} \star \widetilde{\mathrm{ch}}_{k_2}^{[n]} \star \ldots \star \widetilde{\mathrm{ch}}_{k_N}^{[n]}$$

for arbitrary nonnegative integers $k_1, \ldots, k_N$, $N \geq 1$. For partitions λ and μ of n, we define the intersection numbers

$$\langle \lambda, \widetilde{\mathrm{ch}}_{k_1}^{[n]} \cdots \widetilde{\mathrm{ch}}_{k_N}^{[n]}, \mu \rangle_n = \left\langle \tilde{\mathfrak{p}}_{-\lambda}, \widetilde{\mathrm{ch}}_{k_1}^{[n]} \star \widetilde{\mathrm{ch}}_{k_2}^{[n]} \star \ldots \star \widetilde{\mathrm{ch}}_{k_N}^{[n]} \star \tilde{\mathfrak{p}}_{-\mu} \right\rangle_n \tag{12.28}$$

In this section, we will form two types of generating functions (namely, the N-point function and the multi-point trace function) of these intersection numbers and then reformulate them in terms of the operator formalism.

We begin with the N-point function. Given two partitions λ and μ of n, the *N-point function* $G_{\lambda,\mu}(z_1, \ldots, z_N)$ is defined by

$$G_{\lambda,\mu}(z_1, \ldots, z_N) = \sum_{k_1,\ldots,k_N} z_1^{k_1} \ldots z_N^{k_N} \langle \lambda, \widetilde{\mathrm{ch}}_{k_1}^{[n]} \cdots \widetilde{\mathrm{ch}}_{k_N}^{[n]}, \mu \rangle_n. \tag{12.29}$$

This can be reformulated in an operator form:

$$\begin{aligned} G_{\lambda,\mu}(z_1, \ldots, z_N) &= \langle \tilde{\mathfrak{p}}_{-\lambda}, \mathfrak{G}_{z_1} \ldots \mathfrak{G}_{z_N} \tilde{\mathfrak{p}}_{-\mu} \rangle_n \\ &= \langle \tilde{\mathfrak{p}}_{-\lambda}, \mathfrak{H}(z_1) \ldots \mathfrak{H}(z_N) \tilde{\mathfrak{p}}_{-\mu} \rangle_n. \end{aligned}$$

Recall from (12.20) that

$$F^{\bullet}_{\lambda,\mu}(z_1, \ldots, z_N) = \langle \tilde{\mathfrak{p}}_{-\lambda}, \varepsilon_0(z_1) \ldots \varepsilon_0(z_N) \tilde{\mathfrak{p}}_{-\mu} \rangle_n.$$

Note that, in particular for $N = 0$, we have defined $G_{\lambda,\mu}()$ and $F^{\bullet}_{\lambda,\mu}()$. By (12.11), we have

$$G_{\lambda,\mu}() = F^{\bullet}_{\lambda,\mu}() = \frac{\delta_{\lambda,\mu}}{\mathfrak{z}_\lambda}.$$

It follows from (12.26) and Theorem 12.5 that the 1-point function is given by

$$\begin{aligned} G_{\lambda,\mu}(z) &= \frac{1}{\varsigma(z)} \left(F^{\bullet}_{\lambda,\mu}(z) - \frac{1}{\varsigma(z)} F^{\bullet}_{\lambda,\mu}() \right) \\ &= \frac{1}{\varsigma(z)} \left(F^{\bullet}_{\lambda,\mu}(z) - \frac{\delta_{\lambda,\mu}}{\mathfrak{z}_\lambda \varsigma(z)} \right). \end{aligned} \tag{12.30}$$

What remains is computing $F^{\bullet}_{\lambda,\mu}(z)$. The somewhat simpler *connected* series $F^{\circ}_{\lambda,\mu}$ rather than the disconnected series has been computed in [**OP1**, Subsection 3.2].

We introduce some notations on partitions. For two partitions $\lambda = (1^{m_1} 2^{m_2} \ldots)$ and $\mu = (1^{n_1} 2^{n_2} \ldots)$, define the partition

$$\lambda + \mu = (1^{m_1+n_1} 2^{m_2+n_2} \ldots).$$

We say $\lambda \subset \mu$ if $m_i \leq n_i$ for all i, and in this case, define the partition

$$\mu - \lambda = (1^{n_1-m_1} 2^{n_2-m_2} \ldots).$$

For $\lambda = (\lambda_1 \geq \ldots \geq \lambda_r)$ where $r = \ell(\lambda)$, put

$$\varsigma(\lambda, z) = \varsigma(\lambda_1 z) \cdots \varsigma(\lambda_r z).$$

Given a subset $U \subset \underline{r} = \{1, \ldots, r\}$, we denote by λ_U the subpartition of λ which consists of the parts λ_i, $i \in U$.

PROPOSITION 12.6. *Let $\lambda = (\lambda_1 \geq \dots \geq \lambda_r)$ and $\mu = (\mu_1 \geq \dots \geq \mu_s)$ be partitions of n, where $r = \ell(\lambda)$ and $s = \ell(\mu)$. Then,*

$$F^{\bullet}_{\lambda,\mu}(z) = \sum_U \frac{\varsigma(\lambda_U, z)\,\varsigma(\lambda_U + \mu - \lambda, z)}{\mathfrak{z}_\lambda \mathfrak{z}_{\lambda_U+\mu-\lambda}\varsigma(z)}$$

summed over subsets $U \subset \underline{r}$ such that $\lambda \subset \lambda_U + \mu$.

PROOF. Let $[\mathfrak{p}_{\lambda,U}, \varepsilon_0(z)]$ denote the multi-commutator

$$[\cdots[\mathfrak{p}_{\lambda_b}, [\mathfrak{p}_{\lambda_a}, \varepsilon_0(z)]]\cdots],$$

if we write $U = \{a, b, \dots\}$. Note that the multi-commutator is independent of the ordering of elements in U since the $\mathfrak{p}_k$'s $(k > 0)$ commute with each other.

By moving $\varepsilon_0(z)$ to the left whenever possible, we have

$$\begin{aligned} \mathfrak{p}_{\lambda_r}\cdots\mathfrak{p}_{\lambda_1}\varepsilon_0(z) &= \sum_{U\subset\underline{r}}[\mathfrak{p}_{\lambda,U}, \varepsilon_0(z)] \prod_{i\in\underline{r}\backslash U} \mathfrak{p}_{\lambda_i} \\ &= \sum_{U\subset\underline{r}} \varsigma(\lambda_U, z)\varepsilon_{|\lambda_U|}(z) \prod_{i\in\underline{r}\backslash U} \mathfrak{p}_{\lambda_i} \end{aligned} \tag{12.31}$$

where we have repeatedly used (12.23). It follows that

$$\begin{aligned} F^{\bullet}_{\lambda,\mu}(z) &= \frac{1}{\mathfrak{z}_\lambda\mathfrak{z}_\mu}\langle \mathfrak{p}_{\lambda_r}\cdots\mathfrak{p}_{\lambda_1}\varepsilon_0(z)\mathfrak{p}_{-\mu_1}\cdots\mathfrak{p}_{-\mu_s}\rangle \\ &= \frac{1}{\mathfrak{z}_\lambda\mathfrak{z}_\mu}\sum_{U\subset\underline{r}}\varsigma(\lambda_U, z)\left\langle \varepsilon_{|\lambda_U|}(z)\prod_{i\in\underline{r}\backslash U}\mathfrak{p}_{\lambda_i}\cdot\mathfrak{p}_{-\mu_1}\cdots\mathfrak{p}_{-\mu_s}\right\rangle \end{aligned} \tag{12.32}$$

Denote by λ_U^- to be the partition which consists of the parts λ_i, $i \in \underline{r}\backslash U$. Apparently, the vacuum expectation

$$\left\langle \varepsilon_{|\lambda_U|}(z)\prod_{i\in\underline{r}\backslash U}\mathfrak{p}_{\lambda_i}\cdot\mathfrak{p}_{-\mu_1}\cdots\mathfrak{p}_{-\mu_s}\right\rangle = 0 \quad \text{unless } \lambda_U^- \subset \mu.$$

If $\lambda_U^- \subset \mu$, or equivalently if $\lambda \subset \lambda_U + \mu$, then $\mu - \lambda_U^- = (\lambda_U + \mu) - \lambda$, and we can show by induction that

$$\prod_{i\in\underline{r}\backslash U}\mathfrak{p}_{\lambda_i}\cdot\mathfrak{p}_{-\mu_1}\cdots\mathfrak{p}_{-\mu_s}|0\rangle = \frac{\mathfrak{z}_\mu}{\mathfrak{z}_{\lambda_U+\mu-\lambda}}\mathfrak{p}_{-\mu_{a_1}}\cdots\mathfrak{p}_{-\mu_{a_t}}|0\rangle \tag{12.33}$$

where we have denoted $\mu - \lambda_U^- = (\mu_{a_1}, \dots, \mu_{a_t})$.

By (12.32) and (12.33), we have

$$F^{\bullet}_{\lambda,\mu}(z) = \frac{1}{\mathfrak{z}_\lambda\mathfrak{z}_\mu}\sum_{U\subset\underline{r}}\frac{\mathfrak{z}_\mu}{\mathfrak{z}_{\lambda_U+\mu-\lambda}}\varsigma(\lambda_U, z)\left\langle\varepsilon_{|\lambda_U|}(z)\mathfrak{p}_{-\mu_{a_1}}\cdots\mathfrak{p}_{-\mu_{a_t}}\right\rangle. \tag{12.34}$$

Similar to (12.31), we now move $\varepsilon_{|\lambda_U|}(z)$ to the right in $\langle\varepsilon_{|\lambda_U|}(z)\mathfrak{p}_{-\mu_{a_1}}\cdots\mathfrak{p}_{-\mu_{a_t}}\rangle$ whenever possible. Note that if $\varepsilon_K(z)$ $(K > 0)$ results from such a move, then the corresponding vacuum expectation is zero. Therefore

$$\begin{aligned} \langle\varepsilon_{|\lambda_U|}(z)\mathfrak{p}_{-\mu_{a_1}}\cdots\mathfrak{p}_{-\mu_{a_t}}\rangle &= \langle[\cdots[\varepsilon_{|\lambda_U|}(z), \mathfrak{p}_{-\mu_{a_1}}],\cdots\mathfrak{p}_{-\mu_{a_t}}]\rangle \\ &= \varsigma(\lambda_U + \mu - \lambda, z)\langle\varepsilon_0(z)\rangle \\ &= \varsigma(\lambda_U + \mu - \lambda, z)/\varsigma(z). \end{aligned} \tag{12.35}$$

Now the proposition follows from (12.34) and (12.35). □

THEOREM 12.7. *The 1-point function* $G_{\lambda,\mu}(z)$ *is given by*

$$G_{\lambda,\mu}(z) = \frac{1}{\mathfrak{z}_\lambda\, \varsigma(z)^2}\left(\sum_U \frac{\varsigma(\lambda_U, z)\,\varsigma(\lambda_U+\mu-\lambda, z)}{\mathfrak{z}_{\lambda_U+\mu-\lambda}} - \delta_{\lambda,\mu}\right)$$

where U *runs over the subsets of* $\underline{\ell(\lambda)}$ *such that* $\lambda \subset \lambda_U + \mu$.

PROOF. Follows from (12.30) and Proposition 12.6. □

In general, by Theorem 12.5, we have the following.

PROPOSITION 12.8. *We have*

$$G_{\lambda,\mu}(z_1,\ldots,z_N) = \frac{1}{\prod_{i=1}^N \varsigma(z_i)} \sum_{U\subset \underline{N}} \frac{(-1)^{N-|U|}}{\prod_{i\in \underline{N}\setminus U}\varsigma(z_i)} F^\bullet_{\lambda,\mu}(z_U)$$

where

$$F^\bullet_{\lambda,\mu}(z_U) = \left\langle \tilde{\mathfrak{p}}_{-\lambda}, \prod_{i\in U} \varepsilon_0(z_i)\tilde{\mathfrak{p}}_{-\mu}\right\rangle_n.$$

REMARK 12.9. The strategy used in Proposition 12.6 could be generalized to compute the N-point disconnected series $F^\bullet_{\lambda,\mu}(z_1,\ldots,z_N)$, although the notations would be a bit involved. Then the computation of $G_{\lambda,\mu}(z_1,\ldots,z_N)$ follows by Proposition 12.8.

Next, we study the multi-point trace function. Recall from (12.21) the q-trace $\mathrm{Tr}_q(\mathfrak{f})$ of an operator $\mathfrak{f} \in \mathrm{End}(\mathbb{H}_X)$. In particular, for the identity operator Id on $\mathbb{H}_X$, we have

$$\mathrm{Tr}_q(\mathrm{Id}) = \frac{1}{(q;q)_\infty},$$

where

$$(a;q)_n = (1-a)(1-aq)(1-aq^2)\cdots(1-aq^n). \tag{12.36}$$

Our main object here is *the multi-point trace function*

$$\mathrm{Tr}_q(\mathfrak{G}_{z_1}\cdots\mathfrak{G}_{z_N})$$

of a product of the operators $\mathfrak{G}_{z_j}$, which encodes information about the intersection numbers (12.28). Here $z_1,\ldots,z_N$ are independent variables.

We define $\Theta(z) = \Theta(z;q)$ by

$$\Theta(z) = \Theta(z;q) = \eta(q)^{-3}\sum_{m\in\mathbb{Z}}(-1)^m q^{\frac{(m+1/2)^2}{2}} e^{(m+1/2)z}$$

where $\eta(q) = q^{1/24}(q;q)_\infty$ is the Dedekind eta function. By the Jacobi triple product identity, we obtain

$$\Theta(z) = (e^{z/2}-e^{-z/2})(qe^z;q)_\infty(qe^{-z};q)_\infty/(q;q)^2_\infty.$$

We further define

$$\Theta^{(k)}(z) = \frac{d^k}{dz^k}\Theta(z), \qquad k\ge 0.$$

We agree that $\Theta^{(k)}(z) = 0$ for $k<0$. Given a positive integer N, we denote $\underline{N} = \{1,2,\ldots,N\}$. Given a finite set U, we denote by S_U the symmetric group on U. In particular, $S_{\underline{N}} = S_N$. Given $U = \{u_1,\ldots,u_k\} \subset \underline{N}$ with $u_1 < \ldots < u_k$

and a permutation $\sigma \in S_U$, we denote by $M_{U,\sigma}$ the $k \times k$ matrix whose (i,j)-th entry is

$$\Theta^{(j-i+1)}(z_{\sigma u_1} \cdots z_{\sigma u_{k-j}})/(j-i+1)!.$$

Put

$$\Theta_{U,\sigma} = \Theta(z_{\sigma u_1})\Theta(z_{\sigma u_1} z_{\sigma u_2}) \cdots \Theta(z_{\sigma u_1} \cdots z_{\sigma u_k}).$$

By the Theorem 0.5 in [**BO**],

$$\mathrm{Tr}_q(\varepsilon_0(z_1) \cdots \varepsilon_0(z_N)) = \frac{1}{(q;q)_\infty} \sum_{\sigma \in S_N} \frac{\det M_{\underline{N},\sigma}}{\Theta_{\underline{N},\sigma}}. \tag{12.37}$$

In view of (12.22), formula (12.37) also computes the stationary Gromov-Witten invariants of an elliptic curve.

THEOREM 12.10. *We have*

$$\mathrm{Tr}_q(\mathfrak{G}_{z_1} \cdots \mathfrak{G}_{z_N}) = \frac{1}{(q;q)_\infty \prod_{i=1}^N \varsigma(z_i)} \sum_{U \subset \underline{N}} \frac{(-1)^{N-|U|} \sum_{\sigma \in S_U} \frac{\det M_{U,\sigma}}{\Theta_{U,\sigma}}}{\prod_{i \in \underline{N} \setminus U} \varsigma(z_i)}.$$

PROOF. By Lemma 12.4 and Theorem 12.5, we have

$$\mathrm{Tr}_q(\mathfrak{G}_{z_1} \cdots \mathfrak{G}_{z_N}) = \frac{1}{\prod_{i=1}^N \varsigma(z_i)} \sum_{U \subset \underline{N}} \frac{(-1)^{N-|U|}\mathrm{Tr}_q(\prod_{i \in U} \varepsilon_0(z_i))}{\prod_{i \in \underline{N} \setminus U} \varsigma(z_i)}.$$

Now the theorem follows by applying (12.37) by replacing the set $\underline{N}$ by U. □

12.4. Equivariant intersection and τ-functions of 2-Toda hierarchies

The relations between the cohomology rings of Hilbert schemes of points on surfaces and $\mathcal{W}$ algebras (Section 4.4) through vertex operators have led to the following question (Question 5 in [**QW1**, Section 6]): what is the precise connection between Hilbert schemes and integrable hierarchies? The connection between the geometry of Hilbert schemes and vertex operators is made through the Chern character operators. However the Chern character operators are usually nilpotent for (cohomology) degree reasons. This presents a serious difficulty for a sensible answer to the above question in the framework of ordinary cohomology theory. In this section, we will deal with this question in the equivariant setting. We will continue to consider the $\mathbb{T}$-action (12.10) on $X = \mathbb{C}^2$, and define τ-functions as the generating series of certain equivariant intersection numbers on $X^{[n]}$. Then we will prove that these τ-functions satisfy the 2-Toda hierarchies.

Let

$$\begin{aligned} t &= (t_1, t_2, \ldots), \\ s &= (s_1, s_2, \ldots), \\ x &= (x_1, x_2, \ldots) \end{aligned}$$

be three sequences of indeterminates. Define the following half vertex operators:

$$\Gamma_\pm(t) = \exp\left(\sum_{k>0} t_k \mathfrak{p}_{\pm k}/k\right).$$

Given a partition $\mu = (\mu_1 \geq \mu_2 \geq \ldots)$, we write $t_\mu = t_{\mu_1} t_{\mu_2} \cdots$.

Define the τ-*function* $\tau(x,t,s)$ to be the following generating function for the equivariant intersection numbers $\langle \lambda, \widetilde{\mathrm{ch}}_{k_1}^{[n]} \dots \widetilde{\mathrm{ch}}_{k_N}^{[n]}, \mu \rangle_n$ from (12.28):

$$\tau(x,t,s) = \sum_n \sum_{|\lambda|=|\mu|=n} t_\lambda s_\mu \left\langle \lambda, \exp\left(\sum_{k=0}^{+\infty} x_k \widetilde{\mathrm{ch}}_k^{[n]}\right), \mu \right\rangle_n$$

Note that

$$\begin{aligned} \Gamma_-(s) &= \sum_{n\geq 0} \sum_{|\lambda|=n} t_\lambda \tilde{\mathfrak{p}}_{-\lambda}, \\ \Gamma_+(t) &= \Gamma_-(t)^\dagger. \end{aligned}$$

From the definition of $\mathfrak{H}_k$ and Theorem 12.5, we see that the τ-function affords an operator formulation:

$$\tau(x,t,s) = \left\langle \Gamma_+(t) \exp\left(\sum_{k=0}^{+\infty} x_k \mathfrak{H}_k\right) \Gamma_-(s) \right\rangle.$$

Next, we define the Chern character operators from the space $\mathcal{M}(m,n)$. We have a universal exact sequence:

$$0 \to \mathcal{J}_m \to \pi_2^* \mathcal{O}_m \to \mathcal{Q}_m \to 0$$

where π_1, π_2 are the projections of $\mathcal{M}(m,n) \times X$ to the two factors. Denote by $\mathcal{O}_m^{[n]}$ the $\mathbb{T}$-equivariant rank-n vector bundle over $\mathcal{M}(m,n)$ given by the push-forward $\pi_{1*}(\mathcal{Q}_m)$, whose fiber over a point $\xi_\lambda \in \mathcal{M}(m,n) \cong X^{[n]}$ is given by

$$\mathcal{O}_m^{[n]}|_{\xi_\lambda} = \mathcal{O}_m \bigotimes \mathcal{O}^{[n]}|_{\xi_\lambda}. \tag{12.38}$$

In the same way as defining the operators $\mathfrak{G}_z$ and $\mathfrak{G}_k (k \geq 0)$ acting on $\mathbb{H}_X = \mathcal{F}^{(0)}$, we can define the operators $\mathfrak{G}_z^{(m)}$ and $\mathfrak{G}_k^{(m)} (k \geq 0)$ acting on $\mathcal{F}^{(m)}$ using the cup products with $\sum_{k\geq 0} t^{n-k} \mathrm{ch}_k(\mathcal{O}_m^{[n]}) z^k$ and $t^{n-k} \mathrm{ch}_k(\mathcal{O}_m^{[n]})$ respectively on $\mathbb{H}_n^{(m)}$. For technical reasons below, we introduce the following modification

$$\widetilde{\mathrm{ch}}_k(\mathcal{O}_m^{[n]}) = t^{n-k} \mathrm{ch}_k(\mathcal{O}_m^{[n]}) + c_k^{(m)} t^n$$

where the constant $c_k^{(m)}$ is defined by

$$\frac{e^{mz}-1}{\varsigma(z)^2} = mz^{-1} + \sum_{k\geq 0} c_k^{(m)} z^k.$$

Equivalently, if we define

$$\widetilde{\mathfrak{G}}_z^{(m)} = \mathfrak{G}_z^{(m)} + \frac{e^{mz}-1}{\varsigma(z)^2} \mathrm{Id}. \tag{12.39}$$

and further write $\widetilde{\mathfrak{G}}_z^{(m)} = mz^{-1}\mathrm{Id} + \sum_{k\geq 0} \widetilde{\mathfrak{G}}_k^{(m)} z^k$, then

$$\widetilde{\mathfrak{G}}_k^{(m)} = \mathfrak{G}_k^{(m)} + c_k^{(m)} \mathrm{Id}, \quad k \geq 0$$

and $\widetilde{\mathfrak{G}}_k^{(m)}$ acts on $\mathbb{H}_n^{(m)}$ by the product with $\widetilde{\mathrm{ch}}_k(\mathcal{O}_m^{[n]})$. When $m = 0$, we have $c_k^{(0)} = 0$ and $\widetilde{\mathfrak{G}}_k^{(0)} = \mathfrak{G}_k^{(0)} = \mathfrak{G}_k$ for all k.

Similarly we define the equivariant intersection numbers for $\mathcal{M}(m,n)$, denoted by $\langle \cdot \rangle_n^{(m)}$. When $m = 0$, they reduce to the ones defined earlier.

Now define the *τ-function* $\tau(x,t,s,m)$ to be the following generating function of the equivariant intersection numbers on $\mathcal{M}(m,n)$:

$$\tau(x,t,s,m)=\sum_n \sum_{|\lambda|=|\mu|=n} t_\lambda s_\mu \left\langle \lambda, \exp\left(\sum_{k=0}^{+\infty} x_k \widetilde{\mathrm{ch}}_k(\mathcal{O}_m^{[n]})\right), \mu\right\rangle_n^{(m)}$$

In particular we have $\tau(x,t,s,0)=\tau(x,t,s)$.

LEMMA 12.11. *As operators on $\mathcal{F}^{(m)}$, the identification holds:*

$$\widetilde{\mathfrak{G}}_z^{(m)} = \mathfrak{H}(z).$$

PROOF. Under the identification

$$S^m: \quad \mathbb{H}_n = \mathbb{H}_n^{(0)} \xrightarrow{\sim} \mathbb{H}_n^{(m)},$$

we denote by $[\lambda]^{(m)}$ the image of $[\lambda]$. By the identification of toric action (12.38), the same proof as in Lemma 12.3 implies that

$$\widetilde{\mathfrak{G}}_z^{(m)}([\lambda]^{(m)}) \quad = \quad \left(\frac{e^{mz}}{\varsigma(z)}\left(\sum_{i=1}^{+\infty} e^{z(\lambda_i-i+1/2)}-\frac{1}{\varsigma(z)}\right)+\frac{e^{mz}-1}{\varsigma(z)^2}\right)\cdot[\lambda]^{(m)}.$$

By (12.27) and $S^m([\lambda])=[\lambda]^{(m)}$, we have

$$\mathfrak{H}(z)([\lambda]^{(m)})=\mathfrak{H}(z)S^m([\lambda])=e^{mz}S^m\mathfrak{H}(z)([\lambda])+\frac{e^{mz}-1}{\varsigma(z)^2}S^m([\lambda]).$$

This can be rewritten as

$$\mathfrak{H}(z)([\lambda]^{(m)})=\left(\frac{e^{mz}}{\varsigma(z)}\left(\sum_{i=1}^{+\infty} e^{z(\lambda_i-i+1/2)}-\frac{1}{\varsigma(z)}\right)+\frac{e^{mz}-1}{\varsigma(z)^2}\right)\cdot[\lambda]^{(m)}$$

by Lemma 12.3 and Theorem 12.5. This finishes the proof. ☐

THEOREM 12.12.

(i) *The function $\tau(x,t,s,m)$ can be reformulated as:*

$$\tau(x,t,s,m)=\left\langle S^{-m}\Gamma_+(t)\exp\left(\sum_{k=0}^{+\infty} x_k\mathfrak{H}_k\right)\Gamma_-(s)S^m\right\rangle.$$

(ii) *The function $\tau(x,t,s,m)$, $m\in\mathbb{Z}$, satisfies the 2-Toda hierarchy of Ueno-Takasaki* [**UT**, Section 1]. *The lowest equation among the hierarchy reads:*

$$\frac{\partial^2}{\partial t_1\partial s_1}\ln\tau(t,s,x,m)=\frac{\tau(t,s,x,m+1)\,\tau(t,s,x,m-1)}{\tau(t,s,x,m)^2}. \tag{12.40}$$

PROOF. (i) follows from Lemma 12.11 and the definition of $\tau(x,t,s,m)$, while (ii) is standard since the operator $\mathfrak{H}_k$ lies in $\widehat{\mathfrak{sl}}_\infty$. ☐

By setting $t_2=t_3=\cdots=s_2=s_3=\cdots=0$ and $x_2=x_3=\cdots=0$ in $\tau(x,t,s)$, we obtain the following generating function of the intersection numbers on Hilbert schemes:

$$\sum_n (t_1s_1e^{x_0})^n\left\langle \frac{1}{n!}\mathfrak{p}_{-1}^n, \exp(x_1\mathfrak{G}_1)\cdot\frac{1}{n!}\mathfrak{p}_{-1}^n\right\rangle_n$$

thanks to the fact that $\mathfrak{G}_0([\lambda]) = |\lambda| \cdot [\lambda]$. Setting $u = x_0 + \ln(t_1 s_1)$, we denote the above generating function by $\tau(u, x_1)$. A simple computation reduces the Toda equation (12.40) to the following:

$$e^{-u} \frac{\partial^2}{\partial u^2} \ln \tau(u, x_1) = \frac{\tau(u + x_1, x_1)\, \tau(u - x_1, x_1)}{\tau(u, x_1)^2}.$$

It is interesting to observe that this τ function can also be interpreted [**Ok1**, Subsection 2.9] as generating functions of certain Hurwitz numbers.

12.5. Numerical aspects of Hilbert/Gromov-Witten correspondence

In this section, X denotes an arbitrary smooth projective complex surface. We will study the numerical aspect of the Hilbert/Gromov-Witten correspondence discovered in Theorem 12.5. More precisely, we will prove that certain structure constants in the cup products of the cohomology ring of the Hilbert scheme $X^{[n]}$ are related to the 1-point disconnected series $F^{\bullet}_{\lambda,\mu}(z)$ of the stationary Gromov-Witten theory of the projective line $\mathbb{P}^1$ relative to $0, \infty \in \mathbb{P}^1$.

Recall the notations $\mathcal{I}^{[n]}$ and $\mathfrak{a}_{-\lambda}(\alpha)$ from Definition 8.5 and (9.2) respectively. For a partition $\lambda = (1^{m_1} 2^{m_2} \cdots)$ and a cohomology class $\alpha \in H^*(X)$, define

$$\tilde{\mathfrak{a}}_{-\lambda}(\alpha) = \frac{1}{\mathfrak{z}_\lambda}\, \mathfrak{a}_{-\lambda}(\alpha) = \frac{1}{\mathfrak{z}_\lambda} \prod_{r \geq 1} \mathfrak{a}_{-r}(\alpha)^{m_r}. \tag{12.41}$$

The following theorem relates the construct constant $c^{\lambda}_{k,\mu}$ to the 1-point disconnected series $F^{\bullet}_{\lambda,\mu}(z)$.

THEOREM 12.13. *Let $2 \leq k \leq n$, $\mu \vdash n$, and $\mathcal{I} = \bigoplus_{i=1}^{4} H^i(X)$. Then,*

$$\begin{aligned} &\mathfrak{a}_{-(1^{n-k}k)}(1_X)|0\rangle \cup \mathfrak{a}_{-\mu}(1_X)|0\rangle \\ \equiv &\sum_{\substack{\lambda \vdash n \\ \ell(\lambda) = \ell(\mu)+1-k}} c^{\lambda}_{k,\mu}\, \mathfrak{a}_{-\lambda}(1_X)|0\rangle \pmod{\mathcal{I}^{[n]}} \end{aligned} \tag{12.42}$$

where $c^{\lambda}_{k,\mu}$ is the coefficient of z^{k-1} in

$$\frac{(n-k)!\, k!\, \mathfrak{z}_\mu}{\varsigma(z)} F^{\bullet}_{\lambda,\mu}(z).$$

PROOF. In view of (12.41), we note that (12.42) is equivalent to

$$\begin{aligned} &\tilde{\mathfrak{a}}_{-(1^{n-k}k)}(1_X)|0\rangle \cup \tilde{\mathfrak{a}}_{-\mu}(1_X)|0\rangle \\ \equiv &\sum_{\substack{\lambda \vdash n \\ \ell(\lambda) = \ell(\mu)+1-k}} \tilde{c}^{\lambda}_{k,\mu}\, \tilde{\mathfrak{a}}_{-\lambda}(1_X)|0\rangle \pmod{\mathcal{I}^{[n]}} \end{aligned} \tag{12.43}$$

where $\tilde{c}^{\lambda}_{k,\mu}$ is the coefficient of z^{k-1} in

$$\frac{(k-1)!\, \mathfrak{z}_\lambda}{\varsigma(z)} F^{\bullet}_{\lambda,\mu}(z).$$

Next, we outline our main idea used in the proof of (12.43). Using the results in Section 8.3, we will convert (12.43) to a cup product in the cohomology ring $H^*((\mathbb{C}^2)^{[n]})$ of the Hilbert scheme $(\mathbb{C}^2)^{[n]}$. Via the geometric realization of Heisenberg monomial classes given by Proposition 3.16, the cup product in $H^*((\mathbb{C}^2)^{[n]})$

is related to the cup product in $H^*_{\mathbb{T}}((\mathbb{C}^2)^{[n]})$. Then, we will apply (12.30) which is a consequence of Theorem 12.5 where an algebraic relation between the equivariant cohomology ring of $(\mathbb{C}^2)^{[n]}$ and the Gromov-Witten theory of $\mathbb{P}^1$ have been established.

To convert (12.43) to a cup product in $H^*((\mathbb{C}^2)^{[n]})$, define $\tilde{\mathfrak{a}}_{-\mu}(1_{\mathbb{C}^2})$ as in (12.41). By Theorem 8.12, the quotient ring $H^*(X^{[n]})/\mathcal{I}^{[n]}$ is isomorphic to the cohomology ring $H^*((\mathbb{C}^2)^{[n]})$. Moreover, we see from (8.18), (8.19), (8.4) that via this isomorphism, the cohomology classes $\tilde{\mathfrak{a}}_{-(1^{n-k}k)}(1_X)|0\rangle \pmod{\mathcal{I}^{[n]}}$ and $\tilde{\mathfrak{a}}_{-\mu}(1_X)|0\rangle$ $\pmod{\mathcal{I}^{[n]}}$ correspond to the classes $(-1)^{k-1}(k-1)!\ \mathrm{ch}_{k-1}^{[n]}$ and $\tilde{\mathfrak{a}}_{-\mu}(1_{\mathbb{C}^2})|0\rangle$ in $H^*((\mathbb{C}^2)^{[n]})$ respectively. Here, $\mathrm{ch}_{k-1}^{[n]}$ stands for the $(k-1)$-th component of the Chern character of the rank-n tautological vector bundle $(\mathcal{O}_{\mathbb{C}^2})^{[n]}$. Thus, proving (12.43) is equivalent to showing that

$$(12.44)\qquad \begin{aligned} &(-1)^{k-1}(k-1)!\ \mathrm{ch}_{k-1}^{[n]} \cup \tilde{\mathfrak{a}}_{-\mu}(1_{\mathbb{C}^2})|0\rangle \\ = &\sum_{\substack{\lambda\vdash n \\ \ell(\lambda)=\ell(\mu)+1-k}} \tilde{c}_{k,\mu}^{\lambda}\ \tilde{\mathfrak{a}}_{-\lambda}(1_{\mathbb{C}^2})|0\rangle. \end{aligned}$$

Next we rewrite the left-hand-side of (12.44) by using the equivariant cohomology $H^*_{\mathbb{T}}((\mathbb{C}^2)^{[n]})$ where $\mathbb{T}$ acts on $\mathbb{C}^2$ as in (12.10). Recall the equivariant Heisenberg operator $\mathfrak{p}_i(1_{\mathbb{C}^2})$ from Section 11.2. Define $\tilde{\mathfrak{p}}_{-\mu}(1_{\mathbb{C}^2})$ similarly as in (12.41). It is well-known that $H^*_{\mathbb{T}}((\mathbb{C}^2)^{[n]})$ is a $\mathbb{C}[t]$-module, and there is a ring isomorphism

$$H^*_{\mathbb{T}}((\mathbb{C}^2)^{[n]})/\langle t\rangle \cong H^*((\mathbb{C}^2)^{[n]}).$$

Let

$$\Psi : H^*_{\mathbb{T}}((\mathbb{C}^2)^{[n]}) \to H^*_{\mathbb{T}}((\mathbb{C}^2)^{[n]})/\langle t\rangle \cong H^*((\mathbb{C}^2)^{[n]})$$

be the natural projection. Assume that $\mu = (1^{s_1}2^{s_2}\ldots\ell^{s_\ell})$. As in Proposition 3.16, the Heisenberg monomial class

$$\frac{\mathfrak{z}_\mu}{\prod_{r\ge 1} s_r!}\tilde{\mathfrak{a}}_{-\mu}(1_{\mathbb{C}^2})|0\rangle = \left(\prod_{r=1}^{\ell}\frac{\mathfrak{a}_{-r}(1_{\mathbb{C}^2})^{s_r}}{s_r!}\right)|0\rangle \in H^*((\mathbb{C}^2)^{[n]})$$

is represented by the closure $\overline{W^0_\mu} \subset (\mathbb{C}^2)^{[n]}$ of

$$W(\underbrace{1,\mathbb{C}^2;\ldots;1,\mathbb{C}^2}_{s_1\ \text{times}};\ldots;\underbrace{\ell,\mathbb{C}^2;\ldots;\ell,\mathbb{C}^2}_{s_\ell\ \text{times}})^0.$$

Similarly, $\overline{W^0_\mu}$ is a $\mathbb{T}$-equivariant closed subscheme of $(\mathbb{C}^2)^{[n]}$, and

$$\frac{\mathfrak{z}_\mu}{\prod_{r\ge 1} s_r!}\tilde{\mathfrak{p}}_{-\mu}(1_{\mathbb{C}^2})|0\rangle \in H^*_{\mathbb{T}}((\mathbb{C}^2)^{[n]})$$

is the equivariant cohomology class corresponding to $\overline{W^0_\mu}$. So

$$\Psi(\tilde{\mathfrak{p}}_{-\mu}(1_{\mathbb{C}^2})|0\rangle) = \tilde{\mathfrak{a}}_{-\mu}(1_{\mathbb{C}^2})|0\rangle$$

and

$$(12.45)\qquad \begin{aligned} &(-1)^{k-1}(k-1)!\ \mathrm{ch}_{k-1}^{[n]} \cup \tilde{\mathfrak{a}}_{-\mu}(1_{\mathbb{C}^2})|0\rangle \\ = &(-1)^{k-1}(k-1)!\ \Psi\big(\mathrm{ch}_{k-1,\mathbb{T}}^{[n]} \cup \tilde{\mathfrak{p}}_{-\mu}(1_{\mathbb{C}^2})|0\rangle\big). \end{aligned}$$

Note that $1_{\mathbb{C}^2} = [\mathbb{C}^2] = -t^{-1}[\{0\} \times \mathbb{C}]$ in $H^*_{\mathbb{T}}((\mathbb{C}^2)^{[n]})$. It follows that

$$\tilde{\mathfrak{p}}_{-\mu}(1_{\mathbb{C}^2}) = (-t^{-1})^{\ell(\mu)}\,\tilde{\mathfrak{p}}_{-\mu}([\{0\} \times \mathbb{C}]) = (-t)^{-\ell(\mu)}\,\tilde{\mathfrak{p}}_{-\mu} \tag{12.46}$$

where the last $\tilde{\mathfrak{p}}_{-\mu}$ denotes $\tilde{\mathfrak{p}}_{-\mu}([\{0\} \times \mathbb{C}])$ for simplicity. By (12.45),

$$\begin{aligned} &(-1)^{k-1}(k-1)!\ \mathrm{ch}_{k-1}^{[n]} \cup \tilde{\mathfrak{a}}_{-\mu}(1_{\mathbb{C}^2})|0\rangle \\ = \ &(-1)^{k-1-\ell(\mu)}(k-1)!\ \Psi\left(t^{-\ell(\mu)}\ \mathrm{ch}_{k-1,\mathbb{T}}^{[n]} \cup \tilde{\mathfrak{p}}_{-\mu}|0\rangle\right). \end{aligned} \tag{12.47}$$

Recall from Section 11.2 that $\{\tilde{\mathfrak{p}}_{-\lambda}|0\rangle \mid \lambda \vdash n\}$ is a linear basis of $H^{2n}_{\mathbb{T}}((\mathbb{C}^2)^{[n]})$. Put

$$\widetilde{\mathrm{ch}}_{k-1}^{[n]} \star \tilde{\mathfrak{p}}_{-\mu}|0\rangle = \sum_{\lambda \vdash n} d_{k,\mu}^{\lambda}\ \tilde{\mathfrak{p}}_{-\lambda}|0\rangle \tag{12.48}$$

where all the coefficients $d_{k,\mu}^{\lambda}$ are independent of t. So by (12.16) and (11.18),

$$\begin{aligned} \mathrm{ch}_{k-1,\mathbb{T}}^{[n]} \cup \tilde{\mathfrak{p}}_{-\mu}|0\rangle &= t^{-n+(k-1)}\widetilde{\mathrm{ch}}_{k-1,\mathbb{T}}^{[n]} \cup \tilde{\mathfrak{p}}_{-\mu}|0\rangle \\ &= t^{k-1}\widetilde{\mathrm{ch}}_{k-1,\mathbb{T}}^{[n]} \star \tilde{\mathfrak{p}}_{-\mu}|0\rangle \\ &= t^{k-1}\sum_{\lambda \vdash n} d_{k,\mu}^{\lambda}\ \tilde{\mathfrak{p}}_{-\lambda}|0\rangle. \end{aligned} \tag{12.49}$$

Combining this with (12.47) and (12.46), we conclude that

$$\begin{aligned} &(-1)^{k-1}(k-1)!\ \mathrm{ch}_{k-1}^{[n]} \cup \tilde{\mathfrak{a}}_{-\mu}(1_{\mathbb{C}^2})|0\rangle \\ = \ &(-1)^{k-1-\ell(\mu)}(k-1)!\ \Psi\left(t^{k-1-\ell(\mu)}\sum_{\lambda \vdash n} d_{k,\mu}^{\lambda}\ \tilde{\mathfrak{p}}_{-\lambda}|0\rangle\right) \\ = \ &(k-1)!\ \Psi\left(\sum_{\lambda \vdash n}(-t)^{k-1-\ell(\mu)+\ell(\lambda)}\ d_{k,\mu}^{\lambda}\ \tilde{\mathfrak{p}}_{-\lambda}(1_{\mathbb{C}^2})|0\rangle\right). \end{aligned} \tag{12.50}$$

Since $\Psi(\tilde{\mathfrak{p}}_{-\lambda}(1_{\mathbb{C}^2})|0\rangle) = \tilde{\mathfrak{a}}_{-\lambda}(1_{\mathbb{C}^2})|0\rangle \neq 0$ for each $\lambda \vdash n$, we obtain from (12.50) that

$$\begin{aligned} &(-1)^{k-1}(k-1)!\ \mathrm{ch}_{k-1}^{[n]} \cup \tilde{\mathfrak{a}}_{-\mu}(1_{\mathbb{C}^2})|0\rangle \\ = &\sum_{\substack{\lambda \vdash n \\ \ell(\lambda)=\ell(\mu)+1-k}} \left((k-1)!\,d_{k,\mu}^{\lambda}\right)\ \tilde{\mathfrak{a}}_{-\lambda}(1_{\mathbb{C}^2})|0\rangle. \end{aligned} \tag{12.51}$$

By (12.51), to finish the proof of (12.44), it remains to show that the structure constant $d_{k,\mu}^{\lambda}$ in (12.48) is the coefficient of z^{k-1} in the expansion of the series

$$\frac{\mathfrak{z}_\lambda}{\varsigma(z)}F_{\lambda,\mu}^{\bullet}(z). \tag{12.52}$$

By (12.29) and (12.30), we have

$$\sum_i z^i\,\langle\tilde{\mathfrak{p}}_{-\lambda}|0\rangle, \widetilde{\mathrm{ch}}_i^{[n]} \star \tilde{\mathfrak{p}}_{-\mu}|0\rangle\rangle_n = \frac{1}{\varsigma(z)}\left(F_{\lambda,\mu}^{\bullet}(z) - \frac{\delta_{\lambda,\mu}}{\mathfrak{z}_\lambda\varsigma(z)}\right).$$

Combining these observations with (12.48) and (11.30) with $\alpha = \beta = 1$, we get

$$\begin{aligned} d_{k,\mu}^{\lambda} &= \mathfrak{z}_\lambda\langle\tilde{\mathfrak{p}}_{-\lambda}|0\rangle, \widetilde{\mathrm{ch}}_{k-1}^{[n]} \star \tilde{\mathfrak{p}}_{-\mu}|0\rangle\rangle_n \\ &= \mathrm{Coeff}_{z^{k-1}}\left(\frac{\mathfrak{z}_\lambda}{\varsigma(z)}\left(F_{\lambda,\mu}^{\bullet}(z) - \frac{\delta_{\lambda,\mu}}{\mathfrak{z}_\lambda\varsigma(z)}\right)\right). \end{aligned}$$

Finally, since $\ell(\lambda) = \ell(\mu)+1-k$ and $k \geq 2$, we have $\lambda \neq \mu$. Therefore, the structure constant $d^{\lambda}_{k,\mu}$ in (12.51) is the coefficient of z^{k-1} in (12.52). $\square$

REMARK 12.14. Let $n \geq 2$, $\mu \vdash n$, and $\mathcal{I} = \bigoplus_{i=1}^{4} H^i(X)$. Theorem 12.13 handles the cup product $\mathfrak{a}_{-(1^{n-k}k)}(1_X)|0\rangle \cup \mathfrak{a}_{-\mu}(1_X)|0\rangle \pmod{\mathcal{I}^{[n]}}$ for the partition $(1^{n-k}k)$ of n. By Lemma 8.11, the quotient ring $H^*(X^{[n]})/\mathcal{I}^{[n]}$ is generated by the $(n-1)$ classes $\mathfrak{a}_{-(1^{n-k}k)}(1_X)|0\rangle \pmod{\mathcal{I}^{[n]}}$, $2 \leq k \leq n$. It follows that the cup product

$$\mathfrak{a}_{-\nu}(1_X)|0\rangle \cup \mathfrak{a}_{-\mu}(1_X)|0\rangle \pmod{\mathcal{I}^{[n]}}$$

for an arbitrary partition $\nu \vdash n$ can be obtained by repeatedly applying Theorem 12.13. In other words, the structure constants of $\mathfrak{a}_{-\nu}(1_X)|0\rangle \cup \mathfrak{a}_{-\mu}(1_X)|0\rangle \pmod{\mathcal{I}^{[n]}}$ should also be related to the Gromov-Witten theory of $\mathbb{P}^1$.

12.6. Relation to the Hurwitz numbers of $\mathbb{P}^1$

In view of Theorem 8.12, Theorem 12.13 states that in $H^*((\mathbb{C}^2)^{[n]})$, we have

$$\mathfrak{a}_{-(1^{n-k}k)}(1_{\mathbb{C}^2})|0\rangle \cup \mathfrak{a}_{-\mu}(1_{\mathbb{C}^2})|0\rangle = \sum_{\substack{\lambda \vdash n \\ \ell(\lambda)=\ell(\mu)+1-k}} c^{\lambda}_{k,\mu}\, \mathfrak{a}_{-\lambda}(1_{\mathbb{C}^2})|0\rangle$$

where $c^{\lambda}_{k,\mu}$ is the coefficient of z^{k-1} in

$$\frac{(n-k)!\, k!\, \mathfrak{z}_\mu}{\varsigma(z)} F^{\bullet}_{\lambda,\mu}(z).$$

In this section, we will establish a precise identity between some general structure constants under the cup product in $H^*((\mathbb{C}^2)^{[n]})$ and certain Hurwitz numbers for the projective line $\mathbb{P}^1$.

The Hurwitz theory of a nonsingular curve C counts the number of covers of C with specified ramification. The ramifications are determined by the profile of the cover over the branch points. For Hurwitz theory, we will only consider covers

$$\pi : D \to C$$

where D is a nonsingular (possibly disconnected) curve and π is dominant on each component of D. Let $n > 0$ be the degree of π. The profile of π over a point $c \in C$ is the partition λ of n obtained from the multiplicities of $\pi^{-1}(c)$. The profile of π over c is the partition (1^n) if and only if π is unramified over c.

Let $\lambda^1, \ldots, \lambda^k$ be partitions of n assigned to k distinct points $c_1, \ldots, c_k \in C$. A *Hurwitz cover* of C of genus g, degree n and monodromy λ^i at the point c_i is a morphism

$$\pi : D \to C$$

such that

(i) D is a nonsingular (possibly disconnected) curve of genus g;
(ii) π has profile λ^i at the point c_i;
(iii) π is unramified over $C - \{c_1, \ldots, c_k\}$.

The Riemann-Hurwitz formula provides the genus constraint of a Hurwitz cover:

$$2g - 2 + n(2 - 2g_C) = \sum_{i=1}^{k} (n - \ell(\lambda^i)) \tag{12.53}$$

which is valid for both connected and disconnected Hurwitz covers. In disconnected theory, the domain genus may be negative. Two Hurwitz covers $\pi : D \to C$ and $\pi' : D' \to C$ are defined to be isomorphic if there exists an isomorphism of curves $f : D \to D'$ satisfying $\pi' \circ f = \pi$. Up to isomorphism, there are only finitely many Hurwitz covers of C of genus g, degree n and monodromy λ^i at c_i. Each cover π has a finite group of automorphisms $\mathrm{Aut}(\pi)$. The *Hurwitz number*

$$H_n^C(\lambda^1, \dots, \lambda^k) \tag{12.54}$$

is defined to be the weighted count of the distinct, possibly disconnected Hurwitz covers π with the prescribed data. Each such cover is weighted by $1/|\mathrm{Aut}(\pi)|$. We refer to [**GJV**] for more references on (single) Hurwitz numbers and some interesting development which relates double Hurwitz numbers to Gromov-Witten theory.

The enumeration of Hurwitz covers of the projective line $\mathbb{P}^1$ is classically known to be equivalent to multiplication in the class algebra of the symmetric group. Fix $(k+1)$ partitions $\lambda^0, \lambda^1, \dots, \lambda^k$ of n. Hurwitz covers with profile λ^i over $c_i \in \mathbb{P}^1$ canonically yield $(k+1)$-tuples of permutations $(s_0, \dots, s_k)$, in the n-th symmetric group S_n, defined up to conjugation satisfying:

(i) s_i is of cycle type λ^i for each i;
(ii) $s_0 s_1 \dots s_k = 1$.

The elements s_i are determined by the monodromies of π around the points c_i. This leads to the following well-known lemma (e.g. [**OP1**, Subsection 0.3.3]).

LEMMA 12.15. *The Hurwitz number $H_n^{\mathbb{P}^1}(\lambda^0, \dots, \lambda^k)$ is equal to the number of $(k+1)$-tuples of permutations $(s_0, \dots, s_k)$ in the n-th symmetric group S_n satisfying the two conditions below divided by $n!$:*

(i) s_i is of cycle type λ^i for each i;
(ii) $s_0 s_1 \dots s_k = 1$.

Next, we consider the cup product $\mathfrak{a}_{-\lambda^1}(1_{\mathbb{C}^2})|0\rangle \cup \cdots \cup \mathfrak{a}_{-\lambda^k}(1_{\mathbb{C}^2})|0\rangle$ in the cohomology ring of $(\mathbb{C}^2)^{[n]}$. Denote by $c_{\lambda^1,\dots,\lambda^k}^{\lambda^0}$ the coefficient of $\mathfrak{a}_{-\lambda^0}(1_{\mathbb{C}^2})|0\rangle$ in

$$\mathfrak{a}_{-\lambda^1}(1_{\mathbb{C}^2})|0\rangle \cup \cdots \cup \mathfrak{a}_{-\lambda^k}(1_{\mathbb{C}^2})|0\rangle$$

when expressed as a linear combination of the Heisenberg monomial classes. By cohomology degree reasons, we have

$$(n - \ell(\lambda^0)) = \sum_{i=1}^{k} (n - \ell(\lambda^i)). \tag{12.55}$$

Setting $g = 1 - \ell(\lambda^0)$, we see that $n, g, \lambda^0, \lambda^1, \dots, \lambda^k$ satisfy the identity (12.53).

THEOREM 12.16. *Let $\lambda^0, \lambda^1, \dots, \lambda^k$ be partitions of n satisfying (12.55). Then,*

$$c_{\lambda^1,\dots,\lambda^k}^{\lambda^0} = \prod_{i=1}^{k} \mathfrak{z}_{\lambda^i} \cdot H_n^{\mathbb{P}^1}(\lambda^0, \lambda^1, \dots, \lambda^k). \tag{12.56}$$

PROOF. Recall the notation $\tilde{\mathfrak{a}}_{-\lambda}(1_{\mathbb{C}^2})$ from (12.41). Let $\tilde{c}_{\lambda^1,\dots,\lambda^k}^{\lambda^0}$ be the coefficient of $\tilde{\mathfrak{a}}_{-\lambda^0}(1_{\mathbb{C}^2})|0\rangle$ in

$$\tilde{\mathfrak{a}}_{-\lambda^1}(1_{\mathbb{C}^2})|0\rangle \cup \cdots \cup \tilde{\mathfrak{a}}_{-\lambda^k}(1_{\mathbb{C}^2})|0\rangle$$

when expressed as a linear combination of the Heisenberg monomial classes. Then

$$c_{\lambda^1,\ldots,\lambda^k}^{\lambda^0} = \prod_{i=1}^{k} \mathfrak{z}_{\lambda^i} \cdot \frac{1}{\mathfrak{z}_{\lambda^0}} \cdot \tilde{c}_{\lambda^1,\ldots,\lambda^k}^{\lambda^0},$$

and (12.56) is equivalent to

$$\tilde{c}_{\lambda^1,\ldots,\lambda^k}^{\lambda^0} = \mathfrak{z}_{\lambda^0} \cdot H_n^{\mathbb{P}^1}(\lambda^0, \lambda^1, \ldots, \lambda^k). \tag{12.57}$$

Let Ψ_1, Ψ_2 and C_λ be from (8.4), (8.11) and (8.9) respectively. By (8.5) and (8.12), we obtain

$$\Psi_1\Psi_2(C_\lambda) = \tilde{\mathfrak{a}}_{-\lambda}(1_{\mathbb{C}^2})|0\rangle.$$

By the assumption (12.55), we see from Theorem 8.3 that $\tilde{c}_{\lambda^1,\ldots,\lambda^k}^{\lambda^0}$ is equal to the number of conjugacy class C_{λ^0} in the cup product of the conjugacy classes $C_{\lambda^1}, \ldots, C_{\lambda^k}$ in the class algebra $\mathcal{C}(S_n)$, denoted by

$$[C_{\lambda^0}] \prod_{i=1}^{k} C_{\lambda^i}.$$

Since the conjugacy class C_{λ^0} consists of $n!/\mathfrak{z}_{\lambda^0}$ elements, we have

$$[C_{(1^n)}]C_{\lambda^0} \prod_{i=1}^{k} C_{\lambda^i} = \frac{n!}{\mathfrak{z}_{\lambda^0}} \cdot [C_{\lambda^0}] \prod_{i=1}^{k} C_{\lambda^i} = \frac{n!}{\mathfrak{z}_{\lambda^0}} \cdot \tilde{c}_{\lambda^1,\ldots,\lambda^k}^{\lambda^0}.$$

On the other hand, we conclude from Lemma 12.15 that

$$H_n^{\mathbb{P}^1}(\lambda^0, \lambda^1, \ldots, \lambda^k) = \frac{1}{n!} \cdot [C_{(1^n)}]C_{\lambda^0} \prod_{i=1}^{k} C_{\lambda^i}.$$

Now (12.57) follows from putting the above together. □

Part 5

Gromov-Witten theory of the Hilbert schemes of points

CHAPTER 13

Cosection localization for the Hilbert schemes of points

The Heisenberg algebra of Grojnowski and Nakajima offers a basic tool for writing down the cohomology classes of the Hilbert schemes of points on a surface. The geometric representation (Proposition 3.16) of the Heisenberg monomial classes provides a convenient way to analyze the Gromov-Witten invariants of these Hilbert schemes. The study of the Gromov-Witten theory of the Hilbert schemes of points on a surface began with the work [**LQ1**] where the 1-point genus-0 extremal Gromov-Witten invariants are computed. The results in [**LQ1**] give the first evidence to Ruan's Cohomological Crepant Resolution Conjecture [**Rua2**, Section 2]. In [**ELQ**], the genus-0 extremal Gromov-Witten invariants of the Hilbert scheme of 3 points on the complex projective plane $\mathbb{P}^2$ are calculated. Motivated by the Gromov-Witten and Donaldson-Thomas correspondence conjectured in [**MNOP1**, **MNOP2**], Okounkov and Pandharipande [**OP3**, **OP4**] investigated the equivariant Gromov-Witten theory of the Hilbert schemes of points on the affine plane $\mathbb{C}^2$, and proved its equivalence with the local Gromov-Witten theory of the 3-fold $\mathbb{C}^2 \times \mathbb{P}^1$. Using the technique of cosection localization and the equivariant quantum corrected boundary operator of Okounkov and Pandharipande [**OP3**], Jun Li and W.-P. Li [**LL**] determined the 2-point genus-0 extremal Gromov-Witten invariants of the Hilbert scheme $X^{[n]}$ of points on an arbitrary surface X.

In this chapter, we will address the technique of cosection localization developed in [**KL1**, **KL2**], which provides a powerful tool for studying the virtual fundamental cycles. We will apply it to the Gromov-Witten theory of the Hilbert schemes of points on surfaces, and review the reduction results from [**LL**] when the images of the stable maps to the Hilbert schemes are contracted by the Hilbert-Chow morphism. When the smooth projective surface X has a non-trivial holomorphic 2-form, we will prove some vanishing results regarding the Gromov-Witten invariants of the Hilbert scheme $X^{[n]}$. Moreover, we will apply the results to the Hilbert scheme $X^{[2]}$, and calculate its genus-1 Gromov-Witten invariant. Proposition 13.3 is from [**LL**], while Proposition 13.15, Theorem 13.19 and their proofs are from [**HLQ**].

13.1. Cosection localization of Kiem and J. Li

In this section, we will recall from [**KL1**, **KL2**] the technique of cosection localization and its application to the Gromov-Witten theory. This technique was used in [**LL**] to study stable maps to the Hilbert scheme $X^{[n]}$, whose images in $X^{[n]}$ are contracted by the Hilbert-Chow morphism to points in the symmetric product

$X^{(n)}$. We will present the reduction results from [**LL**] regarding the corresponding virtual fundamental cycles.

First of all, we review the technique of cosection localization. Following the concept of a localized Gysin map, a localized virtual fundamental cycle is constructed by Kiem and J. Li [**KL2**, Theorem 1.1].

THEOREM 13.1. *Let $\mathfrak{M}$ be a Deligne-Mumford stack endowed with a perfect obstruction theory. Suppose the obstruction sheaf $\mathcal{O}b_{\mathfrak{M}}$ admits a surjective homomorphism $\sigma : \mathcal{O}b_{\mathfrak{M}}|_U \to \mathcal{O}_U$ over an open $U \subset \mathfrak{M}$. Let*

$$\mathfrak{M}(\sigma) = \mathfrak{M} - U.$$

be the degeneracy locus. Then $(\mathfrak{M}, \sigma)$ has a localized virtual fundamental cycle

$$[\mathfrak{M}]^{\text{vir}}_{\text{loc}} \in A_*\big(\mathfrak{M}(\sigma)\big).$$

This cycle has the usual properties of the virtual fundamental cycles, and relates to the usual virtual fundamental cycle $[\mathfrak{M}]^{\text{vir}}$ via

$$[\mathfrak{M}]^{\text{vir}} = \iota_*[\mathfrak{M}]^{\text{vir}}_{\text{loc}} \in A_*\big(\mathfrak{M}\big)$$

where $\iota : \mathfrak{M}(\sigma) \to \mathfrak{M}$ is the inclusion map.

To apply the technique of cosection localization to Gromov-Witten theory, let Y be a smooth quasi-projective variety with a non-trivial holomorphic 2-form $\tilde{\theta} \in H^0(Y, \Omega_Y^2)$. Let $\mathfrak{M} = \overline{\mathfrak{M}}_{g,r}(Y, \beta)$ be the moduli space of r-pointed genus-g stable maps to Y with homology class β. The 2-form $\tilde{\theta}$ on Y induces a cosection σ of the obstruction sheaf on $\mathfrak{M}$ as follows. Let $f : \mathcal{C} \to Y$ and $\pi : \mathcal{C} \to \mathfrak{M}$ be the universal family of $\mathfrak{M}$. Let $\mathcal{S}$ be the Artin stack of genus-g connected nodal curves. Then the relative obstruction sheaf of the standard relative obstruction theory of $\mathfrak{M}/\mathcal{S}$ is

$$\mathcal{O}b_{\mathfrak{M}/\mathcal{S}} = R^1\pi_* f^*T_Y.$$

Now regard the holomorphic 2-form $\tilde{\theta} \in H^0(Y, \Omega_Y^2)$ as an anti-symmetric homomorphism

$$\hat{\theta} : T_Y \longrightarrow \Omega_Y, \qquad \langle \hat{\theta}(v), v \rangle = 0. \tag{13.1}$$

Then $\hat{\theta}$ defines the first arrow in the following sequence of homomorphisms

$$R^1\pi_* f^*T_Y \longrightarrow R^1\pi_* f^*\Omega_Y \longrightarrow R^1\pi_*\Omega_{\mathcal{C}/\mathfrak{M}} \longrightarrow R^1\pi_*\omega_{\mathcal{C}/\mathfrak{M}} \tag{13.2}$$

where the second is induced by $f^*\Omega_Y \to \Omega_{\mathcal{C}/\mathfrak{M}}$, and the third is induced by the tautological map $\Omega_{\mathcal{C}/\mathfrak{M}} \to \omega_{\mathcal{C}/\mathfrak{M}}$. Since $R^1\pi_*\omega_{\mathcal{C}/\mathfrak{M}} \cong \mathcal{O}_{\mathfrak{M}}$, the composite of this sequence provides

$$\sigma^{\text{rel}} : \quad R^1\pi_* f^*T_Y = \mathcal{O}b_{\mathfrak{M}/\mathcal{S}} \longrightarrow \mathcal{O}_{\mathfrak{M}}. \tag{13.3}$$

The obstruction sheaf of $\mathfrak{M}$ is the cokernel of $p^*T_{\mathcal{S}} \to \mathcal{O}b_{\mathfrak{M}/\mathcal{S}}$ where $p : \mathfrak{M} \to \mathcal{S}$ is the projection. Using the universal family f and $R^1\pi_* f^*T_Y = \mathcal{E}xt^1_\pi(f^*\Omega_Y, \mathcal{O}_{\mathcal{C}})$, we obtain the exact sequence

$$\mathcal{E}xt^1_\pi(\Omega_{\mathcal{C}/\mathfrak{M}}, \mathcal{O}_{\mathcal{C}}) \longrightarrow \mathcal{E}xt^1_\pi(f^*\Omega_Y, \mathcal{O}_{\mathcal{C}}) \longrightarrow \mathcal{O}b_{\mathfrak{M}} \longrightarrow 0 \tag{13.4}$$

where the first arrow is induced by $f^*\Omega_Y \to \Omega_{\mathcal{C}/\mathfrak{M}}$. It can be proved that the composition

$$\mathcal{E}xt^1_\pi(\Omega_{\mathcal{C}/\mathfrak{M}}, \mathcal{O}_{\mathcal{C}}) \longrightarrow \mathcal{E}xt^1_\pi(f^*\Omega_Y, \mathcal{O}_{\mathcal{C}}) \xrightarrow{\sigma^{\text{rel}}} \mathcal{O}_{\mathfrak{M}}$$

is the zero homomorphism. Combining with (13.4), we obtain a cosection of the obstruction sheaf $\mathcal{O}b_{\mathfrak{M}}$:

$$\sigma : \quad \mathcal{O}b_{\mathfrak{M}} \longrightarrow \mathcal{O}_{\mathfrak{M}}. \tag{13.5}$$

The vanishing criterion asserts that the degeneracy locus of σ consists of stable maps $[\varphi : C \to Y]$ satisfying

$$\hat{\theta}(\varphi_* T_{C_{\mathrm{reg}}}) = 0 \tag{13.6}$$

where C_{reg} denotes the smooth locus of the curve C (these stable maps are called $\tilde{\theta}$-null stable maps).

Next, we assume that X denotes a smooth projective complex surface with a non-trivial holomorphic differential 2-form

$$\theta \in H^0(X, \Omega_X^2) = H^0(X, \mathcal{O}_X(K_X)).$$

Using θ, Beauville [**Bea2**, Section 5] (also [**Bea1**]) constructed a holomorphic 2-form on the Hilbert scheme $X^{[n]}$ as follows. Let

$$\begin{aligned}
X^{[n]}{}_* &= \{\xi \in X^{[n]} \,|\, |\mathrm{Supp}(\xi)| \geq n-1\}, \\
X_*^{(n)} &= \rho_n(X^{[n]}{}_*), \\
X_*^n &= \eta_n^{-1}(X_*^{(n)}), \\
\Delta_{n*} &= \{(x_1, \ldots, x_i, \ldots, x_j, \ldots, x_n) \in X_*^n \,|\, x_i = x_j \text{ for some } i < j\}
\end{aligned}$$

where $\rho_n : X^{[n]} \to X^{(n)}$ is the Hilbert-Chow morphism, and $\eta_n : X^n \to X^{(n)}$ is the quotient map. Let

$$\pi_n : \quad \widetilde{X_*^n} \to X_*^n$$

be the blowing-up morphism of X_*^n along the big diagonal Δ_{n*}. Then we have a commutative diagram of morphisms:

$$\begin{array}{ccc}
\widetilde{X_*^n} & \xrightarrow{\pi_n} & X_*^n \\
\downarrow{\scriptstyle \tilde{\eta}_n} & & \downarrow{\scriptstyle \eta_n} \\
X^{[n]}{}_* & \xrightarrow{\rho_n} & X_*^{(n)}
\end{array} \tag{13.7}$$

where $\tilde{\eta}_n : \widetilde{X_*^n} \to \widetilde{X_*^n}/S_n = X^{[n]}{}_*$ is the quotient map. Let $p_i : X^n \to X$ be the i-th projection. Then the holomorphic 2-form $\pi_n^*(\sum_{i=1}^n p_i^*\theta)$ on $\widetilde{X_*^n}$ is S_n-invariant, and thus induces a holomorphic 2-form $\theta_*^{[n]}$ on $X^{[n]}{}_*$. Note that $X^{[n]} - X^{[n]}{}_*$ is a codimension-2 closed subscheme of $X^{[n]}$. By Hartogs' Theorem, $\theta_*^{[n]}$ extends to a holomorphic 2-form $\theta^{[n]}$ on $X^{[n]}$:

$$\theta^{[n]} \in H^0(X^{[n]}, \Omega^2_{X^{[n]}}). \tag{13.8}$$

Now apply the construction of the cosection (13.5) to $\tilde{\theta} = \theta^{[n]}$ and $Y = X^{[n]}$. Put

$$\overline{\mathfrak{M}} = \overline{\mathfrak{M}}_{g,0}(X^{[n]}, d\beta_n).$$

Then, $\theta^{[n]}$ defines a regular cosection of the obstruction sheaf $\mathcal{O}b_{\overline{\mathfrak{M}}}$ of $\overline{\mathfrak{M}}$:

$$\sigma : \mathcal{O}b_{\overline{\mathfrak{M}}} \to \mathcal{O}_{\overline{\mathfrak{M}}}. \tag{13.9}$$

Moreover, if $\overline{\mathfrak{M}}(\sigma) \subset \overline{\mathfrak{M}}$ denotes the degeneracy locus of points over which σ vanishes, then

$$[\overline{\mathfrak{M}}]^{\mathrm{vir}} \in H_*(\overline{\mathfrak{M}}(\sigma)) \tag{13.10}$$

by Theorem 13.1. By abusing notations, denote a point $[\varphi : C \to X^{[n]}] \in \overline{\mathfrak{M}}$ by φ. By the vanishing criterion (13.6), σ vanishes at a point $\varphi \in \overline{\mathfrak{M}}$ if the image of $\varphi_* : T_{C_{\text{reg}}} \to T_{X^{[n]}}$ lies in the null space of

$$\theta^{[n]} : T_{X^{[n]}} \to \Omega_{X^{[n]}}. \tag{13.11}$$

To understand such φ, note that $[\varphi(C)] = d\beta_n$. Thus the composition $\rho_n \circ \varphi$ is a constant map. Let

$$\text{Spt} : \overline{\mathfrak{M}} \to X^{(n)}$$

be the induced map. If $\text{Spt}(\varphi) = \sum_{i=1}^{\ell} n_i x_i \in X^{(n)}$, then the morphism φ factors through the product of punctual Hilbert schemes:

$$\varphi = (\varphi_1, \ldots, \varphi_\ell) : C \to \prod_{i=1}^{\ell} M_{n_i}(x_i) \subset X^{[n]} \tag{13.12}$$

where φ_i is a morphism from C to $M_{n_i}(x_i)$. The collection

$$\varphi = (\varphi_1, \ldots, \varphi_\ell) \tag{13.13}$$

is defined to be the *standard decomposition* of φ, and the point x_i is called the *support* of φ_i. Note that the collection $\{\varphi_1, \ldots, \varphi_\ell\}$ is unique up to the ordering of the φ_i's. The following lemma and its proof are from [**LL**, Lemma 3.2].

LEMMA 13.2. *Let $\Lambda_\theta \subset \overline{\mathfrak{M}} = \overline{\mathfrak{M}}_{g,0}(X^{[n]}, d\beta_n)$ be the subset consisting of the points*

$$\varphi = (\varphi_1, \ldots, \varphi_\ell) \in \overline{\mathfrak{M}}$$

such that for each i, either φ_i is a constant map or its support $x_i = \text{Spt}(\varphi_i)$ lies in the vanishing locus of θ. Then,

$$\overline{\mathfrak{M}}(\sigma) \subset \Lambda_\theta.$$

PROOF. Assume that $\varphi = (\varphi_1, \ldots, \varphi_\ell) \in \overline{\mathfrak{M}}$ lies in the degeneracy locus $\overline{\mathfrak{M}}(\sigma)$ of σ. By the vanishing criterion (13.6), $\varphi_*(T_{C_{\text{reg}}})$ is contained in the null space of $\theta^{[n]}$. Note that if $\xi = \xi_1 + \ldots + \xi_\ell$ with $\xi_i \in M_{n_i}(x_i)$, then

$$T_{\xi, X^{[n]}} \cong \bigoplus_{i=1}^{\ell} T_{\xi_i, X^{[n_i]}}.$$

Thus, $\theta^{[n]}$ is the direct sum of the forms $\theta^{[n_i]}$, and $\varphi_*(T_{C_{\text{reg}}})$ is contained in the null space of $\theta^{[n]}$ if and only if $\varphi_{i*}(T_{C_{\text{reg}}})$ is contained in the null space of $\theta^{[n_i]}$ for each i. According to the work of Beauville, the form $\theta^{[n_i]}$ is non-degenerate along $M_{n_i}(x)$ if $x \notin \theta^{-1}(0)$. Applying this to the support x_i of φ_i, we see that $\varphi \in \Lambda_\theta$. □

Lemma 13.2 is sufficient if we have a regular section $\theta \in H^0(X, \mathcal{O}_X(K_X))$. For a general surface, we take a meromorphic section θ of $\mathcal{O}_X(K_X)$. Let D_0 and D_∞ be the vanishing and pole divisors of θ respectively. Viewed as a meromorphic section of Ω_X^2, θ induces a meromorphic section $\theta^{[n]}$ of $\Omega^2_{X^{[n]}}$ and hence a meromorphic homomorphism

$$\sigma : \mathcal{E} \to \mathcal{O}_{\overline{\mathfrak{M}}}.$$

Here, $\mathcal{E}$ is a certain locally free sheaf admitting a surjection onto $R^1\pi_* f^* T_{X^{[n]}}$, and $f : \mathcal{C} \to X^{[n]}$ and $\pi : \mathcal{C} \to \overline{\mathfrak{M}}$ are the universal family of $\overline{\mathfrak{M}}$. Define *the degeneracy locus* of σ to be the subset $\text{Deg}(\sigma) \subset \overline{\mathfrak{M}}$ consisting of all the points $\varphi \in \overline{\mathfrak{M}}$ such that either σ is undefined or vanishes at φ. Adopting the proof of

Lemma 13.2, one sees that $\mathrm{Deg}(\sigma)$ is contained in the subset of $\overline{\mathfrak{M}}$ consisting of all those $\varphi = (\varphi_1, \ldots, \varphi_\ell) \in \overline{\mathfrak{M}}$ such that either for some i the support of φ_i is contained in $D_0 \cup D_\infty$, or for each i the map φ_i is constant.

The following reduction result states that the virtual fundamental cycle $\overline{\mathfrak{M}}^{\mathrm{vir}}$ is supported in a much smaller subset than $\mathrm{Deg}(\sigma)$.

PROPOSITION 13.3. *Let $\Lambda_\theta \subset \overline{\mathfrak{M}} = \overline{\mathfrak{M}}_{g,0}(X^{[n]}, d\beta_n)$ be the subset consisting of the points*

$$\varphi = (\varphi_1, \ldots, \varphi_\ell) \in \overline{\mathfrak{M}} \tag{13.14}$$

such that for each i, either φ_i is a constant map or the support $x_i = \mathrm{Spt}(\varphi_i)$ lies in $D_0 \cup D_\infty$. Then $\overline{\mathfrak{M}}^{\mathrm{vir}}$ is supported in Λ_θ.

The proof of Proposition 13.3 is based on the observation that when $\varphi \in \overline{\mathfrak{M}}$ is decomposed into ℓ maps into smaller Hilbert schemes as in (13.13), the obstruction sheaf can be decomposed into the direct sum of ℓ factors. We refer to [**LL**, Section 3] for the details.

Proposition 13.3 deals with stable maps whose images are contracted by the Hilbert-Chow morphisms, and only requires the presence of a *meromorphic* section θ of $\mathcal{O}_X(K_X)$. Therefore, it is very useful in studying extremal Gromov-Witten invariants of the Hilbert scheme of points on an arbitrary surface, as we will see in Chapter 15. However, to apply the technique of cosection localization to handle stable maps whose images are not necessarily contracted by the Hilbert-Chow morphisms, we will have to assume the existence of a non-trivial *holomorphic* section of $\mathcal{O}_X(K_X)$. This leads us to the next section.

13.2. Vanishing of Gromov-Witten invariants when $p_g(X) > 0$

In this section, using the technique of cosection localization, we will study the stable maps to the Hilbert scheme $X^{[n]}$, whose images in $X^{[n]}$ are not necessarily contracted by the Hilbert-Chow morphism to points in the symmetric product $X^{(n)}$. When $p_g(X) = h^0(X, \mathcal{O}_X(K_X)) > 0$, we will prove some vanishing results regarding the Gromov-Witten invariants of the Hilbert scheme $X^{[n]}$.

Let X be a smooth projective complex surface. For simplicity, put

$$\overline{\mathfrak{M}} = \overline{\mathfrak{M}}_{g,r}(X^{[n]}, \beta).$$

Assume that the surface X admits a non-trivial holomorphic 2-form

$$\theta \in H^0(X, \Omega_X^2) = H^0(X, \mathcal{O}_X(K_X)).$$

As we have seen in Section 13.1, θ induces a holomorphic 2-form $\theta^{[n]}$ of the Hilbert scheme $X^{[n]}$ which can also be regarded as a map $\theta^{[n]} : T_{X^{[n]}} \to \Omega_{X^{[n]}}$. In turn, $\theta^{[n]}$ induces a regular cosection

$$\sigma : \quad \mathcal{O}b_{\overline{\mathfrak{M}}} \longrightarrow \mathcal{O}_{\overline{\mathfrak{M}}} \tag{13.15}$$

of the obstruction sheaf $\mathcal{O}b_{\overline{\mathfrak{M}}}$ of $\overline{\mathfrak{M}}$. The degeneracy locus

$$\overline{\mathfrak{M}}(\sigma) \tag{13.16}$$

of σ is the subset of $\overline{\mathfrak{M}}$ consisting of all the stable maps $u : \Gamma \to X^{[n]}$ such that the composition

$$u^*(\theta^{[n]}) \circ du : \quad T_{\Gamma_{\mathrm{reg}}} \to u^* T_{X^{[n]}}|_{\Gamma_{\mathrm{reg}}} \to u^* \Omega_{X^{[n]}}|_{\Gamma_{\mathrm{reg}}} \tag{13.17}$$

is trivial over the regular locus Γ_{reg} of Γ. Moreover, the cosection σ is surjective away from the degeneracy locus $\overline{\mathfrak{M}}(\sigma)$, and there exists a localized virtual cycle $[\overline{\mathfrak{M}}]^{\text{vir}}_{\text{loc}} \in A_*(\overline{\mathfrak{M}}(\sigma))$ such that

$$[\overline{\mathfrak{M}}]^{\text{vir}} = \iota_*[\overline{\mathfrak{M}}]^{\text{vir}}_{\text{loc}} \in A_*(\overline{\mathfrak{M}}) \tag{13.18}$$

where $\iota : \overline{\mathfrak{M}}(\sigma) \to \overline{\mathfrak{M}}$ stands for the inclusion map.

Our next lemma is parallel to Lemma 13.2. On one hand, it applies to an arbitrary $\beta \in H_2(X^{[n]})$. On the other hand, it provides less information than Lemma 13.2 does.

LEMMA 13.4. *Let C_0 be the zero divisor of θ. Let $u : \Gamma \to X^{[n]}$ be a stable map in the degeneracy locus $\overline{\mathfrak{M}}(\sigma)$, and let Γ_0 be an irreducible component of Γ with non-constant restriction $u|_{\Gamma_0}$. Then there exists $\xi_1 \in X^{[n_0]}$ for some n_0 such that* $\text{Supp}(\xi_1) \cap C_0 = \emptyset$ *and*

$$u(\Gamma_0) \subset \xi_1 + \big\{\xi_2 | \, \text{Supp}(\xi_2) \subset C_0\big\}. \tag{13.19}$$

PROOF. For notational convenience, we assume that $\Gamma = \Gamma_0$ is irreducible. Then there exist a nonempty open subset $U \subset \Gamma$ and an integer $n_0 \geq 0$ such that U is smooth and for every element $p \in U$, the image $u(p)$ is of the form

$$u(p) = \xi_1(p) + \xi_2(p) \tag{13.20}$$

where $\xi_1(p) \in X^{[n_0]}$ with $\text{Supp}(\xi_1(p)) \cap C_0 = \emptyset$ and $\text{Supp}(\xi_2(p)) \subset C_0$. This induces a decomposition $u|_U = (u_1, u_2)$ where the morphisms $u_1 : U \to X^{[n_0]}$ and $u_2 : U \to X^{[n-n_0]}$ are defined by sending $p \in U$ to $\xi_1(p)$ and $\xi_2(p)$ respectively. Since (13.17) is trivial over the regular locus Γ_{reg} of Γ, the composite

$$u_1^*(\theta^{[n_0]}) \circ du_1 : \quad T_U \to u_1^* T_{X^{[n_0]}}|_U \to u_1^* \Omega_{X^{[n_0]}}|_U \tag{13.21}$$

is trivial. Note that the holomorphic 2-form $\theta^{[n_0]}$ on $X^{[n_0]}$ is non-degenerate at $\xi_1(p), p \in U$ since $\text{Supp}(\xi_1(p)) \cap C_0 = \emptyset$. Thus, $du_1 = 0$ and u_1 is a constant morphism. Setting $\xi_1 = \xi_1(p) = u_1(p), p \in U$ proves the lemma. □

In the rest of this chapter, we will assume that X is simply connected. By (1.30),

$$\text{Pic}(X^{[n]}) \cong \text{Pic}(X) \oplus \mathbb{Z} \cdot (B_n/2).$$

Under this isomorphism, the divisor $D_C \in \text{Pic}(X^{[n]})$ corresponds to $C \in \text{Pic}(X)$. Let $\{\alpha_1, \ldots, \alpha_s\}$ be a linear basis of $H^2(X, \mathbb{C})$. Then,

$$\{D_{\alpha_1}, \ldots, D_{\alpha_s}, B_n\} \tag{13.22}$$

is a linear basis of $H^2(X^{[n]}, \mathbb{C})$. Represent $\alpha_1, \ldots, \alpha_s$ by real-surfaces $C_1, \ldots, C_s \subset X$ respectively. Then a linear basis of $H_2(X^{[n]}, \mathbb{C})$ is given by

$$\{\beta_{C_1}, \ldots, \beta_{C_s}, \beta_n\} \tag{13.23}$$

where β_{C_i} is defined by (2.1).

LEMMA 13.5. *Let the surface X be simply connected. Assume that $C_{0,1}, \ldots, C_{0,s}$ are the irreducible components (with reduced scheme structures) of the zero divisor of θ. Let $\overline{\mathfrak{M}} = \overline{\mathfrak{M}}_{g,r}(X^{[n]}, \beta)$. If the degeneracy locus $\overline{\mathfrak{M}}(\sigma)$ from* (13.16) *is nonempty, then*

$$\beta = \sum_{i=1}^{s} d_i \beta_{C_{0,i}} - d\beta_n$$

for some integers $d_1, \ldots, d_s \geq 0$ and d.

PROOF. Let $u : \Gamma \to X^{[n]}$ be a stable map in $\overline{\mathfrak{M}}(\sigma)$, and let $\Gamma_1, \ldots, \Gamma_t$ be the irreducible components of Γ. Then,

$$\beta = u_*[\Gamma] = \sum_{i=1}^{t} \deg(u|_{\Gamma_i})\, [u(\Gamma_i)]. \tag{13.24}$$

Assume that the restriction $u|_{\Gamma_i}$ is non-constant. Put

$$C_0 = C_{0,1} \cup \cdots \cup C_{0,s}.$$

By Lemma 13.4, there exists $\xi_1 \in X^{[n_0]}$ for some n_0 such that $\mathrm{Supp}(\xi_1) \cap C_0 = \emptyset$ and

$$u(\Gamma_i) \subset \xi_1 + \{\xi_2 |\, \mathrm{Supp}(\xi_2) \subset C_0\}.$$

By Lemma 3.19,

$$[u(\Gamma_i)] = \sum_{i=1}^{s} d_i' \beta_{C_{0,i}} - d'\beta_n$$

for some integers $d_1', \ldots, d_s' \geq 0$ and d'. Now our lemma follows immediately from (13.24). □

THEOREM 13.6. *Let X be a simply connected smooth projective complex surface admitting a holomorphic 2-form θ, and let $C_{0,1}, \ldots, C_{0,s}$ be the irreducible components (with reduced scheme structures) of the zero divisor of θ. If*

$$\beta \neq \sum_{i=1}^{s} d_i \beta_{C_{0,i}} - d\beta_n$$

for some integers $d_1, \ldots, d_s \geq 0$ and d, then all the Gromov-Witten invariants of $X^{[n]}$ defined via the moduli space $\overline{\mathfrak{M}}_{g,r}(X^{[n]}, \beta)$ vanish.

PROOF. By Lemma 13.5, the degeneracy locus $\overline{\mathfrak{M}}(\sigma)$ from (13.16) is empty. It follows from (13.18) that

$$[\overline{\mathfrak{M}}_{g,r}(X^{[n]}, \beta)]^{\mathrm{vir}} = 0.$$

Therefore, all the Gromov-Witten invariants of $X^{[n]}$ defined via the moduli space $\overline{\mathfrak{M}}_{g,r}(X^{[n]}, \beta)$ vanish. □

COROLLARY 13.7. *Let X be a simply connected smooth projective complex surface admitting a holomorphic 2-form with irreducible zero divisor. If $\beta \neq d_0\beta_{K_X} - d\beta_n$ for some integer d and rational number $d_0 \geq 0$, then all the Gromov-Witten invariants of $X^{[n]}$ defined via the moduli space $\overline{\mathfrak{M}}_{g,r}(X^{[n]}, \beta)$ vanish.*

PROOF. Let $\theta \in H^0(X, \Omega_X^2) = H^0(X, \mathcal{O}_X(K_X))$ be the holomorphic 2-form whose zero divisor C_0 is irreducible (but possibly non-reduced). Then,

$$K_X = C_0 = m(C_0)_{\mathrm{red}} \tag{13.25}$$

for some positive integer m, and the corollary follows from Theorem 13.6. □

REMARK 13.8. From the proof of Corollary 13.7, we see that if m is from (13.25), then the rational number d_0 in Corollary 13.7 may be required to be of the form d_0'/m for some integer $d_0' \geq 0$.

Recall from Theorem 1.24 (ii) that $K_{X^{[n]}} = D_{K_X}$. Thus, if $\beta = d_0\beta_{K_X} - d\beta_n$ for some rational number $d_0 \ge 0$ and integer d, then the expected dimension of $\overline{\mathfrak{M}}_{g,r}(X^{[n]}, \beta)$ is

$$(13.26) \qquad \begin{aligned} \mathfrak{d} &= -K_{X^{[n]}} \cdot \beta + (\dim X^{[n]} - 3)(1-g) + r \\ &= -d_0 K_X^2 + (2n-3)(1-g) + r. \end{aligned}$$

Our first corollary deals with the case when X is an elliptic surface.

COROLLARY 13.9. *Let X be a simply connected (minimal) elliptic surface without multiple fibers and with positive geometric genus. Let $n \ge 2$ and $\beta \ne 0$. Then all the Gromov-Witten invariants without descendant insertions defined via the moduli space $\overline{\mathfrak{M}}_{g,r}(X^{[n]}, \beta)$ vanish, except possibly when $0 \le g \le 1$ and $\beta = d_0\beta_{K_X} - d\beta_n$ for some integer d and rational number $d_0 \ge 0$.*

PROOF. Since X is a simply connected elliptic surface without multiple fibers, $K_X = (p_g - 1)f$ where $p_g \ge 1$ is the geometric genus of X and f denotes a smooth fiber of the elliptic fibration. By Corollary 13.7, it remains to consider the case when $\beta = d_0\beta_{K_X} - d\beta_n$ for some integer d and rational number $d_0 \ge 0$. By (13.26) and $K_X^2 = 0$, the expected dimension of the moduli space $\overline{\mathfrak{M}}_{g,r}(X^{[n]}, \beta)$ is equal to

$$\mathfrak{d} = (2n-3)(1-g) + r.$$

By the Fundamental Class Axiom, all the Gromov-Witten invariants without descendant insertions are equal to zero if $g \ge 2$. □

Our second corollary concentrates on the case when X is of general type.

COROLLARY 13.10. *Let X be a simply connected minimal surface of general type admitting a holomorphic 2-form with irreducible zero divisor. Let $n \ge 2$ and $\beta \ne 0$. Then all the Gromov-Witten invariants without descendant insertions defined via $\overline{\mathfrak{M}}_{g,r}(X^{[n]}, \beta)$ vanish, except possibly in the following cases*

(i) *$g = 0$ and $\beta = d\beta_n$ for some integer $d > 0$;*
(ii) *$g = 1$ and $\beta = d\beta_n$ for some integer $d > 0$;*
(iii) *$g = 0$ and $\beta = d_0\beta_{K_X} - d\beta_n$ for some integer d and rational number $d_0 > 0$.*

PROOF. In view of Corollary 13.7, it remains to consider the case when $\beta = d_0\beta_{K_X} - d\beta_n$ for some integer d and rational number $d_0 \ge 0$.

When $d_0 = 0$ and $\beta = d\beta_n$ with $d > 0$, we see from (13.26) that the expected dimension of the moduli space $\overline{\mathfrak{M}}_{g,r}(X^{[n]}, \beta)$ is equal to

$$\mathfrak{d} = (2n-3)(1-g) + r.$$

If $g \ge 2$, then all the Gromov-Witten invariants without descendant insertions defined via $\overline{\mathfrak{M}}_{g,r}(X^{[n]}, \beta)$ vanish by the Fundamental Class Axiom.

Next, assume that $d_0 > 0$. Since $K_X^2 \ge 1$, we see from (13.26) that

$$\mathfrak{d} < (2n-3)(1-g) + r.$$

By the Fundamental Class Axiom, all the Gromov-Witten invariants without descendant insertions vanish except possibly in the case when $g = 0$. □

In Section 13.4 below, for the Hilbert scheme $X^{[2]}$ where X is a simply connected minimal surface of general type admitting a non-trivial holomorphic 2-form with irreducible zero divisor, we will study the exceptional cases listed in Corollary 13.10 (i), (ii) and (iii). The computation will involve certain integrals over some moduli space of genus-1 stable maps. These integrals will be calculated in the next section.

13.3. Intersections on some moduli space of genus-1 stable maps

In this section, we will compute certain intersection numbers on the moduli space of genus-1 stable maps to $\mathbb{P}(V)$ where V is a rank-2 vector bundle over a smooth projective curve C. The results will be used in the next section.

We begin with some notations.

NOTATION 13.11. *Let V be a rank-2 bundle over a smooth projective variety B.*

(i) *f denotes a fiber of the ruling $\pi : \mathbb{P}(V) \to B$ or its cohomology class.*

(ii) *$\mathcal{H} = (f_{1,0})_*\omega$ is the rank-1 Hodge bundle over $\overline{\mathfrak{M}}_{1,0}(\mathbb{P}(V), df)$ where ω is the relative dualizing sheaf for the forgetful map*

$$f_{1,0} : \overline{\mathfrak{M}}_{1,1}(\mathbb{P}(V), df) \to \overline{\mathfrak{M}}_{1,0}(\mathbb{P}(V), df).$$

(iii) $\lambda = c_1(\mathcal{H})$.

Let $d \geq 1$. If $u = [\mu : D \to \mathbb{P}(V)] \in \overline{\mathfrak{M}}_{1,0}(\mathbb{P}(V), df)$, then $\mu(D)$ is a fiber of the ruling $\pi : \mathbb{P}(V) \to B$. Therefore, there exists a natural morphism

$$\phi : \quad \overline{\mathfrak{M}}_{1,0}(\mathbb{P}(V), df) \to B \tag{13.27}$$

whose fiber over $b \in B$ is

$$\overline{\mathfrak{M}}_{1,0}\big(\pi^{-1}(b), d[\pi^{-1}(b)]\big) \cong \overline{\mathfrak{M}}_{1,0}(\mathbb{P}^1, d[\mathbb{P}^1]).$$

So the moduli space $\overline{\mathfrak{M}}_{1,0}(\mathbb{P}(V), df)$ is smooth (as a stack) with dimension

$$\dim \overline{\mathfrak{M}}_{1,0}(\mathbb{P}^1, d[\mathbb{P}^1]) + \dim(B) = 2d + \dim(B).$$

By (12.3), the expected dimension of $\overline{\mathfrak{M}}_{1,0}(\mathbb{P}(V), df)$ is $2d$. Since $d \geq 1$, the sheaf $R^1(f_{1,0})_*\mathrm{ev}_1^*\mathcal{O}_{\mathbb{P}(V)}(-2)$ on $\overline{\mathfrak{M}}_{1,0}(\mathbb{P}(V), df)$ is locally free of rank-$2d$. Moreover,

$$\lambda^2 = 0 \tag{13.28}$$

in view of [**Mum**, (5.2)] regarding the Chern character of the Hodge bundles and the proof of [**FPa**, Proposition 1].

LEMMA 13.12. *Let $d \geq 1$. Let V be a rank-2 bundle over $B_0 \times C$ where B_0 and C are smooth projective curves. Let $V_b = V|_{\{b\} \times C}$ for $b \in B_0$. Then,*

$$\int_{[\overline{\mathfrak{M}}_{1,0}(\mathbb{P}(V_b), df)]} \lambda \cdot c_{2d}\big(R^1(f_{1,0})_*\mathrm{ev}_1^*\mathcal{O}_{\mathbb{P}(V_b)}(-2)\big) \tag{13.29}$$

is independent of the points $b \in B_0$.

PROOF. This follows from the observation that (13.29) is equal to

$$\int_{[\overline{\mathfrak{M}}_{1,0}(\mathbb{P}(V), df)]} \phi^*[\{b\} \times C] \cdot \lambda \cdot c_{2d}\big(R^1(f_{1,0})_*\mathrm{ev}_1^*\mathcal{O}_{\mathbb{P}(V)}(-2)\big)$$

where the morphism $\phi : \overline{\mathfrak{M}}_{1,0}(\mathbb{P}(V), df) \to B_0 \times C$ is from (13.27). □

Formula (13.30) below is probably well-known, but we could not find a reference.

LEMMA 13.13. *Let d be a positive integer. Then, we have*

$$\int_{[\overline{\mathfrak{M}}_{1,0}(\mathbb{P}^1, d[\mathbb{P}^1])]} c_{2d}\big(R^1(f_{1,0})_*\mathrm{ev}_1^*\mathcal{O}_{\mathbb{P}^1}(-2)\big) = 0, \tag{13.30}$$

$$\int_{[\overline{\mathfrak{M}}_{1,0}(\mathbb{P}^1, d[\mathbb{P}^1])]} \lambda \cdot c_{2d-1}\big(R^1(f_{1,0})_*\mathrm{ev}_1^*\mathcal{O}_{\mathbb{P}^1}(-2)\big) = -\frac{1}{12d}. \tag{13.31}$$

PROOF. We begin with the proof of (13.30). Choose a K3 surface S which contains a smooth rational curve C. Then, $C^2 = -2$, $T_S|_C = \mathcal{O}_C(2) \oplus \mathcal{O}_C(-2)$, and dC is the only element in the complete linear system $|dC|$. So we have

$$\begin{aligned}
&\int_{[\overline{\mathfrak{M}}_{1,0}(\mathbb{P}^1, d[\mathbb{P}^1])]} c_{2d}\big(R^1(f_{1,0})_*\mathrm{ev}_1^*\mathcal{O}_{\mathbb{P}^1}(-2)\big)\\
= &\int_{[\overline{\mathfrak{M}}_{1,0}(C, d[C])]} c_{2d}\big(R^1(f_{1,0})_*\mathrm{ev}_1^*(T_S|_C)\big)\\
= &\int_{[\overline{\mathfrak{M}}_{1,0}(S, d[C])]} c_{2d}\big(R^1(f_{1,0})_*\mathrm{ev}_1^*T_S\big).
\end{aligned}$$

Note that $R^1(f_{1,0})_*\mathrm{ev}_1^*T_S$ is a rank-$2d$ bundle on the $2d$-dimensional moduli space $\overline{\mathfrak{M}}_{1,0}(S, d[C])$ whose virtual dimension is 0. By Proposition 12.1,

$$\int_{[\overline{\mathfrak{M}}_{1,0}(\mathbb{P}^1, d[\mathbb{P}^1])]} c_{2d}\big(R^1(f_{1,0})_*\mathrm{ev}_1^*\mathcal{O}_{\mathbb{P}^1}(-2)\big) = \deg\big([\overline{\mathfrak{M}}_{1,0}(S, d[C])]^{\mathrm{vir}}\big).$$

Since $[\overline{\mathfrak{M}}_{g,r}(S, \beta)]^{\mathrm{vir}} = 0$ whenever $\beta \neq 0$, we obtain

$$\int_{[\overline{\mathfrak{M}}_{1,0}(\mathbb{P}^1, d[\mathbb{P}^1])]} c_{2d}\big(R^1(f_{1,0})_*\mathrm{ev}_1^*\mathcal{O}_{\mathbb{P}^1}(-2)\big) = 0.$$

To prove (13.31), we apply $(f_{1,0})_*\mathrm{ev}_1^*$ to the exact sequence

$$0 \to \mathcal{O}_{\mathbb{P}^1}(-2) \to \mathcal{O}_{\mathbb{P}^1}(-1)^{\oplus 2} \to \mathcal{O}_{\mathbb{P}^1} \to 0.$$

Since $(f_{1,0})_*\mathrm{ev}_1^*\mathcal{O}_{\mathbb{P}^1} = \mathcal{O}_{\overline{\mathfrak{M}}_{1,0}(\mathbb{P}^1, d[\mathbb{P}^1])}$ and $R^1(f_{1,0})_*\mathrm{ev}_1^*\mathcal{O}_{\mathbb{P}^1} = \mathcal{H}^*$, we get the exact sequence

$$\begin{aligned}
0 \to \mathcal{O}_{\overline{\mathfrak{M}}_{1,0}(\mathbb{P}^1, d[\mathbb{P}^1])} &\to R^1(f_{1,0})_*\mathrm{ev}_1^*\mathcal{O}_{\mathbb{P}^1}(-2)\\
&\to R^1(f_{1,0})_*\mathrm{ev}_1^*\mathcal{O}_{\mathbb{P}^1}(-1)^{\oplus 2} \to \mathcal{H}^* \to 0.
\end{aligned}$$

Calculating the total Chern class and using (13.28), we see that

$$\begin{aligned}
&c\big(R^1(f_{1,0})_*\mathrm{ev}_1^*\mathcal{O}_{\mathbb{P}^1}(-2)\big)\\
= \;&c\big(R^1(f_{1,0})_*\mathrm{ev}_1^*\mathcal{O}_{\mathbb{P}^1}(-1)^{\oplus 2}\big)/c(\mathcal{H}^*)\\
= \;&c\big(R^1(f_{1,0})_*\mathrm{ev}_1^*\mathcal{O}_{\mathbb{P}^1}(-1)^{\oplus 2}\big) \cdot (1+\lambda).
\end{aligned} \tag{13.32}$$

Thus, the top Chern class $c_{2d}\big(R^1(f_{1,0})_*\mathrm{ev}_1^*\mathcal{O}_{\mathbb{P}^1}(-2)\big)$ is equal to

$$c_{2d}\big(R^1(f_{1,0})_*\mathrm{ev}_1^*\mathcal{O}_{\mathbb{P}^1}(-1)^{\oplus 2}\big) + \lambda \cdot c_{2d-1}\big(R^1(f_{1,0})_*\mathrm{ev}_1^*\mathcal{O}_{\mathbb{P}^1}(-1)^{\oplus 2}\big).$$

By the Proposition 2 in [**GP**] and (13.30), we conclude that

$$\int_{[\overline{\mathfrak{M}}_{1,0}(\mathbb{P}^1, d[\mathbb{P}^1])]} \lambda \cdot c_{2d-1}\big(R^1(f_{1,0})_*\mathrm{ev}_1^*\mathcal{O}_{\mathbb{P}^1}(-1)^{\oplus 2}\big) = -\frac{1}{12d}. \tag{13.33}$$

By (13.32) again,

$$\lambda \cdot c_{2d-1}\big(R^1(f_{1,0})_*\mathrm{ev}_1^*\mathcal{O}_{\mathbb{P}^1}(-2)\big) = \lambda \cdot c_{2d-1}\big(R^1(f_{1,0})_*\mathrm{ev}_1^*\mathcal{O}_{\mathbb{P}^1}(-1)^{\oplus 2}\big).$$

Combining this with (13.33), we obtain our formula (13.31). □

The following lemma deals with the case when V is a rank-2 bundle over $\mathbb{P}^1$. The main idea in its proof is to reduce to the special cases

$$\mathcal{O}_{\mathbb{P}^1} \oplus \mathcal{O}_{\mathbb{P}^1}, \quad \mathcal{O}_{\mathbb{P}^1}(2) \oplus \mathcal{O}_{\mathbb{P}^1}(-1)$$

depending on the parity of the degree of V.

LEMMA 13.14. *Let $d \geq 1$, and V be a rank-2 bundle over $\mathbb{P}^1$. Then,*

$$\int_{[\overline{\mathfrak{M}}_{1,0}(\mathbb{P}(V),df)]} \lambda \cdot c_{2d}\big(R^1(f_{1,0})_*\mathrm{ev}_1^*\mathcal{O}_{\mathbb{P}(V)}(-2)\big) = \frac{\deg(V)}{12d}. \tag{13.34}$$

PROOF. First of all, assume that $\deg(V) = 2k$ for some integer k. Then V can be deformed to

$$\mathcal{O}_{\mathbb{P}^1}(k) \oplus \mathcal{O}_{\mathbb{P}^1}(k) = \big(\mathcal{O}_{\mathbb{P}^1} \oplus \mathcal{O}_{\mathbb{P}^1}\big) \otimes \mathcal{O}_{\mathbb{P}^1}(k).$$

By Lemma 13.12,

$$\begin{aligned} &\int_{[\overline{\mathfrak{M}}_{1,0}(\mathbb{P}(V),df)]} \lambda \cdot c_{2d}\big(R^1(f_{1,0})_*\mathrm{ev}_1^*\mathcal{O}_{\mathbb{P}(V)}(-2)\big) \\ = &\int_{[\overline{\mathfrak{M}}_{1,0}(\mathbb{P}^1\times\mathbb{P}^1,df)]} \lambda \cdot c_{2d}\Big(R^1(f_{1,0})_*\mathrm{ev}_1^*\big(\mathcal{O}_{\mathbb{P}^1\times\mathbb{P}^1}(-2) \otimes \mathcal{O}_{\mathbb{P}^1\times\mathbb{P}^1}(-2kf)\big)\Big). \end{aligned}$$

Note that

$$\overline{\mathfrak{M}}_{1,0}(\mathbb{P}^1 \times \mathbb{P}^1, df) \cong \mathbb{P}^1 \times \overline{\mathfrak{M}}_{1,0}(\mathbb{P}^1, d[\mathbb{P}^1]),$$

Thus, we obtain

$$\begin{aligned} &\int_{[\overline{\mathfrak{M}}_{1,0}(\mathbb{P}(V),df)]} \lambda \cdot c_{2d}\big(R^1(f_{1,0})_*\mathrm{ev}_1^*\mathcal{O}_{\mathbb{P}(V)}(-2)\big) \\ = &\int_{[\mathbb{P}^1\times\overline{\mathfrak{M}}_{1,0}(\mathbb{P}^1,d[\mathbb{P}^1])]} \pi_2^*\lambda \cdot c_{2d}\Big(\pi_2^*\big(R^1(f_{1,0})_*\mathrm{ev}_1^*\mathcal{O}_{\mathbb{P}^1}(-2)\big) \otimes \pi_1^*\mathcal{O}_{\mathbb{P}^1}(-2k)\Big) \end{aligned}$$

where π_1 and π_2 are the projection on $\mathbb{P}^1 \times \overline{\mathfrak{M}}_{1,0}(\mathbb{P}^1, d[\mathbb{P}^1])$. Hence,

$$\begin{aligned} &\int_{[\overline{\mathfrak{M}}_{1,0}(\mathbb{P}(V),df)]} \lambda \cdot c_{2d}\big(R^1(f_{1,0})_*\mathrm{ev}_1^*\mathcal{O}_{\mathbb{P}(V)}(-2)\big) \\ = &\int_{[\mathbb{P}^1\times\overline{\mathfrak{M}}_{1,0}(\mathbb{P}^1,d[\mathbb{P}^1])]} \pi_2^*\lambda \cdot \pi_2^*c_{2d-1}\big(R^1(f_{1,0})_*\mathrm{ev}_1^*\mathcal{O}_{\mathbb{P}^1}(-2)\big) \cdot \pi_1^*c_1\big(\mathcal{O}_{\mathbb{P}^1}(-2k)\big) \\ = &\frac{\deg(V)}{12d} \end{aligned} \tag{13.35}$$

where we have used formula (13.31) in the last step.

Next, assume that $\deg(V) = 2k+1$ for some integer k. Then V can be deformed to $\big(\mathcal{O}_{\mathbb{P}^1}(2) \oplus \mathcal{O}_{\mathbb{P}^1}(-1)\big) \otimes \mathcal{O}_{\mathbb{P}^1}(k)$. As in the previous paragraph, we have

$$\begin{aligned} &\int_{[\overline{\mathfrak{M}}_{1,0}(\mathbb{P}(V),df)]} \lambda \cdot c_{2d}\big(R^1(f_{1,0})_*\mathrm{ev}_1^*\mathcal{O}_{\mathbb{P}(V)}(-2)\big) \\ = &\int_{[\overline{\mathfrak{M}}_{1,0}(S,df)]} \lambda \cdot c_{2d}\Big(R^1(f_{1,0})_*\mathrm{ev}_1^*\big(\mathcal{O}_S(-2) \otimes \mathcal{O}_S(-2kf)\big)\Big) \\ = &\int_{[\overline{\mathfrak{M}}_{1,0}(S,df)]} \lambda \cdot c_{2d}\big(R^1(f_{1,0})_*\mathrm{ev}_1^*\mathcal{O}_S(-2)\big) + \frac{2k}{12d} \end{aligned}$$

where $S = \mathbb{P}(\mathcal{O}_{\mathbb{P}^1}(2) \oplus \mathcal{O}_{\mathbb{P}^1}(-1))$. Let $\mathbb{F}_1$ be the blown-up of $\mathbb{P}^2$ at a point p, and σ be the exceptional curve. Then, $T_{\mathbb{F}_1}|_\sigma \cong \mathcal{O}_{\mathbb{P}^1}(2) \oplus \mathcal{O}_{\mathbb{P}^1}(-1)$. So

$$\begin{aligned} (13.36) \quad & \int_{[\overline{\mathfrak{M}}_{1,0}(\mathbb{P}(V),df)]} \lambda \cdot c_{2d}\big(R^1(f_{1,0})_* \mathrm{ev}_1^* \mathcal{O}_{\mathbb{P}(V)}(-2)\big) \\ = & \int_{[\overline{\mathfrak{M}}_{1,0}(\mathbb{P}(T_{\mathbb{F}_1}|_\sigma),df)]} \lambda \cdot c_{2d}\big(R^1(f_{1,0})_* \mathrm{ev}_1^* \mathcal{O}_{\mathbb{P}(T_{\mathbb{F}_1}|_\sigma)}(-2)\big) + \frac{2k}{12d} \\ = & \int_{[\overline{\mathfrak{M}}_{1,0}(\mathbb{P}(T_{\mathbb{F}_1}),df)]} \phi^*[\sigma] \cdot \lambda \cdot c_{2d}\big(R^1(f_{1,0})_* \mathrm{ev}_1^* \mathcal{O}_{\mathbb{P}(T_{\mathbb{F}_1})}(-2)\big) + \frac{2k}{12d} \end{aligned}$$

where the morphism $\phi : \overline{\mathfrak{M}}_{1,0}(\mathbb{P}(T_{\mathbb{F}_1}), df) \to \mathbb{F}_1$ is from (13.27). Let f_0 be a fiber of the ruling $\mathbb{F}_1 \to \mathbb{P}^1$, and C be a smooth conic in $\mathbb{P}^2$ such that $p \notin C$. We use C to denote its strict transform in $\mathbb{F}_1$. Then, $[\sigma] = [C]/2 - [f_0]$. By (13.36),

$$\begin{aligned} & \int_{[\overline{\mathfrak{M}}_{1,0}(\mathbb{P}(V),df)]} \lambda \cdot c_{2d}\big(R^1(f_{1,0})_* \mathrm{ev}_1^* \mathcal{O}_{\mathbb{P}(V)}(-2)\big) \\ = & \frac{1}{2} \cdot \int_{[\overline{\mathfrak{M}}_{1,0}(\mathbb{P}(T_{\mathbb{F}_1}),df)]} \phi^*[C] \cdot \lambda \cdot c_{2d}\big(R^1(f_{1,0})_* \mathrm{ev}_1^* \mathcal{O}_{\mathbb{P}(T_{\mathbb{F}_1})}(-2)\big) \\ & - \int_{[\overline{\mathfrak{M}}_{1,0}(\mathbb{P}(T_{\mathbb{F}_1}),df)]} \phi^*[f_0] \cdot \lambda \cdot c_{2d}\big(R^1(f_{1,0})_* \mathrm{ev}_1^* \mathcal{O}_{\mathbb{P}(T_{\mathbb{F}_1})}(-2)\big) + \frac{2k}{12d} \\ = & \frac{1}{2} \cdot \int_{[\overline{\mathfrak{M}}_{1,0}(\mathbb{P}(T_{\mathbb{F}_1}|_C),df)]} \lambda \cdot c_{2d}\big(R^1(f_{1,0})_* \mathrm{ev}_1^* \mathcal{O}_{\mathbb{P}(T_{\mathbb{F}_1}|_C)}(-2)\big) \\ & - \int_{[\overline{\mathfrak{M}}_{1,0}(\mathbb{P}(T_{\mathbb{F}_1}|_{f_0}),df)]} \lambda \cdot c_{2d}\big(R^1(f_{1,0})_* \mathrm{ev}_1^* \mathcal{O}_{\mathbb{P}(T_{\mathbb{F}_1}|_{f_0})}(-2)\big) + \frac{2k}{12d}. \end{aligned}$$

Note that $\deg\big(T_{\mathbb{F}_1}|_C\big) = 6$ and $\deg\big(T_{\mathbb{F}_1}|_{f_0}\big) = 2$. By (13.35),

$$\begin{aligned} & \int_{[\overline{\mathfrak{M}}_{1,0}(\mathbb{P}(V),df)]} \lambda \cdot c_{2d}\big(R^1(f_{1,0})_* \mathrm{ev}_1^* \mathcal{O}_{\mathbb{P}(V)}(-2)\big) \\ = & \frac{1}{2} \cdot \frac{6}{12d} - \frac{2}{12d} + \frac{2k}{12d} \\ = & \frac{\deg(V)}{12d}. \end{aligned}$$

This completes the proof of the lemma. □

The following is the main result in this section. The ideas in its proof are to deform the rank-2 vector bundle V over a curve C to a direct sum of two line bundles over C, reduce the situation to a rank-2 bundle over $\mathbb{P}^1$, and then apply Lemma 13.14.

PROPOSITION 13.15. *Let d be a positive integer. Assume that V is a rank-2 vector bundle over a smooth projective curve C. Then, we have*

$$(13.37) \quad \int_{[\overline{\mathfrak{M}}_{1,0}(\mathbb{P}(V),df)]} \lambda \cdot c_{2d}\big(R^1(f_{1,0})_* \mathrm{ev}_1^* \mathcal{O}_{\mathbb{P}(V)}(-2)\big) = \frac{\deg(V)}{12d}.$$

PROOF. It is well-known that there exist a rank-2 bundle $\mathcal{V}$ over $\mathbb{P}^1 \times C$ and two points $b_1, b_2 \in \mathbb{P}^1$ such that

$$\begin{aligned} \mathcal{V}|_{\{b_1\}\times C} &= V, \\ \mathcal{V}|_{\{b_2\}\times C} &= (\mathcal{O}_C \oplus M) \otimes N \end{aligned}$$

where M and N are line bundles on C with M being very ample. As in the proof of Lemma 13.14, we conclude from Lemma 13.12 that

$$\begin{aligned}(13.38)\quad &\int_{[\overline{\mathfrak{M}}_{1,0}(\mathbb{P}(V),df)]} \lambda \cdot c_{2d}\big(R^1(f_{1,0})_*\mathrm{ev}_1^*\mathcal{O}_{\mathbb{P}(V)}(-2)\big) \\ = &\int_{[\overline{\mathfrak{M}}_{1,0}(\mathbb{P}(\mathcal{O}_C\oplus M),df)]} \lambda \cdot c_{2d}\big(R^1(f_{1,0})_*\mathrm{ev}_1^*\mathcal{O}_{\mathbb{P}(\mathcal{O}_C\oplus M)}(-2)\big) + \frac{2\deg(N)}{12d}.\end{aligned}$$

Since M is very ample, there exists a morphism $\alpha : C \to \mathbb{P}^1$ such that $M = \alpha^*\mathcal{O}_{\mathbb{P}^1}(1)$. Then,

$$\mathcal{O}_C \oplus M = \alpha^*\big(\mathcal{O}_{\mathbb{P}^1} \oplus \mathcal{O}_{\mathbb{P}^1}(1)\big).$$

This induces an isomorphism

$$\overline{\mathfrak{M}}_{1,0}(\mathbb{P}(\mathcal{O}_C \oplus M), df) \cong C \times_{\mathbb{P}^1} \overline{\mathfrak{M}}_{1,0}\big(\mathbb{P}(\mathcal{O}_{\mathbb{P}^1} \oplus \mathcal{O}_{\mathbb{P}^1}(1)), df\big).$$

Let

$$\widetilde{\alpha} : \overline{\mathfrak{M}}_{1,0}(\mathbb{P}(\mathcal{O}_C \oplus M), df) \to \overline{\mathfrak{M}}_{1,0}\big(\mathbb{P}(\mathcal{O}_{\mathbb{P}^1} \oplus \mathcal{O}_{\mathbb{P}^1}(1)), df\big)$$

be the projection. Then

$$\begin{aligned}&\int_{[\overline{\mathfrak{M}}_{1,0}(\mathbb{P}(\mathcal{O}_C\oplus M),df)]} \lambda \cdot c_{2d}\big(R^1(f_{1,0})_*\mathrm{ev}_1^*\mathcal{O}_{\mathbb{P}(\mathcal{O}_C\oplus M)}(-2)\big) \\ = &\int_{[\overline{\mathfrak{M}}_{1,0}(\mathbb{P}(\mathcal{O}_C\oplus M),df)]} \widetilde{\alpha}^*\lambda \cdot c_{2d}\Big(\widetilde{\alpha}^*\big(R^1(f_{1,0})_*\mathrm{ev}_1^*\mathcal{O}_{\mathbb{P}\left(\mathcal{O}_{\mathbb{P}^1}\oplus\mathcal{O}_{\mathbb{P}^1}(1)\right)}(-2)\big)\Big) \\ = &\ \deg(\widetilde{\alpha}) \cdot \int_{[\overline{\mathfrak{M}}_{1,0}(\mathbb{P}(\mathcal{O}_{\mathbb{P}^1}\oplus\mathcal{O}_{\mathbb{P}^1}(1)),df)]} \lambda \cdot c_{2d}\big(R^1(f_{1,0})_*\mathrm{ev}_1^*\mathcal{O}_{\mathbb{P}(\mathcal{O}_{\mathbb{P}^1}\oplus\mathcal{O}_{\mathbb{P}^1}(1))}(-2)\big).\end{aligned}$$

By Lemma 13.14 and noticing $\deg(\widetilde{\alpha}) = \deg(\alpha) = \deg(M)$, we get

$$(13.39)\quad \int_{[\overline{\mathfrak{M}}_{1,0}(\mathbb{P}(\mathcal{O}_C\oplus M),df)]} \lambda \cdot c_{2d}\big(R^1(f_{1,0})_*\mathrm{ev}_1^*\mathcal{O}_{\mathbb{P}(\mathcal{O}_C\oplus M)}(-2)\big) = \frac{\deg(M)}{12d}.$$

Now our formula (13.37) follows immediately from (13.38) and (13.39). ☐

13.4. Gromov-Witten invariants of the Hilbert scheme $X^{[2]}$

In this section, a continuation of Section 13.2, we will study the Gromov-Witten invariants of the Hilbert scheme $X^{[2]}$ when X is a simply connected minimal surface of general type admitting a non-trivial holomorphic 2-form with irreducible zero divisor.

To begin with, we obtain the following from Corollary 13.10.

PROPOSITION 13.16. *Let X be a simply connected minimal surface of general type admitting a non-trivial holomorphic 2-form with irreducible zero divisor. Let $\beta \neq 0$. Then all the Gromov-Witten invariants without descendant insertions defined via $\overline{\mathfrak{M}}_{g,r}(X^{[2]}, \beta)$ vanish, except possibly in the following cases*

(i) *$g = 0$ and $\beta = d\beta_2$ for some integer $d > 0$;*
(ii) *$g = 1$ and $\beta = d\beta_2$ for some integer $d \geq 2$;*
(iii) *$K_X^2 = 1, g = 0$ and $\beta = \beta_{K_X} - d\beta_2$ for some integer d.*

PROOF. Cases (i) and (ii) follow from Corollary 13.10 (i) and (ii) respectively. In the case of Corollary 13.10 (iii), we have $g = 0$ and $\beta = d_0\beta_{K_X} - d\beta_2$ for some rational number $d_0 > 0$ and some integer d. We see from (13.26) that the expected dimension of the moduli space $\overline{\mathfrak{M}}_{0,r}(X^{[2]}, \beta)$ is equal to

$$\mathfrak{d} = -d_0 K_X^2 + 1 + r.$$

Since $d_0 K_X^2$ must be a positive integer, we conclude from the Fundamental Class Axiom that all the Gromov-Witten invariants without descendant insertions defined via $\overline{\mathfrak{M}}_{0,r}(X^{[2]}, \beta)$ vanish except possibly when $d_0 K_X^2 = 1$. Now write $K_X = mC_0'$ where C_0' is an irreducible and reduced curve, and $m \geq 1$ is an integer. By Remark 13.8, $d_0 = d_0'/m$ for some integer $d_0' \geq 1$. Therefore, we obtain

$$1 = d_0 K_X^2 = d_0' m (C_0')^2.$$

It follows that $d_0' = m = (C_0')^2 = 1$. Hence $K_X^2 = 1$ and $d_0 = 1$. □

Case (i) in Proposition 13.16 will be dealt with in Example 15.10. By the Divisor Axiom, Case (ii) in Proposition 13.16 can be reduced to the invariant

$$\langle 1 \rangle_{1, d\beta_2}^{X^{[2]}}. \tag{13.40}$$

Similarly, Case (iii) in Proposition 13.16 can be reduced to the invariant

$$\langle 1 \rangle_{0,\ \beta_{K_X} - d\beta_2}^{X^{[2]}}. \tag{13.41}$$

The invariants (13.40) and (13.41) have been computed in [**HLQ**, Section 6].

In the rest of this section, we will present the calculation of (13.40) for an arbitrary smooth projective surface X. Let $d \geq 2$ be an integer. For simplicity, we put

$$\overline{\mathfrak{M}}_{g,r,d} = \overline{\mathfrak{M}}_{g,r}(X^{[2]}, d\beta_2). \tag{13.42}$$

The next lemma determines the obstruction sheaf over the moduli space $\overline{\mathfrak{M}}_{1,0,d}$ and the corresponding virtual fundamental class.

LEMMA 13.17. (i) *The obstruction sheaf* $\mathcal{O}b = R^1(f_{1,0})_*(\mathrm{ev}_1^* T_{X^{[2]}})$ *over the moduli space* $\overline{\mathfrak{M}}_{1,0,d}$ *is locally free of rank* $2d+2$;

(ii) $[\overline{\mathfrak{M}}_{1,0,d}]^{\mathrm{vir}} = c_{2d+2}(\mathcal{O}b) \cap [\overline{\mathfrak{M}}_{1,0,d}]$.

PROOF. (i) Let $\mathrm{ev}_1 : \overline{\mathfrak{M}}_{1,1,d} \to X^{[2]}$ and $f_{1,0} : \overline{\mathfrak{M}}_{1,1,d} \to \overline{\mathfrak{M}}_{1,0,d}$ be the evaluation map and the forgetful map respectively. Let $u = [\mu : D \to X^{[2]}] \in \overline{\mathfrak{M}}_{1,0,d}$. Then

$$H^1\big(f_{1,0}^{-1}(u), (\mathrm{ev}_1^* T_{X^{[2]}})|_{f_{1,0}^{-1}(u)}\big) \cong H^1(D, \mu^* T_{X^{[2]}}) = H^1\big(D, \mu^*(T_{X^{[2]}}|_{\mu(D)})\big).$$

Since $d \geq 1$, $\mu(D) = M_2(x) \cong \mathbb{P}^1$ for some point $x \in X$. By Proposition 2.9 (ii),

$$T_{X^{[2]}}|_{\mu(D)} = \mathcal{O}_{\mu(D)}(2) \oplus \mathcal{O}_{\mu(D)}(-2) \oplus \mathcal{O}_{\mu(D)} \oplus \mathcal{O}_{\mu(D)}.$$

Since the curve D is of genus 1,

$$h^1\big(D, \mu^*(T_{X^{[2]}}|_{\mu(D)})\big) = 2d + 2.$$

It follows that the sheaf $R^1(f_{1,0})_*(\mathrm{ev}_1^* T_{X^{[2]}})$ is locally free of rank $2d+2$.

(ii) First of all, note that there exists a natural morphism

$$\phi : \overline{\mathfrak{M}}_{1,0,d} \to X$$

sending an element $u = [\mu : D \to X^{[2]}] \in \overline{\mathfrak{M}}_{1,0,d}$ to $x \in X$ if $\mu(D) = M_2(x)$. The fiber $\phi^{-1}(x)$ over $x \in X$ is

$$\overline{\mathfrak{M}}_{1,0}(M_2(x), d[M_2(x)] \cong \overline{\mathfrak{M}}_{1,0}(\mathbb{P}^1, d[\mathbb{P}^1]).$$

So the moduli space $\overline{\mathfrak{M}}_{1,0,d}$ is smooth (as a stack) with dimension

$$\dim \overline{\mathfrak{M}}_{1,0,d} = \dim \overline{\mathfrak{M}}_{1,0}(\mathbb{P}^1, d[\mathbb{P}^1]) + 2 = 2d + 2.$$

By (13.26), the expected dimension of $\overline{\mathfrak{M}}_{1,0,d}$ is 0. Thus, the excess dimension of $\overline{\mathfrak{M}}_{1,0,d}$ is $2d+2$. By (i) and Proposition 12.1,

$$[\overline{\mathfrak{M}}_{1,0,d}]^{\mathrm{vir}} = c_{2d+2}(\mathcal{O}b) \cap [\overline{\mathfrak{M}}_{1,0,d}].$$

This completes the proof of the lemma. □

Via the inclusion map $B_2 \hookrightarrow X^{[2]}$, the evaluation map $\mathrm{ev}_1 : \overline{\mathfrak{M}}_{1,1,d} \to X^{[2]}$ factors through a morphism $\widetilde{\mathrm{ev}}_1 : \overline{\mathfrak{M}}_{1,1,d} \to B_2$. Also, $B_2 \cong \mathbb{P}(T_X^*)$. Let $\rho : B_2 \to X$ be the canonical projection. Then, there exists a commutative diagram of morphisms:

$$(13.43) \qquad \begin{array}{ccc} \overline{\mathfrak{M}}_{1,1,d} & \overset{\widetilde{\mathrm{ev}}_1}{\to} & B_2 \\ \downarrow^{f_{1,0}} & & \downarrow^{\rho} \\ \overline{\mathfrak{M}}_{1,0,d} & \overset{\phi}{\to} & X. \end{array}$$

The obstruction sheaf over the moduli space $\overline{\mathfrak{M}}_{1,0,d}$ is further investigated in the following lemma.

LEMMA 13.18. (i) *Let $\mathcal{H}$ be the Hodge bundle over $\overline{\mathfrak{M}}_{1,0,d}$. Then,*

$$R^1(f_{1,0})_*\widetilde{\mathrm{ev}}_1^* T_{B_2} \cong \mathcal{H}^* \otimes \phi^* T_X;$$

(ii) *There exists an exact sequence of locally free sheaves:*

$$(13.44) \quad 0 \to R^1(f_{1,0})_*\widetilde{\mathrm{ev}}_1^* T_{B_2} \to \mathcal{O}b \to R^1(f_{1,0})_*\widetilde{\mathrm{ev}}_1^* \mathcal{O}_{B_2}(-2) \to 0.$$

PROOF. (i) Let $T_{B_2/X}$ be the relative tangent sheaf for the projection $\rho : B_2 \to X$. Applying the functors $\widetilde{\mathrm{ev}}_1^*$ and $(f_{1,0})_*$ to the exact sequence

$$0 \to T_{B_2/X} \to T_{B_2} \to \rho^* T_X \to 0$$

of locally free sheaves, we obtain an exact sequence

$$(13.45) \qquad R^1(f_{1,0})_*\widetilde{\mathrm{ev}}_1^* T_{B_2/X} \to R^1(f_{1,0})_*\widetilde{\mathrm{ev}}_1^* T_{B_2} \to R^1(f_{1,0})_*\widetilde{\mathrm{ev}}_1^*(\rho^* T_X) \to 0.$$

where we have used $R^2(f_{1,0})_*\widetilde{\mathrm{ev}}_1^* T_{B_2/X} = 0$ since $f_{1,0}$ is of relative dimension 1.

We claim that $R^1(f_{1,0})_*\widetilde{\mathrm{ev}}_1^* T_{B_2/X} = 0$. Indeed, let $u = [\mu : D \to X^{[2]}] \in \overline{\mathfrak{M}}_{1,0,d}$, and assume that $\mu(D) = M_2(x)$. Since $T_{B_2/X}|_{M_2(x)} = T_{M_2(x)} = \mathcal{O}_{M_2(x)}(2)$,

$$H^1\big(f_{1,0}^{-1}(u), \widetilde{\mathrm{ev}}_1^* T_{B_2/X}|_{f_{1,0}^{-1}(u)}\big) \cong H^1(D, \mu^*\mathcal{O}_{M_2(x)}(2)) = 0.$$

By (13.45),

$$R^1(f_{1,0})_*\widetilde{\mathrm{ev}}_1^* T_{B_2} \cong R^1(f_{1,0})_*\widetilde{\mathrm{ev}}_1^*(\rho^* T_X).$$

Since $\rho \circ \widetilde{\mathrm{ev}}_1 = \phi \circ f_{1,0}$, we get

$$\begin{aligned} R^1(f_{1,0})_*\widetilde{\mathrm{ev}}_1^* T_{B_2} &\cong R^1(f_{1,0})_*\big(f_{1,0}^*(\phi^* T_X)\big) \\ &\cong R^1(f_{1,0})_*\mathcal{O}_{\overline{\mathfrak{M}}_{1,1,d}} \otimes \phi^* T_X \\ &\cong \mathcal{H}^* \otimes \phi^* T_X. \end{aligned}$$

(ii) Since ev_1 factors through $\widetilde{\mathrm{ev}}_1$, we see from Lemma 13.17 (i) that

$$\mathcal{O}b = R^1(f_{1,0})_*(\mathrm{ev}_1^* T_{X^{[2]}}) = R^1(f_{1,0})_*\big(\widetilde{\mathrm{ev}}_1^*(T_{X^{[2]}}|_{B_2})\big).$$

Since B_2 is a smooth divisor in $X^{[2]}$ and $\mathcal{O}_{B_2}(B_2) = \mathcal{O}_{B_2}(-2)$, we have

$$0 \to T_{B_2} \to T_{X^{[2]}}|_{B_2} \to \mathcal{O}_{B_2}(-2) \to 0.$$

Applying the functors $\widetilde{\mathrm{ev}}_1^*$ and $(f_{1,0})_*$, we obtain an exact sequence

$$(f_{1,0})_*\widetilde{\mathrm{ev}}_1^*\mathcal{O}_{B_2}(-2) \to R^1(f_{1,0})_*\widetilde{\mathrm{ev}}_1^*T_{B_2} \to \mathcal{O}b \tag{13.46}$$

$$\to R^1(f_{1,0})_*\widetilde{\mathrm{ev}}_1^*\mathcal{O}_{B_2}(-2) \to 0$$

where we have used $R^2(f_{1,0})_*\widetilde{\mathrm{ev}}_1^*T_{B_2} = 0$ since $f_{1,0}$ is of relative dimension 1.

We claim that $(f_{1,0})_*\widetilde{\mathrm{ev}}_1^*\mathcal{O}_{B_2}(-2) = 0$. Indeed, let

$$u = [\mu : D \to X^{[2]}] \in \overline{\mathfrak{M}}_{1,0,d},$$

and assume that $\mu(D) = M_2(x)$. Then, since $\mathcal{O}_{B_2}(-2)|_{M_2(x)} = \mathcal{O}_{M_2(x)}(-2)$,

$$H^0\big(f_{1,0}^{-1}(u), \widetilde{\mathrm{ev}}_1^*\mathcal{O}_{B_2}(-2)|_{f_{1,0}^{-1}(u)}\big) \cong H^0(D, \mu^*\mathcal{O}_{M_2(x)}(-2)) = 0.$$

So $(f_{1,0})_*\widetilde{\mathrm{ev}}_1^*\mathcal{O}_{B_2}(B_2) = 0$. It follows that the exact sequence (13.46) is simplified to the exact sequence (13.44). □

THEOREM 13.19. *Let $d \ge 2$. Let X be a smooth projective surface. Then,*

$$\langle 1 \rangle_{1,d\beta_2}^{X^{[2]}} = \frac{K_X^2}{12d}.$$

PROOF. By Lemma 13.17 (ii),

$$\langle 1 \rangle_{1,d\beta_2}^{X^{[2]}} = \deg[\overline{\mathfrak{M}}_{1,0,d}]^{\mathrm{vir}} = \deg\big(c_{2d+2}(\mathcal{O}b) \cap [\overline{\mathfrak{M}}_{1,0,d}]\big).$$

The Hodge bundle $\mathcal{H}$ and $R^1(f_{1,0})_*\widetilde{\mathrm{ev}}_1^*\mathcal{O}_{B_2}(-2)$ on the moduli space $\overline{\mathfrak{M}}_{1,0,d}$ are of ranks 1 and $2d$ respectively. Therefore, by Lemma 13.18 and (13.28),

$$\begin{aligned}
\langle 1 \rangle_{1,d\beta_2}^{X^{[2]}} &= c_2(\mathcal{H}^* \otimes \phi^*T_X) \cdot c_{2d}\big(R^1(f_{1,0})_*\widetilde{\mathrm{ev}}_1^*\mathcal{O}_{B_2}(-2)\big) \\
&= \big(\lambda^2 + \phi^*K_X \cdot \lambda + \phi^*c_2(T_X)\big) \cdot c_{2d}\big(R^1(f_{1,0})_*\widetilde{\mathrm{ev}}_1^*\mathcal{O}_{B_2}(-2)\big) \\
&= \big(\phi^*K_X \cdot \lambda + \phi^*c_2(T_X)\big) \cdot c_{2d}\big(R^1(f_{1,0})_*\widetilde{\mathrm{ev}}_1^*\mathcal{O}_{B_2}(-2)\big).
\end{aligned}$$

Let $\chi(X)$ be the Euler characteristic of X. By (13.30), we obtain

$$\begin{aligned}
&\phi^*c_2(T_X) \cdot c_{2d}\big(R^1(f_{1,0})_*\widetilde{\mathrm{ev}}_1^*\mathcal{O}_{B_2}(-2)\big) \\
= \;& \chi(X) \cdot \int_{[\overline{\mathfrak{M}}_{1,0}(\mathbb{P}^1, d[\mathbb{P}^1])]} c_{2d}\big(R^1(f_{1,0})_*\mathrm{ev}_1^*\mathcal{O}_{\mathbb{P}^1}(-2)\big) \\
= \;& 0.
\end{aligned}$$

Hence,

$$\langle 1 \rangle_{1,d\beta_2}^{X^{[2]}} = \phi^*K_X \cdot \lambda \cdot c_{2d}\big(R^1(f_{1,0})_*\widetilde{\mathrm{ev}}_1^*\mathcal{O}_{B_2}(-2)\big).$$

Choose two smooth irreducible curves C_1 and C_2 satisfying $[C_1]-[C_2]=K_X$. Since $B_2 \cong \mathbb{P}(T_X^*)$, we see that

$$\begin{aligned}
\langle 1\rangle_{1,d\beta_2}^{X^{[2]}} &= \phi^*([C_1]-[C_2])\cdot\lambda\cdot c_{2d}\big(R^1(f_{1,0})_*\widetilde{\mathrm{ev}}_1^*\mathcal{O}_{B_2}(-2)\big)\\
&= \int_{[\overline{\mathfrak{M}}_{1,0}(\mathbb{P}(T_X^*|_{C_1}),df)]}\lambda\cdot c_{2d}\big(R^1(f_{1,0})_*\mathrm{ev}_1^*\mathcal{O}_{\mathbb{P}(T_X^*|_{C_1})}(-2)\big)\\
&\quad -\int_{[\overline{\mathfrak{M}}_{1,0}(\mathbb{P}(T_X^*|_{C_2}),df)]}\lambda\cdot c_{2d}\big(R^1(f_{1,0})_*\mathrm{ev}_1^*\mathcal{O}_{\mathbb{P}(T_X^*|_{C_2})}(-2)\big)\\
&= \frac{\deg(T_X^*|_{C_1})}{12d}-\frac{\deg(T_X^*|_{C_2})}{12d}\\
&= \frac{K_X^2}{12d}
\end{aligned}$$

where we have used Proposition 13.15 in the third step. □

PROBLEM 13.20. *Let $n\geq 3$ and $d\geq 2$. Let X be a smooth projective complex surface. What is the genus-1 extremal Gromov-Witten invariant $\langle 1\rangle_{1,d\beta_n}^{X^{[n]}}$?*

CHAPTER 14

Equivariant quantum operator of Okounkov-Pandharipande

Motivated by the Gromov-Witten and Donaldson-Thomas correspondence conjectured in [**MNOP1**, **MNOP2**], Okounkov and Pandharipande [**OP3**, **OP4**] studied the equivariant Gromov-Witten theory of the Hilbert schemes of points on the affine plane $\mathbb{C}^2$. This chapter surveys their work on the equivariant quantum corrected boundary operator, the equivariant quantum cohomology of the Hilbert schemes of points in the affine plane $\mathbb{C}^2$, and the quantum differential equation for the Hilbert schemes of points. The equivariant quantum corrected boundary operator is the q-deformation of the quantum-mechanical Calogero-Sutherland operator, parallel to those Calogero-Sutherland operators discussed in Section 7.4 for the non-equivariant setting. We will determine the structure of the small equivariant quantum cohomology of the Hilbert scheme $(\mathbb{C}^2)^{[n]}$, and see that the equivariant quantum cohomology of $(\mathbb{C}^2)^{[n]}$ is equivalent to the equivariant Gromov-Witten theory of $\mathbb{C}^2 \times \mathbb{P}^1$. This equivalence is one edge of the tetrahedron of equivalences among four theories: the (equivariant) quantum cohomology of $(\mathbb{C}^2)^{[n]}$, the (equivariant) Gromov-Witten theory of $\mathbb{C}^2 \times \mathbb{P}^1$, the (equivariant) Donaldson-Thomas theory of $\mathbb{C}^2 \times \mathbb{P}^1$, and the (equivariant) quantum orbifold cohomology of the symmetric product $(\mathbb{C}^2)^{(n)}$. The quantum differential equation for the Hilbert schemes of points is defined by the equivariant quantum corrected boundary operator, and closely related to the Gromov-Witten theory, Donaldson-Thomas theory and Macdonald polynomials. Its behavior at the two special values $q = 0$ and $q = -1$ will be discussed. All the results in this chapter are from [**OP3**, **OP4**] which have been generalized to the Hilbert scheme of points on $\mathcal{A}_n$-resolutions by Maulik and Oblomkov [**Mau**, **MO1**, **MO2**].

14.1. Equivariant quantum cohomology of the Hilbert scheme $(\mathbb{C}^2)^{[n]}$

In this section, we will review the equivariant quantum corrected boundary operator of Okounkov and Pandharipande [**OP3**] which will be used in Sections 14.3 and 15.2. We will sketch the calculation of this important operator. Moreover, we will see that the small equivariant quantum cohomology of the Hilbert scheme $(\mathbb{C}^2)^{[n]}$ of points in the affine plane is generated by the boundary divisor.

Define the Fock space $\mathfrak{F}$ which is freely generated over $\mathbb{C}$ by commuting creation operators $\alpha_{-k}, k > 0$, acting on the vacuum vector $|0\rangle$. The annihilation operators $\alpha_k, k > 0$ kill the vacuum $\alpha_k|0\rangle = 0, k > 0$ and satisfy the commutation relations

$$[\alpha_k, \alpha_\ell] = k\delta_{k,-\ell}\ \mathrm{Id}. \tag{14.1}$$

Next, let

$$\mathbb{T} = (\mathbb{C}^*)^2$$

and $H^*_{\mathbb{T}}(\mathrm{pt}) = \mathbb{C}[t_1, t_2]$. Similar to Example 11.2, the 2-dimensional algebraic torus $\mathbb{T}$ acts on $\mathbb{C}^2$ by

$$(s_1, s_2) \cdot (u, v) = (s_1 u, s_2 v), \quad s_1, s_2 \in \mathbb{C}^* \tag{14.2}$$

where u and v denote the coordinate functions of $\mathbb{C}^2$. This action lifts to a $\mathbb{T}$-action on the Hilbert schemes $(\mathbb{C}^2)^{[n]}$. The study of the equivariant cohomology of the Hilbert schemes $(\mathbb{C}^2)^{[n]}$ in terms of the Heisenberg algebra action can be carried out as in Section 11.1 and Section 11.2. The Heisenberg algebra is generated by the operators $\mathfrak{p}_k([Y])$ defined in (11.26) and (11.27), where Y denotes a $\mathbb{T}$-invariant closed subscheme of $\mathbb{C}^2$. The analogue of Proposition 11.6 gives the following commutation relations:

$$[\mathfrak{p}_k([\mathbb{C}^2]), \mathfrak{p}_\ell([x])] = -k\delta_{k,-\ell}\,\mathrm{Id}. \tag{14.3}$$

It allows us to define a linear isomorphism of Fock spaces:

$$\Psi: \quad \mathfrak{F} \otimes_{\mathbb{C}} \mathbb{C}[t_1, t_2] \to \bigoplus_{n=0}^{+\infty} H^*_{\mathbb{T}}\big((\mathbb{C}^2)^{[n]}\big) \tag{14.4}$$

by putting

$$\Psi\big(\alpha_{-k_1} \cdots \alpha_{-k_r}|0\rangle\big) = \mathfrak{p}_{-k_1}([\mathbb{C}^2]) \cdots \mathfrak{p}_{-k_r}([\mathbb{C}^2])|0\rangle \tag{14.5}$$

for $k_1, \ldots, k_r > 0$.

LEMMA 14.1. *Let $k > 0$, and x denote the origin of $\mathbb{C}^2$. Then via the isomorphism Ψ, the operators α_{-k} and α_k correspond to the operators $\mathfrak{p}_{-k}([\mathbb{C}^2]) = 1/(t_1 t_2) \cdot \mathfrak{p}_{-k}([x])$ and $-\mathfrak{p}_k([x])$ respectively.*

PROOF. Note that $[\mathbb{C}^2] = 1/(t_1 t_2) \cdot [x]$. By (14.5), the operator α_{-k} corresponds to the operator $\mathfrak{p}_{-k}([\mathbb{C}^2]) = 1/(t_1 t_2) \cdot \mathfrak{p}_{-k}([x])$. In view of the commutation relations (14.1) and (14.3), α_k corresponds to $-\mathfrak{p}_k([x])$. □

For a partition λ with parts λ_i, put

$$\begin{aligned} |\lambda\rangle &= \frac{1}{\mathfrak{z}_\lambda} \prod_i \alpha_{-\lambda_i} \cdot |0\rangle, \\ |\lambda([x])\rangle &= (t_1 t_2)^{\ell(\lambda)} \cdot |\lambda\rangle \end{aligned}$$

where $\mathfrak{z}_\lambda$ is defined by (1.1). Under the identification (14.4), the elements $|\lambda\rangle, \lambda \vdash n$ form a linear basis of the equivariant cohomology $H^*_{\mathbb{T}}\big((\mathbb{C}^2)^{[n]}\big)$ of $(\mathbb{C}^2)^{[n]}$. By Lemma 14.1, $|\lambda([x])\rangle$ corresponds to the class

$$\frac{1}{\mathfrak{z}_\lambda} \prod_i \mathfrak{p}_{-\lambda_i}([x]) \cdot |0\rangle. \tag{14.6}$$

For simplicity, we will denote the classes $|\lambda\rangle$ and $|\lambda([x])\rangle$ by λ and $\lambda([x])$ respectively if no confusion arises.

The standard inner product on the equivariant cohomology induces the following nonstandard inner product on Fock space after an extension of scalars:

$$\langle \mu, \nu \rangle = \big\langle |\mu\rangle, |\nu\rangle \big\rangle = \frac{(-1)^{|\mu| - \ell(\mu)} \cdot \delta_{\mu,\nu}}{(t_1 t_2)^{\ell(\mu)} \cdot \mathfrak{z}_\mu}.$$

The equivariant Gromov-Witten theory can be defined as in Section 12.1. Let β_n be the homology class of $(\mathbb{C}^2)^{[n]}$ defined in (1.34). The $\mathbb{T}$-fixed points ξ_λ of $(\mathbb{C}^2)^{[n]}$ are in one-to-one correspondence with the partitions $\lambda \vdash n$. Throughout this chapter, as in [**OP3, OP4**], we adopt the convention that ξ_λ is defined by the ideal (12.12). By the localization theorem, the equivariant cohomology classes $[\xi_\lambda]$ form a linear basis of the localized equivariant cohomology $H^*_{\mathbb{T}}((\mathbb{C}^2)^{[n]}) \otimes_{\mathbb{C}} \mathbb{C}(t_1, t_2)$. The locus of the stable maps in $\overline{\mathfrak{M}}_{g,k}((\mathbb{C}^2)^{[n]}, d\beta_n)$ meeting a $\mathbb{T}$-fixed point is compact. The $\mathbb{T}$-equivariant Gromov-Witten invariants of $(\mathbb{C}^2)^{[n]}$ are defined by putting

$$\Big\langle [\xi_{\lambda^1}], \ldots, [\xi_{\lambda^k}] \Big\rangle_{g, d\beta_n} = \int_{[\overline{\mathfrak{M}}_{g,k}((\mathbb{C}^2)^{[n]}, d\beta_n)]^{\rm vir}} {\rm ev}_1^*([\xi_{\lambda^1}]) \cup \cdots \cup {\rm ev}_k^*([\xi_{\lambda^k}])$$

where ${\rm ev}_1, \ldots, {\rm ev}_k$ are the evaluation maps on $\overline{\mathfrak{M}}_{g,k}((\mathbb{C}^2)^{[n]}, d\beta_n)$, and $\lambda^1, \ldots, \lambda^k$ are partitions of n.

The following is [**OP3**, Theorem 1].

THEOREM 14.2. *Let $w_1, w_2 \in H^*_{\mathbb{T}}((\mathbb{C}^2)^{[n]}) \otimes_{\mathbb{C}} \mathbb{C}(t_1, t_2)$, and let $\mathbf{M}(q, t_1, t_2)$ be*

$$(14.7) \qquad (t_1 + t_2) \sum_{k>0} \left(\frac{k}{2} \frac{(-q)^k + 1}{(-q)^k - 1} - \frac{1}{2} \frac{(-q) + 1}{(-q) - 1} \right) \alpha_{-k} \alpha_k + \frac{1}{2} \sum_{k,\ell>0} (t_1 t_2 \alpha_{k+\ell} \alpha_{-k} \alpha_{-\ell} - \alpha_{-k-\ell} \alpha_k \alpha_\ell).$$

Then,

$$\sum_{d \ge 0} \langle w_1, c_1(\mathcal{O}^{[n]}_{\mathbb{C}^2}), w_2 \rangle_{0, d\beta_n} q^d = \langle w_1, \mathbf{M}(q, t_1, t_2)\, w_2 \rangle.$$

Note that $c_1(\mathcal{O}^{[n]}_{\mathbb{C}^2}) = -B_n/2$ where $B_n \subset (\mathbb{C}^2)^{[n]}$ is the boundary divisor defined by (1.26). The operator $\mathbf{M}(q, t_1, t_2)$ is known as the *equivariant quantum corrected boundary operator*. Since $\beta_n \cdot c_1(\mathcal{O}^{[n]}_{\mathbb{C}^2}) = 1$, Theorem 14.2 implies that

$$\sum_{d \ge 1} d \langle w_1, w_2 \rangle_{0, d\beta_n} q^d = \Big\langle w_1, \big(\mathbf{M}(q, t_1, t_2) - \mathbf{M}(0, t_1, t_2)\big) w_2 \Big\rangle$$

which computes the equivariant 2-point genus-0 extremal Gromov-Witten invariants of $(\mathbb{C}^2)^{[n]}$. In view of Lemma 14.1, we have

$$(14.8) \qquad \mathbf{M}(q, t_1, t_2) - \mathbf{M}(0, t_1, t_2) = \widetilde{\mathbf{M}}(q, t_1, t_2) - \widetilde{\mathbf{M}}(0, t_1, t_2)$$

where the operator $\widetilde{\mathbf{M}}(q, t_1, t_2)$ is defined to be

$$(14.9) \qquad -\frac{t_1 + t_2}{t_1 t_2} \sum_{k>0} \left(\frac{k}{2} \frac{(-q)^k + 1}{(-q)^k - 1} - \frac{1}{2} \frac{(-q) + 1}{(-q) - 1} \right) \mathfrak{p}_{-k}([x]) \mathfrak{p}_k([x]).$$

The structure of the small equivariant quantum cohomology ring $QH^*_{\mathbb{T}}((\mathbb{C}^2)^{[n]})$ of $(\mathbb{C}^2)^{[n]}$ is given by the following [**OP3**, Corollary 1].

COROLLARY 14.3. *The divisor class $c_1(\mathcal{O}^{[n]}_{\mathbb{C}^2})$ generates the small equivariant quantum cohomology ring $QH^*_{\mathbb{T}}((\mathbb{C}^2)^{[n]})$ of $(\mathbb{C}^2)^{[n]}$ over $\mathbb{C}(q, t_1, t_2)$.*

PROOF. Since the limiting operator

$$\lim_{t \to \infty} \frac{1}{t} \mathbf{M}(q, t, t^{-1}) = \sum_{k>0} \left(\frac{k}{2} \frac{(-q)^k + 1}{(-q)^k - 1} - \frac{1}{2} \frac{(-q) + 1}{(-q) - 1} \right) \alpha_{-k} \alpha_k$$

is diagonal with distinct eigenvalues, $\mathbf{M}(q, t_1, t_2)$ has distinct eigenvalues for generic values of the parameters. By the Localization Theorem (Theorem 11.1), the equivariant cohomology ring $H^*_{\mathbb{T}}((\mathbb{C}^2)^{[n]})$ of $(\mathbb{C}^2)^{[n]}$ is semisimple after localization.

Thus the equivariant quantum cohomology ring $QH^*_{\mathbb{T}}\big((\mathbb{C}^2)^{[n]}\big)$ of $(\mathbb{C}^2)^{[n]}$ is also semisimple after localization. So

$$QH^*_{\mathbb{T}}\big((\mathbb{C}^2)^{[n]}\big) = \sum_i \mathbb{C}(q, t_1, t_2) \cdot w_i$$

where $\{w_i\}_i$ is the (finite) set of idempotents of $QH^*_{\mathbb{T}}\big((\mathbb{C}^2)^{[n]}\big)$ satisfying $w_i \cdot w_j = \delta_{i,j} \cdot w_i$. Note that the idempotents w_i are eigenvectors of quantum multiplication by $c_1\big(\mathcal{O}_{\mathbb{C}^2}^{[n]}\big)$. So they are eigenvectors under the action of the operator $\mathbf{M}(q, t_1, t_2)$. The element $1/n! \cdot \alpha_{-1}^n|0\rangle$ represents the unit of $QH^*_{\mathbb{T}}\big((\mathbb{C}^2)^{[n]}\big)$. Since the unit is equal to $\sum_i w_i$, the action of $\mathbf{M}(q, t_1, t_2)$ on $1/n! \cdot \alpha_{-1}^n|0\rangle$ generates the n-eigenvalue subspace of Fock space. Hence, $c_1\big(\mathcal{O}_{\mathbb{C}^2}^{[n]}\big)$ generates $QH^*_{\mathbb{T}}\big((\mathbb{C}^2)^{[n]}\big)$ over $\mathbb{C}(q, t_1, t_2)$. □

In the rest of this section, we will outline the proof of Theorem 14.2. The first step is to introduce a $\mathbb{T}$-equivariant reduced obstruction theory for stable maps (of positive degrees) to $(\mathbb{C}^2)^{[n]}$ relative to the moduli space of pointed genus-g curves. The new obstruction theory differs from the standard theory by a 1-dimensional obstruction space L with $c_1(L) = (t_1 + t_2)$. Therefore, when $d > 0$, we have

$$[\overline{\mathfrak{M}}_{g,k}((\mathbb{C}^2)^{[n]}, d\beta_n)]^{\text{vir}} = (t_1 + t_2) \cdot [\overline{\mathfrak{M}}_{g,k}((\mathbb{C}^2)^{[n]}, d\beta_n)]^{\text{vir}}_{\text{red}} \tag{14.10}$$

where $[\overline{\mathfrak{M}}_{g,k}((\mathbb{C}^2)^{[n]}, d\beta_n)]^{\text{vir}}_{\text{red}}$ denotes the virtual fundamental cycle from the reduced theory. Formula (14.10) is the analogue of the reduction lemma (Lemma 13.2) in the equivariant setting.

The second step in the proof of Theorem 14.2 is to establish some additive property. For simplicity, denote the invariant $\langle w_1, \cdots, w_k\rangle_{0,d\beta_n}$ by $\langle w_1, \cdots, w_k\rangle_d$. By (14.10), when $d > 0$, we get

$$\big\langle \mu([x]), c_1\big(\mathcal{O}_{\mathbb{C}^2}^{[n]}\big), \nu\big\rangle_d = (t_1 + t_2) \cdot \big(\mu([x]), c_1\big(\mathcal{O}_{\mathbb{C}^2}^{[n]}\big), \nu\big)_d$$

where $(\cdot, \cdot, \cdot)_d$ on the right-hand-side denotes the reduced invariant of $(\mathbb{C}^2)^{[n]}$. If $\ell(\mu) = \ell(\nu)$, the integral $\big(\mu([x]), c_1\big(\mathcal{O}_{\mathbb{C}^2}^{[n]}\big), \nu\big)_d$ is a nonequivariant constant. Let

$$\xi = \sum_i \mu_i x_i \in (\mathbb{C}^2)^{(n)}$$

where the points $x_1, \ldots, x_{\ell(\mu)} \in \mathbb{C}^2$ are distinct. By Proposition 3.16, we can replace the equivariant class $\mu([x])$ in the integrand by the nonequivariant class

$$\mu(\xi) = \frac{1}{\mathfrak{z}_\lambda} \cdot \rho_n^{-1}(\xi)$$

where ρ_n is the Hilbert-Chow morphism $(\mathbb{C}^2)^{[n]} \to (\mathbb{C}^2)^{(n)}$. Every rational curve in $(\mathbb{C}^2)^{[n]}$ is contracted by ρ_n. The moduli space of maps connecting the locus $\mu(\xi)$ and the cycle $\nu = |\nu\rangle$ is isomorphic to the moduli space of stable maps to the product

$$\prod_{i=1}^{\ell(\mu)} M_{\mu_i}(x_i)$$

if $\mu = \nu$ and empty otherwise, where $M_{\mu_i}(x_i) \subset (\mathbb{C}^2)^{[\mu_i]}$ denotes the punctual Hilbert scheme of $\mathbb{C}^2$ at x_i. The moduli space of maps to the product,

$$\overline{\mathfrak{M}}_{0,3}\left(\prod_{i=1}^{\ell(\mu)} M_{\mu_i}(x_i), d\right) \tag{14.11}$$

has components corresponding to the different distributions of the total degree d among the factors. Consider a component

$$\overline{\mathfrak{M}}[j,k] \subset \overline{\mathfrak{M}}_{0,3}\left(\prod_{i=1}^{\ell(\mu)} M_{\mu_i}(x_i), d\right),$$

for which the degree splitting has at least two non-zero terms corresponding to two different points x_j and x_k. The component $\overline{\mathfrak{M}}[j,k]$ has a standard obstruction theory obtained from the standard obstruction theory of $\overline{\mathfrak{M}}_{0,3}((\mathbb{C}^2)^{[n]}, d\beta_n)$. The standard obstruction theory of $\overline{\mathfrak{M}}[j,k]$ has a 2-dimensional quotient obtained from the 2-dimensional family of holomorphic symplectic forms (at x_j and x_k). So we obtain a doubly reduced obstruction theory by reducing the obstruction space by the 2-dimensional quotient. The nonequivariant integral of the (singly) reduced theory over such a component vanishes since the singly reduced theory contains an additional 1-dimensional trivial factor. Therefore, the only components of (14.11) which contribute to the integral $\big(\mu([x]), c_1\big(\mathcal{O}_{\mathbb{C}^2}^{[n]}\big), \mu\big)_d$ are those for which the degree d is distributed entirely to a single factor of the product.

In summary, the following properties have been verified:

(i) $\big\langle \mu, c_1\big(\mathcal{O}_{\mathbb{C}^2}^{[n]}\big), \nu\big\rangle_d = 0$ if $d > 0$ and $\mu \neq \nu$;

(ii) $\big\langle \mu, c_1\big(\mathcal{O}_{\mathbb{C}^2}^{[n]}\big), \mu\big\rangle_d = \gamma_{n,d} \cdot (t_1t_2)^{-\ell(\mu)} \cdot (t_1 + t_2)$ for some $\gamma_{n,d} \in \mathbb{Q}$;

(iii) the addition formula for $d > 0$,

$$\frac{\big\langle \mu, c_1\big(\mathcal{O}_{\mathbb{C}^2}^{[n]}\big), \mu\big\rangle_d}{\langle \mu, \mu\rangle} = \sum_i \frac{\big\langle \mu_i, c_1\big(\mathcal{O}_{\mathbb{C}^2}^{[n]}\big), \mu_i\big\rangle_d}{\langle \mu_i, \mu_i\rangle} \tag{14.12}$$

where by abusing notations, the μ_i's on the right-hand-side denote the partition with one part μ_i.

Note that the operator $\mathbf{M}(q, t_1, t_2)$ also satisfies the corresponding properties, including the addition formula for $d > 0$,

$$\frac{\big\langle \mu, \mathbf{M}(q, t_1, t_2)\mu\big\rangle_d}{\langle \mu, \mu\rangle} = \sum_i \frac{\big\langle \mu_i, \mathbf{M}(q, t_1, t_2)\mu_i\big\rangle_d}{\langle \mu_i, \mu_i\rangle}. \tag{14.13}$$

Now Theorem 14.2 follows from inductions on both n and d. We refer to [**OP3**, Section 3] for further details.

14.2. Equivalence of four theories

As an application of the results in Section 14.1, Okounkov and Pandharipande [**OP3**, Subsection 4.3] obtained an equivalence between the quantum cohomology of the Hilbert scheme $(\mathbb{C}^2)^{[n]}$ and the local Gromov-Witten theory of $\mathbb{C}^2 \times \mathbb{P}^1$. In this section, we will survey this equivalence which leads to a tetrahedron of equivalences among four theories.

To begin with, let $(\mathbb{P}^1, x_1, \ldots, x_r)$ be the complex projective line with r distinct marked points. Fix partitions $\lambda^1, \ldots, \lambda^r$ of n. Let

$$\overline{\mathfrak{M}}_h^\bullet(\mathbb{P}^1, \lambda^1, \ldots, \lambda^r)$$

denote the moduli space of (possibly disconnected) relative stable maps from genus h curves to $\mathbb{P}^1$ with prescribed ramification λ^i at x_i. The prescribed ramification points on the domain are unmarked, and the maps are required to be nonconstant on all connected components.

The partition function $\mathbf{Z}'_{GW}(\mathbb{C}^2 \times \mathbb{P}^1)_{n[\mathbb{P}^1],\lambda^1,\ldots,\lambda^r}$ of the local Gromov-Witten theory of $\mathbb{C}^2 \times \mathbb{P}^1$ is defined to be

$$\sum_{h\in\mathbb{Z}} u^{2h-2} \cdot \int_{[\overline{\mathfrak{M}}^\bullet_h(\mathbb{P}^1,\lambda^1,\ldots,\lambda^r)]^{\mathrm{vir}}} e\big(-R^\bullet\pi_* f^*(\mathcal{O}_{\mathbb{P}^1} \oplus \mathcal{O}_{\mathbb{P}^1})\big)$$

where π and f are from the universal diagram for $\overline{\mathfrak{M}}^\bullet_h(\mathbb{P}^1, \lambda^1, \ldots, \lambda^r)$:

$$\begin{array}{ccc} \mathfrak{U} & \xrightarrow{f} & \mathbb{P}^1 \\ \pi\big\downarrow & & \\ \overline{\mathfrak{M}}^\bullet_h(\mathbb{P}^1,\lambda^1,\ldots,\lambda^r). & & \end{array}$$

The shifted generating function $\mathbf{GW}^*_n(\mathbb{C}^2 \times \mathbb{P}^1)_{\lambda^1,\ldots,\lambda^r}$ is defined to be

$$(-iu)^{n(2-r)+\sum_{j=1}^r \ell(\lambda^j)} \cdot \mathbf{Z}'_{GW}(\mathbb{C}^2 \times \mathbb{P}^1)_{n[\mathbb{P}^1],\lambda^1,\ldots,\lambda^r}.$$

Fix a point $\xi \in \overline{\mathfrak{M}}_{0,r}$. Define the multipoint invariant $\langle \lambda^1, \ldots, \lambda^r\rangle_\xi^{(\mathbb{C}^2)^{[n]}}$ to be

$$\sum_{d=0}^{+\infty} q^d \cdot \int_{[\overline{\mathfrak{M}}_{0,r}((\mathbb{C}^2)^{[n]},d\beta_n)]^{\mathrm{vir}}} \mathrm{ev}_1^*(\lambda^1) \cup \cdots \cup \mathrm{ev}_r^*(\lambda^r) \cup \tilde{\pi}^*(\xi)$$

where $\tilde{\pi} : \overline{\mathfrak{M}}_{0,r}((\mathbb{C}^2)^{[n]}, d\beta_n) \to \overline{\mathfrak{M}}_{0,r}$ is the $\mathbb{T}$-equivariant forgetful map, and $\mathrm{ev}_1, \ldots, \mathrm{ev}_r$ are the evaluation maps on $\overline{\mathfrak{M}}_{0,r}((\mathbb{C}^2)^{[n]}, d\beta_n)$. The following correspondence theorem [**OP3**, Theorem 2] relates the local Gromov-Witten theory of $\mathbb{C}^2 \times \mathbb{P}^1$ to the genus-0 invariants $\langle \lambda^1, \ldots, \lambda^r\rangle_\xi^{(\mathbb{C}^2)^{[n]}}$ of the Hilbert scheme $(\mathbb{C}^2)^{[n]}$.

THEOREM 14.4. *After the variable change* $e^{iu} = -q$,

$$\mathbf{GW}^*_n(\mathbb{C}^2 \times \mathbb{P}^1)_{\lambda^1,\ldots,\lambda^r} = (-1)^n \cdot \langle \lambda^1, \ldots, \lambda^r\rangle_\xi^{(\mathbb{C}^2)^{[n]}}. \tag{14.14}$$

PROOF. A direct comparison of the formulas of Theorem 14.2 and [**BP**, Theorem 6.5] yields the result in case $r = 3$ and $\lambda^1 = (1^{n-2}2)$. Moreover, a verification shows that the degeneration formula of local Gromov-Witten theory is compatible via the correspondence (14.14) with the splitting formula for genus-0 fixed moduli invariants of $(\mathbb{C}^2)^{[n]}$. So the correspondence (14.14) follows from the observation that by Corollary 14.3 and the reconstruction result in [**BP**, Appendix], both sides of (14.14) are canonically determined by the 3-point case with one 2-cycle $(1^{n-2}2)$. □

It would be interesting to compare the present Hilbert/Gromov-Witten correspondence Theorem 14.4 with the Hilbert/Gromov-Witten correspondence discussed in Chapter 12.

The Gromov-Witten and Donaldson-Thomas theories of $\mathbb{C}^2 \times \mathbb{P}^1$ are related by the correspondence conjectured in [**MNOP1**, **MNOP2**] and refined for the equivariant context in [**BP**]. The proof of this equivalence has been completed in [**OP5**].

In the context of the Crepant Resolution Conjecture, Bryan and Graber [**BG**] proved that the equivariant quantum cohomology of the Hilbert scheme $(\mathbb{C}^2)^{[n]}$ is equivalent to the equivariant quantum orbifold cohomology of the symmetric product $(\mathbb{C}^2)^{(n)}$. Related material will be presented in Chapter 16.

In summary, there exists a tetrahedron of equivalences among the (equivariant) quantum cohomology of $(\mathbb{C}^2)^{[n]}$, the (equivariant) Gromov-Witten theory of $\mathbb{C}^2 \times \mathbb{P}^1$, the (equivariant) Donaldson-Thomas theory of $\mathbb{C}^2 \times \mathbb{P}^1$, and the (equivariant) quantum orbifold cohomology of $(\mathbb{C}^2)^{(n)}$:

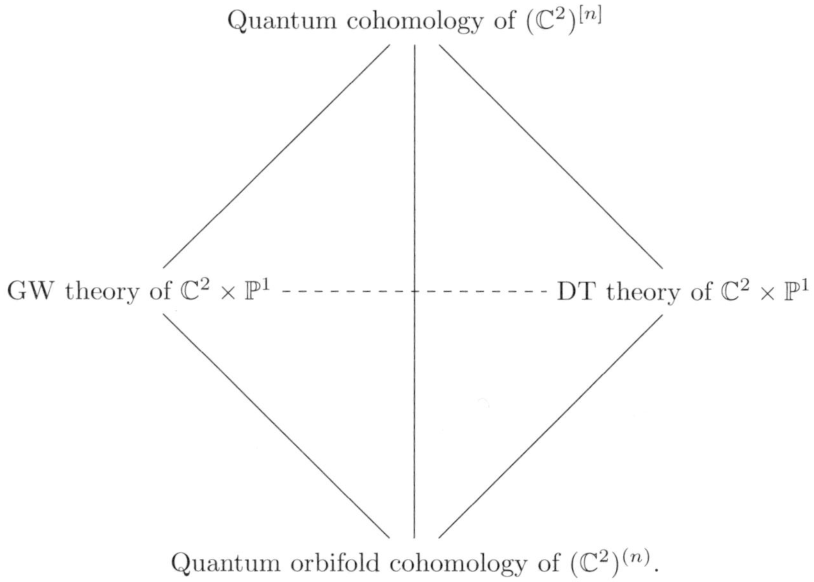

In [**PTse, Tse**], Pandharipande and Tseng generalized this tetrahedron of equivalences to the higher genus case:

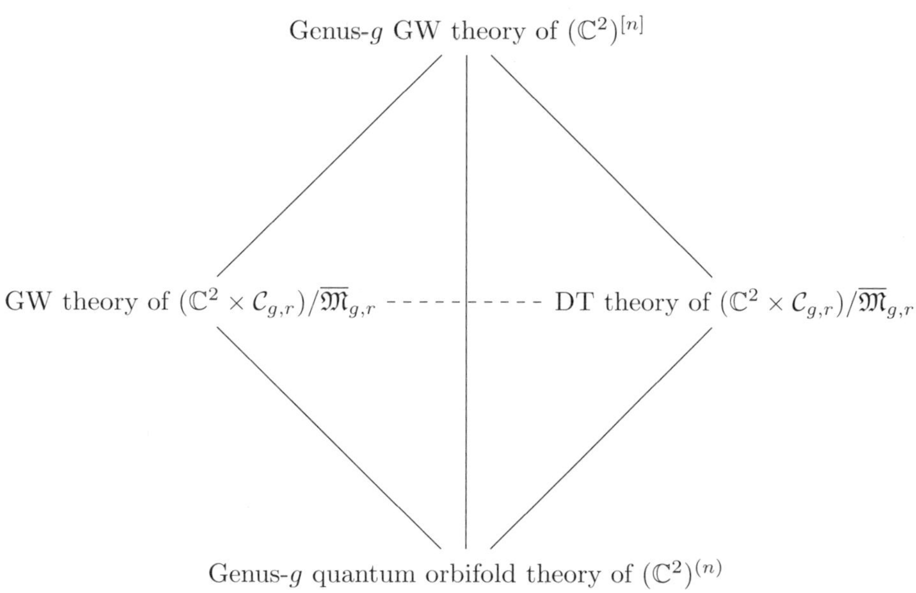

where $\mathcal{C}_{g,r}$ denotes the universal curve over the moduli space $\overline{\mathfrak{M}}_{g,r}$ of genus-g stable curves with r marked points, and $(\mathbb{C}^2 \times \mathcal{C}_{g,r})/\overline{\mathfrak{M}}_{g,r}$ is treated as a family of 3-folds parametrized by $\overline{\mathfrak{M}}_{g,r}$. In establishing the above vertical equivalence between the genus-g Gromov-Witten theory of $(\mathbb{C}^2)^{[n]}$ and the genus-g quantum orbifold theory of $(\mathbb{C}^2)^{(n)}$, a key role was played by the quantum differential equations for which the genus-0 case is the subject of the next section.

14.3. The quantum differential equation of Hilbert schemes of points

This section covers the work of Okounkov and Pandharipande [**OP4**] on the quantum differential equation for the Hilbert schemes of points in the affine plane. The behavior at the special values $q = 0$ and $q = -1$, which are related to the Gromov-Witten and Donaldson-Thomas theories respectively, will be studied. The residue at $q = 0$ is given in terms of the value of the equivariant quantum corrected boundary operator $\mathbf{M}(q, t_1, t_2)$ at $q = 0$, which is essentially the quantum-mechanical Calogero-Sutherland operator. The appearance of the Calogero-Sutherland operator here is the equivariant analogue of those Calogero-Sutherland operators discussed in Section 7.4 in the non-equivariant setting. The monodromy at $q = -1$ will be investigated, and the connection problem between $q = 0$ and $q = -1$ will be solved via Macdonald polynomials.

The *quantum differential equation* (QDE) for the Hilbert scheme of points is the differential equation

$$q\frac{d}{dq}\Psi = \mathbf{M}\,\Psi, \qquad \Psi \in \mathfrak{F} \tag{14.15}$$

where the operator $\mathbf{M} = \mathbf{M}(q, t_1, t_2)$ is the equivariant quantum corrected boundary operator defined by (14.7). An equivalent version is the differential equation

$$q\frac{d}{dq}\Psi = \overline{\mathbf{M}}\,\Psi, \qquad \Psi \in \mathfrak{F} \tag{14.16}$$

where the operator $\overline{\mathbf{M}} = \overline{\mathbf{M}}(q, t_1, t_2)$ is defined by

$$\begin{aligned} &(t_1+t_2)\sum_{k>0}\frac{k}{2}\frac{(-q)^k+1}{(-q)^k-1}\alpha_{-k}\alpha_k + \frac{1}{2}\sum_{k,\ell>0}(t_1t_2\alpha_{k+\ell}\alpha_{-k}\alpha_{-\ell} - \alpha_{-k-\ell}\alpha_k\alpha_\ell) \\ = \quad &\mathbf{M}(q, t_1, t_2) + \frac{t_1+t_2}{2}\sum_{k>0}\frac{(-q)+1}{(-q)-1}\alpha_{-k}\alpha_k. \end{aligned}$$

Since $\overline{\mathbf{M}}$ commutes with the operator $\sum_{k>0}\alpha_{-k}\alpha_k$, the equation splits into a direct sum of finite-dimensional ODEs.

The *quantum-mechanical Calogero-Sutherland operator*

$$\mathbf{H}_{CS} = \frac{1}{2}\sum_i\left(z_i\frac{\partial}{\partial z_i}\right)^2 + \theta(\theta-1)\sum_{i<j}\frac{1}{|z_i-z_j|^2} \tag{14.17}$$

describes particles moving on the torus $|z_i| = 1$ interacting via the potentials $|z_i - z_j|^{-2}$. The parameter θ adjusts the strength of the interaction. The function

$$\phi(z) = \prod_{i<j}(z_i - z_j)^\theta$$

is an eigenfunction of $\mathbf{H}_{CS}$, and the operator $\phi\mathbf{H}_{CS}\phi^{-1}$ preserves the space of symmetric polynomials in the variables $z_1, z_2, z_3, \ldots$. Identify the space of symmetric functions in the variables $z_1, z_2, z_3, \ldots$ with the Fock space $\mathfrak{F}$ via the linear map

$$p_\mu \mapsto \mathfrak{z}_\mu\,|\mu\rangle$$

where p_μ is the power symmetric function defined in Section 1.2. Via this identification, the operator $\phi\mathbf{H}_{CS}\phi^{-1}$ acts on the Fock space $\mathfrak{F}$.

A direct computation shows that the operator $\phi\mathbf{H}_{CS}\phi^{-1}$ is equal to

$$\Delta_{CS} = \frac{1-\theta}{2}\sum_{k>0} k\alpha_{-k}\alpha_k + \frac{1}{2}\sum_{k,\ell>0}(\theta\alpha_{k+\ell}\alpha_{-k}\alpha_{-\ell} + \alpha_{-k-\ell}\alpha_k\alpha_\ell)$$

modulo scalars and a multiple of the momentum operator $\sum_i z_i\frac{\partial}{\partial z_i}$. We have

$$\overline{\mathbf{M}}(0) = -t_1^{\ell(\cdot)+1}\,\Delta_{CS}|_{\theta=-t_2/t_1}\,t_1^{-\ell(\cdot)} \tag{14.18}$$

where $\ell(\cdot)$ is the function mapping the basis $|\mu\rangle$ to $\ell(\mu)$. This implies that the behavior of (14.16) near the regular singularities $q = 0, \infty$ is described by the Schrödinger equation for (14.17).

For Ψ in the n-th component of $\mathfrak{F}$, the equation (14.15) is a linear first order ODE in $P_2(n)$ unknowns, where $P_2(n)$ is the number of partitions of n. In view of (14.7), it has regular singularities. These are $q = 0, \infty$, and the solutions ξ of

$$(-\xi)^k = 1, \qquad k = 2, \ldots, n$$

excluding $q = -1$. Using L'Hospital's rule, we see that for $k > 0$,

$$\lim_{q\to -1}\left(\frac{k}{2}\frac{(-q)^k+1}{(-q)^k-1} - \frac{1}{2}\frac{(-q)+1}{(-q)-1}\right) = 0. \tag{14.19}$$

Therefore, $q = -1$ is a nonsingular point of (14.15).

Two special values of q play a special role. These are $q = -1$ and $q = 0$, which may be called the Gromov-Witten and Donaldson-Thomas points, respectively. On one hand, the point $q = 0$ is a regular singularity of (14.15), and (14.18) relates the residue of (14.15) at $q = 0$ to the Calogero-Sutherland operator. In particular, the eigenfunctions of $\mathbf{M}(0)$ are, up to normalization, the Jack symmetric functions. More precisely, recall the Jack symmetric function $P_\lambda^{(\alpha)} \in \mathfrak{F}$ defined in Section 1.2 with $\alpha = 1/\theta$. Then the eigenfunctions of $\mathbf{M}(0)$ are

$$\mathbf{J}^\lambda = t_2^{|\lambda|}t_1^{\ell(\cdot)}\,P_\lambda^{(\alpha)}|_{\alpha=-t_1/t_2}.$$

Note that geometrically, $\mathbf{M}(0)$ is the equivariant boundary operator

$$\bigoplus_n c_1(\mathcal{O}_{\mathbb{C}^2}^{[n]})\cup,$$

and up to normalization, $\mathbf{J}^\lambda$ corresponds to the class of a $\mathbb{T}$-fixed point in the equivariant cohomology of the Hilbert schemes of points in $\mathbb{C}^2$ (see Chapter 11 and Section 12.2 for details). In particular, we see from (12.13) that the eigenvalue corresponding to the eigenfunction $\mathbf{J}^\lambda$ is

$$-c(\lambda) = -c(\lambda; t_1, t_2) = \sum_{\square\in D_\lambda}(a'(\square)t_1 + \ell'(\square)t_2)$$

where $a'(\square)$ and $\ell'(\square)$ are the arm colength and the leg colength respectively of the cell $\square$ in the Young diagram D_λ, and are defined in Definition 1.4. By the general theory of ODEs, there exists a solution

$$\Psi = \mathbf{Y}^\lambda q^{-c(\lambda)}, \qquad \mathbf{Y}^\lambda \in \mathbb{C}[[q]] \tag{14.20}$$

of (14.15) which converges for $|q| < 1$ and satisfies

$$\mathbf{Y}^\lambda(0) = \mathbf{J}^\lambda.$$

Let $\mathbf{Y}(q)$ denote the matrix formed by the vectors $\mathbf{Y}^\lambda$. Then

$$\mathbf{Y}(q)\, q^{-c(\cdot)} \tag{14.21}$$

is a fundamental solution to the QDE (14.15).

On the other hand, $q = -1$ is a nonsingular point of (14.15), and the monodromy at this point is studied. Let

$$g(x,t) = \frac{x^{tx}}{\Gamma(tx)}, \quad x > 0$$

where Γ denotes the Gamma function. Let $\mathbf{G}_{\mathrm{GW}} = \mathbf{G}_{\mathrm{GW}}(t_1,t_2)$ and $\mathbf{\Gamma} = \mathbf{\Gamma}(t_1,t_2)$ be the diagonal operators defined by

$$\begin{aligned} \mathbf{G}_{\mathrm{GW}}(t_1,t_2)|\mu\rangle &= \prod_j \left(g(\mu_j,t_1)g(\mu_j,t_2)\right) \cdot |\mu\rangle, \\ \mathbf{\Gamma}(t_1,t_2)|\mu\rangle &= \frac{(2\pi i)^{\ell(\mu)}}{\prod_j \mu_j} \mathbf{G}_{\mathrm{GW}}(t_1,t_2)|\mu\rangle. \end{aligned}$$

Define the connection

$$\nabla^{\mathbf{\Gamma}} = \mathbf{\Gamma}\left(q\frac{d}{dq} - \mathbf{M}(q,t_1,t_2)\right)\mathbf{\Gamma}^{-1}.$$

The monodromy at $q = -1$ is given by [**OP4**, Theorem 3].

THEOREM 14.5. *The monodromy of $\nabla^{\mathbf{\Gamma}}$ based at $q = -1$ is a Laurent polynomial in*

$$T_1 = e^{2\pi i t_1}, \quad T_2 = e^{2\pi i t_2}.$$

Moreover, it is unitary with respect to the Hermitian form defined by

$$\langle\langle \mu,\nu\rangle\rangle = \frac{\delta_{\mu,\nu}}{\mathfrak{z}(\mu)} \prod_i \left((T_1^{\mu_i/2} - T_1^{-\mu_i/2})(T_2^{\mu_i/2} - T_2^{-\mu_i/2})\right).$$

Next, the connection problem for the QDE (14.15) between the special points $q = -1$ and $q = 0$ is solved. The connection problem is to find the value of the matrix $\mathbf{Y}(q)$ at $q = -1$. It is proved in [**OP4**] that up to normalization, the value of the matrix $\mathbf{Y}(q)$ at $q = -1$ is given in terms of Macdonald polynomials with parameters $e^{2\pi i t_1}$ and $e^{2\pi i t_2}$. More precisely, let $\mathbf{P}^\lambda \in \mathfrak{F} \otimes \mathbb{Q}(q,t)$ be the monic Macdonald polynomial as defined in [**Mac2**]. In studying the K-theory of the Hilbert scheme, Haiman [**Hai3**] used the polynomial

$$\mathbf{H}^\lambda(q,t) = t^{\mathbf{n}(\lambda)} \prod_{\square\in D_\lambda} \left(1 - q^{a(\square)}t^{-\ell(\square)-1}\right) \cdot \Upsilon \mathbf{P}^\lambda(q,t^{-1})$$

where $\mathbf{n}(\lambda) = \sum_i (i-1)\lambda_i$, and Υ is the diagonal operator defined by

$$\Upsilon|\mu\rangle = \prod_i (1 - t^{-\mu_i})^{-1} \cdot |\mu\rangle.$$

Haiman's parameters q and t are identified with the parameters T_1 and T_2 in Theorem 14.5 respectively. Define the operator $\mathcal{O}(a)$ in K-theory by

$$\mathcal{O}(a)\mathbf{H}^\lambda = e^{-2\pi i a c(\lambda)}\mathbf{H}^\lambda.$$

Let $\mathbf{G}_{\mathrm{DT}} = \mathbf{G}_{\mathrm{DT}}(t_1,t_2)$ be the diagonal operator defined by

$$\mathbf{G}_{\mathrm{DT}}(t_1,t_2)|\lambda\rangle = q^{-c(\lambda)} \prod_{\mathbf{w}} \frac{1}{\Gamma(\mathbf{w}+1)} \cdot |\lambda\rangle$$

where $\mathbf{w}$ runs over the exponents

$$\Big\{t_1\big(a(\square)+1\big)-t_2\ell(\square),\ -t_1a(\square)+t_2\big(\ell(\square)+1\big)\Big\}_{\square\in D_\lambda}$$

(i.e. the tangent weights) in (11.11) by replacing α and β there by t_2 and $-t_1$ respectively. Our settings (14.2) and (12.12) for this chapter are compared with the settings (11.1) and (11.10) respectively.

The following is [**OP4**, Theorem 4].

THEOREM 14.6. *Let* $\mathbf{H}$ *be the matrix with columns* $\mathbf{H}^\lambda$. *Then,*

$$\mathbf{\Gamma}^{-1}\mathbf{Y}\mathbf{G}_{\mathrm{DT}}|_{q=-1}=\frac{1}{(2\pi i)^{|\cdot|}}\mathcal{O}(1/2)\mathbf{H}$$

where $|\cdot|$ *is the degree function mapping* $|\mu\rangle$ *to* $|\mu|$.

Finally, notice that

$$\mathbf{M}(q^{-1})=-(-1)^{\ell(\cdot)}\mathbf{M}(q)(-1)^{\ell(\cdot)}.$$

It follows that

$$(-1)^{\ell(\cdot)}\mathbf{Y}(q^{-1})\,q^{c(\cdot)} \tag{14.22}$$

is another fundamental solution to the QDE (14.15). It is natural to ask for the relation between the two fundamental solutions (14.21) and (14.22). Since, as $q\to 0,\infty$, the QDE becomes the Calogero-Sutherland system, the $q\to q^{-1}$ transformation of the fundamental solution can be naturally interpreted as scattering by the nonstationary terms. Moreover, since $q=-1$ is a fixed point of the involution $q\to q^{-1}$, the evaluation at $q=-1$ can be used to compare the two solutions. By Theorem 14.6, the scattering transformation essentially amounts to the action of the operator $(-1)^{\ell(\cdot)}$.

CHAPTER 15

The genus-0 extremal Gromov-Witten invariants

Motivated by Ruan's Cohomological Crepant Resolution Conjecture [**Rua2**, Section 2], [**LQ1**, **LL**, **LQ5**] studied the genus-0 extremal Gromov-Witten theory of the Hilbert schemes of points on a smooth projective complex surface. This chapter surveys the main results and ideas in these papers. We will begin with the computation of 1-point genus-0 extremal Gromov-Witten invariants by calculating the restriction of the obstruction sheaf to a Zariski open subset of the moduli space of stable maps. Then, we will review the computations of J. Li and W.-P. Li [**LL**] regarding the quantum corrected boundary operator and the 2-point genus-0 extremal Gromov-Witten invariants of the Hilbert schemes of points on surfaces. The establishment of the quantum corrected boundary operator by J. Li and W.-P. Li is one of the major steps toward the proof of Ruan's Cohomological Crepant Resolution Conjecture to be presented in Chapter 16. The main idea of [**LL**] is to use the technique of cosection localization discussed in Section 13.1 and the equivariant quantum corrected boundary operator of Okounkov and Pandharipande discussed in Section 14.1. In addition, we will sketch the argument from [**LQ5**] regarding the structure of the 3-point genus-0 extremal Gromov-Witten invariants of the Hilbert scheme $X^{[n]}$. Once again, the technique of cosection localization plays an essential role. Moreover, the notion of Hilbert schemes of α-points from [**LiJ**, Section 1] enables us to obtain finer analysis of the relevant virtual fundamental cycle. Theorem 15.9 and its proof are from [**LQ1**]. The results of Section 15.2 are from [**LL**], while we will refer to [**LL**] for the detailed proofs (except for the proof of Proposition 15.16 which is slightly different).

15.1. 1-point genus-0 extremal Gromov-Witten invariants

In this section, as a warm-up to the results in later sections, we will compute the 1-point genus-0 extremal Gromov-Witten invariants of the Hilbert scheme $X^{[n]}$ for an arbitrary smooth projective complex surface X. The main idea is to find a non-empty Zariski open (not necessarily dense) subset of the moduli space of stable maps, over which the obstruction sheaf can be determined. The calculation of the obstruction sheaf here is similar to what has been presented in Section 13.4.

15.1.1. Bases of $H_2(X^{[n]})$ and $H^{4n-4}(X^{[n]})$. Let X be a smooth projective complex surface. For simplicity, we will assume that X is simply connected. Recall the homology classes $\beta_n, \beta_C \in H_2(X^{[n]})$ from (1.34) and (2.1) respectively. A basis of the homology group $H_2(X^{[n]})$ is presented by Lemma 2.1 (i). In this subsection, we will write down a basis of the cohomology group $H^{4n-4}(X^{[n]})$.

Note from (3.11) that if $n = \sum_{i=1}^{k} n_i$ and $m = \sum_{i=1}^{k} (2n_i - 2 + |\alpha_i|)$, then

$$\mathfrak{a}_{-n_1}(\alpha_1) \ldots \mathfrak{a}_{-n_k}(\alpha_k)|0\rangle \in H^m(X^{[n]}). \tag{15.1}$$

DEFINITION 15.1. Let $\alpha, \widetilde{\alpha} \in H^2(X^{[n]})$. Then, we define the following classes:

$$\begin{aligned}
\mathfrak{s}_{n,1} &= \mathfrak{a}_{-1}(1_X)\mathfrak{a}_{-1}(x)^{n-1}|0\rangle \\
\mathfrak{s}_{n,2} &= \mathfrak{a}_{-2}(x)\mathfrak{a}_{-2}(x)\mathfrak{a}_{-1}(x)^{n-4}|0\rangle \\
\mathfrak{s}_{n,3} &= \mathfrak{a}_{-3}(x)\mathfrak{a}_{-1}(x)^{n-3}|0\rangle \\
\mathfrak{s}_{\alpha,1} &= \mathfrak{a}_{-1}(\alpha)\mathfrak{a}_{-2}(x)\mathfrak{a}_{-1}(x)^{n-3}|0\rangle \\
\mathfrak{s}_{\alpha,2} &= \mathfrak{a}_{-2}(\alpha)\mathfrak{a}_{-1}(x)^{n-2}|0\rangle \\
\mathfrak{s}_{\alpha,\widetilde{\alpha}} &= \mathfrak{a}_{-1}(\alpha)\mathfrak{a}_{-1}(\widetilde{\alpha})\mathfrak{a}_{-1}(x)^{n-2}|0\rangle
\end{aligned}$$

where x denotes the cohomology class of a point in X.

By (15.1), we have

$$\mathfrak{s}_{n,1}, \mathfrak{s}_{n,2}, \mathfrak{s}_{n,3}, \mathfrak{s}_{\alpha,1}, \mathfrak{s}_{\alpha,2}, \mathfrak{s}_{\alpha,\widetilde{\alpha}} \in H^{4n-4}(X^{[n]}).$$

To discuss the geometric interpretations of $\mathfrak{s}_{n,1}, \mathfrak{s}_{n,2}, \mathfrak{s}_{n,3}, \mathfrak{s}_{\alpha,1}, \mathfrak{s}_{\alpha,2}, \mathfrak{s}_{\alpha,\widetilde{\alpha}}$, we recall Proposition 3.16 and let PD stand for the Poincaré dual. Represent $\alpha, \widetilde{\alpha} \in H^2(X^{[n]})$ respectively by real-2-dimensional cycles $C, \widetilde{C}$ of X in general position. For $x_1, x_2 \in X$ and $\eta \in X^{[n-4]}$ satisfying $x_1 \neq x_2$ and $x_1, x_2 \notin \mathrm{Supp}(\eta)$, we have

$$\mathrm{PD}(\mathfrak{s}_{n,2}) \sim M_2(x_1) + M_2(x_2) + \eta. \tag{15.2}$$

For $x \in X$ and $\eta \in X^{[n-3]}$ with $x \notin C \cup \mathrm{Supp}(\eta)$ and $\mathrm{Supp}(\eta) \cap C = \emptyset$, we get

$$\mathrm{PD}(\mathfrak{s}_{n,3}) \sim M_3(x) + \eta, \tag{15.3}$$

$$\mathrm{PD}(\mathfrak{s}_{\alpha,1}) \sim C + M_2(x) + \eta. \tag{15.4}$$

For a fixed $\eta \in X^{[n-2]}$ satisfying $\mathrm{Supp}(\eta) \cap C = \emptyset$, we have

$$\mathrm{PD}(\mathfrak{s}_{\alpha,2}) \sim M_2(C) + \eta. \tag{15.5}$$

For $\eta = x_1 + \ldots + x_{n-1} \in X^{[n-1]}$ where $x_1, \ldots, x_{n-1}$ are distinct, we obtain

$$\mathrm{PD}(\mathfrak{s}_{n,1}) \sim \text{“the closure of } (X \backslash \mathrm{Supp}(\eta)) + \eta \text{ in } X^{[n]}\text{”}. \tag{15.6}$$

Alternatively, consider the following commutative diagram:

$$\begin{array}{ccccc}
\widetilde{X}_\eta & \subset & X^{[n-1,n]} & \xrightarrow{g_n} & X^{[n]} \\
\downarrow & & \downarrow{\scriptstyle (f_n, \rho)} & & \\
\eta \times X & \subset & X^{[n-1]} \times X & &
\end{array} \tag{15.7}$$

where f_n, g_n are from Notation 1.30, (f_n, ρ) is the blowing-up of $X^{[n-1]} \times X$ along $\mathcal{Z}_{n-1}$ by Theorem 1.31 (ii), and $\widetilde{X}_\eta$ stands for the strict transform of $\eta \times X$. So $\widetilde{X}_\eta$ is isomorphic to the blown-up of X at the $(n-1)$ distinct points $x_1, \ldots, x_{n-1}$. Moreover,

$$g_n|_{\widetilde{X}_\eta} : \quad \widetilde{X}_\eta \to g_n(\widetilde{X}_\eta)$$

is an isomorphism and $g_n(\widetilde{X}_\eta)$ is precisely the closure of $(X \backslash \mathrm{Supp}(\eta)) + \eta$ in the Hilbert scheme $X^{[n]}$. So in view of (15.6), we conclude that

$$\mathrm{PD}(\mathfrak{s}_{n,1}) \sim g_n(\widetilde{X}_\eta). \tag{15.8}$$

Note that the $(n-1)$ exceptional curves in the surface $g_n(\widetilde{X}_\eta)$ are

$$M_2(x_i) + (\eta\backslash\{x_i\}), \qquad i = 1, \ldots, n-1. \tag{15.9}$$

Finally, choose $\eta \in X^{[n-2]}$ such that $\mathrm{Supp}(\eta) \cap (C \cup \widetilde{C}) = \emptyset$. Then according to Proposition 3.16, $\mathfrak{s}_{\alpha,\tilde{\alpha}}$ is represented by the closure of the subset

$$\{x + \tilde{x} + \eta |\ x \in C, \tilde{x} \in \widetilde{C}, \text{ and } x \neq \tilde{x}\} \subset X^{[n]}. \tag{15.10}$$

Via the above notations, the following elementary lemma lists the basis elements of the cohomology groups $H^2(X^{[n]})$, $H^{4n-2}(X^{[n]})$ and $H^{4n-4}(X^{[n]})$.

LEMMA 15.2. *Let $n \geq 2$ and X be simply connected. Let $\{\alpha_1, \ldots, \alpha_s\}$ be a basis of $H^2(X)$ represented by real-2-dimensional cycles $\{C_1, \ldots, C_s\}$ respectively. Then,*

(i) *a basis of $H^2(X^{[n]})$ consists of the cohomology classes*

$$B_n, \quad \mathfrak{a}_{-1}(\alpha_i)\mathbf{1}_{-(n-1)}|0\rangle \quad (i = 1, \ldots, s).$$

(ii) *a basis of $H^{4n-2}(X^{[n]})$ consists of the cohomology classes*

$$\mathfrak{a}_{-2}(x)\mathfrak{a}_{-1}(x)^{n-2}|0\rangle, \quad \mathfrak{a}_{-1}(\alpha_i)\mathfrak{a}_{-1}(x)^{n-1}|0\rangle \quad (i = 1, \ldots, s).$$

(iii) *a basis of $H^{4n-4}(X^{[n]})$ consists of the cohomology classes*

$$\mathfrak{s}_{n,1}, \quad \mathfrak{s}_{n,2}, \quad \mathfrak{s}_{n,3}, \mathfrak{s}_{\alpha_i,1} \ (i = 1, \ldots, s),$$

$$\mathfrak{s}_{\alpha_i,2} \ (i = 1, \ldots, s), \quad \mathfrak{s}_{\alpha_i,\alpha_j} \ (i, j = 1, \ldots, s).$$

PROOF. Note that (i) and (ii) follow from Lemma 2.1. To prove (iii), fix a point $x \in X$. Expand the basis $\{\alpha_1, \ldots, \alpha_s\}$ of $H^2(X)$ to the basis

$$\{\alpha_0 = 1_X, \alpha_1, \ldots, \alpha_s, \alpha_{s+1} = x\}$$

of $H^*(X) = H^0(X) \oplus H^2(X) \oplus H^4(X)$. By (15.1), a basis of $H^{4n-4}(X^{[n]})$ consists of the classes

$$\mathfrak{a}_{-n_1}(\alpha_{m_1}) \cdots \mathfrak{a}_{-n_k}(\alpha_{m_k})|0\rangle \tag{15.11}$$

satisfying $n_i \geq 1$, $\sum_{i=1}^{k} n_i = n$, and

$$\sum_{i=1}^{k} (2n_i - 2 + |\alpha_{m_i}|) = 4n - 4.$$

Without loss of generality, we may assume that $|\alpha_{m_1}| \leq \ldots \leq |\alpha_{m_k}|$. We have

$$\sum_{i=1}^{k} (-2 + |\alpha_{m_i}|) = 2n - 4 \geq 2k - 4.$$

Thus,

$$\sum_{i=1}^{k} |\alpha_{m_i}| \geq 4k - 4.$$

Since X is simply connected, $|\alpha_{m_i}| \in \{0, 2, 4\}$ for every i. It follows that $|\alpha_{m_i}| = 4$ for at least $(k-2)$ many i's, i.e., $\alpha_{m_3} = \ldots = \alpha_{m_k} = x$. Then,

$$|\alpha_{m_1}| + |\alpha_{m_2}| = 2(n - k) + 4 \geq 4.$$

Assume that $|\alpha_{m_2}| = 4$, that is, $\alpha_{m_2} = x$. Then, $|\alpha_{m_1}| = 2(n-k)$. If $|\alpha_{m_1}| = 4$, then $\alpha_{m_1} = x$, $k = n-2$, and $\sum_{i=1}^{n-2} n_i = n$. In this case, (15.11) is equal to either $\mathfrak{s}_{n,2}$ or $\mathfrak{s}_{n,3}$. If $|\alpha_{m_1}| = 2$, i.e., $\alpha_{m_1} = \alpha_i$ for some $1 \le i \le s$, then $k = n-1$ and $\sum_{i=1}^{n-1} n_i = n$. In this case, (15.11) is equal to either $\mathfrak{s}_{\alpha_i,1}$ or $\mathfrak{s}_{\alpha_i,2}$. If $|\alpha_{m_1}| = 0$, then $\alpha_{m_1} = 1_X$, $k = n$, and $\sum_{i=1}^{n} n_i = n$. In this case, (15.11) is equal to $\mathfrak{s}_{n,1}$.

Finally, assume that $|\alpha_{m_2}| \le 2$. Since $|\alpha_{m_1}| \le |\alpha_{m_2}|$ and

$$|\alpha_{m_1}| + |\alpha_{m_2}| = 2(n-k) + 4 \ge 4,$$

we must have $|\alpha_{m_1}| = |\alpha_{m_2}| = 2$. So $k = n$ and $\sum_{i=1}^{n} n_i = n$. In this case, (15.11) is equal to $\mathfrak{s}_{\alpha_i,\alpha_j}$ for some i and j satisfying $1 \le i, j \le s$. □

15.1.2. The obstruction sheaf. In this subsection, we will construct a non-empty Zariski open subset $\mathfrak{U}_0$ of the moduli space $\overline{\mathfrak{M}}_{0,0}(X^{[n]}, d\beta_n)$. Then we will determine the restriction of the obstruction sheaf to the open subset $\mathfrak{U}_0$.

We begin with several notations.

NOTATION 15.3. *Let $\sigma : X^n \to X^{(n)}$ be the natural quotient map. Put*

$$\begin{aligned} {X^{[n]}}_* &= \{\xi \in X^{[n]} \,|\, |\mathrm{Supp}(\xi)| \ge n-1\}, \\ X^{(n)}_* &= \rho_n({X^{[n]}}_*), \\ X^n_* &= \sigma^{-1}(X^{(n)}_*), \\ B &= B_n = \{\xi \in X^{[n]} \,|\, |\mathrm{Supp}(\xi)| < n\}, \\ B_* &= \{\xi \in X^{[n]} \,|\, |\mathrm{Supp}(\xi)| = n-1\}, \\ X^{(n)}_{s*} &= \rho_n(B_*), \\ \Delta_{n*} &= \sigma^{-1}(\rho_n(B)) \cap X^n_* = \coprod_{1 \le i < j \le n} \Delta^{i,j}_{n*} \end{aligned}$$

where $\Delta^{i,j}_{n*} = \{(x_1, \ldots, x_i, \ldots, x_j, \ldots, x_n) \in X^n_* \,|\, x_i = x_j\}$ for $1 \le i < j \le n$.

When we compute the 1-point Gromov-Witten invariants $\langle \alpha \rangle_{0,d\beta_n}$, only ${X^{[n]}}_*$ is involved in most of the cases. Even though the Hilbert scheme $X^{[n]}$ is very complicated, the open subset ${X^{[n]}}_*$ has a very simple description given below. Let

$$\pi_n : \widetilde{X^n_*} \to X^n_* \tag{15.12}$$

be the blowing-up morphism of X^n_* along the big diagonal Δ_{n*}. The action of S_n on X^n_* lifts to an action on $\widetilde{X^n_*}$ and

$${X^{[n]}}_* = \widetilde{X^n_*}/S_n.$$

Let $\tilde{\sigma} : \widetilde{X^n_*} \to {X^{[n]}}_*$ be the quotient map. Let $E^{i,j}_* \subset \widetilde{X^n_*}$ be the exceptional locus over $\Delta^{i,j}_{n*}$. Consider the following morphisms:

$$p_{1,2} \quad : \Delta^{1,2}_{n*} \longrightarrow X, \quad (x, x, x_3, \ldots, x_n) \to x, \tag{15.13}$$

$$j_2 \quad : X^{(n)}_{s*} \longrightarrow X, \quad 2x + x_3 + \ldots + x_n \to x. \tag{15.14}$$

Since the normal bundle of $\Delta_{n*}^{1,2}$ in X_*^n is isomorphic to $p_{1,2}^*T_X$, we have

$$E_*^{1,2} \cong \mathbb{P}(p_{1,2}^*T_X^*).$$

The subgroup $S_2 \times S_{n-2} \subset S_n$ acts on $\Delta_{n*}^{1,2}$ with the S_2-factor acting trivially on $\Delta_{n*}^{1,2}$. The action of $S_2 \times S_{n-2}$ on $\Delta_{n*}^{1,2}$ lifts to an action on $E_*^{1,2}$. It is easy to see that

$$X_{s*}^{(n)} = \Delta_{n*}^{1,2}/(S_2 \times S_{n-2})$$

and

$$B_* = E_*^{1,2}/(S_2 \times S_{n-2}).$$

Regard $p_{1,2}\colon \Delta_{n*}^{1,2} \to X$ as an $S_2 \times S_{n-2}$-equivariant morphism where $S_2 \times S_{n-2}$ acts on X trivially. Then, $S_2 \times S_{n-2}$ acts on $p_{1,2}^*T_X^*$, and the isomorphism $E_*^{1,2} \cong \mathbb{P}(p_{1,2}^*T_X^*)$ is $S_2 \times S_{n-2}$-equivariant. So we get an isomorphism

$$j_1\colon B_* = E_*^{1,2}/(S_2 \times S_{n-2}) \cong \mathbb{P}(p_{1,2}^*T_X^*)/(S_2 \times S_{n-2}) \cong \mathbb{P}(j_2^*T_X^*).$$

where the last isomorphism is due to the fact that the S_2-factor acts trivially on $p_{1,2}^*T_X$ and the S_{n-2}-factor commutes with the morphism $p_{1,2}$.

Next, we study $\mathcal{O}_{B_*}(B_*)$. Since

$$\tilde{\sigma}^*\mathcal{O}_{X^{[n]}_*}(B_*) \cong \mathcal{O}_{X^{[n]}_*}\left(2\sum_{1\le i<j\le n} E_*^{i,j}\right)$$

and $E_*^{i,j} \cap E_*^{1,2} \ne \emptyset$ if and only $i = 1$ and $j = 2$, we conclude that

$$(15.15)\qquad \begin{aligned} (\tilde{\sigma}|_{E_*^{1,2}})^*\mathcal{O}_{B_*}(B_*) &\cong \tilde{\sigma}^*\mathcal{O}_{X^{[n]}_*}(B_*)|_{E_*^{1,2}} \\ &\cong \mathcal{O}_{E_*^{1,2}}(2E_*^{1,2}) \\ &\cong \mathcal{O}_{\mathbb{P}(p_{1,2}^*T_X^*)}(-2) \end{aligned}$$

where we have used the fact that $\mathcal{O}_{E_*^{1,2}}(E_*^{1,2}) \cong \mathcal{O}_{\mathbb{P}(p_{1,2}^*T_X^*)}(-1)$ via the isomorphism $E_*^{1,2} \cong \mathbb{P}(p_{1,2}^*T_X^*)$. Note that

$$\mathcal{O}_{\mathbb{P}(p_{1,2}^*T_X^*)}(-2) = \tau^*\left(\mathcal{O}_{\mathbb{P}(j_2^*T_X^*)}(-2)\right)$$

where $\tau : \mathbb{P}(p_{1,2}^*T_X^*) \to \mathbb{P}(j_2^*T_X^*)$ is the natural morphism. Moreover, $j_1 \circ (\tilde{\sigma}|_{E_*^{1,2}}) = \tau$ via the isomorphism $E_*^{1,2} \cong \mathbb{P}(p_{1,2}^*T_X^*)$. Combining with (15.15), we obtain

$$(\tilde{\sigma}|_{E_*^{1,2}})^*\mathcal{O}_{B_*}(B_*) \cong (\tilde{\sigma}|_{E_*^{1,2}})^*\left(j_1^*\mathcal{O}_{\mathbb{P}(j_2^*T_X^*)}(-2)\right).$$

Since $\mathrm{Pic}(B_*)$ has no torsion, we conclude that

$$(15.16)\qquad \mathcal{O}_{B_*}(B_*) \cong j_1^*\mathcal{O}_{\mathbb{P}(j_2^*T_X^*)}(-2).$$

Consider the open subset $\mathfrak{U}_0$ of $\overline{\mathfrak{M}}_{0,0}(X^{[n]}, d\beta_n)$ consisting of stable maps $[\mu\colon D \to X^{[n]}]$ such that $\mu(D) \subset X^{[n]}_*$. Similarly, take the open subset $\mathfrak{U}_1$ of $\overline{\mathfrak{M}}_{0,1}(X^{[n]}, d\beta_n)$ consisting of stable maps $[\mu\colon (D;p) \to X^{[n]}]$ such that $\mu(D) \subset X^{[n]}_*$. Clearly $\mathfrak{U}_1 = f_{1,0}^{-1}(\mathfrak{U}_0)$. Let $[\mu\colon (D;p) \to X^{[n]}] \in \mathfrak{U}_1$. Since $\mu_*(D) \sim d\beta_n$, we must have

$$\mu(D) = M_2(x_2) + x_3 + \ldots + x_n$$

for some distinct points $x_2, \ldots, x_n \in X$. Hence $\mu(D) \subset B_*$. Moreover, the composite $\rho_n \circ \mathrm{ev}_1$ sends the stable map $[\mu\colon (D;p) \to X^{[n]}]$ to the point $2x_2 + x_3 + \ldots + x_n$, which is independent of the marked point p on D. Hence ev_1 induces a morphism ϕ

from $\mathfrak{U}_0$ to $\rho_n(B_*)$. Putting $\tilde{\mathrm{ev}}_1 = \mathrm{ev}_1|_{\mathfrak{U}_1}$ and $\tilde{f}_{1,0} = f_{1,0}|_{\mathfrak{U}_1}$, we have the following commutative diagram:

$$\begin{array}{ccccccc} \mathfrak{U}_1 & \overset{\tilde{\mathrm{ev}}_1}{\to} & B_* & \overset{j_1}{\cong} & \mathbb{P}(j_2^*T_X^*) & & \\ \downarrow \tilde{f}_{1,0} & & \downarrow \rho_n & & \downarrow \pi & & \\ \mathfrak{U}_0 & \overset{\phi}{\to} & \rho_n(B_*) & = & X_{s*}^{(n)} & \overset{j_2}{\to} & X \end{array} \tag{15.17}$$

where $\pi\colon \mathbb{P}(j_2^*T_X^*) \to X_{s*}^{(n)}$ is the natural projection of the $\mathbb{P}^1$-bundle.

Note that the fiber $\phi^{-1}(2x_2+x_3+\ldots+x_n)$ over a fixed point $2x_2+x_3+\ldots+x_n \in \rho_n(B_*)$ is simply

$$\overline{\mathfrak{M}}_{0,0}\big(M_2(x_2) + x_3 + \ldots + x_n, d[M_2(x_2) + x_3 + \ldots + x_n]\big)$$

which is isomorphic to the moduli space $\overline{\mathfrak{M}}_{0,0}(\mathbb{P}^1, d[\mathbb{P}^1])$ via the isomorphism

$$M_2(x_2) + x_3 + \ldots + x_n \cong \mathbb{P}^1.$$

Hence the complex dimension of $\mathfrak{U}_0$ is equal to

$$\dim \overline{\mathfrak{M}}_{0,0}(\mathbb{P}^1, d[\mathbb{P}^1]) + 2(n-1) = 2n - 3 + 2d - 1.$$

Since $K_{X^{[n]}} \cdot d\beta_n = 0$, the expected dimension of $\mathfrak{M}_{0,0}(X^{[n]}, d\beta_n)$ is $2n-3$ according to (12.3) . Hence the excess dimension of $\mathfrak{U}_0$ is $e = (2d-1)$.

LEMMA 15.4. *With notations as above, the restriction of the direct image sheaf* $R^1(f_{1,0})_*(\mathrm{ev}_1^*T_{X^{[n]}})$ *to* $\mathfrak{U}_0$ *is a locally free sheaf of rank* $(2d-1)$.

PROOF. Take a stable map $u = [\mu\colon D \to X^{[n]}]$ in $\mathfrak{U}_0$, and consider

$$H^1(f_{1,0}^{-1}(u), (\mathrm{ev}_1^*T_{X^{[n]}})|_{f_{1,0}^{-1}(u)}) \cong H^1(D, \mu^*T_{X^{[n]}}).$$

Since $\mu(D) = M_2(x_2) + x_3 + \ldots + x_n \cong \mathbb{P}^1$ for some distinct points $x_2, \ldots, x_n$, we have

$$T_{X^{[n]}}|_{\mu(D)} = \mathcal{O}_{\mathbb{P}^1}(2) \oplus \mathcal{O}_{\mathbb{P}^1}(-2) \oplus \mathcal{O}_{\mathbb{P}^1}^{(2n-2)}$$

by Proposition 2.9 (ii). Thus

$$H^1(D, \mu^*T_{X^{[n]}}) \cong H^1(D, \mu^*\mathcal{O}_{\mathbb{P}^1}(-2))$$

which has dimension equal to the excess dimension $e = (2d-1)$. Hence the direct image sheaf $R^1(f_{1,0})_*(\mathrm{ev}_1^*T_{X^{[n]}})$ over $\mathfrak{U}_0$ is locally free of rank $(2d-1)$. □

Suppose that $\mathfrak{M}_1$ is a closed subset of $\overline{\mathfrak{M}}_{0,1}(X^{[n]}, d\beta_n)$ contained in $\mathfrak{U}_1$ and

$$\mathfrak{M}_0 = f_{1,0}(\mathfrak{M}_1) \subset \mathfrak{U}_0 \subset \overline{\mathfrak{M}}_{0,0}(X^{[n]}, d\beta_n).$$

By Proposition 12.1 (i) and (iii), we have

$$[\overline{\mathfrak{M}}_{0,1}(Y,\beta)]^{\mathrm{vir}}|_{\mathfrak{M}_1} = (\tilde{f}_{1,0})^* c_{2d-1}\big((R^1(f_{1,0})_*(\mathrm{ev}_1)^*T_{X^{[n]}})|_{\mathfrak{M}_0}\big). \tag{15.18}$$

Hence it is crucial to determine the sheaf $\mathcal{V} = R^1(f_{1,0})_*(\mathrm{ev}_1)^*T_{X^{[n]}}$ over $\mathfrak{U}_0$.

LEMMA 15.5. *Let* $\mathcal{V}$ *denote the restriction of* $R^1(f_{1,0})_*(\mathrm{ev}_1)^*T_{X^{[n]}}$ *to* $\mathfrak{U}_0$. *Then,*

(i) $\mathcal{V} \cong R^1(\tilde{f}_{1,0})_*(\tilde{\mathrm{ev}}_1)^*\mathcal{O}_{B_*}(B_*) \cong R^1(\tilde{f}_{1,0})_*(j_1 \circ \tilde{\mathrm{ev}}_1)^*\mathcal{O}_{\mathbb{P}(j_2^*T_X^*)}(-2)$.

(ii) *the locally free sheaf* $\mathcal{V}$ *sits in the exact sequence*

$$\begin{aligned} 0 &\to (j_2 \circ \phi)^*\mathcal{O}_X(-K_X) \to \mathcal{V} \\ &\to R^1(\tilde{f}_{1,0})_*(j_1 \circ \tilde{\mathrm{ev}}_1)^*((j_2 \circ \pi)^*T_X \otimes \mathcal{O}_{\mathbb{P}(j_2^*T_X^*)}(-1)) \to 0. \end{aligned} \tag{15.19}$$

PROOF. (i) Since $\mathrm{ev}_1(\mathfrak{U}_1) \subset B_*$, we have

$$((\mathrm{ev}_1)^* T_{X^{[n]}})|_{\mathfrak{U}_1} = (\widetilde{\mathrm{ev}}_1)^* (T_{X_*^{[n]}}|_{B_*})$$

and

$$\begin{aligned} \mathcal{V} &= (R^1(f_{1,0})_* (\mathrm{ev}_1)^* T_{X^{[n]}})|_{\mathfrak{U}_0} \\ &= R^1(\tilde{f}_{1,0})_* (((\mathrm{ev}_1)^* T_{X^{[n]}})|_{\mathfrak{U}_1}) \\ &= R^1(\tilde{f}_{1,0})_* (\widetilde{\mathrm{ev}}_1)^* (T_{X^{[n]}}|_{B_*}). \end{aligned}$$

Since B_* is a smooth codimension-1 subvariety of $X^{[n]}$, we obtain the exact sequence

$$0 \to T_{B_*} \to T_{X^{[n]}}|_{B_*} \to \mathcal{O}_{B_*}(B_*) \to 0. \tag{15.20}$$

Applying $(\widetilde{\mathrm{ev}}_1)^*$ and $(\tilde{f}_{1,0})_*$ to the exact sequence (15.20), we get

$$R^1(\tilde{f}_{1,0})_* (\widetilde{\mathrm{ev}}_1)^* T_{B_*} \to \mathcal{V} \to R^1(\tilde{f}_{1,0})_* (\widetilde{\mathrm{ev}}_1)^* \mathcal{O}_{B_*}(B_*) \to 0$$

where we have used $R^2(\tilde{f}_{1,0})_* (\widetilde{\mathrm{ev}}_1)^* T_{B_*} = 0$ since $\tilde{f}_{1,0}$ is of relative dimension 1.

If $[\mu\colon D \to X^{[n]}]$ is a stable map in $\mathfrak{U}_0$, then

$$\mu(D) = M_2(x_2) + x_3 + \ldots + x_n.$$

Hence the normal bundle of $\mu(D)$ in B_* is trivial since $\mu(D)$ is a fiber of the $\mathbb{P}^1$-bundle $\mathbb{P}(j_2^* T_X^*)$ over $X_{s*}^{(n)}$. Thus

$$T_{B_*}|_{\mu(D)} \cong \mathcal{O}_{\mu(D)}(2) \oplus \mathcal{O}_{\mu(D)}^{\oplus(2n-2)}.$$

Therefore,

$$H^1(D, \mu^* T_{B_*}) \cong H^1(D, \mu^*(\mathcal{O}_{\mu(D)}(2) \oplus \mathcal{O}_{\mu(D)}^{\oplus(2n-2)})) = 0,$$

and $R^1(\tilde{f}_{1,0})_* (\widetilde{\mathrm{ev}}_1)^* T_{B_*} = 0$. So in view of (15.16), we obtain

$$\mathcal{V} \cong R^1(\tilde{f}_{1,0})_* (\widetilde{\mathrm{ev}}_1)^* \mathcal{O}_{B_*}(B_*) \cong R^1(\tilde{f}_{1,0})_* (j_1 \circ \widetilde{\mathrm{ev}}_1)^* \mathcal{O}_{\mathbb{P}(j_2^* T_X^*)}(-2).$$

(ii) For simplicity, we denote $\mathbb{P}(j_2^* T_X^*)$ by $\mathbb{P}$. Consider the natural surjection

$$\pi^*(j_2^* T_X^*) \to \mathcal{O}_{\mathbb{P}}(1) \to 0.$$

The kernel of this surjection is a line bundle. By comparing the first Chern classes, we get the following exact sequence:

$$0 \to \pi^* \mathcal{O}_{X_{s*}^{(n)}}(j_2^* K_X) \otimes \mathcal{O}_{\mathbb{P}}(-1) \to \pi^*(j_2^* T_X^*) \to \mathcal{O}_{\mathbb{P}}(1) \to 0. \tag{15.21}$$

Tensoring (15.21) with $\pi^* \mathcal{O}_{X_{s*}^{(n)}}(-j_2^* K_X) \otimes \mathcal{O}_{\mathbb{P}}(-1)$, we get

$$\begin{aligned} 0 \quad \to \quad & \mathcal{O}_{\mathbb{P}}(-2) \to (j_2 \circ \pi)^*(T_X^* \otimes \mathcal{O}_X(-K_X)) \otimes \mathcal{O}_{\mathbb{P}}(-1) \\ & \to (j_2 \circ \pi)^* \mathcal{O}_X(-K_X) \to 0. \end{aligned} \tag{15.22}$$

Note that $T_X^* \otimes \mathcal{O}_X(-K_X) \cong T_X$. Applying $(j_1 \circ \widetilde{\mathrm{ev}}_1)^*$ to (15.22) yields

$$\begin{aligned} 0 \quad \to \quad & (j_1 \circ \widetilde{\mathrm{ev}}_1)^* \mathcal{O}_{\mathbb{P}}(-2) \to (j_1 \circ \widetilde{\mathrm{ev}}_1)^*((j_2 \circ \pi)^* T_X \otimes \mathcal{O}_{\mathbb{P}}(-1)) \\ & \to (j_2 \circ \pi \circ j_1 \circ \widetilde{\mathrm{ev}}_1)^* \mathcal{O}_X(-K_X) \to 0. \end{aligned} \tag{15.23}$$

By (15.17), we have

$$(j_2 \circ \pi \circ j_1 \circ \widetilde{\mathrm{ev}}_1)^* = (j_2 \circ \phi \circ \tilde{f}_{1,0})^* = (\tilde{f}_{1,0})^* \circ (j_2 \circ \phi)^*.$$

So rewriting the 3rd term in the exact sequence (15.23), we obtain

$$(15.24)\quad 0 \to (j_1 \circ \widetilde{\text{ev}}_1)^* \mathcal{O}_{\mathbb{P}}(-2) \to (j_1 \circ \widetilde{\text{ev}}_1)^*((j_2 \circ \pi)^* T_X \otimes \mathcal{O}_{\mathbb{P}}(-1)) \to (\tilde{f}_{1,0})^*((j_2 \circ \phi)^* \mathcal{O}_X(-K_X)) \to 0.$$

Applying $(\tilde{f}_{1,0})_*$ to the above exact sequence and using part (i), we have

$$\begin{aligned} 0 &\to (\tilde{f}_{1,0})_*(\tilde{f}_{1,0})^*((j_2 \circ \phi)^* \mathcal{O}_X(-K_X)) \to \mathcal{V} \\ &\to R^1(\tilde{f}_{1,0})_*(j_1 \circ \widetilde{\text{ev}}_1)^*((j_2 \circ \pi)^* T_X \otimes \mathcal{O}_{\mathbb{P}}(-1)) \\ &\to R^1(\tilde{f}_{1,0})_*(\tilde{f}_{1,0})^*((j_2 \circ \phi)^* \mathcal{O}_X(-K_X)) \end{aligned}$$

where we have used

$$(\tilde{f}_{1,0})_*(j_1 \circ \widetilde{\text{ev}}_1)^*((j_2 \circ \pi)^* T_X \otimes \mathcal{O}_{\mathbb{P}}(-1)) = 0.$$

Note that $(\tilde{f}_{1,0})_* \mathcal{O}_{\mathfrak{U}_1} \cong \mathcal{O}_{\mathfrak{U}_0}$ and $R^1(\tilde{f}_{1,0})_* \mathcal{O}_{\mathfrak{U}_1} = 0$. So we get

$$\begin{aligned} (\tilde{f}_{1,0})_*(\tilde{f}_{1,0})^*((j_2 \circ \phi)^* \mathcal{O}_X(-K_X)) &\cong (j_2 \circ \phi)^* \mathcal{O}_X(-K_X) \otimes (\tilde{f}_{1,0})_* \mathcal{O}_{\mathfrak{U}_1} \\ &\cong (j_2 \circ \phi)^* \mathcal{O}_X(-K_X) \end{aligned}$$

by the projection formula. Similarly,

$$R^1(\tilde{f}_{1,0})_*(\tilde{f}_{1,0})^*((j_2 \circ \phi)^* \mathcal{O}_X(-K_X)) = 0.$$

Therefore, the locally free sheaf $\mathcal{V}$ sits in the exact sequence (15.19). □

REMARK 15.6. Fix distinct points $x_2, \ldots, x_n$ on X. Via the isomorphism

$$\phi^{-1}(2x_2 + x_3 + \ldots + x_n) \cong \overline{\mathfrak{M}}_{0,0}(\mathbb{P}^1, d[\mathbb{P}^1]),$$

the restriction of

$$R^1(\tilde{f}_{1,0})_*(j_1 \circ \widetilde{\text{ev}}_1)^*\big((j_2 \circ \pi)^* T_X \otimes \mathcal{O}_{\mathbb{P}}(-1)\big)$$

to $\phi^{-1}(2x_2 + x_3 + \ldots + x_n)$ is isomorphic to

$$R^1(f_{1,0})_*(\text{ev}_1)^*(\mathcal{O}_{\mathbb{P}^1}(-1) \oplus \mathcal{O}_{\mathbb{P}^1}(-1))$$

where by abusing notations, we still use $f_{1,0}$ and ev_1 to denote the forgetful map and the evaluation map from $\overline{\mathfrak{M}}_{0,1}(\mathbb{P}^1, d[\mathbb{P}^1])$ to $\overline{\mathfrak{M}}_{0,0}(\mathbb{P}^1, d[\mathbb{P}^1])$ and $\mathbb{P}^1$ respectively.

15.1.3. The 1-point genus-0 extremal Gromov-Witten invariants. In this subsection, we will compute all the genus-0 1-point extremal Gromov-Witten invariants $\langle \gamma \rangle_{0,d\beta_n}$ for the Hilbert schemes $X^{[n]}$. Recall from (12.4) and (12.3) that $|\gamma| = 4n - 4$. In view of Lemma 15.2 (iii), we need only to compute $\langle \gamma \rangle_{0,d\beta_n}$ when γ is one of the cohomology classes

$$\mathfrak{s}_{n,1}, \mathfrak{s}_{n,2}, \mathfrak{s}_{n,3}, \mathfrak{s}_{\alpha_1,1}, \mathfrak{s}_{\alpha_1,2}, \mathfrak{s}_{\alpha_1,\alpha_2}$$

where $\alpha_1, \alpha_2 \in H^2(X)$. These six cases will be treated in two lemmas.

LEMMA 15.7. *Let $d \geq 1$, and $\alpha_1, \alpha_2 \in H^2(X)$.*
(i) If $\gamma = \mathfrak{s}_{n,1}$, $\mathfrak{s}_{\alpha_1,\alpha_2}$ or $\mathfrak{s}_{\alpha_1,1}$, then $\langle \gamma \rangle_{0,d\beta_n} = 0$.
(ii) If $\gamma = \mathfrak{s}_{\alpha_1,2}$, then $\langle \gamma \rangle_{0,d\beta_n} = 2\langle K_X, \alpha_1 \rangle / d^2$.

PROOF. (i) Represent the cohomology classes $\alpha_1, \alpha_2 \in H^2(X)$ respectively by real-2-dimensional cycles $C_1, C_2 \subset X$ in general position. Fix distinct points $x_1, \ldots, x_{n-1} \in X$ which are not contained in $C_1 \cup C_2$.

Let $\gamma = \mathfrak{s}_{n,1}$. By (15.8),

$$\text{PD}(\mathfrak{s}_{n,1}) \sim g_n(\widetilde{X_\eta}) \cong \widetilde{X_\eta}$$

where $\widetilde{X_\eta}$ is the blow up of X along $\eta = x_1 + \ldots + x_{n-1}$. Moreover, the exceptional curves in $g_n(\widetilde{X_\eta})$ are

$$\rho_n^{-1}(x_1 + \ldots + x_{i-1} + 2x_i + x_{i+1} + \ldots + x_{n-1})$$

where $1 \le i \le n-1$. Let $\mathfrak{M}_1$ be the subset of $\overline{\mathfrak{M}}_{0,1}(X^{[n]}, d\beta_n)$ consisting of all the stable maps $[\mu\colon (D;p) \to X^{[n]}]$ such that $\mu(p) \in g_n(\widetilde{X_\eta})$. In this case, $\mu(D)$ is one of the exceptional curves in $g_n(\widetilde{X_\eta}) \subset B_*$. In particular, the stable maps $[\mu\colon (D;p) \to X^{[n]}]$ are contained in $\mathfrak{U}_1$, and

$$\mathfrak{M}_1 = \coprod_{1 \le i \le n-1} (\tilde{f}_{1,0})^{-1}(\phi^{-1}(x_1 + \ldots + x_{i-1} + 2x_i + x_{i+1} + \ldots + x_{n-1})).$$

So as algebraic cycles, we have

$$[\mathfrak{M}_1] = \sum_{i=1}^{n-1} (\tilde{f}_{1,0})^* \phi^* [x_1 + \ldots + x_{i-1} + 2x_i + x_{i+1} + \ldots + x_{n-1}].$$

By Lemma 15.5 (ii), we get $c_{2d-1}(\mathcal{V}) = -(j_2 \circ \phi)^* K_X \cdot c_{2d-2}(\mathcal{E})$ where

$$\mathcal{E} = R^1(\tilde{f}_{1,0})_* (j_1 \circ \widetilde{\mathrm{ev}}_1)^* ((j_2 \circ \pi)^* T_X \otimes \mathcal{O}_{\mathbb{P}(j_2^* T_X^*)}(-1)).$$

In view of (12.5) and (15.18), we conclude that

$$\begin{aligned}
\langle \gamma \rangle_{0, d\beta_n} &= \int_{[\overline{\mathfrak{M}}_{0,1}(Y,\beta)]^{\mathrm{vir}}} (\mathrm{ev}_1)^* \gamma \\
&= [\mathfrak{M}_1] \cdot [\overline{\mathfrak{M}}_{0,1}(Y,\beta)]^{\mathrm{vir}} \\
&= [\mathfrak{M}_1] \cdot [\overline{\mathfrak{M}}_{0,1}(Y,\beta)]^{\mathrm{vir}}|_{\mathfrak{M}_1} \\
&= [\mathfrak{M}_1] \cdot (\tilde{f}_{1,0})^* (c_{2d-1}(\mathcal{V})) \\
&= -\sum_{i=1}^{n-1} (\tilde{f}_{1,0})^* \Big(\phi^* ([x_1 + \ldots + 2x_i + \ldots + x_{n-1}] \cdot j_2^* K_X) \cdot c_{2d-2}(\mathcal{E}) \Big) \\
&= 0.
\end{aligned}$$

Next let $\gamma = \mathfrak{s}_{\alpha_1,\alpha_2}$. Assume that C_1 and C_2 intersect transversally at the points $y_1, \ldots, y_m$. By (15.10), $\mathfrak{s}_{\alpha_1,\alpha_2}$ is represented by the closure of

$$\{x + \tilde{x} + x_1 + \ldots + x_{n-2} |\ x \in C_1, \tilde{x} \in C_2, \text{ and } x \ne \tilde{x}\} \subset X^{[n]}.$$

Let $\mathfrak{M}_1' \subset \overline{\mathfrak{M}}_{0,1}(X^{[n]}, d\beta_n)$ consist of all the stable maps $[\mu\colon (D;p) \to X^{[n]}]$ such that $\mu(p) \in \mathfrak{s}_{C_1,C_2}$. In this case,

$$\rho_n(\mu(D)) = 2y_k + x_1 + \ldots + x_{n-2}$$

for some k with $1 \le k \le m$. Therefore,

$$\mu(D) = \rho_n^{-1}(2y_k + x_1 + \ldots + x_{n-2}).$$

Hence the stable map $[\mu\colon (D;p) \to X^{[n]}]$ is contained in $\mathfrak{U}_1$. So $\mathfrak{M}_1' \subset \mathfrak{U}_1$ is the disjoint union of

$$(\tilde{f}_{1,0})^{-1}(\phi^{-1}(2y_k + x_1 + \ldots + x_{n-2})), \quad 1 \le k \le m$$

with $\pm$ orientations. By the same computations as in the previous paragraph, we obtain $\langle \gamma \rangle_{0,d\beta_n} = 0$.

For the case of $\mathfrak{s}_{\alpha_1,1}$, the proof is similar to the cases of $\mathfrak{s}_{n,1}$ and $\mathfrak{s}_{\alpha_1,\alpha_2}$.

(ii) Let $\tilde{\eta} = x_1 + \ldots + x_{n-2}$. By (15.5),

$$\mathrm{PD}(\mathfrak{s}_{\alpha_1,2}) \sim M_2(C_1) + \tilde{\eta} = \rho_n^{-1}(2C_1 + \tilde{\eta}).$$

Thus, $\gamma = \mathrm{PD}(\rho_n^{-1}(2C_1 + \tilde{\eta}))$. So we see from (15.18) and Lemma 15.5 (ii) that

$$(15.25) \quad \langle \gamma \rangle_{0,d\beta_n}$$

$$\begin{aligned}
&= \int_{[\overline{\mathfrak{M}}_{0,1}(Y,\beta)]^{\mathrm{vir}}} (\mathrm{ev}_1)^*\gamma \\
&= \int_{-(\tilde{f}_{1,0})^*(j_2\circ\phi)^*K_X\cdot(\tilde{f}_{1,0})^*c_{2d-2}(\mathcal{E})} (\mathrm{ev}_1)^*\gamma \\
&= -\int_{(\tilde{\mathrm{ev}}_1)^*(\rho_n^* j_2^* K_X)\cdot(\tilde{f}_{1,0})^*c_{2d-2}(\mathcal{E})} (\mathrm{ev}_1)^*\gamma \\
&= -\int_{(\tilde{\mathrm{ev}}_1)^*(\rho_n^* j_2^* K_X)\cdot(\tilde{f}_{1,0})^*c_{2d-2}(\mathcal{E})} (\tilde{\mathrm{ev}}_1)^*\mathrm{PD}(\rho_n^{-1}(2C_1 + \tilde{\eta}) \cdot c_1(\mathcal{O}_{B_*}(B_*))) \\
&= -\int_{(\tilde{f}_{1,0})^*c_{2d-2}(\mathcal{E})} (\tilde{\mathrm{ev}}_1)^*\mathrm{PD}\left(\rho_n^{-1}\left((j_2^*K_X)\cdot(2C_1+\tilde{\eta})\right)\cdot c_1(\mathcal{O}_{B_*}(B_*))\right) \\
&= 2\langle K_X, C_1\rangle \cdot \int_{(\tilde{f}_{1,0})^*c_{2d-2}(\mathcal{E})} (\tilde{\mathrm{ev}}_1)^*\mathrm{PD}(\xi)
\end{aligned}$$

where $\xi \in \rho_n^{-1}(2x + \tilde{\eta}) = \rho_n^{-1}(2x + x_1 + \ldots + x_{n-2})$ is a fixed point for some fixed point $x \in C_1$. Also, we have used the isomorphism (15.16) in the last step.

Let $\mathfrak{M}_1'' \subset \overline{\mathfrak{M}}_{0,1}(X^{[n]}, d\beta_n)$ be the subset consisting of all the stable maps $[\mu\colon (D;p) \to X^{[n]}]$ with $\mu(p) = \xi$. If $[\mu\colon (D;p) \to X^{[n]}] \in \mathfrak{M}_1''$, then

$$\rho_n(\mu(D)) = \rho_n(\mu(p)) = 2x + x_1 + \ldots + x_{n-2}.$$

So $\mu(D) = \rho_n^{-1}(2x + x_1 + \ldots + x_{n-2})$. Thus the restriction of the forgetful map $\tilde{f}_{1,0}$ to $\mathfrak{M}_1$ gives a degree-d morphism from $\mathfrak{M}_1''$ to $\phi^{-1}(2x + x_1 + \ldots + x_{n-2})$. Put

$$\mathfrak{M}_0'' = \phi^{-1}(2x + x_1 + \ldots + x_{n-2}).$$

Now, as algebraic cycles, we have

$$(\tilde{f}_{1,0})_*[\mathfrak{M}_1''] = d[\mathfrak{M}_0''] = d \cdot \phi^*[2x + x_1 + \ldots + x_{n-2}].$$

By (15.25), we obtain

$$\begin{aligned}
(15.26) \quad \langle \gamma \rangle_{0,d\beta_2} &= 2\langle K_X, C_1\rangle \cdot [\mathfrak{M}_1''] \cdot (\tilde{f}_{1,0})^*c_{2d-2}(\mathcal{E}) \\
&= 2\langle K_X, C_1\rangle \cdot (\tilde{f}_{1,0})_*[\mathfrak{M}_1''] \cdot c_{2d-1}(\mathcal{V}) \\
&= 2d\langle K_X, C_1\rangle \cdot \phi^*[2x + x_1 + \ldots + x_{n-2}] \cdot c_{2d-2}(\mathcal{E}) \\
&= 2d\langle K_X, C_1\rangle \cdot c_{2d-2}(\mathcal{E}|_{\phi^{-1}(2x+x_1+\ldots+x_{n-2})}).
\end{aligned}$$

By Remark 15.6,

$$\mathcal{E}|_{\phi^{-1}(2x+x_1+\ldots+x_{n-2})} \cong R^1(f_{1,0})_*(\mathrm{ev}_1)^*(\mathcal{O}_{\mathbb{P}^1}(-1) \oplus \mathcal{O}_{\mathbb{P}^1}(-1))$$

where $f_{1,0}$ and ev_1 denote the forgetful map and the evaluation map from the moduli space $\overline{\mathfrak{M}}_{0,1}(\mathbb{P}^1, d[\mathbb{P}^1])$ to $\overline{\mathfrak{M}}_{0,0}(\mathbb{P}^1, d[\mathbb{P}^1])$ and $\mathbb{P}^1$ respectively. By the Theorem 9.2.3 in [**CK**],

$$c_{2d-2}(R^1(f_{1,0})_*(\mathrm{ev}_1)^*(\mathcal{O}_{\mathbb{P}^1}(-1) \oplus \mathcal{O}_{\mathbb{P}^1}(-1))) = 1/d^3.$$

So we have

$$\begin{aligned} & c_{2d-2}(\mathcal{E}|_{\phi^{-1}(2x+x_1+\ldots+x_{n-2})}) \\ = \; & c_{2d-2}(R^1(f_{1,0})_*(\mathrm{ev}_1)^*(\mathcal{O}_{\mathbb{P}^1}(-1) \oplus \mathcal{O}_{\mathbb{P}^1}(-1)) \\ = \; & 1/d^3. \end{aligned}$$

Combining this with (15.26), we conclude that $\langle \gamma \rangle_{0,d\beta_n} = 2\langle K_X, C_1\rangle/d^2$. □

LEMMA 15.8. *Let $d \geq 1$. If $\gamma = \mathfrak{s}_{n,2}$ or $\mathfrak{s}_{n,3}$, then $\langle \gamma \rangle_{0,d\beta_n} = 0$.*

PROOF. Since similar argument works for $\mathfrak{s}_{n,2}$, we will only prove the lemma for $\mathfrak{s}_{n,3}$. So assume $\gamma = \mathfrak{s}_{n,3}$. Let $x_1, \ldots, x_{n-2} \in X$ be fixed distinct points on X contained in a small analytic open subset U of X. We may assume that U is independent of the smooth surface X. Let $\mathfrak{U}_1' \subset \overline{\mathfrak{M}}_{0,1}(X^{[n]}, d\beta_n)$ be the analytic open subset consisting of all stable maps $[\mu\colon (D;p) \to X^{[n]}]$ with $\mu(p) \in U^{[n]}$. Since $\mu_*(D) \sim d\beta_n$, we see that $\mathrm{Supp}(\mu(D)) = \mathrm{Supp}(\mu(p))$ for

$$[\mu\colon (D;p) \to X^{[n]}] \in \overline{\mathfrak{M}}_{0,1}(X^{[n]}, d\beta_n).$$

So $\mu(D) \subset U^{[n]}$, and $\mathfrak{U}_1'$ is independent of X.

Next, recall from (15.3) that $\mathfrak{s}_{n,3}$ is represented by $M_3(x_1) + x_2 + \ldots + x_{n-2}$. Let $\mathfrak{M}_1 \subset \overline{\mathfrak{M}}_{0,1}(X^{[n]}, d\beta_n)$ be the closed subset consisting of all stable maps $[\mu\colon (D;p) \to X^{[n]}]$ with $\mu(p) \in M_3(x_1) + x_2 + \ldots + x_{n-2}$. Then, $\mathfrak{M}_1 \subset \mathfrak{U}_1'$ since $M_3(x_1) + x_2 + \ldots + x_{n-2} \subset U^{[n]}$. In addition, since $\mathrm{Supp}(\mu(D)) = \mathrm{Supp}(\mu(p))$, we must have

$$\mu(D) \subset M_3(x_1) + x_2 + \ldots + x_{n-2}$$

for every $[\mu\colon (D;p) \to X^{[n]}] \in \mathfrak{M}_1$. So $\mathfrak{M}_1$ is independent of X. Thus the pull-back $\mathrm{ev}_1^*(\gamma)$ is also independent of X.

In summary, $\mathfrak{M}_1 \subset \mathfrak{U}_1'$, $\mathfrak{U}_1'$ is analytic open in $\overline{\mathfrak{M}}_{0,1}(X^{[n]}, d\beta_n)$, and $\mathfrak{M}_1$ and $\mathfrak{U}_1'$ are independent of X. It follows from the constructions of the virtual fundamental cycle in [**LT2, LT3**] that the restriction $[\overline{\mathfrak{M}}_{0,1}(X^{[n]}, d\beta_n)]^{\mathrm{vir}}|_{\mathfrak{M}_1}$ is independent of the smooth surface X. So the 1-point Gromov-Witten invariant $\langle \gamma \rangle_{0,d\beta_n}$, which is defined to be $[\overline{\mathfrak{M}}_{0,1}(X^{[n]}, d\beta_n)]^{\mathrm{vir}} \cdot \mathrm{ev}_1^*(\gamma)$ with $\mathrm{ev}_1^*(\gamma)$ being independent of X, is independent of X as well. Since all the Gromov-Witten invariants $\langle \gamma_1, \ldots, \gamma_k \rangle_{0,\beta}$ with $\beta \neq 0$ for a $K3$ surface are zero, we conclude that $\langle \gamma \rangle_{0,d\beta_n} = 0$ for $d \geq 1$. □

The following is the main result in this section.

THEOREM 15.9. *Let X be a simply connected smooth projective surface. Let $n \geq 2$, $d \geq 1$, and $\alpha_1, \alpha_2 \in H^2(X)$.*

(i) *If $\gamma = \mathfrak{s}_{n,1}, \mathfrak{s}_{\alpha_1,\alpha_2}, \mathfrak{s}_{\alpha_1,1}, \mathfrak{s}_{n,2}$ or $\mathfrak{s}_{n,3}$, then*

$$\langle \gamma \rangle_{0,d\beta_n} = 0.$$

(ii) *If $\gamma = \mathfrak{s}_{\alpha_1,2}$, then*

$$\langle \gamma \rangle_{0,d\beta_n} = \frac{2}{d^2} \langle K_X, \alpha_1 \rangle.$$

PROOF. Follows immediately from Lemma 15.7 and Lemma 15.8. □

EXAMPLE 15.10. Let X be a simply connected smooth projective surface. Let us consider all the genus-0 extremal Gromov-Witten invariants of $X^{[2]}$:

$$\langle \gamma_1, \ldots, \gamma_k \rangle_{0,d\beta_2} \tag{15.27}$$

where $k \geq 0$, $d > 0$, and $\gamma_1, \ldots, \gamma_k \in H^*(X^{[2]})$ are homogeneous classes. By the Fundamental Class Axiom (12.6) and the Divisor Axiom (12.7), we may assume $|\gamma_i| \geq 4$ for every i. In view of the degree constraint (12.4) and the expected dimension (12.3), we have

$$\frac{1}{2}\sum_{i=1}^{k} |\gamma_i| = -\big(K_{X^{[2]}} \cdot d\beta_2\big) + \big(\dim(X^{[2]}) - 3\big) + k = 1 + k.$$

It follows that $k = 1$. In other words, all the genus-0 extremal Gromov-Witten invariants of $X^{[2]}$ are reduced to the 1-point genus-0 extremal Gromov-Witten invariants $\langle \gamma \rangle_{0,d\beta_n}$ which is given by Theorem 15.9.

The above calculation of the 1-point genus-0 extremal Gromov-Witten invariants of the Hilbert scheme $X^{[n]}$ follows a somewhat standard process. However, for $k \geq 2$, the determination of the k-point genus-0 extremal Gromov-Witten invariants of $X^{[n]}$ requires completely new ideas. The case $r = 2$ is worked out in [**LL**] which will be presented in the next section.

15.2. 2-point genus-0 extremal invariants of J. Li and W.-P. Li

In this section, X denotes a simply connected smooth projective complex surface. We will survey the results in [**LL**] regarding the 2-point genus-0 extremal Gromov-Witten invariants of the Hilbert scheme $X^{[n]}$ and the quantum corrected boundary operator. We will refer to [**LL**] for the detailed proofs of these results except for the proof of Proposition 15.16, which is presented here in a slightly different way.

We begin with some notations. As in the paragraph after Lemma 13.2, we fix a non-trivial meromorphic section θ of $\mathcal{O}_X(K_X)$ whose vanishing and pole divisors are denoted by D_0 and D_∞ respectively. Let (μ^1, μ^2, μ^3) denote a triple of partitions with $|\mu^1| + |\mu^2| + |\mu^3| = n$. Let $r = \ell(\mu^1)$, $s = \ell(\mu^2)$ and $t = \ell(\mu^3)$. For cohomology classes $\mathbf{c}_1, \ldots, \mathbf{c}_s \in H^2(X)$, define the class $A^\mu_{\mathbf{c}} \in H^*(X^{[n]})$ by

$$A^\mu_{\mathbf{c}} = \prod_{i=1}^{r} \mathfrak{a}_{-\mu^1_i}(x) \cdot \prod_{j=1}^{s} \mathfrak{a}_{-\mu^2_j}(\mathbf{c}_j) \cdot \prod_{k=1}^{t} \mathfrak{a}_{-\mu^3_k}(1_X)|0\rangle. \tag{15.28}$$

Note from (3.11) that the cohomology degree of $A^\mu_{\mathbf{c}}$ is given by

$$\begin{aligned} |A^\mu_{\mathbf{c}}| &= \sum_{i=1}^{r}(2(\mu^1_i - 1) + 4) + \sum_{j=1}^{s}(2(\mu^2_j - 1) + 2) + \sum_{k=1}^{t} 2(\mu^3_k - 1) \\ &= 2(n + \ell(\mu^1) - \ell(\mu^3)). \end{aligned} \tag{15.29}$$

By Proposition 3.16, up to a scalar multiple, the cohomology class $A^\mu_{\mathbf{c}}$ is geometrically represented by the closure $W^\mu_{\mathbf{c}}$ of the subset

$$W(\mu^1_1, q_1; \ldots; \mu^1_r, q_r; \mu^2_1, C_1; \ldots; \mu^2_s, C_s; \mu^3_1, X; \ldots; \mu^3_t, X)^0 \tag{15.30}$$

where $q_1, \ldots, q_r$ are points and $C_1, \ldots, C_s$ are Riemann surfaces Poincaré dual to $\mathbf{c}_1, \ldots, \mathbf{c}_s \in H^2(X)$ respectively such that

$$D_0, D_\infty, q_1, \ldots, q_r, C_1, \ldots, C_s$$

are in general position. Similarly, we have the class $A^\lambda_{\mathbf{e}} \in H^*(X^{[n]})$ and its geometric representation $W^\lambda_{\mathbf{e}}$ for another triple $(\lambda^1, \lambda^2, \lambda^3)$ of partitions and cohomology classes $\mathbf{e}_1, \ldots, \mathbf{e}_b \in H^2(X)$ with $|\lambda^1| + |\lambda^2| + |\lambda^3| = n$, $a = \ell(\lambda^1)$,

$b = \ell(\lambda^2)$ and $c = \ell(\lambda^3)$. The analogues to $q_1, \ldots, q_r$ and $C_1, \ldots, C_s$ are denoted by $p_1, \ldots, p_a$ and $E_1, \ldots, E_b$, where $E_1, \ldots, E_b$ are Riemann surfaces Poincaré dual to $\mathbf{e}_1, \ldots, \mathbf{e}_b \in H^2(X)$ respectively. We assume that

$$D_0, D_\infty, q_1, \ldots, q_r, C_1, \ldots, C_s, p_1, \ldots, p_a, E_1, \ldots, E_b$$

are in general position.

The goal is to calculate the 2-point genus-0 extremal Gromov-Witten invariants

$$\langle A_{\mathbf{e}}^{\lambda}, A_{\mathbf{c}}^{\mu} \rangle_{0,d\beta_n}.$$

Assume that $\langle A_{\mathbf{e}}^{\lambda}, A_{\mathbf{c}}^{\mu} \rangle_{0,d\beta_n} \neq 0$. By (12.3), the expected dimension of the moduli space $\overline{\mathfrak{M}}_{0,2}(X^{[n]}, d\beta_n)$ is $2n-1$. So

$$2(n + \ell(\lambda^1) - \ell(\lambda^3)) + 2(n + \ell(\mu^1) - \ell(\mu^3)) = 2(2n-1)$$

in view of (12.4) and (15.29). Therefore, we must have

$$(\ell(\lambda^3) - \ell(\mu^1)) + (\ell(\mu^3) - \ell(\lambda^1)) = 1. \tag{15.31}$$

Moreover, by Proposition 13.3, $\langle A_{\mathbf{e}}^{\lambda}, A_{\mathbf{c}}^{\mu} \rangle_{0,d\beta_n} \neq 0$ implies that

$$\Lambda_\theta \cap \mathrm{ev}_1^{-1}(W_{\mathbf{e}}^{\lambda}) \cap \mathrm{ev}_2^{-1}(W_{\mathbf{c}}^{\mu}) \neq \emptyset \tag{15.32}$$

where ev_i is the i-th evaluation map from (12.2). A detailed analysis involving (15.32), (15.30) and the definition of Λ_θ in Proposition 13.3 leads to the following.

PROPOSITION 15.11. *Let $d \geq 1$. Assume that $\langle A_{\mathbf{e}}^{\lambda}, A_{\mathbf{c}}^{\mu} \rangle_{0,d\beta_n} \neq 0$. Then,*

$$\ell(\lambda^3) = \ell(\mu^1) + \delta \tag{15.33}$$

$$\ell(\mu^3) = \ell(\lambda^1) + (1-\delta) \tag{15.34}$$

where $\delta = 0$ or 1. If $\delta = 0$, then $\lambda^3 = \mu^1$, and there exists an integer $\ell = \mu_i^3 = \lambda_j^2$ for some i and j such that the partition λ^1 is obtained from μ^3 with ℓ deleted, and the partition μ^2 is obtained from λ^2 with ℓ deleted.

Let $A_{\mathbf{e}}^{\lambda-\lambda_j^2}$ be the cohomology class in $H^*(X^{[n-\lambda_j^2]})$ obtained from $A_{\mathbf{e}}^{\lambda}$ with the factor $\mathfrak{a}_{-\lambda_j^2}(\mathbf{e}_j)$ deleted. We similarly define $A_{\mathbf{e}}^{\lambda-\lambda_j^1}$, $A_{\mathbf{e}}^{\lambda-\lambda_i^2-\lambda_j^2}$, etc.

LEMMA 15.12. *Let notations be as in Proposition 15.11 with $\delta = 0$. Then there exists a universal function $c_{\ell,d}$ depending only on ℓ and d such that*

$$\langle A_{\mathbf{e}}^{\lambda}, A_{\mathbf{c}}^{\mu} \rangle_{0,d\beta_n} = \sum_{\ell=\mu_i^3=\lambda_j^2} \langle A_{\mathbf{e}}^{\lambda-\lambda_j^2}, A_{\mathbf{c}}^{\mu-\mu_i^3} \rangle \cdot \langle K_X, \mathbf{e}_j \rangle \cdot c_{\ell,d}. \tag{15.35}$$

In fact, the proof of Proposition 15.11 shows that $\langle A_{\mathbf{e}}^{\lambda}, A_{\mathbf{c}}^{\mu} \rangle_{0,d\beta_n}$ is equal to

$$\sum_{\ell=\mu_i^3=\lambda_j^2} \langle A_{\mathbf{e}}^{\lambda-\lambda_j^2}, A_{\mathbf{c}}^{\mu-\mu_i^3} \rangle \cdot \langle \mathfrak{a}_{-\ell}(\mathbf{e}_j)|0\rangle, \mathfrak{a}_{-\ell}(1_X)|0\rangle \rangle_{0,d\beta_\ell}. \tag{15.36}$$

By the Lemma 4.3 in [**LL**], there exists a universal function $c_{\ell,d}$ depending only on ℓ and d such that for every $\mathbf{e} \in H^2(X)$, we have

$$\langle \mathfrak{a}_{-\ell}(\mathbf{e})|0\rangle, \mathfrak{a}_{-\ell}(1_X)|0\rangle \rangle_{0,d\beta_\ell} = \langle K_X, \mathbf{e} \rangle \cdot c_{\ell,d}. \tag{15.37}$$

Combining (15.36) and (15.37) yields formula (15.35).

By the Divisor Axiom (12.7) and Theorem 1.24 (i), we have

$$\begin{aligned}\langle A^\lambda_{\mathbf{e}}, c_1(\mathcal{O}^{[n]}_X), A^\mu_{\mathbf{c}}\rangle_{0,d\beta_n} &= \big(c_1(\mathcal{O}^{[n]}_X)\cdot d\beta_n\big)\cdot \langle A^\lambda_{\mathbf{e}}, A^\mu_{\mathbf{c}}\rangle_{0,d\beta_n}\\ &= \big(-B_n/2\cdot d\beta_n\big)\cdot \langle A^\lambda_{\mathbf{e}}, A^\mu_{\mathbf{c}}\rangle_{0,d\beta_n}\\ &= d\cdot \langle A^\lambda_{\mathbf{e}}, A^\mu_{\mathbf{c}}\rangle_{0,d\beta_n}\end{aligned}$$

if $d \ge 1$. Hence we obtain the following corollary from Lemma 15.12.

COROLLARY 15.13. *Let λ and μ be as in Proposition 15.11 with $\delta = 0$. Then*

$$\begin{aligned}&\sum_{d\ge 0} \langle A^\lambda_{\mathbf{e}}, c_1(\mathcal{O}^{[n]}_X), A^\mu_{\mathbf{c}}\rangle_{0,d\beta_n}\, q^d\\ =\ & \langle A^\lambda_{\mathbf{e}}, c_1(\mathcal{O}^{[n]}_X), A^\mu_{\mathbf{c}}\rangle + \sum_{\substack{1\le i\le t\\ 1\le j\le b\\ \mu^3_i=\lambda^2_j=\ell}} \langle A^{\lambda-\lambda^2_j}_{\mathbf{e}}, A^{\mu-\mu^3_i}_{\mathbf{c}}\rangle \cdot \langle K_X, \mathbf{e}_j\rangle \cdot \sum_{d\ge 1} d\, c_{\ell,d}\, q^d.\end{aligned}$$

Next, recall the notation τ_{k*} from Definition 3.9. Consider the operator

$$\begin{aligned}&\mathbf{M}(q) \qquad (15.38)\\ =\ & \sum_{k>0}\left(\frac{k}{2}\frac{(-q)^k+1}{(-q)^k-1} - \frac{1}{2}\frac{(-q)+1}{(-q)-1}\right)\mathfrak{a}_{-k}\mathfrak{a}_k(\tau_{2*}K_X)\\ & -\frac{1}{2}\sum_{k,\ell>0}\big(\mathfrak{a}_{-k}\mathfrak{a}_{-\ell}\mathfrak{a}_{k+\ell} + \mathfrak{a}_{-k-\ell}\mathfrak{a}_k\mathfrak{a}_\ell\big)(\tau_{3*}[X])\\ =\ & \sum_{k>0}\left(\frac{k(-q)^k}{(-q)^k-1} - \frac{q}{1+q}\right)\mathfrak{a}_{-k}\mathfrak{a}_k(\tau_{2*}K_X) - \sum_{k>0}\frac{k-1}{2}\mathfrak{a}_{-k}\mathfrak{a}_k(\tau_{2*}K_X)\\ & -\frac{1}{2}\sum_{k,\ell>0}\big(\mathfrak{a}_{-k}\mathfrak{a}_{-\ell}\mathfrak{a}_{k+\ell} + \mathfrak{a}_{-k-\ell}\mathfrak{a}_k\mathfrak{a}_\ell\big)(\tau_{3*}[X]).\end{aligned}$$

By (4.19),

$$\mathbf{M}(0) = \mathfrak{d}.$$

In view of this and Theorem 15.17 below, $\mathbf{M}(q)$ is defined to be the *quantum corrected boundary operator*. The operator $\mathbf{M}(q)$ is closely related to the operator $\mathbf{M}(q, t_1, t_2)$ introduced in Theorem 14.2.

To compute $\mathbf{M}(q)(A^\mu_{\mathbf{c}})$, it suffices to calculate

$$\mathfrak{a}_{-k}\mathfrak{a}_k(\tau_{2*}K_X)(A^\mu_{\mathbf{c}}), \tag{15.39}$$

$$\mathfrak{a}_{-k}\mathfrak{a}_{-\ell}\mathfrak{a}_{k+\ell}(\tau_{3*}1_X)(A^\mu_{\mathbf{c}}), \tag{15.40}$$

$$\mathfrak{a}_{-k-\ell}\mathfrak{a}_k\mathfrak{a}_\ell(\tau_{3*}1_X)(A^\mu_{\mathbf{c}}) \tag{15.41}$$

where $k, \ell > 0$. Using Lemma 3.12, we see that $\mathfrak{a}_{-k}\mathfrak{a}_k(\tau_{2*}K_X)(A^\mu_{\mathbf{c}})$ is equal to

$$\sum_{\substack{1\le i\le s\\ \mu^2_i=k}} (-\mu^2_i)\mathfrak{a}_{-k}(K_X c_i)A^{\mu-\mu^2_i}_{\mathbf{c}} + \sum_{\substack{1\le i\le t\\ \mu^3_i=k}} (-\mu^3_i)\mathfrak{a}_{-k}(K_X)A^{\mu-\mu^3_i}_{\mathbf{c}}. \tag{15.42}$$

PROPOSITION 15.14. *If $\langle A^\lambda_{\mathbf{e}}, \mathbf{M}(q)(A^\mu_{\mathbf{c}})\rangle$ is not a constant function of q, then*

$$\ell(\lambda^3) = \ell(\mu^1) + \delta \tag{15.43}$$

$$\ell(\mu^3) = \ell(\lambda^1) + (1-\delta) \tag{15.44}$$

where $\delta = 0$ *or* 1. *If* $\delta = 0$, *then* $\lambda^3 = \mu^1$, *and there exists an integer* $\ell = \mu_i^3 = \lambda_j^2$ *for some* i *and* j *such that the partition* λ^1 *is obtained from* μ^3 *with* ℓ *deleted, and the partition* μ^2 *is obtained from* λ^2 *with* ℓ *deleted.*

Proposition 15.14 is parallel to Proposition 15.11. While the proof of Proposition 15.11 is geometric, the proof of Proposition 15.14 is algebraic, and follows from (15.42) and the Heisenberg commutation relation (3.9). The following is parallel to Corollary 15.13.

COROLLARY 15.15. *Let the partitions* λ *and* μ *be as in Proposition* 15.14 *with* $\delta = 0$. *Then* $\langle A_{\mathbf{e}}^{\lambda}, (\mathbf{M}(q) - \mathbf{M}(0))(A_{\mathbf{c}}^{\mu})\rangle$ *is equal to*

$$\sum_{\substack{1 \le i \le t \\ 1 \le j \le b \\ \mu_i^3 = \lambda_j^2 = \ell}} \langle A_{\mathbf{e}}^{\lambda - \lambda_j^2}, A_{\mathbf{c}}^{\mu - \mu_i^3}\rangle \cdot \langle K_X, \mathbf{e}_j\rangle \cdot (-1)^{\ell}\ell^2 \left(\frac{\ell(-q)^{\ell}}{(-q)^{\ell} - 1} - \frac{q}{1+q}\right).$$

Corollary 15.13 and Corollary 15.15 will be connected via the complex projective plane $\mathbb{P}^2$. To achieve this, let $X = \mathbb{P}^2$, $\mathbb{T} = (\mathbb{C}^*)^3$ and $H_{\mathbb{T}}^*(\mathrm{pt}) = \mathbb{C}[t_1, t_2, t_3]$. The group $\mathbb{T}$ acts on a point $[z_1, z_2, z_3] \in X$ by

$$(s_1, s_2, s_3) \cdot [z_1, z_2, z_3] = [s_1 z_1, s_2 z_2, s_3 z_3]$$

for $(s_1, s_2, s_3) \in \mathbb{T}$. There are three $\mathbb{T}$-fixed points:

$$q_1 = [1, 0, 0], \quad q_2 = [0, 1, 0], \quad q_3 = [0, 0, 1].$$

At q_1, under the identification $[z_1, z_2, z_3] = [1, z_2/z_1, z_3/z_1]$, $\mathbb{T}$ acts as

$$(s_1, s_2, s_3) \cdot (z_2/z_1, z_3/z_1) = (s_1^{-1}s_2\, z_2/z_1, s_1^{-1}s_3\, z_3/z_1).$$

Let N_i be the normal bundle of q_i in X. Then as a $\mathbb{T}$-module,

$$N_1 \cong T_1^{-1}T_2 \oplus T_1^{-1}T_3$$

where T_i is the 1-dimensional representation given by $(s_1, s_2, s_3) \to s_i$. Similarly,

$$\begin{aligned} N_2 &\cong T_2^{-1}T_1 \oplus T_2^{-1}T_3, \\ N_3 &\cong T_3^{-1}T_1 \oplus T_3^{-1}T_1. \end{aligned}$$

Arguments parallel to (11.22) show that for $1 \le i, j \le 3$, we have

$$[q_i] \cdot [q_j] = \delta_{i,j}[q_i] \cdot \prod_{k \in \{1,2,3\} - \{i\}} (t_k - t_i). \tag{15.45}$$

Let $L_1 \subset X$ be the line passing through q_2 and q_3, $L_2 \subset X$ be the line passing through q_1 and q_3, and $L_3 \subset X$ be the line passing through q_1 and q_2. The line L_3 is defined by the equation $z_3/z_1 = 0$ near q_1 and the equation $z_3/z_2 = 0$ near q_2. Therefore by the localization theorem, we have

$$\begin{aligned} [L_3] &= \frac{(t_2 - t_1)[q_1]}{(t_2 - t_1)(t_3 - t_1)} + \frac{(t_1 - t_2)[q_2]}{(t_1 - t_2)(t_3 - t_2)} \\ &= \frac{[q_1]}{t_3 - t_1} + \frac{[q_2]}{t_3 - t_2} \end{aligned} \tag{15.46}$$

in $H^*_{\mathbb{T}}(X)' = H^*_{\mathbb{T}}(X) \otimes_{\mathbb{C}[t_1,t_2,t_3]} \mathbb{C}(t_1,t_2,t_3)$. Similarly, we obtain

$$[L_1] = \frac{[q_2]}{t_1 - t_2} + \frac{[q_3]}{t_1 - t_3}, \tag{15.47}$$

$$[L_2] = \frac{[q_1]}{t_2 - t_1} + \frac{[q_3]}{t_2 - t_3}, \tag{15.48}$$

$$[X] = \frac{[q_1]}{(t_2 - t_1)(t_3 - t_1)} + \frac{[q_2]}{(t_1 - t_2)(t_3 - t_2)} + \frac{[q_3]}{(t_1 - t_3)(t_2 - t_3)}. \tag{15.49}$$

Using $-K_X = [L_1] + [L_2] + [L_3]$ and the above formulas, we see that $-K_X$ is equal to

$$\frac{(t_2 + t_3 - 2t_1)[q_1]}{(t_2 - t_1)(t_3 - t_1)} + \frac{(t_1 + t_3 - 2t_2)[q_2]}{(t_1 - t_2)(t_3 - t_2)} + \frac{(t_1 + t_2 - 2t_3)[q_3]}{(t_1 - t_3)(t_2 - t_3)}. \tag{15.50}$$

Let $Y_1, \ldots, Y_k$ be $\mathbb{T}$-equivariant subvarities of X. Put

$$\mathfrak{a}_{n_1} \cdots \mathfrak{a}_{n_k}([Y_1 \times \ldots \times Y_k]) = \mathfrak{a}_{n_1}([Y_1]) \cdots \mathfrak{a}_{n_k}([Y_k]).$$

PROPOSITION 15.16. *Let L be a line in the projective plane $X = \mathbb{P}^2$. Then,*

$$\begin{aligned} &\sum_{d \ge 0} \big\langle \mathfrak{a}_{-n}([L])|0\rangle, c_1(\mathcal{O}_X^{[n]}), \mathfrak{a}_{-n}([X])|0\rangle \big\rangle_{0,d\beta_n} q^d \\ = \;& \big\langle \mathfrak{a}_{-n}([L])|0\rangle, \mathbf{M}(q)(\mathfrak{a}_{-n}([X])|0\rangle)\big\rangle. \end{aligned} \tag{15.51}$$

PROOF. Note that $\mathbf{M}(0) = \mathfrak{d} = c_1(\mathcal{O}_X^{[n]}) \cup$. Thus, we have

$$\big\langle \mathfrak{a}_{-n}([L])|0\rangle, c_1(\mathcal{O}_X^{[n]}), \mathfrak{a}_{-n}([X])|0\rangle \big\rangle = \big\langle \mathfrak{a}_{-n}([L])|0\rangle, \mathbf{M}(0)(\mathfrak{a}_{-n}([X])|0\rangle)\big\rangle.$$

Therefore (15.51) is equivalent to the following:

$$\begin{aligned} &\sum_{d \ge 1} \big\langle \mathfrak{a}_{-n}([L])|0\rangle, c_1(\mathcal{O}_X^{[n]}), \mathfrak{a}_{-n}([X])|0\rangle \big\rangle_{0,d\beta_n} q^d \\ = \;& \big\langle \mathfrak{a}_{-n}([L])|0\rangle, (\mathbf{M}(q) - \mathbf{M}(0))(\mathfrak{a}_{-n}([X])|0\rangle)\big\rangle. \end{aligned} \tag{15.52}$$

Recall from Section 11.2 that the equivariant Heisenberg operators are denoted by $\mathfrak{p}_i(\alpha)$. Take $L = L_3$. For $d \ge 1$, we have

$$\begin{aligned} &\big\langle \mathfrak{a}_{-n}([L_3])|0\rangle, c_1(\mathcal{O}_X^{[n]}), \mathfrak{a}_{-n}([X])|0\rangle \big\rangle_{0,d\beta_n} \\ = \;& d\big\langle \mathfrak{a}_{-n}([L_3])|0\rangle, \mathfrak{a}_{-n}([X])|0\rangle \big\rangle_{0,d\beta_n} \\ = \;& d\big\langle \mathfrak{p}_{-n}([L_3])|0\rangle, \mathfrak{p}_{-n}([X])|0\rangle \big\rangle_{0,d\beta_n}. \end{aligned}$$

By (15.46) and (15.49), $\big\langle \mathfrak{a}_{-n}([L_3])|0\rangle, c_1(\mathcal{O}_X^{[n]}), \mathfrak{a}_{-n}([X])|0\rangle \big\rangle_{0,d\beta_n}$ is equal to

$$\begin{aligned} d\Big\langle &\mathfrak{p}_{-n}\Big(\frac{[q_1]}{t_3 - t_1} + \frac{[q_2]}{t_3 - t_2}\Big)|0\rangle, \\ &\mathfrak{p}_{-n}\Big(\frac{[q_1]}{(t_2 - t_1)(t_3 - t_1)} + \frac{[q_2]}{(t_1 - t_2)(t_3 - t_2)} + \frac{[q_3]}{(t_1 - t_3)(t_2 - t_3)}\Big)|0\rangle\Big\rangle_{0,d\beta_n}. \end{aligned}$$

A simple geometric argument shows that $\big\langle \mathfrak{p}_{-n}([q_i])|0\rangle, \mathfrak{p}_{-n}([q_j])|0\rangle \big\rangle_{0,d\beta_n} = 0$ if $i \ne j$. Thus for $d > 0$, $\big\langle \mathfrak{a}_{-n}([L_3])|0\rangle, c_1(\mathcal{O}_X^{[n]}), \mathfrak{a}_{-n}([X])|0\rangle \big\rangle_{0,d\beta_n}$ is equal to

$$\frac{d}{(t_2 - t_1)(t_3 - t_1)^2} \big\langle \mathfrak{p}_{-n}([q_1])|0\rangle, \mathfrak{p}_{-n}([q_1])|0\rangle \big\rangle_{0,d\beta_n}$$

$$+\frac{d}{(t_1-t_2)(t_3-t_2)^2}\big\langle \mathfrak{p}_{-n}([q_2])|0\rangle, \mathfrak{p}_{-n}([q_2])|0\rangle\big\rangle_{0,d\beta_n}. \tag{15.53}$$

The term $\big\langle \mathfrak{p}_{-n}([q_i])|0\rangle, \mathfrak{p}_{-n}([q_i])|0\rangle\big\rangle_{0,d\beta_n}$ has been computed by Theorem 14.2, (14.8) and (14.9):

$$\begin{aligned}
&\sum_{d\ge 1} d\big\langle \mathfrak{p}_{-n}([q_i])|0\rangle, \mathfrak{p}_{-n}([q_i])|0\rangle\big\rangle_{0,d\beta_n} q^d \\
= &\sum_{d\ge 1} \big\langle \mathfrak{p}_{-n}([q_i])|0\rangle, c_1(\mathcal{O}^{[n]}_{\mathbb{C}^2}), \mathfrak{p}_{-n}([q_i])|0\rangle\big\rangle_{0,d\beta_n} q^d \\
= &\big\langle \mathfrak{p}_{-n}([q_i])|0\rangle, \big(\widetilde{\mathbf{M}}_i(q)-\widetilde{\mathbf{M}}_i(0)\big)\mathfrak{p}_{-n}([q_i])|0\rangle\big\rangle
\end{aligned} \tag{15.54}$$

where the operators $\widetilde{\mathbf{M}}_i(q)$ are defined by

$$\begin{aligned}
\widetilde{\mathbf{M}}_1(q) &= -\frac{(t_2+t_3-2t_1)}{(t_2-t_1)(t_3-t_1)}\sum_{k>0}\left(\frac{k}{2}\frac{(-q)^k+1}{(-q)^k-1}-\frac{1}{2}\frac{(-q)+1}{(-q)-1}\right)\mathfrak{p}_{-k}([q_1])\mathfrak{p}_k([q_1]), \\
\widetilde{\mathbf{M}}_2(q) &= -\frac{(t_1+t_3-2t_2)}{(t_1-t_2)(t_3-t_2)}\sum_{k>0}\left(\frac{k}{2}\frac{(-q)^k+1}{(-q)^k-1}-\frac{1}{2}\frac{(-q)+1}{(-q)-1}\right)\mathfrak{p}_{-k}([q_2])\mathfrak{p}_k([q_2]), \\
\widetilde{\mathbf{M}}_3(q) &= -\frac{(t_1+t_2-2t_3)}{(t_1-t_3)(t_2-t_3)}\sum_{k>0}\left(\frac{k}{2}\frac{(-q)^k+1}{(-q)^k-1}-\frac{1}{2}\frac{(-q)+1}{(-q)-1}\right)\mathfrak{p}_{-k}([q_3])\mathfrak{p}_k([q_3]).
\end{aligned}$$

By (15.53) and (15.54),

$$\begin{aligned}
&\sum_{d\ge 1} \big\langle \mathfrak{a}_{-n}([L])|0\rangle, c_1(\mathcal{O}^{[n]}_X), \mathfrak{a}_{-n}([X])|0\rangle\big\rangle_{0,d\beta_n}\, q^d \\
= &\frac{1}{(t_2-t_1)(t_3-t_1)^2}\big\langle \mathfrak{p}_{-n}([q_1])|0\rangle, \big(\widetilde{\mathbf{M}}_1(q)-\widetilde{\mathbf{M}}_1(0)\big)\mathfrak{p}_{-n}([q_1])|0\rangle\big\rangle \\
&+\frac{1}{(t_1-t_2)(t_3-t_2)^2}\big\langle \mathfrak{p}_{-n}([q_2])|0\rangle, \big(\widetilde{\mathbf{M}}_2(q)-\widetilde{\mathbf{M}}_2(0)\big)\mathfrak{p}_{-n}([q_2])|0\rangle\big\rangle.
\end{aligned} \tag{15.55}$$

On the other hand, by (15.50), $\mathfrak{p}_{-k}\mathfrak{p}_k(\tau_{2*}K_X)$ is equal to

$$\begin{aligned}
&-\frac{(t_2+t_3-2t_1)}{(t_2-t_1)(t_3-t_1)}\mathfrak{p}_{-k}([q_1])\mathfrak{p}_k([q_1]) \\
&-\frac{(t_1+t_3-2t_2)}{(t_1-t_2)(t_3-t_2)}\mathfrak{p}_{-k}([q_2])\mathfrak{p}_k([q_2]) - \frac{(t_1+t_2-2t_3)}{(t_1-t_3)(t_2-t_3)}\mathfrak{p}_{-k}([q_3])\mathfrak{p}_k([q_3])
\end{aligned} \tag{15.56}$$

noting that $\tau_{2*}([q_i]) = [q_i]\times[q_i]$. By (15.46) and (15.49),

$$\begin{aligned}
&\big\langle \mathfrak{a}_{-n}([L])|0\rangle, (\mathbf{M}(q)-\mathbf{M}(0))(\mathfrak{a}_{-n}([X])|0\rangle)\big\rangle \\
= &\big\langle \mathfrak{p}_{-n}([L_3])|0\rangle, (\mathbf{M}(q)-\mathbf{M}(0))(\mathfrak{p}_{-n}([X])|0\rangle)\big\rangle \\
= &\left\langle \mathfrak{p}_{-n}\left(\frac{[q_1]}{t_3-t_1}+\frac{[q_2]}{t_3-t_2}\right)|0\rangle, (\mathbf{M}(q)-\mathbf{M}(0))\mathfrak{p}_{-n}\left(\frac{[q_1]}{(t_2-t_1)(t_3-t_1)}\right.\right. \\
&\left.\left.+\frac{[q_2]}{(t_1-t_2)(t_3-t_2)}+\frac{[q_3]}{(t_1-t_3)(t_2-t_3)}\right)|0\rangle\right\rangle.
\end{aligned}$$

Combining this with (15.38) and (15.56), we conclude that

$$\begin{aligned}
&\big\langle \mathfrak{a}_{-n}([L])|0\rangle, (\mathbf{M}(q) - \mathbf{M}(0))(\mathfrak{a}_{-n}([X])|0\rangle)\big\rangle \\
=\;& \frac{1}{(t_2-t_1)(t_3-t_1)^2}\big\langle \mathfrak{p}_{-n}([q_1])|0\rangle, \big(\widetilde{\mathbf{M}}_1(q) - \widetilde{\mathbf{M}}M_1(0)\big)\mathfrak{p}_{-n}([q_1])|0\rangle\big\rangle \\
&+ \frac{1}{(t_1-t_2)(t_3-t_2)^2}\big\langle \mathfrak{p}_{-n}([q_2])|0\rangle, \big(\widetilde{\mathbf{M}}_2(q) - \widetilde{\mathbf{M}}_2(0)\big)\mathfrak{p}_{-n}([q_2])|0\rangle\big\rangle.
\end{aligned}$$

In view of (15.55), we have proved the desired formula (15.52). □

The following determines all the 2-point genus-0 extremal Gromov-Witten invariants of $X^{[n]}$.

THEOREM 15.17. *Let X be a simply connected smooth projective complex surface. Then*

$$\sum_{d\geq 0}\big\langle A^{\lambda}_{\mathbf{e}}, c_1(\mathcal{O}_X^{[n]}), A^{\mu}_{\mathbf{c}}\big\rangle_{0,d\beta_n}\, q^d = \big\langle A^{\lambda}_{\mathbf{e}}, \mathbf{M}(q) A^{\mu}_{\mathbf{c}}\big\rangle. \tag{15.57}$$

We now sketch the proof of Theorem 15.17, and refer to [**LL**, Section 7] for the details. Since

$$\mathbf{M}(0) = \mathfrak{d} = \bigoplus_n c_1(\mathcal{O}_X^{[n]})\cup,$$

Proposition 15.11 and Proposition 15.14 reduce (15.57) to the case when $\lambda^3 = \mu^1$ and $\ell(\mu^3) = \ell(\lambda^1)+1$. By Corollary 15.13 and Corollary 15.15, the proof of (15.57) is further reduced to the proof of the following identity:

$$\sum_{d\geq 1} d\, c_{\ell,d}\, q^d = (-1)^{\ell}\ell^2\left(\frac{\ell(-q)^{\ell}}{(-q)^{\ell}-1} - \frac{q}{1+q}\right). \tag{15.58}$$

To prove (15.58), consider the special case when $X = \mathbb{P}^2$, $A^{\lambda}_{\mathbf{e}} = \mathfrak{a}_{-\ell}([L])$ and $A^{\mu}_{\mathbf{c}} = \mathfrak{a}_{-\ell}([X])$ where L denotes a line in X. By Corollary 15.13,

$$\sum_{d\geq 0}\big\langle A^{\lambda}_{\mathbf{e}}, c_1(\mathcal{O}_X^{[n]}), A^{\mu}_{\mathbf{c}}\big\rangle_{0,d\beta_n}\, q^d = \big\langle A^{\lambda}_{\mathbf{e}}, c_1(\mathcal{O}_X^{[n]}), A^{\mu}_{\mathbf{c}}\big\rangle - 3\sum_{d\geq 1} d\, c_{\ell,d}\, q^d.$$

Combining with Proposition 15.16, we conclude that

$$\begin{aligned}
&-3\sum_{d\geq 1} d\, c_{\ell,d}\, q^d \\
=\;& \big\langle \mathfrak{a}_{-\ell}([L])|0\rangle, \big(\mathbf{M}(q) - \mathbf{M}(0)\big)(\mathfrak{a}_{-\ell}([X])|0\rangle)\big\rangle \\
=\;& \left(\frac{\ell(-q)^{\ell}}{(-q)^{\ell}-1} - \frac{q}{1+q}\right)\big\langle \mathfrak{a}_{-\ell}([L])|0\rangle, \mathfrak{a}_{-\ell}\mathfrak{a}_{\ell}(\tau_{2*}K_X)(\mathfrak{a}_{-\ell}([X])|0\rangle)\big\rangle \\
=\;& 3\ell\left(\frac{\ell(-q)^{\ell}}{(-q)^{\ell}-1} - \frac{q}{1+q}\right)\big\langle \mathfrak{a}_{-\ell}([L])|0\rangle, \mathfrak{a}_{-\ell}([L])|0\rangle)\big\rangle.
\end{aligned}$$

Since $\big\langle \mathfrak{a}_{-\ell}([L])|0\rangle, \mathfrak{a}_{-\ell}([L])|0\rangle)\big\rangle = (-1)^{\ell-1}\ell$, formula (15.58) follows.

REMARK 15.18. Note that Theorem 15.9 is a special case of Theorem 15.17. However unlike the proof of Theorem 15.17, the proof of Theorem 15.9 is independent of Theorem 14.2 and is based on partial computation of the obstruction sheaf on the moduli space of stable maps to the Hilbert scheme $X^{[n]}$.

15.3. The structure of the genus-0 extremal Gromov-Witten invariants

In this section, we will sketch the argument in [**LQ5**] which leads to the structure of 3-point genus-0 extremal Gromov-Witten invariants of the Hilbert scheme $X^{[n]}$. It involves finer analysis of the virtual fundamental cycle using the method in [**LiJ**] and the cosection localization technique in Section 13.1.

First of all, we recall the concept of Hilbert schemes of α-points from [**LiJ**, Section 1]. Let Λ be a finite set with $|\Lambda| = n$. We define

$$\begin{aligned} X^{[\Lambda]} &= X^{[n]}, \\ X^{(\Lambda)} &= X^{(n)}, \end{aligned}$$

and for accounting purpose, denote

$$X^{\Lambda} = \{(x_a)_{a\in\Lambda} | \, x_a \in X\}.$$

Let

$$\rho_\Lambda = \rho_n : X^{[\Lambda]} \to X^{(\Lambda)}$$

be the Hilbert-Chow morphism. Using ρ_Λ, we define the *Hilbert scheme of* Λ*-points* to be

$$X^{[\![\Lambda]\!]} = X^{[\Lambda]} \times_{X^{(\Lambda)}} X^{\Lambda}. \tag{15.59}$$

These spaces $X^{[\![\Lambda]\!]}$ can be thought of as Hilbert schemes of ordered points.

Let $\mathcal{P}_\Lambda$ be the set of partitions or equivalence relations on Λ. When $\alpha \in \mathcal{P}_\Lambda$ consists of l equivalence classes $\alpha_1, \ldots, \alpha_l$, we write $\alpha = (\alpha_1, \ldots, \alpha_l)$. For such α, put

$$\begin{aligned} X^{(\alpha)} &= \prod_{i=1}^{l} X^{(\alpha_i)}, \\ X^{[\alpha]} &= \prod_{i=1}^{l} X^{[\alpha_i]}. \end{aligned}$$

Define the *Hilbert scheme* $X^{[\![\alpha]\!]}$ *of* α*-points* by

$$X^{[\![\alpha]\!]} = \prod_{i=1}^{l} X^{[\![\alpha_i]\!]}. \tag{15.60}$$

Note that

$$X^{[\![\alpha]\!]} = X^{[\alpha]} \times_{X^{(\alpha)}} X^{\Lambda}.$$

The spaces $X^{[\![\alpha]\!]}$ and $X^{[\![\beta]\!]}$ are birational. To make this precise, we first fix the convention on a partial ordering on $\mathcal{P}_\Lambda$. We define

$$\text{“}\alpha \ge \beta\text{”} \iff \text{“}a \sim_\beta b \Rightarrow a \sim_\alpha b\text{”}.$$

Namely, $\alpha \ge \beta$ if β is finer than α. When $\beta = (\beta_1, \ldots, \beta_r)$, we put

$$\alpha \wedge \beta = (\alpha_1 \cap \beta_1, \ldots, \alpha_l \cap \beta_r)$$

which is the largest element among all that are less than or equal to both α and β. Note that $\mathcal{P}_\Lambda$ contains a maximal and a minimal element. The maximal element is Λ consisting of a single equivalence class Λ; the minimal element is 1^Λ whose equivalence classes are single element sets.

For $\alpha > \beta \in \mathcal{P}_\Lambda$, define

$$\Xi^\alpha_\beta = \{x \in X^\Lambda \mid \exists\, a, b \in \Lambda \text{ so that } x_a = x_b,\ a \sim_\alpha b,\ a \not\sim_\beta b\}.$$

For $\alpha \ne \beta \in \mathcal{P}_\Lambda$, define $\Xi^\alpha_\beta = \Xi^\alpha_{\alpha\wedge\beta} \cup \Xi^\beta_{\alpha\wedge\beta}$. The discrepancy between $X^{[\![\alpha]\!]}$ and $X^{[\![\beta]\!]}$ (in $X^{[\![\alpha]\!]}$) and its complement are defined to be

$$\begin{aligned} \Xi^{[\![\alpha]\!]}_\beta &= X^{[\![\alpha]\!]} \times_{X^\Lambda} \Xi^\alpha_\beta, \\ X^{[\![\alpha]\!]}_\beta &= X^{[\![\alpha]\!]} - \Xi^{[\![\alpha]\!]}_\beta \end{aligned}$$

respectively. More precisely, by Lemma 1.2 in [**LiJ**], there exists a functorial open embedding

$$\zeta^\beta_\alpha : \quad X^{[\![\alpha]\!]}_\beta \to X^{[\![\beta]\!]}$$

induced by the universal property of the respective moduli spaces such that

$$\mathrm{Im}(\zeta^\beta_\alpha) = X^{[\![\beta]\!]}_\alpha.$$

Thus we obtain an isomorphism (equivalence) $\zeta^\beta_\alpha : X^{[\![\alpha]\!]}_\beta \xrightarrow{\cong} X^{[\![\beta]\!]}_\alpha$. We define

$$X^{[\![\le\alpha]\!]} = \Big(\coprod_{\beta\le\alpha} X^{[\![\beta]\!]}\Big)/\sim, \tag{15.61}$$

where the equivalence is by identifying $X^{[\![\beta]\!]}_\gamma \subset X^{[\![\beta]\!]}$ and $X^{[\![\gamma]\!]}_\beta \subset X^{[\![\gamma]\!]}$ via ζ^γ_β for all $\beta, \gamma \le \alpha$. Note that $X^{[\![\le\alpha]\!]}$ is non-separated (except when $\alpha = 1^\Lambda$), and contains the spaces $X^{[\![\le\beta]\!]}$, $\beta \le \alpha$, as open subschemes.

Next, let

$$X^{[n,d]} = \overline{\mathfrak{M}}_{0,3}(X^{[n]}, d\beta_n)$$

be the moduli space of 3-pointed genus-0 degree-$d\beta_n$ stable maps to $X^{[n]}$. Recall from (13.12) that every stable map $(\varphi, C) \in X^{[n,d]}$ has a standard decomposition

$$\varphi = (\varphi_1, \ldots, \varphi_l) \in X^{[n,d]}$$

where the stable reduction φ_i^{st} is contained in $X^{[n_i,d_i]}$ for some n_i and d_i,

$$\rho_{n_i}(\mathrm{Im}(\varphi_i)) = n_i x_i,$$

the points $x_1, \ldots, x_l$ are distinct, and $\varphi(p) = \sum_i \varphi_i(p)$ for all $p \in C$. We use the ideas from [**LiJ**, Section 1] to index the support of

$$\rho_n(\mathrm{Im}(\varphi)) = \sum_i n_i x_i \in X^{(n)}.$$

This is done by introducing the notion of 3-pointed genus-0 degree-δ α-stable maps to $X^{[n]}$, where $\alpha = (\alpha_1, \cdots, \alpha_l)$ denotes a partition of the set $[n] = \{1, \ldots, n\}$ and $\delta = (\delta_1, \cdots, \delta_l)$ with δ_i's being nonnegative integers. The set of such pairs (α, δ) with $\sum_i \delta_i = d$ is denoted by $\mathcal{P}_{[n],d}$. The techniques in [**LiJ**, Section 1] and the product formula [**Beh2**, (1)] for Gromov-Witten invariants enable us to express the virtual fundamental cycle $[X^{[n,d]}]^{\mathrm{vir}}$ in terms of certain discrepancy cycles

$$[\Theta^{[\![\alpha,\delta]\!]}], \quad (\alpha, \delta) \in \mathcal{P}_{[n],d}.$$

In fact, one of the key points in [**LQ5**] is to study such decomposition of $[X^{[n,d]}]^{\mathrm{vir}}$ as a sum of cycles indexed by the partition type of $\rho_n(\mathrm{Im}(\varphi)) \in X^{(n)}$. However, this cannot be done on the moduli space $X^{[n,d]}$. The technique to overcome this impasse is to use the Hilbert scheme $X^{[\![\alpha]\!]}$ of α-points defined in (15.60) and the space $X^{[\![\le\alpha]\!]}$ defined in (15.61). Then the cycle $\mathrm{ev}_*([X^{[n,d]}]^{\mathrm{vir}})$ is a sum of various $[\Theta^{[\![\alpha,\delta]\!]}]$ in $(X^{[\![\le[n]]\!]})^3$. Combining with the cosection localization theory in Section 13.1, pairings with $[\Theta^{[\![\alpha,\delta]\!]}]$ can be studied via C^∞-maps from X to the Grassmannians. For $d \geq 1$, we assemble those

$$[\Theta^{[\![\alpha,\delta]\!]}], \quad (\alpha, \delta) \in \mathcal{P}_{[n],d}$$

with $\delta_i > 0$ for every i into a homology class

$$\mathfrak{Z}_{n,d} \in H_*((X^{[n]})^3).$$

Note that we are back to the original Hilbert scheme $X^{[n]}$. Now the structure of the 3-point genus-0 extremal Gromov-Witten invariants of $X^{[n]}$ is given by the following two theorems from [**LQ5**, Section 1].

THEOREM 15.19. *Let X be a simply connected smooth projective surface. Let $A_1, A_2, A_3 \in H^*(X^{[n]})$ be Heisenberg monomial classes, and $\pi_{m,i}$ be the i-th projection on $(X^{[m]})^3$. Then, $\langle A_1, A_2, A_3\rangle_{0,d\beta_n}$ is equal to*

$$\sum_{m \leq n} \sum_{\substack{A_{1,1}\circ A_{1,2}=A_1 \\ A_{2,1}\circ A_{2,2}=A_2 \\ A_{3,1}\circ A_{3,2}=A_3}} \langle A_{1,1}, A_{2,1}, A_{3,1}\rangle \cdot \left\langle \mathfrak{Z}_{m,d}, \prod_{i=1}^{3} \pi_{m,i}^* A_{i,2} \right\rangle. \tag{15.62}$$

THEOREM 15.20. *Let X be a simply connected smooth projective surface. Let $A_1, A_2, A_3 \in H^*(X^{[n]})$ be Heisenberg monomial classes.*

(i) *If A_i contains a factor $\mathfrak{a}_{-j}(x)$ for some i, then*

$$\left\langle \mathfrak{Z}_{n,d}, \prod_{i=1}^{3} \pi_{n,i}^* A_i \right\rangle = 0.$$

(ii) *For $1 \leq i \leq 3$, let*

$$A_i = \mathfrak{a}_{-\lambda^{(i)}}(1_X)\mathfrak{a}_{-n_{i,1}}(\alpha_{i,1}) \cdots \mathfrak{a}_{-n_{i,u_i}}(\alpha_{i,u_i})|0\rangle$$

where $u_i \geq 0$ and $|\alpha_{i,1}| = \ldots = |\alpha_{i,u_i}| = 2$. Then,

$$\left\langle \mathfrak{Z}_{n,d}, \prod_{i=1}^{3} \pi_{n,i}^* A_i \right\rangle = \prod_{i=1}^{3} \prod_{j=1}^{u_i} \langle K_X, \alpha_{i,j}\rangle \cdot p \tag{15.63}$$

where p is a polynomial in $\langle K_X, K_X\rangle$ whose degree is at most

$$(n - \sum_{i,j} n_{i,j})/2,$$

and whose coefficients depend only on $d, n, \lambda^{(i)}, n_{i,j}$ (and hence are independent of the surface X and the classes $\alpha_{i,j}$).

We refer to Definition 3.13 for the operation $\circ$ appearing in (15.62), and to Definition 3.28 (i) for the notation $\mathfrak{a}_{-\lambda^{(i)}}(1_X)$ appearing in Theorem 15.20 (ii). Geometrically, we may think of the pairing

$$\left\langle \mathfrak{Z}_{m,d}, \prod_{i=1}^{3} \pi_{m,i}^* A_{i,2} \right\rangle$$

in (15.62) as the contributions of the non-constant components φ_i in the standard decomposition of $\varphi = (\varphi_1, \ldots, \varphi_l) \in X^{[n,d]}$, while those constant components φ_i contribute to the factor $\langle A_{1,1}, A_{2,1}, A_{3,1}\rangle$ in (15.62).

COROLLARY 15.21. *Let $d \geq 1$, and let $A_1, A_2, A_3 \in H^*(X^{[n]})$ be Heisenberg monomial classes.*

(i) *If $A_1 = \mathfrak{a}_{-1}(1_X)^{n-1}\mathfrak{a}_{-1}(\alpha)|0\rangle$, then*

$$\left\langle \mathfrak{Z}_{n,d}, \prod_{i=1}^{3} \pi_{n,i}^* A_i \right\rangle = 0.$$

(ii) *If $A_1 = \mathfrak{a}_{-1}(1_X)^{n-1-|\lambda|}\mathfrak{a}_{-1}(\alpha)\mathfrak{a}_{-\lambda}(x)|0\rangle$ for some λ, then*

$$\langle A_1, A_2, A_3 \rangle_{0,d\beta_n} = 0.$$

PROOF. (i) First of all, if $\alpha = x$, then

$$\left\langle \mathfrak{Z}_{n,d}, \prod_{i=1}^{3} \pi_{n,i}^* A_i \right\rangle = 0.$$

by Theorem 15.20 (i).

Next, let $\alpha = 1_X$. Use induction on n. Since $d \geq 1$, the conclusion is trivially true when $n = 1$. Let $n > 1$. Recall that $1/n! \cdot A_1$ is the fundamental class $1_{X^{[n]}}$ of $X^{[n]}$. By Theorem 15.19 and the Fundamental Class Axiom of Gromov-Witten theory,

$$\sum_{m=2}^{n} \sum_{\substack{A_{1,1} \circ A_{1,2} = A_1 \\ A_{2,1} \circ A_{2,2} = A_2 \\ A_{3,1} \circ A_{3,2} = A_3}} \langle A_{1,1}, A_{2,1}, A_{3,1} \rangle \cdot \left\langle \mathfrak{Z}_{m,d}, \prod_{i=1}^{3} \pi_{m,i}^* A_{i,2} \right\rangle = 0.$$

Since $A_1 = \mathfrak{a}_{-1}(1_X)^n|0\rangle$, we have $A_{1,2} = \mathfrak{a}_{-1}(1_X)^m|0\rangle$. By induction,

$$\left\langle \mathfrak{Z}_{m,d}, \prod_{i=1}^{3} \pi_{m,i}^* A_{i,2} \right\rangle = 0$$

if $2 \leq m \leq n-1$. It follows that

$$\left\langle \mathfrak{Z}_{n,d}, \prod_{i=1}^{3} \pi_{n,i}^* A_i \right\rangle = 0.$$

Now let $|\alpha| = 2$. By $\langle A_1, \beta_n \rangle = 0$ and the Divisor Axiom of Gromov-Witten theory, we have $\langle A_1, A_2, A_3 \rangle_{0,d\beta_n} = 0$. Using an argument similar to the one in the previous paragraph, we conclude that

$$\left\langle \mathfrak{Z}_{n,d}, \prod_{i=1}^{3} \pi_{n,i}^* A_i \right\rangle = 0.$$

(ii) We compute $\langle A_1, A_2, A_3 \rangle_{0,d\beta_n}$ by using (15.62). Note that the class $A_{1,2}$ in (15.62) is equal to $\mathfrak{a}_{-1}(1_X)^m|0\rangle$, or is equal to $\mathfrak{a}_{-1}(1_X)^{m-1}\mathfrak{a}_{-1}(\alpha)|0\rangle$, or contains a factor $\mathfrak{a}_{-i}(x)$ for some $i > 0$. By (i) and Theorem 15.20 (i), we get $\langle A_1, A_2, A_3 \rangle_{0,d\beta_n} = 0$. □

PROBLEM 15.22. *Let $n \geq 3$, $k \geq 4$ and $d \geq 1$. Let X be a simply connected smooth projective complex surface, and let $A_1, \ldots, A_k \in H^*(X^{[n]})$ be Heisenberg monomial classes. What is the structure of the genus-0 extremal Gromov-Witten invariant $\langle A_1, \ldots, A_k \rangle_{0,d\beta_n}$?*

We remark that when $k \geq 4$, the arguments in [**LQ5**, Section 5] do not go through since the moduli space $\overline{\mathfrak{M}}_{0,k}$ of genus-0 k-pointed stable curves has positive dimension.

CHAPTER 16

Ruan's Cohomological Crepant Resolution Conjecture

The celebrated Crepant Resolution Conjecture of Bryan-Graber [**BG**, Conjecture 1.2], Coates-Corti-Iritani-Tseng [**CCIT**, Conjecture 1.3], Coates-Ruan [**CR**, Conjecture 4.1], and Ruan [**Rua2**, Section 2] (also Batyrev-Borisov [**BB**], Etingof-Ginzburg [**EtG**, Section 1] and Reid [**Reid**]) has been extensively investigated from various aspects in recent years. It originated from the orbifold string theory in physics (notably, the work of Dixon-Harvey-Vafa-Witten [**DHVW**, **Zas**]). An orbifold Z is a topological space locally modeled on the quotient of Euclidean space by a finite group. Physicists attempted to formulate a string theory on Calabi-Yau orbifolds which could be equivalent to the string theory on its crepant resolutions (here a resolution $\rho : W \to Z$ is crepant if $K_W = \rho^* K_Z$). A consequence of this orbifold string theory is the prediction that orbifold quantum cohomology of an orbifold is isomorphic to the quantum cohomology of its crepant resolutions. Using the orbifold cohomology $H^*_{\rm orb}(Z)$ developed in [**CR1**, **CR2**] (we also refer to [**AGV**] for the algebraic setting), Ruan [**Rua2**, Section 2] formulated a more specific consequence of this orbifold string theory conjecture, known as the *Cohomological Crepant Resolution Conjecture*. It eventually involved into the Crepant Resolution Conjecture after the work of Bryan-Graber [**BG**, Conjecture 1.2], Coates-Corti-Iritani-Tseng [**CCIT**, Conjecture 1.3] and Coates-Ruan [**CR**, Conjecture 4.1]. Roughly speaking, *the Crepant Resolution Conjecture* predicts that the orbifold Gromov-Witten theory of an orbifold Z is ring isomorphic to (in the sense of analytic continuations, symplectic transformations and change of variables of type $q = ce^{iu}$) the ordinary cohomology ring of its crepant resolution W plus those quantum corrections on W which are related to curves contracted by the crepant resolution ρ. We refer to [**Coa**] and the references there for excellent examples confirming this conjecture.

In this chapter, we will concentrate on Ruan's Cohomological Crepant Resolution Conjecture for the Hilbert scheme $X^{[n]}$ where X denotes a smooth projective complex surface. We will prove Ruan's conjecture for the Hilbert-Chow morphism

$$\rho_n : X^{[n]} \to X^{(n)}$$

which is a crepant resolution of the global orbifold $X^{(n)}$ by Remark 1.25. The orbifold cohomology ring $H^*_{\rm orb}(X^{(n)})$ has been studied in Chapter 10. We will define the quantum corrected cohomology ring $H^*_{\rho_n}(X^{[n]})$ for the Hilbert scheme $X^{[n]}$. As vector spaces, $H^*_{\rho_n}(X^{[n]})$ and $H^*(X^{[n]})$ are the same. We will introduce the cohomology classes $\widetilde{G}_k(\alpha, n) \in H^*_{\rho_n}(X^{[n]})$ and the operators $\widetilde{\mathfrak{G}}_k(\alpha)$, which are the analogues of the cohomology classes $G_k(\alpha, n) \in H^*(X^{[n]})$ and the operators $\mathfrak{G}_k(\alpha)$ defined in Definition 4.1. From previous chapters, it is clear that the cohomology classes $G_k(\alpha, n) \in H^*(X^{[n]})$ and the operators $\mathfrak{G}_k(\alpha)$ play significant roles in understanding the ordinary cohomology ring $H^*(X^{[n]})$. Similarly, the classes $\widetilde{G}_k(\alpha, n) \in H^*_{\rho_n}(X^{[n]})$ and the operators $\widetilde{\mathfrak{G}}_k(\alpha)$ will play essential roles in studying the quantum corrected cohomology ring $H^*_{\rho_n}(X^{[n]})$. We will prove the identity

$$[\widetilde{\mathfrak{G}}_k(\alpha), \mathfrak{a}_{-1}(\beta)] = \frac{1}{k!}\, \mathfrak{a}_{-1}^{\{k\}}(\alpha\beta).$$

Following [**LQ5**, Section 6], an axiomatization originated from [**Leh1**, **LQW1**] and formulated in [**QW1**, Subsection 2.6] completes the proof of Ruan's conjecture. Theorem 16.5 and its proof are from [**LL**], while Theorem 16.20, Theorem 16.22 and their proofs are from [**LQ5**].

16.1. The quantum corrected cohomology ring $H^*_{\rho_n}(X^{[n]})$

In this section, we will define the quantum corrected product $w_1 \cdot_{\rho_n} w_2$ and the quantum corrected cohomology ring $H^*_{\rho_n}(X^{[n]})$ for the Hilbert scheme $X^{[n]}$ of points on a smooth projective complex surface X. Moreover, we will define the cohomology classes $\widetilde{G}_k(\alpha, n) \in H^*_{\rho_n}(X^{[n]})$ and the operators $\widetilde{\mathfrak{G}}_k(\alpha)$.

Recall the homology class $\beta_n \in H_2(X^{[n]}; \mathbb{Z})$ from (1.34). An irreducible curve $C \subset X^{[n]}$ is contracted to a point by the Hilbert-Chow morphism

$$\rho_n : X^{[n]} \to X^{(n)}$$

if and only if $C \sim d\beta_n$ for some positive integer d. Let q be a formal variable. Define a 3-point function with variable q for cohomology classes $w_1, w_2, w_3 \in H^*(X^{[n]})$:

$$\langle w_1, w_2, w_3 \rangle_{\rho_n}(q) = \sum_{d \ge 0} \langle w_1, w_2, w_3 \rangle_{0, d\beta_n}\, q^d \tag{16.1}$$

where $\langle w_1, w_2, w_3 \rangle_{0, d\beta_n}$ is the 3-point Gromov-Witten invariant defined by (12.5).

The following definition is adopted from [**Rua2**, Section 2] for the setting of the Hilbert schemes $X^{[n]}$.

DEFINITION 16.1. The *quantum corrected cohomology ring* $H^*_{\rho_n}(X^{[n]})$ is the group $H^*(X^{[n]})$ together with the *quantum corrected product* $w_1 \cdot_{\rho_n} w_2$ defined by

$$\langle w_1 \cdot_{\rho_n} w_2, w_3 \rangle = \langle w_1, w_2, w_3 \rangle_{\rho_n}(-1). \tag{16.2}$$

REMARK 16.2. A priori, it is not clear why the power series (16.1) should be convergent at $q = -1$ so that (16.2) makes sense. The convergence of (16.1) at $q = -1$ will follow from the proofs in Subsection 16.2 which reduce the convergence for an arbitrary surface X to the case of smooth toric surfaces. The convergence of (16.1) at $q = -1$ for smooth toric surfaces is a consequence of [**Cheo2**, Theorem 0.2] where a stronger version of Ruan's Cohomological Crepant Resolution Conjecture was proved for smooth toric surfaces.

Note that as complex vector spaces, $H^*_{\rho_n}(X^{[n]})$ and $H^*(X^{[n]})$ are the same. Thus,

$$\mathbb{H}_X = \bigoplus_{n=0}^{+\infty} H^*(X^{[n]}) = \bigoplus_{n=0}^{+\infty} H^*_{\rho_n}(X^{[n]})$$

as complex vector spaces. Moreover, we see from (16.2), (16.1) and the Fundamental Class Axiom (12.6) that

$$w \cdot_{\rho_n} 1_{X^{[n]}} = w$$

for every $w \in H^*_{\rho_n}(X^{[n]})$.

In the setting of the quantum corrected product, the Chern character operator $\mathfrak{G}_i(\alpha) \in \mathrm{End}(\mathbb{H}_X)$ and the cohomology class $G_i(\alpha, n) \in H^*(X^{[n]})$ defined in Definition 4.1 (iii) must be modified as follows.

DEFINITION 16.3. Let $k \geq 0$ and $\alpha \in H^*(X)$.

(i) Define the cohomology class $\widetilde{G}_k(\alpha, n) \in H^*_{\rho_n}(X^{[n]})$ to be

$$(16.3)\qquad \sum_{0 \leq j \leq k} \sum_{\substack{\lambda \vdash (j+1) \\ \ell(\lambda) = k-j+1}} \frac{(-1)^{|\lambda|-1}}{\lambda^! \cdot |\lambda|!} \cdot \mathbf{1}_{-(n-j-1)}\mathfrak{a}_{-\lambda}(\tau_*\alpha)|0\rangle$$

$$+ \sum_{0 \leq j \leq k} \sum_{\substack{\lambda \vdash (j+1) \\ \ell(\lambda) = k-j-1}} \frac{(-1)^{|\lambda|}}{\lambda^! \cdot |\lambda|!} \cdot \frac{|\lambda| + s(\lambda) - 2}{24} \cdot \mathbf{1}_{-(n-j-1)}\mathfrak{a}_{-\lambda}(\tau_*(e_X\alpha))|0\rangle.$$

(ii) Define $\widetilde{\mathfrak{G}}_k(\alpha) \in \mathrm{End}(\mathbb{H}_X)$ to be the operator on the space

$$\mathbb{H}_X = \bigoplus_{n=0}^{+\infty} H^*_{\rho_n}(X^{[n]})$$

which acts on the n-th component $H^*_{\rho_n}(X^{[n]})$ by the quantum corrected product $\widetilde{G}_k(\alpha, n)\cdot_{\rho_n}$.

REMARK 16.4. We have

$$\widetilde{G}_0(\alpha, n) = \mathbf{1}_{-(n-1)}\mathfrak{a}_{-1}(\alpha)|0\rangle = G_0(\alpha, n),$$
$$\widetilde{G}_1(\alpha, n) = -\frac{1}{2}\mathbf{1}_{-(n-2)}\mathfrak{a}_{-2}(\alpha)|0\rangle = G_1(\alpha, n).$$

In general, we see from Theorem 4.10 that the class $\widetilde{G}_k(\alpha, n)$ consists of those terms in $G_k(\alpha, n)$ which do not contain the canonical divisor K_X.

Note from the definition of the operator $\widetilde{\mathfrak{G}}_k(\alpha)$ that

$$(16.4)\qquad \langle \widetilde{\mathfrak{G}}_k(\alpha)w_1, w_2\rangle = \langle \widetilde{G}_k(\alpha, n) \cdot_{\rho_n} w_1, w_2\rangle = \langle \widetilde{G}_k(\alpha, n), w_1, w_2\rangle_{\rho_n}(-1)$$

for $w_1, w_2 \in H^*_{\rho_n}(X^{[n]})$. For convenience, we introduce the operator $\widetilde{\mathfrak{G}}_k(\alpha; q)$ by

$$(16.5)\qquad \langle \widetilde{\mathfrak{G}}_k(\alpha; q)w_1, w_2\rangle = \sum_{d \geq 0} \langle \widetilde{G}_k(\alpha, n), w_1, w_2\rangle_{0, d\beta_n}\, q^d.$$

From now on, let the surface X be simply connected. By Remark 16.4,

$$\widetilde{G}_1(1_X, n) = G_1(1_X, n) = c_1(\mathcal{O}_X^{[n]}).$$

The following was proved in [**LL**, Subsection 7.2].

THEOREM 16.5. *We have*

$$\widetilde{\mathfrak{G}}_1(1_X) = \bigoplus_n c_1(\mathcal{O}_X^{[n]})\cdot_{\rho_n} = -\frac{1}{6} : \mathfrak{a}^3 :_0 (\tau_{3*}1_X).$$

PROOF. First of all, recall from (14.19) that

$$(16.6)\qquad \lim_{q \to -1} \left(\frac{k}{2}\frac{(-q)^k + 1}{(-q)^k - 1} - \frac{1}{2}\frac{(-q) + 1}{(-q) - 1} \right) = 0.$$

Next, by (16.2) and Theorem 15.17,

$$\bigoplus_n c_1(\mathcal{O}_X^{[n]})\cdot_{\rho_n} = \mathbf{M}(-1).$$

So it remains to prove that

$$\mathbf{M}(-1) = -\frac{1}{6} : \mathfrak{a}^3 :_0 (\tau_{3*} 1_X).$$

In view of (4.17), this is equivalent to proving

$$\mathbf{M}(-1) = -\frac{1}{2} \sum_{k,\ell>0} \big(\mathfrak{a}_{-k}\mathfrak{a}_{-\ell}\mathfrak{a}_{k+\ell} + \mathfrak{a}_{-k-\ell}\mathfrak{a}_k\mathfrak{a}_\ell\big)(\tau_{3*}1_X). \tag{16.7}$$

Finally, (16.7) follows from (15.38) and (16.6). □

The operator $\widetilde{\mathfrak{G}}_1(1_X)$ for the quantum corrected product on $H^*_{\rho_n}(X^{[n]})$ is parallel to the boundary operator $\mathfrak{d}$ for the ordinary cup product on $H^*(X^{[n]})$. The following definition is parallel to Definition 3.23.

DEFINITION 16.6. For a linear operator $\mathfrak{f} \in \mathrm{End}(\mathbb{H}_X)$, put $\mathfrak{f}^{\{0\}} = \mathfrak{f}$. For $k \ge 1$, define $\mathfrak{f}^{\{k\}}$ inductively by

$$\mathfrak{f}^{\{k\}} = [\widetilde{\mathfrak{G}}_1(1_X), \mathfrak{f}^{\{k-1\}}].$$

The next lemma follows from Theorem 16.5 and the same proof of Theorem 3.31.

LEMMA 16.7. *Let $k \ge 0$, $m \in \mathbb{Z}$, and $\alpha \in H^*(X)$. Then, $\mathfrak{a}_m^{\{k\}}(\alpha)$ is equal to*

$$(-m)^k k! \left(\sum_{\ell(\lambda)=k+1, |\lambda|=m} \frac{1}{\lambda^!}\mathfrak{a}_\lambda(\tau_*\alpha) - \sum_{\ell(\lambda)=k-1, |\lambda|=m} \frac{s(\lambda)-1}{24\lambda^!}\mathfrak{a}_\lambda(\tau_*(e_X\alpha)) \right).$$

Comparing with Theorem 3.31, we see that the operator $\mathfrak{a}_m^{\{k\}}(\alpha)$ consists of those terms in $\mathfrak{a}_m^{(k)}(\alpha)$ which do not contain the canonical divisor K_X.

16.2. The commutator $[\widetilde{\mathfrak{G}}_k(\alpha), \mathfrak{a}_{-1}(\beta)]$

In this section, we will prove

$$[\widetilde{\mathfrak{G}}_k(\alpha), \mathfrak{a}_{-1}(\beta)] = \frac{1}{k!}\mathfrak{a}_{-1}^{\{k\}}(\alpha\beta) \tag{16.8}$$

for every simply connected smooth projective complex surface X. First of all, using [**Cheo2**], we will show that (16.8) holds for a smooth projective toric surface X. Then for an arbitrary surface X, using the results from Section 15.3 about the structure of 3-point genus-0 extremal Gromov-Witten invariants of $X^{[n]}$, we will prove (16.8) in three separate cases: $\alpha = x$ is the cohomology class of a point, $|\alpha| = 2$, and $\alpha = 1_X$.

We begin with the following lemma which verifies (16.8) when X is a smooth projective toric surface.

LEMMA 16.8. *Let X be a smooth projective toric surface. Then*

$$[\widetilde{\mathfrak{G}}_k(\alpha), \mathfrak{a}_{-1}(\beta)] = \frac{1}{k!}\mathfrak{a}_{-1}^{\{k\}}(\alpha\beta). \tag{16.9}$$

PROOF. Recall that $\mathbb{P}^2$ and the Hirzebruch surfaces $\mathbb{F}_a$ are smooth toric surfaces, and admit $\mathbb{T} = (\mathbb{C}^*)^2$-actions. By the Proposition in Subsection 2.5 of [**Ful**], X is obtained from $\mathbb{P}^2$ or $\mathbb{F}_a$ by a succession of blow-ups at $\mathbb{T}$-fixed points.

Now let $\mathfrak{a}_m^{\mathbb{T}}(\alpha), H^{*,\mathbb{T}}_{\rho_n}(X^{[n]})$ and $\mathfrak{p}_m^{\mathbb{T}}(\alpha), H^{*,\mathbb{T}}_{\text{orb}}(X^{(n)})$ be the equivariant versions of $\mathfrak{a}_m(\alpha), H^*_{\rho_n}(X^{[n]})$ and $\mathfrak{p}_m(\alpha), H^*_{\text{orb}}(X^{(n)})$ respectively. Following [**Cheo2**, Subsection 3.1], there exists a ring isomorphism

$$\Psi_n^{\mathbb{T}} : H^{*,\mathbb{T}}_{\text{orb}}(X^{(n)}) \to H^{*,\mathbb{T}}_{\rho_n}(X^{[n]})$$

defined by

$$\begin{aligned} & \Psi_n^{\mathbb{T}}\big(\sqrt{-1}^{n_1+\ldots+n_s-s}\mathfrak{p}^{\mathbb{T}}_{-n_1}(\alpha_1)\cdots\mathfrak{p}^{\mathbb{T}}_{-n_s}(\alpha_s)|0\rangle\big) \\ = \ & \mathfrak{a}^{\mathbb{T}}_{-n_1}(\alpha_1)\cdots\mathfrak{a}^{\mathbb{T}}_{-n_s}(\alpha_s)|0\rangle. \end{aligned}$$

Note that up to a scalar factor which depends only on the partition $\lambda = (n_1, \ldots, n_s)$ and the tuple $\overrightarrow{\alpha} = (\alpha_1, \ldots, \alpha_s)$, our notation $\mathfrak{p}^{\mathbb{T}}_{-n_1}(\alpha_1)\cdots\mathfrak{p}^{\mathbb{T}}_{-n_s}(\alpha_s)|0\rangle$ coincides with the notation $\lambda(\overrightarrow{\alpha})$ used in [**Cheo2**]. Also, our notation $\mathfrak{a}_m^{\mathbb{T}}(\alpha)$ coincides with the notation $\mathfrak{p}_m(\alpha)$ used in [**Cheo2**]. The integer $n_1 + \ldots + n_s - s$ is the age. Passing the map $\Psi_n^{\mathbb{T}}$ to the ordinary cohomology, we obtain a ring isomorphism

$$\Psi_n : H^*_{\text{orb}}(X^{(n)}) \to H^*_{\rho_n}(X^{[n]})$$

with

$$\begin{aligned} & \Psi_n\big(\sqrt{-1}^{n_1+\ldots+n_s-s}\mathfrak{p}_{-n_1}(\alpha_1)\cdots\mathfrak{p}_{-n_s}(\alpha_s)|0\rangle\big) \\ = \ & \mathfrak{a}_{-n_1}(\alpha_1)\cdots\mathfrak{a}_{-n_s}(\alpha_s)|0\rangle. \end{aligned}$$

Using (10.8) and (16.3), we see that $\Psi_n\big(\sqrt{-1}^k O_k(\alpha, n)\big) = \widetilde{G}_k(\alpha, n)$.

Next, let $A = \mathfrak{a}_{-n_1}(\alpha_1)\cdots\mathfrak{a}_{-n_s}(\alpha_s)|0\rangle \in H^*(X^{[n-1]})$. By definition,

$$\begin{aligned} [\widetilde{\mathfrak{G}}_k(\alpha), \mathfrak{a}_{-1}(\beta)]A &= \widetilde{\mathfrak{G}}_k(\alpha)\mathfrak{a}_{-1}(\beta)A - \mathfrak{a}_{-1}(\beta)\widetilde{\mathfrak{G}}_k(\alpha)A \\ &= \widetilde{\mathfrak{G}}_k(\alpha, n) \cdot_{\rho_n} \mathfrak{a}_{-1}(\beta)A - \mathfrak{a}_{-1}(\beta)\big(\widetilde{\mathfrak{G}}_k(\alpha, n-1) \cdot_{\rho_{n-1}} A\big). \end{aligned}$$

Put $P = \mathfrak{p}_{-n_1}(\alpha_1)\cdots\mathfrak{p}_{-n_s}(\alpha_s)|0\rangle$ and $a = n_1 + \ldots + n_s - s$. Let $\bullet$ denote the orbifold ring product. Then,

$$\begin{aligned} & \Psi_n\Big(\mathfrak{p}_{-1}(\beta)\big(\sqrt{-1}^k O_k(\alpha, n-1) \bullet \sqrt{-1}^a P\big)\Big) \\ = \ & \mathfrak{a}_{-1}(\beta)\Psi_n\Big(\sqrt{-1}^k O_k(\alpha, n-1) \bullet \sqrt{-1}^a P\Big) \\ = \ & \mathfrak{a}_{-1}(\beta)\big(\widetilde{\mathfrak{G}}_k(\alpha, n-1) \cdot_{\rho_{n-1}} A\big), \end{aligned}$$

and $\Psi_n\big(\sqrt{-1}^a \mathfrak{p}_{-1}(\beta)P\big) = \mathfrak{a}_{-1}(\beta)A$. So $[\widetilde{\mathfrak{G}}_k(\alpha), \mathfrak{a}_{-1}(\beta)]A$ is equal to

$$\begin{aligned} & \Psi_n\big(\sqrt{-1}^k O_k(\alpha, n) \bullet \sqrt{-1}^a \mathfrak{p}_{-1}(\beta)P\big) \\ - \ & \Psi_n\big(\mathfrak{p}_{-1}(\beta)\big(\sqrt{-1}^k O_k(\alpha, n-1) \bullet \sqrt{-1}^a P\big)\big). \end{aligned}$$

Since $O_k(\alpha, n) \bullet \mathfrak{p}_{-1}(\beta)P = \mathfrak{O}_k(\alpha)\mathfrak{p}_{-1}(\beta)P$, we obtain

$$[\widetilde{\mathfrak{G}}_k(\alpha), \mathfrak{a}_{-1}(\beta)]A = \sqrt{-1}^{k+a} \cdot \Psi_n\big([\mathfrak{O}_k(\alpha), \mathfrak{p}_{-1}(\beta)]P\big).$$

By Theorem 10.1 (iii), we conclude that $[\widetilde{\mathfrak{G}}_k(\alpha), \mathfrak{a}_{-1}(\beta)]A$ is equal to

$$\sqrt{-1}^{k+a} \cdot (-1)^k \cdot \Psi_n\left(\sum_{\substack{\ell(\lambda)=k+1 \\ |\lambda|=-1}} \frac{1}{\lambda^!}\mathfrak{p}_\lambda(\tau_*(\alpha\beta))P + \sum_{\substack{\ell(\lambda)=k-1 \\ |\lambda|=-1}} \frac{s(\lambda)-1}{24\lambda^!}\mathfrak{p}_\lambda(\tau_*(e_X\alpha\beta))P\right).$$

Finally, by the definition of Ψ_n and Lemma 16.7, $[\widetilde{\mathfrak{G}}_k(\alpha), \mathfrak{a}_{-1}(\beta)]A$ is equal to

$$\sum_{\substack{\ell(\lambda)=k+1 \\ |\lambda|=-1}} \frac{1}{\lambda!}\mathfrak{a}_\lambda(\tau_*(\alpha\beta))A - \sum_{\substack{\ell(\lambda)=k-1 \\ |\lambda|=-1}} \frac{s(\lambda)-1}{24\lambda!}\mathfrak{a}_\lambda(\tau_*(e_X\alpha\beta))A = \frac{1}{k!}\mathfrak{a}_{-1}^{\{k\}}(\alpha\beta)A.$$

Therefore,

$$[\widetilde{\mathfrak{G}}_k(\alpha), \mathfrak{a}_{-1}(\beta)] = \frac{1}{k!}\cdot \mathfrak{a}_{-1}^{\{k\}}(\alpha\beta).$$

This completes the proof of the lemma. □

To prove (16.8) for every simply connected smooth projective surface X, for simplicity, we put

$$\langle w_1, w_2, w_3\rangle_d = \langle w_1, w_2, w_3\rangle_{0,d\beta_n}.$$

In addition, w_1, w_2 and w_3 will stand for Heisenberg monomial classes.

Let $\alpha, \beta \in H^*(X)$. Then (16.8) is equivalent to

$$(16.10) \qquad \big\langle [\widetilde{\mathfrak{G}}_k(\alpha), \mathfrak{a}_{-1}(\beta)]w_1, w_2\big\rangle = \frac{1}{k!}\big\langle \mathfrak{a}_{-1}^{\{k\}}(\alpha\beta)w_1, w_2\big\rangle.$$

for all cohomology classes $w_1 \in H^*_{\rho_n}(X^{[n-1]}) = H^*(X^{[n-1]})$ and $w_2 \in H^*_{\rho_n}(X^{[n]}) = H^*(X^{[n]})$. Put

$$(16.11) \qquad D^\alpha_\beta(w_1, w_2; q) = \big\langle [\widetilde{\mathfrak{G}}_k(\alpha; q), \mathfrak{a}_{-1}(\beta)]w_1, w_2\big\rangle - \frac{1}{k!}\big\langle \mathfrak{a}_{-1}^{\{k\}}(\alpha\beta)w_1, w_2\big\rangle$$

where we have omitted k in $D^\alpha_\beta(w_1, w_2; q)$ since it will be clear from the context.

The following lemma gives an explicit expression to $D^\alpha_\beta(w_1, w_2; q)$, which enables us to analyze $D^\alpha_\beta(w_1, w_2; q)$.

LEMMA 16.9. *The difference $D^\alpha_\beta(w_1, w_2; q)$ is equal to*

$$\begin{aligned}
&\sum_{0\le j\le k} \sum_{\substack{\lambda\vdash(j+1) \\ \ell(\lambda)=k-j+1}} \frac{(-1)^{|\lambda|-1}}{\lambda!\cdot|\lambda|!} \sum_{d\ge 1} \Big(\big\langle \mathbf{1}_{-(n-j-1)}\mathfrak{a}_{-\lambda}(\tau_*\alpha)|0\rangle, \mathfrak{a}_{-1}(\beta)w_1, w_2\big\rangle_d \\
&\qquad - \big\langle \mathbf{1}_{-(n-j-2)}\mathfrak{a}_{-\lambda}(\tau_*\alpha)|0\rangle, w_1, \mathfrak{a}_{-1}(\beta)^\dagger w_2\big\rangle_d\Big) q^d \\
&+ \sum_{\epsilon\in\{K_X, K_X^2\}} \sum_{\substack{\ell(\lambda)=k+1-|\epsilon|/2 \\ |\lambda|=-1}} \tilde{f}_{|\epsilon|}(\lambda)\cdot \big\langle \mathfrak{a}_\lambda(\tau_*(\epsilon\alpha\beta))w_1, w_2\big\rangle \\
&- \sum_{\substack{\epsilon\in\{K_X, K_X^2\} \\ 0\le j\le k}} \sum_{\substack{\lambda\vdash(j+1) \\ \ell(\lambda)=k-j+1-|\epsilon|/2}} \tilde{g}_{|\epsilon|}(\lambda)\cdot \Big(\big\langle \mathbf{1}_{-(n-j-1)}\mathfrak{a}_{-\lambda}(\tau_*(\epsilon\alpha))|0\rangle, \mathfrak{a}_{-1}(\beta)w_1, w_2\big\rangle \\
&\qquad - \big\langle \mathbf{1}_{-(n-j-2)}\mathfrak{a}_{-\lambda}(\tau_*(\epsilon\alpha))|0\rangle, w_1, \mathfrak{a}_{-1}(\beta)^\dagger w_2\big\rangle\Big)
\end{aligned}$$

where $\mathfrak{a}_{-1}(\beta)^\dagger = -\mathfrak{a}_1(\beta)$ is the adjoint operator of $\mathfrak{a}_{-1}(\beta)$, and the functions $\tilde{f}_{|\epsilon|}(\lambda)$ and $\tilde{g}_{|\epsilon|}(\lambda)$ depend only on $k, |\epsilon|$ and λ.

PROOF. By (16.5), $\big\langle [\widetilde{\mathfrak{G}}_k(\alpha; q), \mathfrak{a}_{-1}(\beta)] w_1, w_2 \big\rangle$ is equal to

$$
\begin{aligned}
(16.12) \qquad & \big\langle \widetilde{\mathfrak{G}}_k(\alpha;q)\big(\mathfrak{a}_{-1}(\beta)w_1\big), w_2\big\rangle - \big\langle \mathfrak{a}_{-1}(\beta)\widetilde{\mathfrak{G}}_k(\alpha;q)(w_1), w_2\big\rangle \\
= \; & \big\langle \widetilde{\mathfrak{G}}_k(\alpha;q)\big(\mathfrak{a}_{-1}(\beta)w_1\big), w_2\big\rangle - \big\langle \widetilde{\mathfrak{G}}_k(\alpha;q)(w_1), \mathfrak{a}_{-1}(\beta)^\dagger w_2\big\rangle \\
= \; & \sum_{d \ge 0} \Big(\big\langle \widetilde{G}_k(\alpha, n), \mathfrak{a}_{-1}(\beta) w_1, w_2 \big\rangle_d \\
& - \big\langle \widetilde{G}_k(\alpha, n-1), w_1, \mathfrak{a}_{-1}(\beta)^\dagger w_2\big\rangle_d \Big) q^d.
\end{aligned}
$$

If $d \ge 1$, then we see from (16.3) and Corollary 15.21 (ii) that

$$
\begin{aligned}
(16.13) \qquad & \big\langle \widetilde{G}_k(\alpha, n), \mathfrak{a}_{-1}(\beta) w_1, w_2 \big\rangle_d \\
= & \sum_{0 \le j \le k} \sum_{\substack{\lambda \vdash (j+1) \\ \ell(\lambda) = k-j+1}} \frac{(-1)^{|\lambda|-1}}{\lambda^! \cdot |\lambda|!} \big\langle \mathbf{1}_{-(n-j-1)} \mathfrak{a}_{-\lambda}(\tau_* \alpha)|0\rangle, \mathfrak{a}_{-1}(\beta) w_1, w_2 \big\rangle_d.
\end{aligned}
$$

Similarly, if $d \ge 1$, then $\big\langle \widetilde{G}_k(\alpha, n-1), w_1, \mathfrak{a}_{-1}(\beta)^\dagger w_2\big\rangle_d$ is equal to

$$
(16.14) \qquad \sum_{0 \le j \le k} \sum_{\substack{\lambda \vdash (j+1) \\ \ell(\lambda) = k-j+1}} \frac{(-1)^{|\lambda|-1}}{\lambda^! \cdot |\lambda|!} \big\langle \mathbf{1}_{-(n-j-2)} \mathfrak{a}_{-\lambda}(\tau_* \alpha)|0\rangle, w_1, \mathfrak{a}_{-1}(\beta)^\dagger w_2 \big\rangle_d.
$$

Next, we study the two terms with $d = 0$ in (16.12). By (16.3) and Theorem 4.10, $\widetilde{G}_k(\alpha, n)$ is equal to

$$
G_k(\alpha, n) - \sum_{\substack{\epsilon \in \{K_X, K_X^2\} \\ 0 \le j \le k}} \sum_{\substack{\lambda \vdash (j+1) \\ \ell(\lambda) = k-j+1-|\epsilon|/2}} \tilde{g}_{|\epsilon|}(\lambda) \cdot \mathbf{1}_{-(n-j-1)} \mathfrak{a}_{-\lambda}(\tau_*(\epsilon\alpha))|0\rangle
$$

where $\tilde{g}_{|\epsilon|}(\lambda)$ depends only on k, $|\epsilon|$ and λ. Note that

$$
\begin{aligned}
& \langle G_k(\alpha, n), \mathfrak{a}_{-1}(\beta) w_1, w_2 \rangle - \langle G_k(\alpha, n-1), w_1, \mathfrak{a}_{-1}(\beta)^\dagger w_2 \rangle \\
= \; & \langle G_k(\alpha, n) \cdot \mathfrak{a}_{-1}(\beta) w_1, w_2 \rangle - \langle G_k(\alpha, n-1) \cdot w_1, \mathfrak{a}_{-1}(\beta)^\dagger w_2 \rangle \\
= \; & \langle \mathfrak{G}_k(\alpha) \mathfrak{a}_{-1}(\beta) w_1, w_2 \rangle - \langle \mathfrak{a}_{-1}(\beta) \mathfrak{G}_k(\alpha) w_1, w_2 \rangle \\
= \; & \big\langle [\mathfrak{G}_k(\alpha), \mathfrak{a}_{-1}(\beta)] w_1, w_2 \big\rangle.
\end{aligned}
$$

By (4.7), Theorem 3.31 and Lemma 16.7, we obtain

$$
\begin{aligned}
& \langle G_k(\alpha, n), \mathfrak{a}_{-1}(\beta) w_1, w_2 \rangle - \langle G_k(\alpha, n-1), w_1, \mathfrak{a}_{-1}(\beta)^\dagger w_2 \rangle \\
= \; & \frac{1}{k!} \langle \mathfrak{a}_{-1}^{(k)}(\alpha\beta) w_1, w_2 \rangle \\
= \; & \frac{1}{k!} \langle \mathfrak{a}_{-1}^{\{k\}}(\alpha\beta) w_1, w_2 \rangle + \sum_{\epsilon \in \{K_X, K_X^2\}} \sum_{\substack{\ell(\lambda) = k+1-|\epsilon|/2 \\ |\lambda| = -1}} \tilde{f}_{|\epsilon|}(\lambda) \cdot \big\langle \mathfrak{a}_\lambda(\tau_*(\epsilon\alpha\beta)) w_1, w_2 \big\rangle.
\end{aligned}
$$

Thus, $\big\langle \widetilde{G}_k(\alpha, n), \mathfrak{a}_{-1}(\beta) w_1, w_2 \big\rangle - \big\langle \widetilde{G}_k(\alpha, n-1), w_1, \mathfrak{a}_{-1}(\beta)^\dagger w_2 \big\rangle$ is equal to

$$
\begin{aligned}
& \frac{1}{k!} \langle \mathfrak{a}_{-1}^{\{k\}}(\alpha\beta) w_1, w_2 \rangle + \sum_{\epsilon \in \{K_X, K_X^2\}} \sum_{\substack{\ell(\lambda) = k+1-|\epsilon|/2 \\ |\lambda| = -1}} \tilde{f}_{|\epsilon|}(\lambda) \cdot \big\langle \mathfrak{a}_\lambda(\tau_*(\epsilon\alpha\beta)) w_1, w_2 \big\rangle \\
- & \sum_{\substack{\epsilon \in \{K_X, K_X^2\} \\ 0 \le j \le k}} \sum_{\substack{\lambda \vdash (j+1) \\ \ell(\lambda) = k-j+1-|\epsilon|/2}} \tilde{g}_{|\epsilon|}(\lambda) \cdot \Big(\big\langle \mathbf{1}_{-(n-j-1)} \mathfrak{a}_{-\lambda}(\tau_*(\epsilon\alpha))|0\rangle, \mathfrak{a}_{-1}(\beta) w_1, w_2 \big\rangle \\
& - \big\langle \mathbf{1}_{-(n-j-2)} \mathfrak{a}_{-\lambda}(\tau_*(\epsilon\alpha))|0\rangle, w_1, \mathfrak{a}_{-1}(\beta)^\dagger w_2 \big\rangle \Big).
\end{aligned}
$$

Combining this with (16.12), (16.13) and (16.14), we prove the lemma. $\square$

In the following, the proof of (16.8) will be divided into three cases:

- $\alpha = x$ is the cohomology class of a point,
- $|\alpha| = 2$,
- $\alpha = 1_X$.

We start with the simplest case when $\alpha = x$ and β is arbitrary.

PROPOSITION 16.10. *If $\alpha = x$ is the cohomology class of a point, then (16.8) is true.*

PROOF. By Corollary 15.21 (ii), every term in Lemma 16.9 is equal to zero. So $D^x_\beta(w_1, w_2; q) = 0$. In particular, $D^x_\beta(w_1, w_2; -1) = 0$. By (16.11),

$$[\widetilde{\mathfrak{G}}_k(x; -1), \mathfrak{a}_{-1}(\beta)] = \frac{1}{k!}\,\mathfrak{a}_{-1}^{\{k\}}(x\beta).$$

In view of (16.4) and (16.5), $\widetilde{\mathfrak{G}}_k(x) = \widetilde{\mathfrak{G}}_k(x; -1)$. Hence (16.8) is true when $\alpha = x$. $\square$

Next, we deal with the case $|\alpha| = 2$. We will study $D^\alpha_\beta(w_1, w_2; q)$ when $|\alpha| = 2$. We begin with two lemmas about the structures of the intersections in $H^*(X^{[n]})$.

LEMMA 16.11. *Let λ be a partition with $|\lambda| \le n$. For $i = 1$ and 2, let*

$$w_i = \mathfrak{a}_{-\lambda^{(i)}}(x)\mathfrak{a}_{-\mu^{(i)}}(1_X)\mathfrak{a}_{-n_{i,1}}(\alpha_{i,1}) \cdots \mathfrak{a}_{-n_{i,u_i}}(\alpha_{i,u_i})|0\rangle \tag{16.15}$$

where $|\alpha_{i,j}| = 2$ for all i and j. Then,

$$\begin{aligned} &\big\langle \mathfrak{a}_{-1}(1_X)^{n-|\lambda|}\mathfrak{a}_{-\lambda}(x)|0\rangle, w_1, w_2\big\rangle \\ &= \delta_{u_1,u_2} \cdot \sum_{\sigma \in \mathrm{Perm}\{1,\ldots,u_1\}} \prod_{i=1}^{u_1} \langle \alpha_{1,i}, \alpha_{2,\sigma(i)}\rangle \cdot p(\sigma) \end{aligned} \tag{16.16}$$

where $p(\sigma)$ depends only on σ, n, λ and all the $\lambda^{(i)}, \mu^{(i)}, n_{i,j}$.

PROOF. By Remark 8.10 (ii-a), $\mathfrak{a}_{-1}(1_X)^{n-|\lambda|}\mathfrak{a}_{-\lambda}(x)|0\rangle$ is a polynomial of the classes $G_k(x, n), k \ge 0$ whose coefficients are independent of X. In addition, the integers k involved depend only on λ. Note that

$$\begin{aligned} &\big\langle G_{k_1}(x, n) \cdots G_{k_l}(x, n), w_1, w_2\big\rangle \\ =\ &\big\langle \mathfrak{G}_{k_1}(x) \cdots \mathfrak{G}_{k_l}(x) w_1, w_2\big\rangle \\ =\ &\big\langle \mathfrak{a}_{-\lambda^{(1)}}(x)\mathfrak{a}_{-n_{1,1}}(\alpha_{1,1}) \cdots \mathfrak{a}_{-n_{1,u_1}}(\alpha_{1,u_1})\mathfrak{G}_{k_1}(x) \cdots \mathfrak{G}_{k_l}(x)\mathfrak{a}_{-\mu^{(1)}}(1_X)|0\rangle, w_2\big\rangle. \end{aligned}$$

So by Theorem 4.7 and Theorem 3.8, $\big\langle G_{k_1}(x, n) \cdots G_{k_l}(x, n), w_1, w_2\big\rangle$ is equal to

$$\delta_{u_1,u_2} \cdot \sum_{\sigma \in \mathrm{Perm}\{1,\ldots,u_1\}} \prod_{i=1}^{u_1} \langle \alpha_{1,i}, \alpha_{2,\sigma(i)}\rangle \cdot \tilde{p}(\sigma) \tag{16.17}$$

where $\tilde{p}(\sigma)$ depends only on $\sigma, n, k_1, \ldots, k_l$ and all the $\lambda^{(i)}, \mu^{(i)}, n_{i,j}$. $\square$

LEMMA 16.12. *Let $n_0 \geq 1$, $|\alpha| = 2$, and λ be a partition. Let w_1 and w_2 be given by (16.15). Then, $\langle \mathbf{1}_{-(n-|\lambda|-n_0)}\mathfrak{a}_{-\lambda}(x)\mathfrak{a}_{-n_0}(\alpha)|0\rangle, w_1, w_2\rangle$ is equal to*

$$\begin{aligned}
(16.18) \qquad & \langle K_X, \alpha\rangle \cdot \delta_{u_1,u_2} \cdot \sum_{\sigma_1 \in \mathrm{Perm}\{1,\ldots,u_1\}} \prod_{i=1}^{u_1} \langle \alpha_{1,i}, \alpha_{2,\sigma_1(i)}\rangle \cdot p_1(\sigma_1) \\
& + \sum_{j=1}^{u_1} \langle \alpha, \alpha_{1,j}\rangle \cdot \delta_{u_1-1,u_2} \cdot \sum_{\sigma_2} \prod_{i\neq j} \langle \alpha_{1,i}, \alpha_{2,\sigma_2(i)}\rangle \cdot p_2(\sigma_2) \\
& + \sum_{j=1}^{u_2} \langle \alpha, \alpha_{2,j}\rangle \cdot \delta_{u_1,u_2-1} \cdot \sum_{\sigma_3} \prod_{i=1}^{u_1} \langle \alpha_{1,i}, \alpha_{2,\sigma_3(i)}\rangle \cdot p_3(\sigma_3)
\end{aligned}$$

where σ_2 runs over all bijections $\{1,\ldots,u_1\} - \{j\} \to \{1,\ldots,u_2\}$, σ_3 runs over all bijections $\{1,\ldots,u_1\} \to \{1,\ldots,u_2\} - \{j\}$, and $p_1(\sigma_1)$ (respectively, $p_2(\sigma_2)$, $p_3(\sigma_3)$) depend only on σ_1 (respectively, σ_2, σ_3), n, n_0, λ and all the $\lambda^{(i)}, \mu^{(i)}, n_{i,j}$.

PROOF. By Remark 8.10 (ii-b), $\mathbf{1}_{-(n-|\lambda|-n_0)}\mathfrak{a}_{-\lambda}(x)\mathfrak{a}_{-n_0}(\alpha)|0\rangle$ can be written as

$$\langle K_X, \alpha\rangle \cdot F_1(n) + \sum_i G_{k_i}(\alpha, n) \cdot F_{2,i}(n)$$

where $F_1(n)$ and $F_{2,i}(n)$ are polynomials of $G_k(x,n), k \geq 0$ whose coefficients are independent of n and α. Moreover, the integers k and k_i depend only on λ and n_0. Thus,

$$\begin{aligned}
(16.19) \qquad & \langle \mathbf{1}_{-(n-|\lambda|-n_0)}\mathfrak{a}_{-\lambda}(x)\mathfrak{a}_{-n_0}(\alpha)|0\rangle, w_1, w_2\rangle \\
= \;& \langle K_X, \alpha\rangle \cdot \langle F_1(n), w_1, w_2\rangle + \sum_i \langle G_{k_i}(\alpha, n) \cdot F_{2,i}(n), w_1, w_2\rangle.
\end{aligned}$$

As in the proof of Lemma 16.11, $\langle F_1(n), w_1, w_2\rangle$ is of the form

$$(16.20) \qquad \delta_{u_1,u_2} \cdot \sum_{\sigma_1 \in \mathrm{Perm}\{1,\ldots,u_1\}} \prod_{i=1}^{u_1} \langle \alpha_{1,i}, \alpha_{2,\sigma_1(i)}\rangle \cdot \tilde{p}_{1,1}(\sigma_1)$$

where $\tilde{p}_{1,1}(\sigma_1)$ depends only on $\sigma_1, n, n_0, \lambda$ and all the $\lambda^{(i)}, \mu^{(i)}, n_{i,j}$. Also,

$$\begin{aligned}
& \langle G_{k_i}(\alpha, n) G_{s_1}(x,n) \cdots G_{s_l}(x,n), w_1, w_2\rangle \\
= \;& \langle \mathfrak{G}_{s_1}(x) \cdots \mathfrak{G}_{s_l}(x) \mathfrak{G}_{k_i}(\alpha) w_1, w_2\rangle.
\end{aligned}$$

By Theorem 4.7 and Lemma 3.12, $\langle G_{k_i}(\alpha, n) G_{s_1}(x,n) \cdots G_{s_l}(x,n), w_1, w_2\rangle$ is equal to

$$\begin{aligned}
(16.21) \qquad & \langle K_X, \alpha\rangle \cdot \delta_{u_1,u_2} \cdot \sum_{\sigma_1 \in \mathrm{Perm}\{1,\ldots,u_1\}} \prod_{i=1}^{u_1} \langle \alpha_{1,i}, \alpha_{2,\sigma_1(i)}\rangle \cdot \tilde{p}_{1,2}(\sigma_1) \\
& + \sum_{j=1}^{u_1} \langle \alpha, \alpha_{1,j}\rangle \cdot \delta_{u_1-1,u_2} \cdot \sum_{\sigma_2} \prod_{i\neq j} \langle \alpha_{1,i}, \alpha_{2,\sigma_2(i)}\rangle \cdot \tilde{p}_2(\sigma_2) \\
& + \sum_{j=1}^{u_2} \langle \alpha, \alpha_{2,j}\rangle \cdot \delta_{u_1,u_2-1} \cdot \sum_{\sigma_3} \prod_{i=1}^{u_1} \langle \alpha_{1,i}, \alpha_{2,\sigma_3(i)}\rangle \cdot \tilde{p}_3(\sigma_3)
\end{aligned}$$

where σ_2 runs over all the bijections $\{1, \ldots, u_1\} - \{j\} \to \{1, \ldots, u_2\}$, and σ_3 runs over all the bijections $\{1, \ldots, u_1\} \to \{1, \ldots, u_2\} - \{j\}$. Hence

$$\sum_i \big\langle G_{k_i}(\alpha, n) \cdot F_{2,i}(n), w_1, w_2 \big\rangle$$

is of the form (16.21) as well. Combining with (16.19) and (16.20), we obtain formula (16.18). □

Now we introduce the notion of universal polynomials

$$P(K_X, S_1, S_2)$$

in $\langle K_X, K_X \rangle$ of degree at most m and of type (u_1, u_2), and prove a vanishing lemma.

DEFINITION 16.13. Fix three integers $m, u_1, u_2 \geq 0$. Then *a universal polynomial $P(K_X, S_1, S_2)$ in $\langle K_X, K_X \rangle$ of degree at most m and of type (u_1, u_2)* is of the form

$$\sum_{\substack{1 \leq j_1 < \ldots < j_s \leq u_1 \\ 1 \leq l_1 < \ldots < l_s \leq u_2}} \prod_{i \notin \{j_1, \ldots, j_s\}} \langle K_X, \alpha_{1,i} \rangle \cdot \prod_{i \notin \{l_1, \ldots, l_s\}} \langle K_X, \alpha_{2,i} \rangle \tag{16.22}$$

$$\cdot \sum_{\sigma \in \mathrm{Perm}\{l_1, \ldots, l_s\}} \prod_{i=1}^{s} \langle \alpha_{1,j_i}, \alpha_{2,\sigma(l_i)} \rangle \cdot p(j_1, \ldots, j_s; l_1, \ldots, l_s; \sigma)$$

where $S_i = \{\alpha_{i,1}, \ldots, \alpha_{i,u_i}\} \subset H^2(X)$, and $p(j_1, \ldots, j_s; l_1, \ldots, l_s; \sigma)$ is a polynomial in $\langle K_X, K_X \rangle$ whose degree is at most m and whose coefficients are independent of X and the classes $\alpha_{i,j}$.

LEMMA 16.14. *Fix $m, u_1, u_2 \geq 0$. Let $P(K_X, S_1, S_2)$ be a universal polynomial in $\langle K_X, K_X \rangle$ of degree at most m and of type (u_1, u_2). Assume that*

$$P(K_X, S_1, S_2) = 0$$

whenever X is a smooth projective toric surface. Then $P(K_X, S_1, S_2) = 0$ for every smooth projective surface X.

PROOF. Let $r \gg m + u_1 + u_2$, and let X_r be a smooth toric surface obtained from $\mathbb{P}^2$ as an r-fold blown-up. Let L_0 be a line in $\mathbb{P}^2$, and let $e_1, \ldots, e_r$ be the exceptional divisors. Then,

$$K_{X_r} = -3L_0 + e_1 + \ldots + e_r.$$

For fixed $j_1, \ldots, j_s, l_1, \ldots, l_s$ and σ, let

$$\begin{aligned} \big\{\alpha_{1,i} \big| i \in \{1, \ldots, u_1\} - \{j_1, \ldots, j_s\}\big\} &= \big\{-e_1, \ldots, -e_{u_1 - s}\big\}, \\ \big\{\alpha_{2,i} \big| i \in \{1, \ldots, u_2\} - \{l_1, \ldots, l_s\}\big\} &= \big\{-e_{u_1 - s + 1}, \ldots, -e_{u_1 - s + u_2 - s}\big\}, \end{aligned}$$

and

$$\alpha_{1,j_i} = \alpha_{2,\sigma(l_i)} = e_{u_1 - s + u_2 - s + 2i} - e_{u_1 - s + u_2 - s + 2i - 1}$$

for $i = 1, \ldots, s$. Then,

$$0 = P(K_{X_r}, S_1, S_2) = (-2)^s \cdot p(j_1, \ldots, j_s; l_1, \ldots, l_s; \sigma) \tag{16.23}$$

by (16.22). It follows that $p(j_1, \ldots, j_s; l_1, \ldots, l_s; \sigma) = 0$ for all the surfaces X_r with $r \gg m + u_1 + u_2$. Since $p(j_1, \ldots, j_s; l_1, \ldots, l_s; \sigma)$ is a polynomial in $\langle K_{X_r}, K_{X_r} \rangle$ whose degree is at most m, we conclude that as polynomials,

$$p(j_1, \ldots, j_s; l_1, \ldots, l_s; \sigma) = 0.$$

Therefore, $P(K_X, S_1, S_2) = 0$ for every smooth projective surface X. □

Our next lemma is about the structure of certain 3-pointed extremal Gromov-Witten invariants, and provides the motivation for Definition 16.13.

LEMMA 16.15. *Let $d, n_0 \ge 1$ and $|\alpha| = 2$. Let w_1 and w_2 be given by (16.15). Then,*

$$\begin{aligned} &\big\langle \mathbf{1}_{-(n-|\lambda|-n_0)}\mathfrak{a}_{-\lambda}(x)\mathfrak{a}_{-n_0}(\alpha)|0\rangle, w_1, w_2\big\rangle_d \\ =\ & \langle K_X, \alpha\rangle \cdot P(K_X, S_1, S_2) \end{aligned} \tag{16.24}$$

where $S_1 = \{\alpha_{1,1}, \ldots, \alpha_{1,u_1}\}$, $S_2 = \{\alpha_{2,1}, \ldots, \alpha_{2,u_2}\}$, and $P(K_X, S_1, S_2)$ is a universal polynomial in $\langle K_X, K_X\rangle$ of degree at most $(n-n_0)/2$ and of type (u_1, u_2).

PROOF. For simplicity, let

$$w_0 = \mathbf{1}_{-(n-|\lambda|-n_0)}\mathfrak{a}_{-\lambda}(x)\mathfrak{a}_{-n_0}(\alpha)|0\rangle.$$

Also, for $i = 1$ and 2, let

$$\widetilde{w}_i = \mathfrak{a}_{-\mu^{(i)}}(1_X)\mathfrak{a}_{-n_{i,1}}(\alpha_{i,1})\cdots\mathfrak{a}_{-n_{i,u_i}}(\alpha_{i,u_i})|0\rangle.$$

We compute $\langle w_0, w_1, w_2\rangle_d$ by using (15.62). Consider the following term from (15.62):

$$\langle B_0, B_1, B_2\rangle \cdot \left\langle \mathfrak{Z}_{m,d},\ \pi_{m,1}^*\left(\frac{w_0}{B_0}\right)\cdot \pi_{m,2}^*\left(\frac{w_1}{B_1}\right)\cdot \pi_{m,3}^*\left(\frac{w_2}{B_2}\right)\right\rangle \tag{16.25}$$

where $m \le n$, $B_0, B_1, B_2 \in H^*(X^{[n-m]})$, $B_0 \subset w_0$, $B_1 \subset w_1$, and $B_2 \subset w_2$. By Theorem 15.20 (i) and Corollary 15.21 (i), such a term is nonzero only if

$$\begin{aligned} B_0 &= \mathfrak{a}_{-1}(1_X)^j\mathfrak{a}_{-\lambda}(x)|0\rangle, \\ B_1 &= \mathfrak{a}_{-\lambda^{(1)}}(x)\widetilde{B}_1, \\ B_2 &= \mathfrak{a}_{-\lambda^{(2)}}(x)\widetilde{B}_2 \end{aligned}$$

with $j \le (n-|\lambda|-n_0)$, $\widetilde{B}_1 \subset \widetilde{w}_1$, and $\widetilde{B}_2 \subset \widetilde{w}_2$. In this situation, (16.25) can be rewritten as

$$\big\langle \mathfrak{a}_{-1}(1_X)^j\mathfrak{a}_{-\lambda}(x)|0\rangle, \mathfrak{a}_{-\lambda^{(1)}}(x)\widetilde{B}_1, \mathfrak{a}_{-\lambda^{(2)}}(x)\widetilde{B}_2\big\rangle \tag{16.26}$$

$$\cdot\left\langle \mathfrak{Z}_{m,d},\ \pi_{m,1}^*\left(\frac{\mathbf{1}_{-(n-|\lambda|-n_0)}\mathfrak{a}_{-n_0}(\alpha)|0\rangle}{\mathfrak{a}_{-1}(1_X)^j|0\rangle}\right)\cdot \pi_{m,2}^*\left(\frac{\widetilde{w}_1}{\widetilde{B}_1}\right)\cdot \pi_{m,3}^*\left(\frac{\widetilde{w}_2}{\widetilde{B}_2}\right)\right\rangle. \tag{16.27}$$

Note that

$$\widetilde{B}_1 = \mathfrak{a}_{-\nu^{(1)}}(1_X)\mathfrak{a}_{-n_{1,j_1}}(\alpha_{1,j_1})\cdots\mathfrak{a}_{-n_{1,j_s}}(\alpha_{1,j_s})|0\rangle$$

for some $1 \le j_1 < \ldots < j_s \le u_1$ and some sub-partition $\nu^{(1)}$ of $\mu^{(1)}$ (i.e., every part of $\nu^{(1)}$ is a part of $\mu^{(1)}$). Similarly,

$$\widetilde{B}_2 = \mathfrak{a}_{-\nu^{(2)}}(1_X)\mathfrak{a}_{-n_{2,l_1}}(\alpha_{2,l_1})\cdots\mathfrak{a}_{-n_{2,l_t}}(\alpha_{2,l_t})|0\rangle$$

for some $1 \le l_1 < \ldots < l_t \le u_2$ and some sub-partition $\nu^{(2)}$ of $\mu^{(2)}$. By Lemma 16.11, (16.26) is equal to

$$\delta_{s,t}\cdot \sum_{\sigma\in \mathrm{Perm}\{l_1,\ldots,l_s\}} \prod_{i=1}^{s}\langle \alpha_{1,j_i}, \alpha_{2,\sigma(l_i)}\rangle \cdot p_1(j_1,\ldots,j_s; l_1,\ldots,l_s;\sigma) \tag{16.28}$$

where p_1 is a number independent of the surface X and the classes $\alpha_{i,j}$. By Theorem 15.20, we see that the factor (16.27) is equal to

$$\langle K_X, \alpha\rangle \cdot \prod_{i \notin \{j_1,\ldots,j_s\}} \langle K_X, \alpha_{1,i}\rangle \cdot \prod_{i \notin \{l_1,\ldots,l_s\}} \langle K_X, \alpha_{2,i}\rangle \cdot p_2(j_1,\ldots,j_s; l_1,\ldots,l_s;\sigma)$$

where p_2 is a polynomial in $\langle K_X, K_X\rangle$ whose degree is at most $(m-n_0)/2 \le (n-n_0)/2$, and whose coefficients are independent of the surface X and the classes $\alpha_{i,j}$. Combining this with (16.25), (16.26), (16.27) and (16.28), we obtain (16.24). □

PROPOSITION 16.16. *If $|\alpha| = 2$, then (16.8) is true.*

PROOF. Recall that (16.8) is equivalent to (16.10), and $D^\alpha_\beta(w_1, w_2; q)$ from (16.11) is computed by Lemma 16.9. Let w_1 and w_2 be given by (16.15). Let $u_1' = \delta_{2,|\beta|} + u_1$ and $S_2 = \{\alpha_{2,1}, \ldots, \alpha_{2,u_2}\}$. Let $S_1 = \{\alpha_{1,1}, \ldots, \alpha_{1,u_1}\}$ if $|\beta| \ne 2$, and $S_1 = \{\beta, \alpha_{1,1}, \ldots, \alpha_{1,u_1}\}$ if $|\beta| = 2$.

By Lemma 16.8 and Lemma 16.14, it suffices to prove that

$$D^\alpha_\beta(w_1, w_2; -1) = \langle K_X, \alpha\rangle \cdot P(K_X, S_1, S_2) \tag{16.29}$$

where $P(K_X, S_1, S_2)$ is a universal polynomial in $\langle K_X, K_X\rangle$ of degree at most $(n-1)/2$ and of type (u_1', u_2). This follows if we can prove that

$$D^\alpha_\beta(w_1, w_2; q) = \langle K_X, \alpha\rangle \cdot \sum_{d \ge 0} P(K_X, S_1, S_2; d)\, q^d \tag{16.30}$$

where every $P(K_X, S_1, S_2; d)$ is a universal polynomial in $\langle K_X, K_X\rangle$ of degree at most $(n-1)/2$ and of type (u_1', u_2). We remark that d has been inserted into the notation $P(K_X, S_1, S_2; d)$ to emphasis its dependence on d.

In the rest of the proof, we will show that the contribution of every term in Lemma 16.9 is of the form $P(K_X, S_1, S_2; d)$ for a suitable $d \ge 0$. Note that in $H^*(X^i)$,

$$\tau_{i*}(\alpha) = \alpha \otimes x \otimes \cdots \otimes x + x \otimes \alpha \otimes x \otimes \cdots \otimes x + \ldots + x \otimes \cdots \otimes x \otimes \alpha.$$

Thus, by Lemma 16.15, $\big\langle \mathbf{1}_{-(n-j-1)}\mathfrak{a}_{-\lambda}(\tau_*\alpha)|0\rangle, \mathfrak{a}_{-1}(\beta)w_1, w_2\big\rangle_d$ is equal to

$$\langle K_X, \alpha\rangle \cdot P_1(K_X, S_1, S_2; d)$$

where $P_1(K_X, S_1, S_2; d)$ is a universal polynomial in $\langle K_X, K_X\rangle$ of degree at most $(n-1)/2$ and of type (u_1', u_2). Similarly, since $\mathfrak{a}_{-1}(\beta)^\dagger w_2 = -\mathfrak{a}_1(\beta)w_2$, we see from Theorem 3.8 and Lemma 16.15 that

$$\big\langle \mathbf{1}_{-(n-j-2)}\mathfrak{a}_{-\lambda}(\tau_*\alpha)|0\rangle, w_1, \mathfrak{a}_{-1}(\beta)^\dagger w_2\big\rangle_d = \langle K_X, \alpha\rangle \cdot P_2(K_X, S_1, S_2; d).$$

Next, we move to the term $\big\langle \mathfrak{a}_\lambda(\tau_*(\epsilon\alpha\beta))w_1, w_2\big\rangle$ in Lemma 16.9, where $\epsilon \in \{K_X, K_X^2\}$. Such a term is zero unless $\epsilon = K_X$ and $|\beta| = 0$. In this case, we may assume that $\beta = 1_X$. So let $\epsilon = K_X$ and $\beta = 1_X$. Then,

$$\begin{aligned}\big\langle \mathfrak{a}_\lambda(\tau_*(\epsilon\alpha\beta))w_1, w_2\big\rangle &= \langle K_X, \alpha\rangle \cdot \big\langle \mathfrak{a}_\lambda(x)w_1, w_2\big\rangle \\ &= \langle K_X, \alpha\rangle \cdot P_3(K_X, S_1, S_2; 0).\end{aligned}$$

by Theorem 3.8, where $P_3(K_X, S_1, S_2; 0)$ is a universal polynomial in $\langle K_X, K_X\rangle$ of degree 0 (i.e., $\langle K_X, K_X\rangle$ does not appear) and of type (u_1', u_2).

Finally, $\tau_*(\epsilon\alpha)$ is zero unless $\epsilon = K_X$. Let $\epsilon = K_X$. By Lemma 16.11,

$$\begin{aligned} &\big\langle \mathbf{1}_{-(n-j-1)}\mathfrak{a}_{-\lambda}(\tau_*(\epsilon\alpha))|0\rangle, \mathfrak{a}_{-1}(\beta)w_1, w_2\big\rangle \\ = \quad &\langle K_X, \alpha\rangle \cdot \big\langle \mathbf{1}_{-(n-j-1)}\mathfrak{a}_{-\lambda}(x)|0\rangle, \mathfrak{a}_{-1}(\beta)w_1, w_2\big\rangle \\ = \quad &\langle K_X, \alpha\rangle \cdot P_4(K_X, S_1, S_2; 0) \end{aligned}$$

where $P_4(K_X, S_1, S_2; 0)$ is a universal polynomial in $\langle K_X, K_X\rangle$ of degree 0 and of type (u_1', u_2). Similarly, since $\mathfrak{a}_{-1}(\beta)^\dagger w_2 = -\mathfrak{a}_1(\beta)w_2$, we obtain

$$\big\langle \mathbf{1}_{-(n-j-2)}\mathfrak{a}_{-\lambda}(\tau_*(\epsilon\alpha))|0\rangle, w_1, \mathfrak{a}_{-1}(\beta)^\dagger w_2\big\rangle = \langle K_X, \alpha\rangle \cdot P_5(K_X, S_1, S_2; 0).$$

This completes the proof of our proposition. □

We are left with the case $\alpha = 1_X$. We start with two lemmas.

LEMMA 16.17. *Let $d, n_0 \geq 1$. Let w_1 and w_2 be given by (16.15). Then,*

$$\big\langle \mathbf{1}_{-(n-|\lambda|-n_0)}\mathfrak{a}_{-\lambda}(x)\mathfrak{a}_{-n_0}(1_X)|0\rangle, w_1, w_2\big\rangle_d = P(K_X, S_1, S_2) \tag{16.31}$$

where $S_1 = \{\alpha_{1,1}, \ldots, \alpha_{1,u_1}\}$, $S_2 = \{\alpha_{2,1}, \ldots, \alpha_{2,u_2}\}$, and $P(K_X, S_1, S_2)$ is a universal polynomial in $\langle K_X, K_X\rangle$ of degree at most $n/2$ and of type (u_1, u_2).

PROOF. This follows from the proof of Lemma 16.15 by replacing α by 1_X (and then by noticing that the factor $\langle K_X, \alpha\rangle$ there will not appear here). □

LEMMA 16.18. *Let $d \geq 1$ and $|\lambda| \leq n$. Let w_1 and w_2 be given by (16.15). Then,*

$$\big\langle \mathbf{1}_{-(n-|\lambda|)}\mathfrak{a}_{-\lambda}(\tau_* 1_X)|0\rangle, w_1, w_2\big\rangle_d = P(K_X, S_1, S_2) \tag{16.32}$$

where $S_1 = \{\alpha_{1,1}, \ldots, \alpha_{1,u_1}\}$, $S_2 = \{\alpha_{2,1}, \ldots, \alpha_{2,u_2}\}$, and $P(K_X, S_1, S_2)$ is a universal polynomial in $\langle K_X, K_X\rangle$ of degree at most $n/2$ and of type (u_1, u_2).

PROOF. For $i = 1$ and 2, let

$$\widetilde{w}_i = \mathfrak{a}_{-\mu^{(i)}}(1_X)\mathfrak{a}_{-n_{i,1}}(\alpha_{i,1}) \cdots \mathfrak{a}_{-n_{i,u_i}}(\alpha_{i,u_i})|0\rangle.$$

Note that if the Künneth decomposition of $\tau_{2*}1_X \in H^*(X^2)$ is given by

$$\tau_{2*}1_X = x \otimes 1_X + 1_X \otimes x + \sum_j \gamma_{j,1} \otimes \gamma_{j,2}$$

where $|\gamma_{j,1}| = |\gamma_{j,2}| = 2$, then up to permutations of factors, a typical term in the Künneth decomposition of $\tau_{i*}1_X \in H^*(X^i)$ with $i \geq 3$ is either $x \otimes \cdots \otimes x \otimes 1_X$ or $x \otimes \cdots \otimes x \otimes \gamma_{j,1} \otimes \gamma_{j,2}$. In view of Lemma 16.17, it suffices to verify that

$$\sum_j \big\langle \mathbf{1}_{-\tilde{n}}\mathfrak{a}_{-\tilde{\lambda}}(x)\mathfrak{a}_{-n_1}(\gamma_{j,1})\mathfrak{a}_{-n_2}(\gamma_{j,2})|0\rangle, w_1, w_2\big\rangle_d = P_1(K_X, S_1, S_2) \tag{16.33}$$

where $\tilde{n} = n - |\tilde{\lambda}| - n_1 - n_2$, and $P_1(K_X, S_1, S_2)$ is a universal polynomial in $\langle K_X, K_X\rangle$ of degree at most $n/2$ and of type (u_1, u_2). For simplicity, let

$$w_0 = \mathbf{1}_{-\tilde{n}}\mathfrak{a}_{-\tilde{\lambda}}(x)\mathfrak{a}_{-n_1}(\gamma_{j,1})\mathfrak{a}_{-n_2}(\gamma_{j,2})|0\rangle.$$

We see from (15.62) that to prove (16.33), it suffices to show that

$$\sum_j \langle B_0, B_1, B_2\rangle \cdot \left\langle \mathfrak{Z}_{m,d}, \ \pi_{m,1}^*\left(\frac{w_0}{B_0}\right) \cdot \pi_{m,2}^*\left(\frac{w_1}{B_1}\right) \cdot \pi_{m,3}^*\left(\frac{w_2}{B_2}\right)\right\rangle \tag{16.34}$$

is equal to $P_2(K_X, S_1, S_2)$, where $m \leq n$, $B_0, B_1, B_2 \in H^*(X^{[n-m]})$, $B_0 \subset w_0$, $B_1 \subset w_1$, and $B_2 \subset w_2$. By Theorem 15.20 (i) and Corollary 15.21 (i), such a term is nonzero only if $B_1 = \mathfrak{a}_{-\lambda^{(1)}}(x)\widetilde{B}_1$ with $\widetilde{B}_1 \subset \widetilde{w}_1$, $B_2 = \mathfrak{a}_{-\lambda^{(2)}}(x)\widetilde{B}_2$ with $\widetilde{B}_2 \subset \widetilde{w}_2$, and $B_0 = \mathfrak{a}_{-1}(1_X)^s\mathfrak{a}_{-\tilde{\lambda}}(x)|0\rangle$ or $\mathfrak{a}_{-1}(1_X)^s\mathfrak{a}_{-\tilde{\lambda}}(x)\mathfrak{a}_{-n_1}(\gamma_{j,1})|0\rangle$ or $\mathfrak{a}_{-1}(1_X)^s\mathfrak{a}_{-\tilde{\lambda}}(x)\mathfrak{a}_{-n_2}(\gamma_{j,2})|0\rangle$ where $s \le \tilde{n}$. In the following, we assume that (16.34) is nonzero. By symmetry, we need only to consider two cases for B_0:

$$B_0 = \mathfrak{a}_{-1}(1_X)^s\mathfrak{a}_{-\tilde{\lambda}}(x)|0\rangle$$

or

$$B_0 = \mathfrak{a}_{-1}(1_X)^s\mathfrak{a}_{-\tilde{\lambda}}(x)\mathfrak{a}_{-n_1}(\gamma_{j,1})|0\rangle.$$

We begin with the case $B_0 = \mathfrak{a}_{-1}(1_X)^s\mathfrak{a}_{-\tilde{\lambda}}(x)|0\rangle$. Then (16.34) becomes

$$\sum_j \big\langle \mathfrak{a}_{-1}(1_X)^s\mathfrak{a}_{-\tilde{\lambda}}(x)|0\rangle, \mathfrak{a}_{-\lambda^{(1)}}(x)\widetilde{B}_1, \mathfrak{a}_{-\lambda^{(2)}}(x)\widetilde{B}_2\big\rangle$$
$$\cdot \left\langle \mathfrak{Z}_{m,d},\ \pi_{m,1}^*\left(\frac{\mathbf{1}_{-\tilde{n}}\mathfrak{a}_{-n_1}(\gamma_{j,1})\mathfrak{a}_{-n_2}(\gamma_{j,2})|0\rangle}{\mathfrak{a}_{-1}(1_X)^s|0\rangle}\right)\cdot \pi_{m,2}^*\left(\frac{\widetilde{w}_1}{\widetilde{B}_1}\right)\cdot \pi_{m,3}^*\left(\frac{\widetilde{w}_2}{\widetilde{B}_2}\right)\right\rangle.$$

Applying the same arguments as in the computations of (16.26) and (16.27), we conclude that the term (16.34) is equal to

$$\sum_j \langle K_X, \gamma_{j,1}\rangle \cdot \langle K_X, \gamma_{j,2}\rangle \cdot P_3(K_X, S_1, S_2)$$

where $P_3(K_X, S_1, S_2)$ is a universal polynomial in $\langle K_X, K_X\rangle$ of degree at most $(m - n_1 - n_2)/2$ and of type (u_1, u_2). Note that for $\beta_1, \beta_2 \in H^2(X)$, we have

$$\sum_j \langle \beta_1, \gamma_{j,1}\rangle \cdot \langle \beta_2, \gamma_{j,2}\rangle = \langle \beta_1, \beta_2\rangle. \tag{16.35}$$

Therefore, (16.34) is equal to $\langle K_X, K_X\rangle \cdot P_3(K_X, S_1, S_2)$ which is a universal polynomial in $\langle K_X, K_X\rangle$ of degree at most $m/2 \le n/2$ and of type (u_1, u_2).

Next, let $B_0 = \mathfrak{a}_{-1}(1_X)^s\mathfrak{a}_{-\tilde{\lambda}}(x)\mathfrak{a}_{-n_1}(\gamma_{j,1})|0\rangle$. This time, (16.34) becomes

$$\sum_j \big\langle \mathfrak{a}_{-1}(1_X)^s\mathfrak{a}_{-\tilde{\lambda}}(x)\mathfrak{a}_{-n_1}(\gamma_{j,1})|0\rangle, \mathfrak{a}_{-\lambda^{(1)}}(x)\widetilde{B}_1, \mathfrak{a}_{-\lambda^{(2)}}(x)\widetilde{B}_2\big\rangle$$
$$\cdot \left\langle \mathfrak{Z}_{m,d},\ \pi_{m,1}^*\left(\frac{\mathbf{1}_{-\tilde{n}}\mathfrak{a}_{-n_2}(\gamma_{j,2})|0\rangle}{\mathfrak{a}_{-1}(1_X)^s|0\rangle}\right)\cdot \pi_{m,2}^*\left(\frac{\widetilde{w}_1}{\widetilde{B}_1}\right)\cdot \pi_{m,3}^*\left(\frac{\widetilde{w}_2}{\widetilde{B}_2}\right)\right\rangle.$$

Using Lemma 16.12, Theorem 15.20 and (16.35), we conclude that (16.34) is equal to $P_4(K_X, S_1, S_2)$ which is a universal polynomial in $\langle K_X, K_X\rangle$ of degree at most

$$(m - n_2)/2 + 1 \le ((n - n_1) - n_2)/2 + 1 \le n/2$$

and of type (u_1, u_2). This completes the proof of (16.32). □

PROPOSITION 16.19. *If $\alpha = 1_X$, then (16.8) is true.*

PROOF. We adopt the same notations and approaches as in the proof of Proposition 16.16. By Lemma 16.8 and Lemma 16.14, it suffices to prove that

$$D_\beta^{1_X}(w_1, w_2; -1) = P(K_X, S_1, S_2) \tag{16.36}$$

where $P(K_X, S_1, S_2)$ is a universal polynomial in $\langle K_X, K_X\rangle$ of degree at most $(n+1)/2$ and of type (u_1', u_2). This follows if we can prove that

$$D_\beta^{1_X}(w_1, w_2; q) = \sum_{d\ge 0} P(K_X, S_1, S_2; d)\, q^d \tag{16.37}$$

where $P(K_X, S_1, S_2; d)$ is a universal polynomial in $\langle K_X, K_X\rangle$ of degree at most $(n+1)/2$ and of type (u_1', u_2). In the following, we will show that the contribution of every term in Lemma 16.9 is of the form $P(K_X, S_1, S_2; d)$ for a suitable $d \geq 0$.

First of all, when $d \geq 1$, we conclude from Lemma 16.18 that

$$\begin{aligned} &\big\langle \mathbf{1}_{-(n-j-1)}\mathfrak{a}_{-\lambda}(\tau_* 1_X)|0\rangle, \mathfrak{a}_{-1}(\beta)w_1, w_2\big\rangle_d \\ - &\big\langle \mathbf{1}_{-(n-j-2)}\mathfrak{a}_{-\lambda}(\tau_* 1_X)|0\rangle, w_1, \mathfrak{a}_{-1}(\beta)^\dagger w_2\big\rangle_d \end{aligned}$$

is equal to $P_1(K_X, S_1, S_2; d)$ which is a universal polynomial in $\langle K_X, K_X\rangle$ of degree at most $n/2$ and of type (u_1', u_2).

Next, consider

$$\big\langle \mathfrak{a}_\lambda(\tau_*(\epsilon\alpha\beta))w_1, w_2\big\rangle = \big\langle \mathfrak{a}_\lambda(\tau_*(\epsilon\beta))w_1, w_2\big\rangle$$

from Lemma 16.9, where $\epsilon \in \{K_X, K_X^2\}$. It is zero unless $\epsilon = K_X^2$ and $\beta = 1_X$ (when $|\beta| = 0$, we let $\beta = 1_X$), or $\epsilon = K_X$ and $|\beta| = 2$, or $\epsilon = K_X$ and $\beta = 1_X$. If $\epsilon = K_X^2$ and $\beta = 1_X$, then

$$\begin{aligned} \big\langle \mathfrak{a}_\lambda(\tau_*(\epsilon\beta))w_1, w_2\big\rangle &= \langle K_X, K_X\rangle \cdot \big\langle \mathfrak{a}_\lambda(x)w_1, w_2\big\rangle \\ &= \langle K_X, K_X\rangle \cdot P_2(K_X, S_1, S_2; 0) \end{aligned}$$

by Theorem 3.8, where $P_2(K_X, S_1, S_2; 0)$ is a universal polynomial in $\langle K_X, K_X\rangle$ of degree 0 and of type (u_1', u_2). If $\epsilon = K_X$ and $|\beta| = 2$, then

$$\begin{aligned} \big\langle \mathfrak{a}_\lambda(\tau_*(\epsilon\beta))w_1, w_2\big\rangle &= \langle K_X, \beta\rangle \cdot \big\langle \mathfrak{a}_\lambda(x)w_1, w_2\big\rangle \\ &= \langle K_X, \beta\rangle \cdot P_3(K_X, S_1, S_2; 0) \end{aligned}$$

which is a universal polynomial in $\langle K_X, K_X\rangle$ of degree 0 and of type (u_1', u_2). If $\epsilon = K_X$ and $\beta = 1_X$, then we obtain

$$\big\langle \mathfrak{a}_\lambda(\tau_*(\epsilon\beta))w_1, w_2\big\rangle = \big\langle \mathfrak{a}_\lambda(\tau_* K_X)w_1, w_2\big\rangle$$

which again is a universal polynomial in $\langle K_X, K_X\rangle$ of degree 0 and of type (u_1', u_2).

Finally, let $\epsilon \in \{K_X, K_X^2\}$. We have $\tau_*(\epsilon\alpha) = \tau_*\epsilon$. Let I_ϵ be the difference

$$\begin{aligned} &\big\langle \mathbf{1}_{-(n-j-1)}\mathfrak{a}_{-\lambda}(\tau_*\epsilon)|0\rangle, \mathfrak{a}_{-1}(\beta)w_1, w_2\big\rangle \\ - &\big\langle \mathbf{1}_{-(n-j-2)}\mathfrak{a}_{-\lambda}(\tau_*\epsilon)|0\rangle, w_1, \mathfrak{a}_{-1}(\beta)^\dagger w_2\big\rangle \end{aligned}$$

from Lemma 16.9. When $\epsilon = K_X^2$, we see from Lemma 16.11 that

$$\begin{aligned} I_\epsilon &= \langle K_X, K_X\rangle \cdot \big\langle \mathbf{1}_{-(n-j-1)}\mathfrak{a}_{-\lambda}(x)|0\rangle, \mathfrak{a}_{-1}(\beta)w_1, w_2\big\rangle \\ &\quad - \langle K_X, K_X\rangle \cdot \big\langle \mathbf{1}_{-(n-j-2)}\mathfrak{a}_{-\lambda}(x)|0\rangle, w_1, \mathfrak{a}_{-1}(\beta)^\dagger w_2\big\rangle \\ &= \langle K_X, K_X\rangle \cdot P_4(K_X, S_1, S_2; 0) \end{aligned}$$

where $P_4(K_X, S_1, S_2; 0)$ is a universal polynomial in $\langle K_X, K_X\rangle$ of degree 0 and of type (u_1', u_2). When $\epsilon = K_X$, we see from Lemma 16.12 that

$$\begin{aligned} I_\epsilon &= \big\langle \mathbf{1}_{-(n-j-1)}\mathfrak{a}_{-\lambda}(\tau_* K_X)|0\rangle, \mathfrak{a}_{-1}(\beta)w_1, w_2\big\rangle \\ &\quad - \big\langle \mathbf{1}_{-(n-j-2)}\mathfrak{a}_{-\lambda}(\tau_* K_X)|0\rangle, w_1, \mathfrak{a}_{-1}(\beta)^\dagger w_2\big\rangle \end{aligned}$$

is a universal polynomial in $\langle K_X, K_X\rangle$ of degree at most 1 and of type (u_1', u_2). □

The following is the main result in this section.

THEOREM 16.20. *Let X be a simply connected smooth projective complex surface. Then*

$$[\widetilde{\mathfrak{G}}_k(\alpha), \mathfrak{a}_{-1}(\beta)] = \frac{1}{k!} \mathfrak{a}_{-1}^{\{k\}}(\alpha\beta).$$

PROOF. Follows from Propositions 16.10, 16.16 and 16.19. □

At this point, all the necessary properties (Theorem 3.8, Theorem 16.5 and Theorem 16.20) on the Hilbert scheme side in order to prove Ruan's Cohomological Crepant Resolution Conjecture for the Hilbert-Chow morphisms have been established. We are therefore led to the final section of the book.

16.3. Ruan's Cohomological Crepant Resolution Conjecture

In this section, we will review Ruan's Cohomological Crepant Resolution Conjecture for the Hilbert-Chow morphisms, and present a proof of it following the axiomatization approach formulated in [**QW1**, Subsection 2.6].

Let X denote a smooth projective complex surface. By Remark 1.25, the Hilbert-Chow morphism

$$\rho_n : X^{[n]} \to X^{(n)}$$

is a crepant resolution of the global orbifold $X^{(n)}$. In this setting, Ruan's Cohomological Crepant Resolution Conjecture states as follows.

CONJECTURE 16.21. (Ruan's Cohomological Crepant Resolution Conjecture) *Let X be a smooth projective complex surface. Then, the quantum corrected cohomology ring $H^*_{\rho_n}(X^{[n]})$ is ring isomorphic to $H^*_{\text{orb}}(X^{(n)})$.*

Conjecture 16.21 was first confirmed for $n = 2$, following Theorem 15.9 and Example 15.10. When the smooth projective surface X has a numerically trivial canonical class K_X, Conjecture 16.21 was proved in [**FG1, LS2, Uri**] (the case of certain quasi-projective surfaces X was also considered in [**LQW5**, Section 5] and [**QW3**, Subsection 3.4]). In [**Cheo2**, Corollary 0.3], Conjecture 16.21 was verified when X is a smooth projective toric surface (in fact, a stronger version of Conjecture 16.21 was proved there). The main idea in [**Cheo2**] is to take a toric action on X with finitely many fixed points $x_1, \ldots, x_s$. Each point x_i has an open neighborhood X_i such that $X_i \cong \mathbb{C}^2$ and $x_i \in X_i$ is identified with the origin of $\mathbb{C}^2$. In terms of the Heisenberg monomial classes (3.10), a product formula is proved, which reduces the 3-point genus-0 extremal Gromov-Witten invariants of $X^{[n]}$ to those of $X_i \cong \mathbb{C}^2$. There is also a similar product formula on the orbifold side. Therefore, when X is a smooth projective toric surface, Conjecture 16.21 is reduced to the case of the affine plane $\mathbb{C}^2$ in the equivariant setting. The case of the affine plane $\mathbb{C}^2$ in the equivariant setting has been verified earlier in [**BG**].

An axiomatization originated from [**Leh1, LQW1**] and formulated in [**QW1**, Subsection 2.6] was used in [**LQ5**, Section 6] to prove Conjecture 16.21. This axiomatization involves Heisenberg algebra actions and vertex operator techniques pioneered in [**Groj, Nak3**]. We recall from Definition 7.12 that a graded Frobenius algebra over $\mathbb{C}$ is a finite dimensional graded (complex) vector space A with a graded associative multiplication $A \otimes A \to A$ and unit element 1_A together with a linear form $T : A \to \mathbb{C}$ such that the induced bilinear form $\langle a, b \rangle = T(ab)$ is nondegenerate. For $k \geq 1$, the k-th co-product

$$\tau_{k*} : A \to A^{\otimes k}$$

is defined by requiring

$$\langle \tau_{k*}(a), b_1 \otimes \cdots \otimes b_k \rangle = T(ab_1 \cdots b_k).$$

Now the axiomatization in [**QW1**, Subsection 2.6] states that the algebra structure on each $A^{[n]}$ in a sequence of graded Frobenius algebras $A^{[n]}$ $(n \geq 0)$ is determined if

(A1) the direct sum

$$\bigoplus_{n=0}^{+\infty} A^{[n]}$$

affords the structure of the Fock space of a Heisenberg algebra modeled on A which denotes $A^{[1]}$.

(A2) There exists a sequence of elements $\widetilde{G}_k(\alpha, n) \in A^{[n]}$ depending on $\alpha \in A$ (linearly) and a non-negative integer k. Define the operators $\widetilde{\mathfrak{G}}_k(\alpha)$ on

$$\bigoplus_{n=0}^{+\infty} A^{[n]}$$

which act on the component $A^{[n]}$ via multiplication by $\widetilde{G}_k(\alpha, n) \in A^{[n]}$. The operators $\widetilde{\mathfrak{G}}_k(\alpha)$ and the Heisenberg generators satisfy:

$$\widetilde{\mathfrak{G}}_1(1_A) = -\frac{1}{6} : \mathfrak{a}^3 :_0 (\tau_{3*}1_A), \tag{16.38}$$

$$[\widetilde{\mathfrak{G}}_k(\alpha), \mathfrak{a}_{-1}(\beta)] = \frac{1}{k!}\, \mathfrak{a}_{-1}^{\{k\}}(\alpha\beta) \tag{16.39}$$

where $: \mathfrak{a}^3 :_0$ is the zero mode in the normally ordered product $: \mathfrak{a}^3 :$, and $\mathfrak{a}_{-1}^{\{k\}}(\alpha)$ denotes the k-th derivative with $\mathfrak{a}_{-1}^{\{0\}}(\alpha) = \mathfrak{a}_{-1}(\alpha)$ and

$$\mathfrak{a}_{-1}^{\{k\}}(\alpha) = [\widetilde{\mathfrak{G}}_1(1_A), \mathfrak{a}_{-1}^{\{k-1\}}(\alpha)]$$

for $k \geq 1$.

When (A1) and (A2) are satisfied, the algebra $A^{[n]}$ is generated by the elements

$$\widetilde{G}_k(\alpha, n) \in A^{[n]}, \quad \alpha \in A, k \geq 0.$$

In addition, the product is determined by (16.38) and (16.39).

On one hand, with

$$A^{[n]} = H^*_{\mathrm{orb}}(X^{(n)})$$

(viewed as an algebra over $\mathbb{C}$) and the sequence of classes $O_k(\alpha, n) \in H^*_{\mathrm{orb}}(X^{(n)})$, we see from Theorem 10.1, Theorem 10.6 and Theorem 10.8 that (A1) and (A2) hold for the orbifold cohomology rings $H^*_{\mathrm{orb}}(X^{(n)})$. On the other hand, by Theorem 3.8, Theorem 16.5 and Theorem 16.20, the quantum corrected cohomology rings

$$A^{[n]} = H^*_{\rho_n}(X^{[n]})$$

also satisfy (A1) and (A2) with $A = A^{[1]} = H^*(X)$ and the sequence of classes $\widetilde{G}_k(\alpha, n) \in H^*_{\rho_n}(X^{[n]})$. This leads to the following result.

THEOREM 16.22. *Conjecture* 16.21 *holds when X is a simply connected smooth projective surface.*

PROOF. Recall the operators $\mathfrak{p}_n(\alpha)$ from Section 10.2. Note that the shift number (or the age) of $\mathfrak{p}_{-n_1}(\alpha_1)\cdots\mathfrak{p}_{-n_s}(\alpha_s)|0\rangle$ is equal to $n_1+\ldots+n_s-s$. Define a linear isomorphism

$$\Psi: \bigoplus_{n=0}^{+\infty} H^*_{\text{orb}}(X^{(n)}) \longrightarrow \bigoplus_{n=0}^{+\infty} H^*_{\rho_n}(X^{[n]}) \tag{16.40}$$

by

$$\Psi_n\Big(\sqrt{-1}^{\,n_1+\ldots+n_s-s}\mathfrak{p}_{-n_1}(\alpha_1)\cdots\mathfrak{p}_{-n_s}(\alpha_s)|0\rangle\Big) = \mathfrak{a}_{-n_1}(\alpha_1)\cdots\mathfrak{a}_{-n_s}(\alpha_s)|0\rangle.$$

This induces a linear isomorphism

$$\Psi_n: \quad H^*_{\text{orb}}(X^{(n)}) \longrightarrow H^*_{\rho_n}(X^{[n]}) = H^*(X^{[n]})$$

for each n. Moreover, Ψ_1 is simply the identity map on the cohomology group of the surface X.

By Theorem 16.5 and Theorem 16.20, the two formulas (16.38) and (16.39) hold for $A^{[n]} = H^*_{\rho_n}(X^{[n]})$. By the proof of Theorem 4.7, we have

$$\widetilde{\mathfrak{G}}_k(\alpha) = -\sum_{\ell(\lambda)=k+2,|\lambda|=0}\frac{1}{\lambda^!}\mathfrak{a}_\lambda(\tau_*\alpha) + \sum_{\ell(\lambda)=k,|\lambda|=0}\frac{s(\lambda)-2}{24\lambda^!}\mathfrak{a}_\lambda(\tau_*(e_X\alpha)).$$

Combining this with Theorem 10.15, we check directly that

$$\begin{aligned}
&\Psi_n\Big(\sqrt{-1}^{\,k}O_k(\alpha,n)\bullet\sqrt{-1}^{\,n_1+\ldots+n_s-s}\mathfrak{p}_{-n_1}(\alpha_1)\cdots\mathfrak{p}_{-n_s}(\alpha_s)|0\rangle\Big)\\
=\;&\Psi_n\Big(\sqrt{-1}^{\,k+n_1+\ldots+n_s-s}\mathfrak{O}_k(\alpha)\mathfrak{p}_{-n_1}(\alpha_1)\cdots\mathfrak{p}_{-n_s}(\alpha_s)|0\rangle\Big)\\
=\;&\widetilde{\mathfrak{G}}_k(\alpha)\mathfrak{a}_{-n_1}(\alpha_1)\cdots\mathfrak{a}_{-n_s}(\alpha_s)|0\rangle\\
=\;&\widetilde{G}_k(\alpha,n)\cdot_{\rho_n}\big(\mathfrak{a}_{-n_1}(\alpha_1)\cdots\mathfrak{a}_{-n_s}(\alpha_s)|0\rangle\big)
\end{aligned}$$

where $n_1+\ldots+n_s = n$. In particular, letting $s = n$, $n_1 = \ldots = n_s = 1$ and $\alpha_1 = \ldots = \alpha_s = 1_X$, we obtain

$$\Psi_n(\sqrt{-1}^{\,k}O_k(\alpha,n)) = \widetilde{G}_k(\alpha,n).$$

Thus,

$$\begin{aligned}
&\Psi_n\Big(\sqrt{-1}^{\,k}O_k(\alpha,n)\bullet\sqrt{-1}^{\,n_1+\ldots+n_s-s}\mathfrak{p}_{-n_1}(\alpha_1)\cdots\mathfrak{p}_{-n_s}(\alpha_s)|0\rangle\Big)\\
=\;&\Psi_n(\sqrt{-1}^{\,k}O_k(\alpha,n))\cdot_{\rho_n}\big(\mathfrak{a}_{-n_1}(\alpha_1)\cdots\mathfrak{a}_{-n_s}(\alpha_s)|0\rangle\big)\\
=\;&\Psi_n(\sqrt{-1}^{\,k}O_k(\alpha,n))\cdot_{\rho_n}\Psi_n\big(\sqrt{-1}^{\,n_1+\ldots+n_s-s}\mathfrak{p}_{-n_1}(\alpha_1)\cdots\mathfrak{p}_{-n_s}(\alpha_s)|0\rangle\big).
\end{aligned}$$

Since the classes $O_k(\alpha,n)$ with $k\ge 0, \alpha\in H^*(X)$ generate the ring $H^*_{\text{orb}}(X^{(n)})$, we conclude that

$$\Psi_n: \quad H^*_{\text{orb}}(X^{(n)}) \longrightarrow H^*_{\rho_n}(X^{[n]})$$

is a ring isomorphism. □

REMARK 16.23. Using Heisenberg monomial classes, one checks that the ring isomorphism Ψ_n preserves the pairings on $H^*_{\text{orb}}(X^{(n)})$ and $H^*_{\rho_n}(X^{[n]}) = H^*(X^{[n]})$.

REMARK 16.24. The assumption in Theorem 16.22 that the smooth projective complex surface X is simply connected is not essential. This assumption is for the convenience that no odd degree cohomology classes appear. The proof of Theorem 16.22 should work for every smooth projective complex surface X.

Bibliography

[AGV] D. Abramovich, T. Graber, and A. Vistoli, *Algebraic orbifold quantum products*, Orbifolds in mathematics and physics (Madison, WI, 2001), Contemp. Math., vol. 310, Amer. Math. Soc., Providence, RI, 2002, pp. 1–24, DOI 10.1090/conm/310/05397. MR1950940

[AGT] L. F. Alday, D. Gaiotto, and Y. Tachikawa, *Liouville correlation functions from four-dimensional gauge theories*, Lett. Math. Phys. **91** (2010), no. 2, 167–197, DOI 10.1007/s11005-010-0369-5. MR2586871

[AIK] A. B. Altman, A. Iarrobino, and S. L. Kleiman, *Irreducibility of the compactified Jacobian*, Real and complex singularities (Proc. Ninth Nordic Summer School/NAVF Sympos. Math., Oslo, 1976), Sijthoff and Noordhoff, Alphen aan den Rijn, 1977, pp. 1–12. MR0498546

[And] G. E. Andrews, *The theory of partitions*, Addison-Wesley Publishing Co., Reading, Mass.-London-Amsterdam, 1976. Encyclopedia of Mathematics and its Applications, Vol. 2. MR0557013

[ACGH] E. Arbarello, M. Cornalba, P. A. Griffiths, and J. Harris, *Geometry of algebraic curves. Vol. I*, Grundlehren der Mathematischen Wissenschaften [Fundamental Principles of Mathematical Sciences], vol. 267, Springer-Verlag, New York, 1985. MR770932

[ABCH] D. Arcara, A. Bertram, I. Coskun, and J. Huizenga, *The minimal model program for the Hilbert scheme of points on* $\mathbb{P}^2$ *and Bridgeland stability*, Adv. Math. **235** (2013), 580–626, DOI 10.1016/j.aim.2012.11.018. MR3010070

[AH] M. F. Atiyah and F. Hirzebruch, *Vector bundles and homogeneous spaces*, Proc. Sympos. Pure Math., Vol. III, American Mathematical Society, Providence, R.I., 1961, pp. 7–38. MR0139181

[BBC] D. H. Bailey, J. M. Borwein, and R. E. Crandall, *Computation and theory of extended Mordell-Tornheim-Witten sums*, Math. Comp. **83** (2014), no. 288, 1795–1821, DOI 10.1090/S0025-5718-2014-02768-3. MR3194130

[BB] V. V. Batyrev and L. A. Borisov, *Mirror duality and string-theoretic Hodge numbers*, Invent. Math. **126** (1996), no. 1, 183–203, DOI 10.1007/s002220050093. MR1408560

[BBM] P. Baum, J.-L. Brylinski, and R. MacPherson, *Cohomologie équivariante délocalisée* (French, with English summary), C. R. Acad. Sci. Paris Sér. I Math. **300** (1985), no. 17, 605–608. MR791098

[BC] P. Baum and A. Connes, *Chern character for discrete groups*, A fête of topology, Academic Press, Boston, MA, 1988, pp. 163–232, DOI 10.1016/B978-0-12-480440-1.50015-0. MR928402

[Bea1] A. Beauville, *Variétés Kähleriennes dont la première classe de Chern est nulle* (French), J. Differential Geom. **18** (1983), no. 4, 755–782 (1984). MR730926

[Bea2] ———, *Variétés kählériennes compactes avec* $c_1 = 0$ (French), Astérisque **126** (1985), 181–192. Geometry of $K3$ surfaces: moduli and periods (Palaiseau, 1981/1982). MR785234

[Beh1] K. Behrend, *Gromov-Witten invariants in algebraic geometry*, Invent. Math. **127** (1997), no. 3, 601–617, DOI 10.1007/s002220050132. MR1431140

[Beh2] ———, *The product formula for Gromov-Witten invariants*, J. Algebraic Geom. **8** (1999), no. 3, 529–541. MR1689355

[BF1] K. Behrend and B. Fantechi, *The intrinsic normal cone*, Invent. Math. **128** (1997), no. 1, 45–88, DOI 10.1007/s002220050136. MR1437495

[BF2] K. Behrend and B. Fantechi, *Symmetric obstruction theories and Hilbert schemes of points on threefolds*, Algebra Number Theory **2** (2008), no. 3, 313–345, DOI 10.2140/ant.2008.2.313. MR2407118

[BSG] M. Beltrametti and A. J. Sommese, *Zero cycles and kth order embeddings of smooth projective surfaces*, Problems in the theory of surfaces and their classification (Cortona, 1988), Sympos. Math., XXXII, Academic Press, London, 1991, pp. 33–48. With an appendix by Lothar Göttsche. MR1273371

[BT] A. Bertram and M. Thaddeus, *On the quantum cohomology of a symmetric product of an algebraic curve*, Duke Math. J. **108** (2001), no. 2, 329–362, DOI 10.1215/S0012-7094-01-10825-9. MR1833394

[BO] S. Bloch and A. Okounkov, *The character of the infinite wedge representation*, Adv. Math. **149** (2000), no. 1, 1–60, DOI 10.1006/aima.1999.1845. MR1742353

[Boi1] S. Boissière, *Chern classes of the tangent bundle on the Hilbert scheme of points on the affine plane*, J. Algebraic Geom. **14** (2005), no. 4, 761–787, DOI 10.1090/S1056-3911-05-00412-1. MR2147349

[Boi2] ———, *On the McKay correspondences for the Hilbert scheme of points on the affine plane*, Math. Ann. **334** (2006), no. 2, 419–438, DOI 10.1007/s00208-005-0738-z. MR2207706

[BN] S. Boissière and M. A. Nieper-Wisskirchen, *Generating series in the cohomology of Hilbert schemes of points on surfaces*, LMS J. Comput. Math. **10** (2007), 254–270, DOI 10.1112/S146115700000139X. MR2320831

[Bor] R. E. Borcherds, *Vertex algebras, Kac-Moody algebras, and the Monster*, Proc. Nat. Acad. Sci. U.S.A. **83** (1986), no. 10, 3068–3071, DOI 10.1073/pnas.83.10.3068. MR843307

[Bra1] D. M. Bradley, *Multiple q-zeta values*, J. Algebra **283** (2005), no. 2, 752–798, DOI 10.1016/j.jalgebra.2004.09.017. MR2111222

[Bra2] ———, *On the sum formula for multiple q-zeta values*, Rocky Mountain J. Math. **37** (2007), no. 5, 1427–1434, DOI 10.1216/rmjm/1194275927. MR2382894

[Bri] J. Briançon, *Description de $H\mathrm{ilb}^n C\{x,y\}$*, Invent. Math. **41** (1977), no. 1, 45–89, DOI 10.1007/BF01390164. MR0457432

[BG] J. Bryan and T. Graber, *The crepant resolution conjecture*, Algebraic geometry—Seattle 2005. Part 1, Proc. Sympos. Pure Math., vol. 80, Amer. Math. Soc., Providence, RI, 2009, pp. 23–42, DOI 10.1090/pspum/080.1/2483931. MR2483931

[BP] J. Bryan and R. Pandharipande, *The local Gromov-Witten theory of curves*, J. Amer. Math. Soc. **21** (2008), no. 1, 101–136, DOI 10.1090/S0894-0347-06-00545-5. With an appendix by Bryan, C. Faber, A. Okounkov and Pandharipande. MR2350052

[Can] M. B. Can, *Nested Hilbert schemes and the nested q,t-Catalan series* (English, with English and French summaries), 20th Annual International Conference on Formal Power Series and Algebraic Combinatorics (FPSAC 2008), Discrete Math. Theor. Comput. Sci. Proc., AJ, Assoc. Discrete Math. Theor. Comput. Sci., Nancy, 2008, pp. 61–69. MR2721442

[Car1] E. Carlsson, *Vertex operators and moduli spaces of sheaves*, ProQuest LLC, Ann Arbor, MI, 2008. Thesis (Ph.D.)–Princeton University. MR2711527

[Car2] ———, *Vertex operators and quasimodularity of Chern numbers on the Hilbert scheme*, Adv. Math. **229** (2012), no. 5, 2888–2907, DOI 10.1016/j.aim.2011.10.003. MR2889150

[CO] E. Carlsson and A. Okounkov, *Exts and vertex operators*, Duke Math. J. **161** (2012), no. 9, 1797–1815, DOI 10.1215/00127094-1593380. MR2942794

[dCM] M. A. A. de Cataldo and L. Migliorini, *The Douady space of a complex surface*, Adv. Math. **151** (2000), no. 2, 283–312, DOI 10.1006/aima.1999.1896. MR1758249

[CG] F. Catanese and L. Gœttsche, *d-very-ample line bundles and embeddings of Hilbert schemes of 0-cycles*, Manuscripta Math. **68** (1990), no. 3, 337–341, DOI 10.1007/BF02568768. MR1065935

[CL] H.-l. Chang and J. Li, *Semi-perfect obstruction theory and Donaldson-Thomas invariants of derived objects*, Comm. Anal. Geom. **19** (2011), no. 4, 807–830, DOI 10.4310/CAG.2011.v19.n4.a6. MR2880216

[Che1] J. Cheah, *On the cohomology of Hilbert schemes of points*, J. Algebraic Geom. **5** (1996), no. 3, 479–511. MR1382733

[Che2] ———, *Cellular decompositions for nested Hilbert schemes of points*, Pacific J. Math. **183** (1998), no. 1, 39–90, DOI 10.2140/pjm.1998.183.39. MR1616606

[Che3] ———, *The virtual Hodge polynomials of nested Hilbert schemes and related varieties*, Math. Z. **227** (1998), no. 3, 479–504, DOI 10.1007/PL00004387. MR1612677

[CR1] W. Chen and Y. Ruan, *Orbifold Gromov-Witten theory*, Orbifolds in mathematics and physics (Madison, WI, 2001), Contemp. Math., vol. 310, Amer. Math. Soc., Providence, RI, 2002, pp. 25–85, DOI 10.1090/conm/310/05398. MR1950941

[CR2] ———, *A new cohomology theory of orbifold*, Comm. Math. Phys. **248** (2004), no. 1, 1–31, DOI 10.1007/s00220-004-1089-4. MR2104605

[Cheo1] Wan Keng Cheong, *Orbifold quantum cohomology of the symmetric product of $\mathcal{A}_r$*. Preprint. arXiv:0910.0629

[Cheo2] W. K. Cheong, *Strengthening the cohomological crepant resolution conjecture for Hilbert-Chow morphisms*, Math. Ann. **356** (2013), no. 1, 45–72, DOI 10.1007/s00208-012-0833-x. MR3038121

[ChG] W. K. Cheong and A. Gholampour, *Orbifold Gromov-Witten theory of the symmetric product of $\mathcal{A}_r$*, Geom. Topol. **16** (2012), no. 1, 475–527, DOI 10.2140/gt.2012.16.475. MR2916292

[Coa] T. Coates, *On the crepant resolution conjecture in the local case*, Comm. Math. Phys. **287** (2009), no. 3, 1071–1108, DOI 10.1007/s00220-008-0715-y. MR2486673

[CCIT] T. Coates, H. Iritani, and H.-H. Tseng, *Wall-crossings in toric Gromov-Witten theory. I. Crepant examples*, Geom. Topol. **13** (2009), no. 5, 2675–2744, DOI 10.2140/gt.2009.13.2675. MR2529944

[CR] T. Coates and Y. Ruan, *Quantum cohomology and crepant resolutions: a conjecture* (English, with English and French summaries), Ann. Inst. Fourier (Grenoble) **63** (2013), no. 2, 431–478. MR3112518

[Cos] K. Costello, *Higher genus Gromov-Witten invariants as genus zero invariants of symmetric products*, Ann. of Math. (2) **164** (2006), no. 2, 561–601, DOI 10.4007/annals.2006.164.561. MR2247968

[CoG] K. Costello, I. Grojnowski, *Hilbert schemes, Hecke algebras and the Calogero-Sutherland system*. Preprint. math.AG/0310189.

[CK] D. A. Cox and S. Katz, *Mirror symmetry and algebraic geometry*, Mathematical Surveys and Monographs, vol. 68, American Mathematical Society, Providence, RI, 1999. MR1677117

[DHVW] L. Dixon, J. A. Harvey, C. Vafa, and E. Witten, *Strings on orbifolds*, Nuclear Phys. B **261** (1985), no. 4, 678–686, DOI 10.1016/0550-3213(85)90593-0. MR818423

[ELQ] D. Edidin, W.-P. Li, and Z. Qin, *Gromov-Witten invariants of the Hilbert scheme of 3-points on $\mathbb{P}^2$*, Asian J. Math. **7** (2003), no. 4, 551–574, DOI 10.4310/AJM.2003.v7.n4.a7. MR2074891

[EG] G. Ellingsrud and L. Göttsche, *Hilbert schemes of points and Heisenberg algebras*, School on Algebraic Geometry (Trieste, 1999), ICTP Lect. Notes, vol. 1, Abdus Salam Int. Cent. Theoret. Phys., Trieste, 2000, pp. 59–100. MR1795861

[EGL] G. Ellingsrud, L. Göttsche, and M. Lehn, *On the cobordism class of the Hilbert scheme of a surface*, J. Algebraic Geom. **10** (2001), no. 1, 81–100. MR1795551

[EL] G. Ellingsrud and M. Lehn, *Irreducibility of the punctual quotient scheme of a surface*, Ark. Mat. **37** (1999), no. 2, 245–254, DOI 10.1007/BF02412213. MR1714770

[ES1] G. Ellingsrud and S. A. Strømme, *On the homology of the Hilbert scheme of points in the plane*, Invent. Math. **87** (1987), no. 2, 343–352, DOI 10.1007/BF01389419. MR870732

[ES2] ———, *Towards the Chow ring of the Hilbert scheme of $\mathbf{P}^2$*, J. Reine Angew. Math. **441** (1993), 33–44. MR1228610

[ES3] ———, *An intersection number for the punctual Hilbert scheme of a surface*, Trans. Amer. Math. Soc. **350** (1998), no. 6, 2547–2552, DOI 10.1090/S0002-9947-98-01972-2. MR1432198

[EtG] P. Etingof and V. Ginzburg, *Symplectic reflection algebras, Calogero-Moser space, and deformed Harish-Chandra homomorphism*, Invent. Math. **147** (2002), no. 2, 243–348, DOI 10.1007/s002220100171. MR1881922

[FPa] C. Faber and R. Pandharipande, *Hodge integrals and Gromov-Witten theory*, Invent. Math. **139** (2000), no. 1, 173–199, DOI 10.1007/s002229900028. MR1728879

[FG1] B. Fantechi and L. Göttsche, *Orbifold cohomology for global quotients*, Duke Math. J. **117** (2003), no. 2, 197–227, DOI 10.1215/S0012-7094-03-11721-4. MR1971293

[FG2] ———, *Local properties and Hilbert schemes of points*, Fundamental algebraic geometry, Math. Surveys Monogr., vol. 123, Amer. Math. Soc., Providence, RI, 2005, pp. 139–178. MR2223408

[Fog1] J. Fogarty, *Algebraic families on an algebraic surface*, Amer. J. Math **90** (1968), 511–521, DOI 10.2307/2373541. MR0237496

[Fog2] J. Fogarty, *Algebraic families on an algebraic surface. II. The Picard scheme of the punctual Hilbert scheme*, Amer. J. Math. **95** (1973), 660–687, DOI 10.2307/2373734. MR0335512

[FU] D. S. Freed and K. K. Uhlenbeck, *Instantons and four-manifolds*, 2nd ed., Mathematical Sciences Research Institute Publications, vol. 1, Springer-Verlag, New York, 1991. MR1081321

[FB] E. Frenkel and D. Ben-Zvi, *Vertex algebras and algebraic curves*, Mathematical Surveys and Monographs, vol. 88, American Mathematical Society, Providence, RI, 2001. MR1849359

[FKRW] E. Frenkel, V. Kac, A. Radul, and W. Wang, *$\mathcal{W}_{1+\infty}$ and $\mathcal{W}(\mathfrak{gl}_N)$ with central charge N*, Comm. Math. Phys. **170** (1995), no. 2, 337–357. MR1334399

[FJW] I. B. Frenkel, N. Jing, and W. Wang, *Vertex representations via finite groups and the McKay correspondence*, Internat. Math. Res. Notices **4** (2000), 195–222, DOI 10.1155/S107379280000012X. MR1747618

[FW] I. B. Frenkel and W. Wang, *Virasoro algebra and wreath product convolution*, J. Algebra **242** (2001), no. 2, 656–671, DOI 10.1006/jabr.2001.8860. MR1848963

[Fu] Y. Fu, *Quantum cohomology of a Hilbert scheme of a Hirzebruch surface*, ProQuest LLC, Ann Arbor, MI, 2010. Thesis (Ph.D.)–University of Illinois at Urbana-Champaign. MR2941780

[FN] A. Fujiki and S. Nakano, *Supplement to "On the inverse of monoidal transformation"*, Publ. Res. Inst. Math. Sci. **7** (1971/72), 637–644, DOI 10.2977/prims/1195193401. MR0294712

[FO] K. Fukaya and K. Ono, *Arnold conjecture and Gromov-Witten invariant for general symplectic manifolds*, The Arnoldfest (Toronto, ON, 1997), Fields Inst. Commun., vol. 24, Amer. Math. Soc., Providence, RI, 1999, pp. 173–190. MR1733575

[Ful] W. Fulton, *Introduction to toric varieties*, Annals of Mathematics Studies, vol. 131, Princeton University Press, Princeton, NJ, 1993. The William H. Roever Lectures in Geometry. MR1234037

[FP] W. Fulton and R. Pandharipande, *Notes on stable maps and quantum cohomology*, Algebraic geometry—Santa Cruz 1995, Proc. Sympos. Pure Math., vol. 62, Amer. Math. Soc., Providence, RI, 1997, pp. 45–96, DOI 10.1090/pspum/062.2/1492534. MR1492534

[Get] E. Getzler, *Intersection theory on $\overline{\mathcal{M}}_{1,4}$ and elliptic Gromov-Witten invariants*, J. Amer. Math. Soc. **10** (1997), no. 4, 973–998, DOI 10.1090/S0894-0347-97-00246-4. MR1451505

[GSY1] A. Gholampour, A. Sheshmani, S.-T. Yau, *Nested Hilbert schemes on surfaces: virtual fundamental class.* Preprint. arXiv.1701.08899.

[GSY2] A. Gholampour, A. Sheshmani, S.-T. Yau, *Localized Donaldson-Thomas theory of surfaces.* Preprint. arXiv.1701.08902.

[Got1] L. Göttsche, *The Betti numbers of the Hilbert scheme of points on a smooth projective surface*, Math. Ann. **286** (1990), no. 1-3, 193–207, DOI 10.1007/BF01453572. MR1032930

[Got2] ———, *Hilbert schemes of zero-dimensional subschemes of smooth varieties*, Lecture Notes in Mathematics, vol. 1572, Springer-Verlag, Berlin, 1994. MR1312161

[Got3] ———, *On the motive of the Hilbert scheme of points on a surface*, Math. Res. Lett. **8** (2001), no. 5-6, 613–627, DOI 10.4310/MRL.2001.v8.n5.a3. MR1879805

[GS] L. Göttsche and W. Soergel, *Perverse sheaves and the cohomology of Hilbert schemes of smooth algebraic surfaces*, Math. Ann. **296** (1993), no. 2, 235–245, DOI 10.1007/BF01445104. MR1219901

[Gou] I. P. Goulden, *A differential operator for symmetric functions and the combinatorics of multiplying transpositions*, Trans. Amer. Math. Soc. **344** (1994), no. 1, 421–440, DOI 10.2307/2154724. MR1249468

[GJV] I. Goulden, D. Jackson, R. Vakil, *Towards the geometry of double Hurwitz numbers*, math.AG/0309440.

[GP] T. Graber and R. Pandharipande, *Localization of virtual classes*, Invent. Math. **135** (1999), no. 2, 487–518, DOI 10.1007/s002220050293. MR1666787

[Groj] I. Grojnowski, *Instantons and affine algebras. I. The Hilbert scheme and vertex operators*, Math. Res. Lett. **3** (1996), no. 2, 275–291, DOI 10.4310/MRL.1996.v3.n2.a12. MR1386846

[Grot] A. Grothendieck, *Techniques de construction et théorèmes d'existence en géométrie algébrique. IV. Les schémas de Hilbert* (French), Séminaire Bourbaki, Vol. 6, Soc. Math. France, Paris, 1995, pp. Exp. No. 221, 249–276. MR1611822

[Hai1] M. Haiman, *Hilbert schemes, polygraphs and the Macdonald positivity conjecture*, J. Amer. Math. Soc. **14** (2001), no. 4, 941–1006, DOI 10.1090/S0894-0347-01-00373-3. MR1839919

[Hai2] ———, *Vanishing theorems and character formulas for the Hilbert scheme of points in the plane*, Invent. Math. **149** (2002), no. 2, 371–407, DOI 10.1007/s002220200219. MR1918676

[Hai3] ———, *Combinatorics, symmetric functions, and Hilbert schemes*, Current developments in mathematics, 2002, Int. Press, Somerville, MA, 2003, pp. 39–111. MR2051783

[Harr] J. Harris, *Algebraic geometry*, Graduate Texts in Mathematics, vol. 133, Springer-Verlag, New York, 1992. A first course. MR1182558

[Hart] R. Hartshorne, *Algebraic geometry*, Springer-Verlag, New York-Heidelberg, 1977. Graduate Texts in Mathematics, No. 52. MR0463157

[Hern] D. Hernandez, *An introduction to affine Kac-Moody algebras*. DEA. Lecture notes from CTQM Master Class, Aarhus University, Denmark, October 2006.

[HLQ] J. Hu, W.-P. Li, and Z. Qin, *The Gromov-Witten invariants of the Hilbert schemes of points on surfaces with* $p_g > 0$, Internat. J. Math. **26** (2015), no. 1, 1550009, 26, DOI 10.1142/S0129167X15500093. MR3313654

[Iar1] A. A. Iarrobino, *Punctual Hilbert schemes*, Mem. Amer. Math. Soc. **10** (1977), no. 188, viii+112, DOI 10.1090/memo/0188. MR0485867

[Iar2] A. Iarrobino, *Hilbert scheme of points: overview of last ten years*, Algebraic geometry, Bowdoin, 1985 (Brunswick, Maine, 1985), Proc. Sympos. Pure Math., vol. 46, Amer. Math. Soc., Providence, RI, 1987, pp. 297–320. MR927986

[IN] Y. Ito and I. Nakamura, *McKay correspondence and Hilbert schemes*, Proc. Japan Acad. Ser. A Math. Sci. **72** (1996), no. 7, 135–138. MR1420598

[Juc] A.-A. A. Jucys, *Symmetric polynomials and the center of the symmetric group ring*, Rep. Mathematical Phys. **5** (1974), no. 1, 107–112. MR0419576

[Kac] V. Kac, *Vertex algebras for beginners*, 2nd ed., University Lecture Series, vol. 10, American Mathematical Society, Providence, RI, 1998. MR1651389

[Kap] S. Kapfer, *Computing cup products in integral cohomology of Hilbert schemes of points on K3 surfaces*, LMS J. Comput. Math. **19** (2016), no. 1, 78–97, DOI 10.1112/S1461157016000012. MR3542518

[KaM] S. Kapfer, G. Menet, *Integral cohomology of the generalized Kummer fourfold*. Preprint.

[KMM] Y. Kawamata, K. Matsuda, and K. Matsuki, *Introduction to the minimal model problem*, Algebraic geometry, Sendai, 1985, Adv. Stud. Pure Math., vol. 10, North-Holland, Amsterdam, 1987, pp. 283–360. MR946243

[KL1] Y. Kiem, J. Li, *Gromov-Witten invariants of varieties with holomorphic 2-forms*. Preprint. arXiv:0707.2986

[KL2] Y.-H. Kiem and J. Li, *Localizing virtual cycles by cosections*, J. Amer. Math. Soc. **26** (2013), no. 4, 1025–1050, DOI 10.1090/S0894-0347-2013-00768-7. MR3073883

[King] A. D. King, *Moduli of representations of finite-dimensional algebras*, Quart. J. Math. Oxford Ser. (2) **45** (1994), no. 180, 515–530, DOI 10.1093/qmath/45.4.515. MR1315461

[KM] M. Kontsevich and Yu. Manin, *Gromov-Witten classes, quantum cohomology, and enumerative geometry*, Comm. Math. Phys. **164** (1994), no. 3, 525–562. MR1291244

[Kuhn] N. J. Kuhn, *Character rings in algebraic topology*, Advances in homotopy theory (Cortona, 1988), London Math. Soc. Lecture Note Ser., vol. 139, Cambridge Univ. Press, Cambridge, 1989, pp. 111–126. MR1055872

[LT] A. Lascoux and J.-Y. Thibon, *Vertex operators and the class algebras of symmetric groups* (English, with English and Russian summaries), Zap. Nauchn. Sem. S.-Peterburg. Otdel. Mat. Inst. Steklov. (POMI) **283** (2001), no. Teor. Predst. Din. Sist. Komb. i Algoritm. Metody. 6, 156–177, 261, DOI 10.1023/B:JOTH.0000024619.77778.3d; English transl., J. Math. Sci. (N. Y.) **121** (2004), no. 3, 2380–2392. MR1879068

[LY] H. B. Lawson Jr. and S. S.-T. Yau, *Holomorphic symmetries*, Ann. Sci. École Norm. Sup. (4) **20** (1987), no. 4, 557–577. MR932798

[Leh1] M. Lehn, *Chern classes of tautological sheaves on Hilbert schemes of points on surfaces*, Invent. Math. **136** (1999), no. 1, 157–207, DOI 10.1007/s002220050307. MR1681097

[Leh2] ———, *Lectures on Hilbert schemes*, Algebraic structures and moduli spaces, CRM Proc. Lecture Notes, vol. 38, Amer. Math. Soc., Providence, RI, 2004, pp. 1–30. MR2095898

[LS1] M. Lehn and C. Sorger, *Symmetric groups and the cup product on the cohomology of Hilbert schemes*, Duke Math. J. **110** (2001), no. 2, 345–357, DOI 10.1215/S0012-7094-01-11026-0. MR1865244

[LS2] ———, *The cup product of Hilbert schemes for $K3$ surfaces*, Invent. Math. **152** (2003), no. 2, 305–329, DOI 10.1007/s00222-002-0270-7. MR1974889

[LZ] C. Li, X. Zhao, *The MMP for deformations of Hilbert schemes of points on the projective plane*. Algebraic Geometry (to appear).

[LiJ] J. Li, *Zero dimensional Donaldson-Thomas invariants of threefolds*, Geom. Topol. **10** (2006), 2117–2171, DOI 10.2140/gt.2006.10.2117. MR2284053

[LL] J. Li and W.-P. Li, *Two point extremal Gromov-Witten invariants of Hilbert schemes of points on surfaces*, Math. Ann. **349** (2011), no. 4, 839–869, DOI 10.1007/s00208-010-0542-2. MR2777035

[LT1] J. Li and G. Tian, *Virtual moduli cycles and Gromov-Witten invariants of algebraic varieties*, J. Amer. Math. Soc. **11** (1998), no. 1, 119–174, DOI 10.1090/S0894-0347-98-00250-1. MR1467172

[LT2] ———, *Virtual moduli cycles and Gromov-Witten invariants of general symplectic manifolds*, Topics in symplectic 4-manifolds (Irvine, CA, 1996), First Int. Press Lect. Ser., I, Int. Press, Cambridge, MA, 1998, pp. 47–83. MR1635695

[LT3] ———, *Comparison of algebraic and symplectic Gromov-Witten invariants*, Asian J. Math. **3** (1999), no. 3, 689–728, DOI 10.4310/AJM.1999.v3.n3.a7. MR1793677

[LQ1] W.-P. Li and Z. Qin, *On 1-point Gromov-Witten invariants of the Hilbert schemes of points on surfaces*, Turkish J. Math. **26** (2002), no. 1, 53–68. MR1892799

[LQ2] W.-P. Li, Z. Qin, *Incidence Hilbert schemes and infinite dimensional Lie algebras*. Proceedings of Fourth International Congress of Chinese Mathematicians (Hangzhou, 2007), 408-441, Vol II.

[LQ3] ———, *Integral cohomology of Hilbert schemes of points on surfaces*, Comm. Anal. Geom. **16** (2008), no. 5, 969–988, DOI 10.4310/CAG.2008.v16.n5.a3. MR2471364

[LQ4] ———, *Equivariant cohomology of incidence Hilbert schemes and infinite dimensional Lie algebras*, Manuscripta Math. **133** (2010), no. 3-4, 519–544, DOI 10.1007/s00229-010-0387-8. MR2729266

[LQ5] ———, *The cohomological crepant resolution conjecture for the Hilbert-Chow morphisms*, J. Differential Geom. **104** (2016), no. 3, 499–557. MR3568629

[LQW1] W.-p. Li, Z. Qin, and W. Wang, *Vertex algebras and the cohomology ring structure of Hilbert schemes of points on surfaces*, Math. Ann. **324** (2002), no. 1, 105–133, DOI 10.1007/s002080200330. MR1931760

[LQW2] ———, *Generators for the cohomology ring of Hilbert schemes of points on surfaces*, Internat. Math. Res. Notices **20** (2001), 1057–1074, DOI 10.1155/S1073792801000502. MR1857595

[LQW3] W.-P. Li, Z. Qin, and W. Wang, *Stability of the cohomology rings of Hilbert schemes of points on surfaces*, J. Reine Angew. Math. **554** (2003), 217–234, DOI 10.1515/crll.2003.007. MR1952174

[LQW4] ———, *Hilbert schemes and $\mathcal{W}$ algebras*, Int. Math. Res. Not. **27** (2002), 1427–1456, DOI 10.1155/S1073792802110129. MR1908477

[LQW5] ———, *Ideals of the cohomology rings of Hilbert schemes and their applications*, Trans. Amer. Math. Soc. **356** (2004), no. 1, 245–265, DOI 10.1090/S0002-9947-03-03422-6. MR2020031

[LQW6] ———, *Hilbert schemes, integrable hierarchies, and Gromov-Witten theory*, Int. Math. Res. Not. **40** (2004), 2085–2104, DOI 10.1155/S1073792804140452. MR2064317

[LQW7] ———, *The cohomology rings of Hilbert schemes via Jack polynomials*, Algebraic structures and moduli spaces, CRM Proc. Lecture Notes, vol. 38, Amer. Math. Soc., Providence, RI, 2004, pp. 249–258. MR2096149

[LQW8] ———, *Hilbert scheme intersection numbers, Hurwitz numbers, and Gromov-Witten invariants*, Infinite-dimensional aspects of representation theory and applications, Contemp. Math., vol. 392, Amer. Math. Soc., Providence, RI, 2005, pp. 67–81, DOI 10.1090/conm/392/07354. MR2189871

[LQZ] W.-P. Li, Z. Qin, and Q. Zhang, *Curves in the Hilbert schemes of points on surfaces*, Vector bundles and representation theory (Columbia, MO, 2002), Contemp. Math., vol. 322, Amer. Math. Soc., Providence, RI, 2003, pp. 89–96, DOI 10.1090/conm/322/05681. MR1987741

[Mac1] I. G. Macdonald, *The Poincaré polynomial of a symmetric product*, Proc. Cambridge Philos. Soc. **58** (1962), 563–568. MR0143204

[Mac2] ———, *Symmetric functions and Hall polynomials*, 2nd ed., Oxford Mathematical Monographs, The Clarendon Press, Oxford University Press, New York, 1995. With contributions by A. Zelevinsky; Oxford Science Publications. MR1354144

[Mar1] E. Markman, *Generators of the cohomology ring of moduli spaces of sheaves on symplectic surfaces*, J. Reine Angew. Math. **544** (2002), 61–82, DOI 10.1515/crll.2002.028. MR1887889

[Mar2] ———, *Integral generators for the cohomology ring of moduli spaces of sheaves over Poisson surfaces*, Adv. Math. **208** (2007), no. 2, 622–646, DOI 10.1016/j.aim.2006.03.006. MR2304330

[Mat] Y. Matsuo, *Matrix theory, Hilbert scheme and integrable system*, Modern Phys. Lett. A **13** (1998), no. 34, 2731–2742, DOI 10.1142/S0217732398002904. MR1660859

[Mau] D. Maulik, *Gromov-Witten theory of $\mathcal{A}_n$-resolutions*, Geom. Topol. **13** (2009), no. 3, 1729–1773, DOI 10.2140/gt.2009.13.1729. MR2496055

[MNOP1] D. Maulik, N. Nekrasov, A. Okounkov, and R. Pandharipande, *Gromov-Witten theory and Donaldson-Thomas theory. I*, Compos. Math. **142** (2006), no. 5, 1263–1285, DOI 10.1112/S0010437X06002302. MR2264664

[MNOP2] ———, *Gromov-Witten theory and Donaldson-Thomas theory. II*, Compos. Math. **142** (2006), no. 5, 1286–1304, DOI 10.1112/S0010437X06002314. MR2264665

[MO1] D. Maulik and A. Oblomkov, *Quantum cohomology of the Hilbert scheme of points on $\mathcal{A}_n$-resolutions*, J. Amer. Math. Soc. **22** (2009), no. 4, 1055–1091, DOI 10.1090/S0894-0347-09-00632-8. MR2525779

[MO2] ———, *Donaldson-Thomas theory of $\mathcal{A}_n \times P^1$*, Compos. Math. **145** (2009), no. 5, 1249–1276, DOI 10.1112/S0010437X09003972. MR2551996

[MJD] T. Miwa, M. Jimbo, and E. Date, *Solitons*, Cambridge Tracts in Mathematics, vol. 135, Cambridge University Press, Cambridge, 2000. Differential equations, symmetries and infinite-dimensional algebras; Translated from the 1993 Japanese original by Miles Reid. MR1736222

[Mod] L. J. Mordell, *On the evaluation of some multiple series*, J. London Math. Soc. **33** (1958), 368–371, DOI 10.1112/jlms/s1-33.3.368. MR0100181

[Mum] D. Mumford, *Towards an enumerative geometry of the moduli space of curves*, Arithmetic and geometry, Vol. II, Progr. Math., vol. 36, Birkhäuser Boston, Boston, MA, 1983, pp. 271–328. MR717614

[MF] D. Mumford and J. Fogarty, *Geometric invariant theory*, 2nd ed., Ergebnisse der Mathematik und ihrer Grenzgebiete [Results in Mathematics and Related Areas], vol. 34, Springer-Verlag, Berlin, 1982. MR719371

[Mu] G. E. Murphy, *A new construction of Young's seminormal representation of the symmetric groups*, J. Algebra **69** (1981), no. 2, 287–297, DOI 10.1016/0021-8693(81)90205-2. MR617079

[Mur] R. Murphy, *A transition matrix for two bases of the integral cohomology of the Hilbert scheme of points in the projective plane*, ProQuest LLC, Ann Arbor, MI, 2010. Thesis (Ph.D.)–University of Missouri - Columbia. MR2982352

[Nag] K. Nagao, *Quiver varieties and Frenkel-Kac construction*, J. Algebra **321** (2009), no. 12, 3764–3789, DOI 10.1016/j.jalgebra.2009.03.012. MR2517812

[Nak1] H. Nakajima, *Instantons on ALE spaces, quiver varieties, and Kac-Moody algebras*, Duke Math. J. **76** (1994), no. 2, 365–416, DOI 10.1215/S0012-7094-94-07613-8. MR1302318

[Nak2] ———, *Gauge theory on resolutions of simple singularities and simple Lie algebras*, Internat. Math. Res. Notices **2** (1994), 61–74, DOI 10.1155/S1073792894000085. MR1264929

[Nak3] ———, *Heisenberg algebra and Hilbert schemes of points on projective surfaces*, Ann. of Math. (2) **145** (1997), no. 2, 379–388, DOI 10.2307/2951818. MR1441880

[Nak4] ———, *Quiver varieties and Kac-Moody algebras*, Duke Math. J. **91** (1998), no. 3, 515–560, DOI 10.1215/S0012-7094-98-09120-7. MR1604167

[Nak5] ———, *Lectures on Hilbert schemes of points on surfaces*, University Lecture Series, vol. 18, American Mathematical Society, Providence, RI, 1999. MR1711344

[Nak6] ———, *The Heisenberg algebra and Hilbert schemes of points on surfaces* (Japanese), Sūgaku **50** (1998), no. 4, 385–398. MR1690690

[Nak7] ———, *More lectures on Hilbert schemes of points on surfaces*, Development of moduli theory—Kyoto 2013, Adv. Stud. Pure Math., vol. 69, Math. Soc. Japan, [Tokyo], 2016, pp. 173–205. MR3586508

[Naka] S. Nakano, *On the inverse of monoidal transformation*, Publ. Res. Inst. Math. Sci. **6** (1970/71), 483–502, DOI 10.2977/prims/1195193917. MR0294710

[Neg] A. Neguţ, *Exts and the AGT relations*, Lett. Math. Phys. **106** (2016), no. 9, 1265–1316, DOI 10.1007/s11005-016-0865-3. MR3533570

[NO] N. A. Nekrasov and A. Okounkov, *Seiberg-Witten theory and random partitions*, The unity of mathematics, Progr. Math., vol. 244, Birkhäuser Boston, Boston, MA, 2006, pp. 525–596, DOI 10.1007/0-8176-4467-9_15. MR2181816

[Nie1] M. A. Nieper-Wisskirchen, *Characteristic classes of the Hilbert schemes of points on non-compact simply-connected surfaces*, JP J. Geom. Topol. **8** (2008), no. 1, 7–21. MR2444822

[Nie2] ———, *Twisted cohomology of the Hilbert schemes of points on surfaces*, Doc. Math. **14** (2009), 749–770. MR2578804

[Obe] G. Oberdieck, *Gromov-Witten invariants of the Hilbert scheme of points of a K3 surface*. Preprint. arXiv:1406.1139

[Ok1] A. Okounkov, *Toda equations for Hurwitz numbers*, Math. Res. Lett. **7** (2000), no. 4, 447–453, DOI 10.4310/MRL.2000.v7.n4.a10. MR1783622

[Ok2] A. Yu. Okun′kov, *Hilbert schemes and multiple q-zeta values* (Russian, with Russian summary), Funktsional. Anal. i Prilozhen. **48** (2014), no. 2, 79–87; English transl., Funct. Anal. Appl. **48** (2014), no. 2, 138–144. MR3288178

[OP1] A. Okounkov and R. Pandharipande, *Gromov-Witten theory, Hurwitz theory, and completed cycles*, Ann. of Math. (2) **163** (2006), no. 2, 517–560, DOI 10.4007/annals.2006.163.517. MR2199225

[OP2] ———, *The equivariant Gromov-Witten theory of* $\mathbf{P}^1$, Ann. of Math. (2) **163** (2006), no. 2, 561–605, DOI 10.4007/annals.2006.163.561. MR2199226

[OP3] ———, *Quantum cohomology of the Hilbert scheme of points in the plane*, Invent. Math. **179** (2010), no. 3, 523–557, DOI 10.1007/s00222-009-0223-5. MR2587340

[OP4] ———, *The quantum differential equation of the Hilbert scheme of points in the plane*, Transform. Groups **15** (2010), no. 4, 965–982, DOI 10.1007/s00031-010-9116-3. MR2753265

[OP5] ———, *The local Donaldson-Thomas theory of curves*, Geom. Topol. **14** (2010), no. 3, 1503–1567, DOI 10.2140/gt.2010.14.1503. MR2679579

[OT] J.-i. Okuda and Y. Takeyama, *On relations for the multiple q-zeta values*, Ramanujan J. **14** (2007), no. 3, 379–387, DOI 10.1007/s11139-007-9053-5. MR2357443

[PTse] R. Pandharipande, H.-H. Tseng, *Higher genus Gromov-Witten theory of the Hilbert scheme of points of the plane and* CohFTs *associated to local curves*. Preprint. arXiv:1707.01406

[QT] Z. Qin and Y. Tu, *The nef cones of and minimal-degree curves in the Hilbert schemes of points on certain surfaces*, Pacific J. Math. **284** (2016), no. 2, 439–453, DOI 10.2140/pjm.2016.284.439. MR3544309

[QW1] Z. Qin and W. Wang, *Hilbert schemes and symmetric products: a dictionary*, Orbifolds in mathematics and physics (Madison, WI, 2001), Contemp. Math., vol. 310, Amer. Math. Soc., Providence, RI, 2002, pp. 233–257, DOI 10.1090/conm/310/05407. MR1950950

[QW2] ———, *Integral operators and integral cohomology classes of Hilbert schemes*, Math. Ann. **331** (2005), no. 3, 669–692, DOI 10.1007/s00208-004-0602-6. MR2122545

[QW3] ———, *Hilbert schemes of points on the minimal resolution and soliton equations*, Lie algebras, vertex operator algebras and their applications, Contemp. Math., vol. 442, Amer. Math. Soc., Providence, RI, 2007, pp. 435–462, DOI 10.1090/conm/442/08541. MR2372578

[QY] Z. Qin, F. Yu, *On Okounkov's conjecture connecting Hilbert schemes of points and multiple q-zeta values.* Intern. Math. Res. Notices (to appear). Published online: December 16, 2016.

[Reid] M. Reid, *La correspondance de McKay*, Astérisque **276** (2002), 53–72. Séminaire Bourbaki, Vol. 1999/2000. MR1886756

[Rua1] Y. Ruan, *Topological sigma model and Donaldson-type invariants in Gromov theory*, Duke Math. J. **83** (1996), no. 2, 461–500, DOI 10.1215/S0012-7094-96-08316-7. MR1390655

[Rua2] ———, *The cohomology ring of crepant resolutions of orbifolds*, Gromov-Witten theory of spin curves and orbifolds, Contemp. Math., vol. 403, Amer. Math. Soc., Providence, RI, 2006, pp. 117–126, DOI 10.1090/conm/403/07597. MR2234886

[Seg] G. Segal, *Equivariant K-theory and symmetric products.* Lecture at Cambridge, July 1996.

[Sim] C. T. Simpson, *Moduli of representations of the fundamental group of a smooth projective variety. I*, Inst. Hautes Études Sci. Publ. Math. **79** (1994), 47–129. MR1307297

[Tik] A. S. Tikhomirov, *Standard bundles on a Hilbert scheme of points on a surface*, Algebraic geometry and its applications (Yaroslavl′, 1992), Aspects Math., E25, Friedr. Vieweg, Braunschweig, 1994, pp. 183–203. MR1282029

[Tor] L. Tornheim, *Harmonic double series*, Amer. J. Math. **72** (1950), 303–314, DOI 10.2307/2372034. MR0034860

[Tse] H.-H. Tseng, *A tale of four theories.* Talk at the conference "Geometry of Moduli Spaces". University of California at San Diego, May 2017.

[UT] K. Ueno and K. Takasaki, *Toda lattice hierarchy*, Group representations and systems of differential equations (Tokyo, 1982), Adv. Stud. Pure Math., vol. 4, North-Holland, Amsterdam, 1984, pp. 1–95. MR810623

[Uri] B. Uribe, *Orbifold cohomology of the symmetric product*, Comm. Anal. Geom. **13** (2005), no. 1, 113–128. MR2154668

[VW] C. Vafa and E. Witten, *A strong coupling test of S-duality*, Nuclear Phys. B **431** (1994), no. 1-2, 3–77, DOI 10.1016/0550-3213(94)90097-3. MR1305096

[Vas] E. Vasserot, *Sur l'anneau de cohomologie du schéma de Hilbert de* $\mathbf{C}^2$ (French, with English and French summaries), C. R. Acad. Sci. Paris Sér. I Math. **332** (2001), no. 1, 7–12, DOI 10.1016/S0764-4442(00)01766-3. MR1805619

[Wan1] W. Wang, *Equivariant K-theory, wreath products, and Heisenberg algebra*, Duke Math. J. **103** (2000), no. 1, 1–23, DOI 10.1215/S0012-7094-00-10311-0. MR1758236

[Wan2] ———, *Algebraic structures behind Hilbert schemes and wreath products*, Recent developments in infinite-dimensional Lie algebras and conformal field theory (Charlottesville, VA, 2000), Contemp. Math., vol. 297, Amer. Math. Soc., Providence, RI, 2002, pp. 271–295, DOI 10.1090/conm/297/05102. MR1919822

[Wan3] ———, *The Farahat-Higman ring of wreath products and Hilbert schemes*, Adv. Math. **187** (2004), no. 2, 417–446, DOI 10.1016/j.aim.2003.09.003. MR2078343

[Wan4] ———, *Vertex algebras and the class algebras of wreath products*, Proc. London Math. Soc. (3) **88** (2004), no. 2, 381–404, DOI 10.1112/S0024611503014382. MR2032512

[Wit] E. Witten, *On quantum gauge theories in two dimensions*, Comm. Math. Phys. **141** (1991), no. 1, 153–209. MR1133264

[Zas] E. Zaslow, *Topological orbifold models and quantum cohomology rings*, Comm. Math. Phys. **156** (1993), no. 2, 301–331. MR1233848

[Zhao] J. Zhao, *Uniform approach to double shuffle and duality relations of various q-analogs of multiple zeta values via Rota-Baxter algebras.* Preprint. arXiv:1412.8044

[Zud] V. V. Zudilin, *Algebraic relations for multiple zeta values* (Russian, with Russian summary), Uspekhi Mat. Nauk **58** (2003), no. 1(349), 3–32, DOI 10.1070/RM2003v058n01ABEH000592; English transl., Russian Math. Surveys **58** (2003), no. 1, 1–29. MR1992130

Index

Selected Published Titles in This Series

228 **Zhenbo Qin,** Hilbert Schemes of Points and Infinite Dimensional Lie Algebras, 2018
227 **Roberto Frigerio,** Bounded Cohomology of Discrete Groups, 2017
226 **Marcelo Aguiar and Swapneel Mahajan,** Topics in Hyperplane Arrangements, 2017
225 **Mario Bonk and Daniel Meyer,** Expanding Thurston Maps, 2017
224 **Ruy Exel,** Partial Dynamical Systems, Fell Bundles and Applications, 2017
223 **Guillaume Aubrun and Stanisław J. Szarek,** Alice and Bob Meet Banach, 2017
222 **Alexandru Buium,** Foundations of Arithmetic Differential Geometry, 2017
221 **Dennis Gaitsgory and Nick Rozenblyum,** A Study in Derived Algebraic Geometry, 2017
220 **A. Shen, V. A. Uspensky, and N. Vereshchagin,** Kolmogorov Complexity and Algorithmic Randomness, 2017
219 **Richard Evan Schwartz,** The Projective Heat Map, 2017
218 **Tushar Das, David Simmons, and Mariusz Urbański,** Geometry and Dynamics in Gromov Hyperbolic Metric Spaces, 2017
217 **Benoit Fresse,** Homotopy of Operads and Grothendieck–Teichmüller Groups, 2017
216 **Frederick W. Gehring, Gaven J. Martin, and Bruce P. Palka,** An Introduction to the Theory of Higher-Dimensional Quasiconformal Mappings, 2017
215 **Robert Bieri and Ralph Strebel,** On Groups of PL-homeomorphisms of the Real Line, 2016
214 **Jared Speck,** Shock Formation in Small-Data Solutions to 3D Quasilinear Wave Equations, 2016
213 **Harold G. Diamond and Wen-Bin Zhang (Cheung Man Ping),** Beurling Generalized Numbers, 2016
212 **Pandelis Dodos and Vassilis Kanellopoulos,** Ramsey Theory for Product Spaces, 2016
211 **Charlotte Hardouin, Jacques Sauloy, and Michael F. Singer,** Galois Theories of Linear Difference Equations: An Introduction, 2016
210 **Jason P. Bell, Dragos Ghioca, and Thomas J. Tucker,** The Dynamical Mordell–Lang Conjecture, 2016
209 **Steve Y. Oudot,** Persistence Theory: From Quiver Representations to Data Analysis, 2015
208 **Peter S. Ozsváth, András I. Stipsicz, and Zoltán Szabó,** Grid Homology for Knots and Links, 2015
207 **Vladimir I. Bogachev, Nicolai V. Krylov, Michael Röckner, and Stanislav V. Shaposhnikov,** Fokker–Planck–Kolmogorov Equations, 2015
206 **Bennett Chow, Sun-Chin Chu, David Glickenstein, Christine Guenther, James Isenberg, Tom Ivey, Dan Knopf, Peng Lu, Feng Luo, and Lei Ni,** The Ricci Flow: Techniques and Applications: Part IV: Long-Time Solutions and Related Topics, 2015
205 **Pavel Etingof, Shlomo Gelaki, Dmitri Nikshych, and Victor Ostrik,** Tensor Categories, 2015
204 **Victor M. Buchstaber and Taras E. Panov,** Toric Topology, 2015
203 **Donald Yau and Mark W. Johnson,** A Foundation for PROPs, Algebras, and Modules, 2015
202 **Shiri Artstein-Avidan, Apostolos Giannopoulos, and Vitali D. Milman,** Asymptotic Geometric Analysis, Part I, 2015
201 **Christopher L. Douglas, John Francis, André G. Henriques, and Michael A. Hill, Editors,** Topological Modular Forms, 2014